D1029272

VANADIUM IN THE ENVIRONMENT

Volume

30

in the Wiley Series in

Advances in Environmental Science and Technology

JEROME O. NRIAGU, Series Editor

VANADIUM IN THE ENVIRONMENT

Part 1: Chemistry and Biochemistry

Edited by

Jerome O. Nriagu
Department of Environmental and Industrial Health
School of Public Health, The University of Michigan
Ann Arbor, Michigan

A WILEY-INTERSCIENCE PUBLICATION
JOHN WILEY & SONS, INC.
New York • Chichester • Weinheim • Brisbane • Singapore • Toronto

Copyright © 1998 by John Wiley & Sons, Inc. All rights reserved.

Published simultaneously in Canada.

Library of Congress Cataloging-in-Publication Data:

Vanadium in the environment / edited by Jerome O. Nriagu.
    p. cm.—(Advances in environmental science and technology ; v. 30)
    Includes index.
    1. Vanadium—Environmental aspects. I. Nriagu, Jerome O.
II. Series.
TD180.A38 vol. 30
[TD196.V35]
628 s—dc21
[363.738]   97-14872
ISBN 0-471-17778-4 (cloth : alk. paper) (Part 1)
ISBN 0-471-17776-8 (cloth : alk. paper) (Part 2)

Printed in the United States of America

10 9 8 7 6 5 4 3 2 1

CONTRIBUTORS

S. A. ABBASI, Centre for Pollution Control and Energy Technology, Pondicherry University, Kalapet, Pondicherry–605 014, India

SEAN S. AMIN, Department of Chemistry, Colorado State University, Fort Collins, CO 80523

MANUEL AURELIANO,* Centro de Neurociências de Coimbra, Departamento de Engenharia Química, Universidade de Coimbra, Largo Marquês de Pombal, 3000 Coimbra, Portugal

ANA M. CORTIZO, Cátedra de Bioquímica Patológica, Facultad de Ciencias Exactas, Universidad Nacional de La Plata, Calle 47 y 115, 1900 La Plata, Argentina

DEBBIE C. CRANS, Department of Chemistry, Colorado State University, Fort Collins, CO 80523

SUSANA B. ETCHEVERRY, Cátedra de Bioquímica Patológica and CEQUINOR, Facultad de Ciencias Exactas, Universidad Nacional de La Plata, Calle 47 y 115, 1900 La Plata, Argentina

SCOTT W. FOWLER, International Atomic Energy Agency, Marine Environment Laboratory, P.O. Box 800, MC 98012, Monaco

TATSUO HAMADA, Department of Animal Science, Tokyo University of Agriculture, Sakuragaoka 1-1-1, Setagaya, Tokyo 156, Japan

W. HEMRIKA, E. C. Slater Institute, Laboratory of Biochemistry, University of Amsterdam, Plantage Muidergracht 12, 1018 TV Amsterdam, The Netherlands

TOSHIAKI ISHII, Division of Marine Radioecology, National Institute of Radiological Sciences, Hitachinaka 311-12, Japan

SVEN JANTZEN, Institut für Anorganische und Angewandte Chemie der Universität Hamburg, D-20146 Hamburg, Germany

KAN KANAMORI, Department of Chemistry, Faculty of Science, Toyama University, Gofuku 3190, Toyama 930, Japan

* *Present address:* Universidade do Algarve, U.C.E.H.—Campus de Gambelas, 8000 Faro, Portugal.

ANASTASIOS D. KERAMIDAS, Department of Chemistry, Colorado State University, Fort Collins, CO 80523

VÍTOR M. C. MADEIRA, Departamento de Bioquímica, Universidade de Coimbra, 3000 Coimbra, Portugal

Y. MAMANE, Technion–Israel Institute of Technology, Haifa 32000, Israel

GEORGE L. MENDZ, School of Biochemistry and Molecular Genetics, The University of New South Wales, Sydney, NSW 2052, Australia

GIOVANNI MICERA, Dipartimento di Chimica, Università di Sassari, Via Vienna 2, 07100 Sassari, Italy

HITOSHI MICHIBATA, Mukaishima Marine Biological Laboratory, Faculty of Science and Laboratory of Marine Molecular Biology, Graduate School of Science, Hiroshima University, Mukaishima-cho 2445, Hiroshima 722, Japan

PIERRE MIRAMAND, Laboratoire de Biologie et Biochimie Marines (EA 1220), Université de La Rochelle, Pôle Sciences et Technologie, Avenue Marillac, 17042 La Rochelle, Cedex 1, France

JEROME O. NRIAGU, Department of Environmental and Industrial Health, School of Public Health, University of Michigan, Ann Arbor, MI 48109

NICOLA PIRRONE, CNR Institute for Atmospheric Pollution, c/o DEIS, University of Calabria, 87036 Rende, Italy

DIETER REHDER, Institut für Anorganische und Angewandte Chemie der Universität Hamburg, D-20146 Hamburg, Germany

DANIELE SANNA, Dipartimento di Chimica, Università di Sassari, Via Vienna 2, 07100 Sassari, Italy

R. WEVER, E. C. Slater Institute, Laboratory of Biochemistry, University of Amsterdam, Plantage Muidergracht 12, 1018 TV Amsterdam, The Netherlands

CONTENTS

INTRODUCTION
TO THE SERIES

The deterioration of environmental quality, which began when mankind first congregated into villages, has existed as a serious problem since the industrial revolution. In the second half of the twentieth century, under the ever-increasing impacts of exponentially growing population and of industrializing society, environmental contamination of the air, water, soil, and food has become a threat to the continued existence of many plant and animal communities of various ecosystems and may ultimately threaten the very survival of the human race. Understandably, many scientific, industrial, and governmental communities have recently committed large resources of money and human power to the problems of environmental pollution and pollution abatement by effective control measures.

Advances in Environmental Science and Technology deals with creative reviews and critical assessments of all studies pertaining to the quality of the environment and to the technology of its conservation. The volumes published in the series are expected to service several objectives: (1) stimulate interdisciplinary cooperation and understanding among the environmental scientists; (2) provide the scientists with a periodic overview of environmental developments that are of general concern or that are of relevance to their own work or interests; (3) provide the graduate student with a critical assessment of past accomplishment, which may help stimulate him or her toward the career opportunities in this vital area; and (4) provide the research manager and the legislative or administrative official with an assured awareness of newly developing research work on the critical pollutants and with the background information important to their responsibility.

As the skills and techniques of many scientific disciplines are brought to bear on the fundamental and applied aspects of the environmental issues, there is a heightened need to draw together the numerous threads and to present a coherent picture of the various research endeavors. This need and the recent tremendous growth in the field of environmental studies have clearly made some editorial adjustments necessary. Apart from the changes in style and format, each future volume in the series will focus on one particular theme or timely topic, starting with Volume 12. The author(s) of each pertinent section will be expected to critically review the literature and the most important recent developments in the particular field; to critically evaluate new concepts, methods, and data; and to focus attention on important unresolved or controversial questions and on probable future trends. Monographs embodying the results of unusually extensive and well-rounded investigations will also be published in the series. The net result of the new editorial policy should be more integrative and comprehensive volumes on key environmental issues and pollutants. Indeed, the development of realistic standards of environmental quality for many pollutants often entails such a holistic treatment.

JEROME O. NRIAGU, Series Editor

PREFACE

Vanadium is widely distributed and the 21st most abundant element in the earth's crust. Although vanadium is more abundant than many of the more common metals such as iron, zinc, and copper, there is no convincing evidence that it is an essential element for man. It is, however, believed to be essential for a number of species such as chicken and rats; deficiency symptoms in such species include retarded growth, impairment of reproduction, disturbance of lipid metabolism and inhibition of Na^+/K^+-ATPase activity in the kidney, brain, and heart. In spite of being a nutritional element, vanadium is not accumulated by the biota, the only organisms known to bioaccumulate it to any significant degree being some mushrooms, tunicates, and sea squirts.

Vanadium has two different catalytic functions in biological systems. It can exist in many oxidation states and the oxyanions and oxycations are known to be powerful oxidants under physiological conditions. Secondly, vanadium forms sulfur-containing anionic and cationic centers and iron-vanadium-sulfur clusters associated in biology with nitrogen fixation. The multiple oxidation states, ready hydrolysis and polymerization confer a level of complexity to the chemistry of vanadium well above that of many elements. Although its current biological role seems to be diminished, the high accumulation of vanadium in some oils suggests that vanadium was involved in early photosynthetic processes.

Vanadium is used widely in industrial processes including the production of special vanadium–iron steels, in the production of hard metals and temperature-resistant alloys, in iron and steel refining and tempering, in glass industry, in the manufacture of pigments, paints, and printing inks, for lining arc welding electrodes, and as catalysts in the pharmaceutical industry. Its use with non-ferrous metals is of particular importance in the atomic energy industry, aircraft construction, and space technology. The high temperature industrial processes as well as the combustion of fossil fuels, especially oils, now release large quantities of vanadium into the environment. The behavior and effects of vanadium pollution in most ecosystems remain poorly understood. A number of studies have even associated the ambient levels of vanadium in the air with cardiovascular diseases, bronchitis, and lung carcinoma in the general population. Vanadium also interferes with a multitude of biochemical processes, can penetrate the blood–brain and placental barriers and is present in milk. The reproductive toxicology of vanadium must be of concern and a fertile field for study.

Vanadium is an element of considerable environmental and scientific interest because of its wide industrial applications and large releases into the environment, complex chemistry, and narrow thresholds between essential and toxic doses. While there are now a number of review papers and reports on vanadium toxicity in animal species including human beings, few attempts have been made to relate the toxicological aspects to the physical-chemical features of the element in the ecosystem. An objective of this volume has been to provide a comprehensive picture of current biological, chemical, and clinical research on vanadium in various environmental media. Individual chapters cover the sources, distribution, transformations, mechanisms of transport, fate as well as human and ecosystem effects. The chapters are written by acknowledged experts from a wide variety of scientific disciplines; indeed, the literature on vanadium is so disparate that no single scientist can provide a detailed account of all the recent developments. The authors have been asked to focus on general principles of vanadium behavior rather than on systematic compilation of published data. The volume is intended to be of interest to graduate students and practicing scientists in the fields of environmental science and engineering, ecology, nutrition, toxicology, public health, and environmental control. More importantly, it is addressed to everyone who is concerned about the impact of metallic pollutants on our health and our life support system.

JEROME O. NRIAGU

Ann Arbor, Michigan

ADVANCES IN ENVIRONMENTAL SCIENCE AND TECHNOLOGY

Jerome O. Nriagu, Series Editor

VANADIUM IN THE ENVIRONMENT

1

HISTORY, OCCURRENCE, AND USES OF VANADIUM

Jerome O. Nriagu

Department of Environmental and Industrial Health, School of Public Health, University of Michigan, Ann Arbor, MI 48109

Vanadium in the Environment. Part 1: Chemistry and Biochemistry, Edited by Jerome O. Nriagu.
ISBN 0-471-17778-4.   © 1998 John Wiley & Sons, Inc.

1. INTRODUCTION

Few of the elements discovered during the last two centuries have had a more romantic history than vanadium. Even the name itself provides allegorical links between chemistry, mythology, and music. Because of its beautiful and multicolored compounds, Nils Sefstrom (1831) named the element he discovered in honor of Vanadis, a surname of the beautiful Scandinavian goddess of youth and love, who is none other than the Freya of Richard Wagner's "Rheingold" (Alexander, 1929). Vanadium in its naked state is a soft and ductile metal that enters into amorous (or congenial) relationships with iron, aluminum, titanium, and other metals. Small amounts of vanadium (0.02–0.06%) result in steels that possess (a) higher strength without heat treatment, (b) aging suppression, (c) improved mechanical properties, and (d) more desirable grain size (Beliles, 1991). Its alloys with nonferrous metals are used extensively in the atomic energy industry, aircraft construction, space technology, and other high-tech industries. Vanadium is endowed with many oxidation states (-1 to $+5$) and thus can form a wide array of stable complex ions, coordination complexes, and solid phases, all of which can be put to commercial use. In less than 100 years, vanadium has gone from being a rare and obscure metal to become one of strategic military importance and a pillar of modern technology. As scientific and technological developments expand its horizon, vanadium is clearly poised to become the element for the twenty-first century.

2. HISTORY

In 1802, Andres Manuel del Rio (1764–1849), a professor of mineralogy in the School of Mines of Mexico, discovered that the brown lead mineral, *plumbo pardo de Zimapan,* from the Cardonal Mine in Hidalgo contained what he thought was a new metal similar to chromium and uranium. He first called it *panchromium* because of the varied colors of its salts, but he later changed the name to *erythronium* ("red")—in allusion to the red color of its salts when treated with acids (Weeks, 1968). When Baron Alexander von Humboldt visited Mexico in 1803, del Rio gave him several specimens of the brown lead ore with a copy of his experiments in order that he might publish them. However, del Rio's claim to the discovery was thwarted by the loss in a shipwreck of the detailed notes he had given to von Humboldt and by a report by H. V. Collect-Descotils, a French chemist and friend of Vauquelin, who declared that the Mexican ore sample contained nothing but basic lead chromate.

In 1831 Nils Gabriel Sefstrom (1787–1845), in collaboration with Jons Jacob Berzelius (1799–1848), discovered a new element from iron ores of Taberg, Sweden, which he named vanadium, a cognomen for Vanadis, the legendary Norse goddess of beauty, in recognition of the richness of the colors of its derivatives. Sefstrom provided the following account of the discovery:

I dissolved a considerable quantity of iron in muriatic (hydrochloric) acid and I noticed that while it was dissolving, a few particles of iron, mainly those which deposit the black powder, dissolved more rapidly than the others, in such a way that there remained hollow veins in the midst of the iron bar. Upon examining this black powder, I found silica, iron, alumina, lime copper and, among other things, uranium. I could not discover in what condition this substance was, because the small quantity of powder did not exceed two decigrams, and moreover more than half of it was silica. After several experiments, I saw that it certainly was not uranium. I had sought to compare the highest degrees of oxidation, but I must remark that vanadium is found partly in the lower degree. (Sefstrom, 1831)

After Sefstrom announced the discovery of vanadium, F. Wohler reanalyzed the "brown lead from Zimapan," which del Rio had sent to Europe through von Humbolt, and was able to show that the ore really contained vanadium instead of chromium; the ore is now known to contain the mineral, vanadinite, $Pb_5(VO_4)_3Cl$. In his reply to a January 22, 1831, letter from Berzelius to announce the new metal, Wohler agonized over the priority of discovery of vanadium:

Anticipatory as it may seem yet, because of the slowness of the mails, it is time to ask whether, when I publish a notice of the mineral, I ought to give its earlier history, the supposed discovery by del Rio of a new metal in it, the refutation by Descotils, that Humbolt brought with him, etc.? I would not want in the least to take away from Sefstrom anything of his priority of discovery, especially since such indecision is repugnant in cases like this; on the other hand one must not expose oneself to charge by the public or especially by one's opponents that one through partisanship concealed earlier claims. In any case Humboldt shall be named, since he alone brought it with him, and with that the rest seems unavoidably linked. Do not laugh at me because of my diplomatic question. (Weeks, 1968, p. 366)

In 1831, J. J. Berzelius published a memoir, *Om Vanadin och dess Egenskaper,* giving details about the properties of the new element (Mellor, 1952). He apparently never realized that what he perceived as the element in fact was vanadous oxide (VO) and that the trichloride he described was an oxychloride. The next major breakthrough that solidified the chemical knowledge about vanadium came from Sir Henry Enfield Roscoe (1833–1915), a former professor at the University of Manchester. In 1865 he came into possession of a plentiful source of vanadium in copper-bearing beds of the lower Keuper Sandston of the Trias at Alderley Edge, Cheshire. After careful studies, he was able to announce that "the substance supposed by Berzelius to be vanadium (atomic weight = 68.5) is not the metal but an oxide, and that the true atomic weight of the metal is $68.5 - 16 = 52.5$. . . . The highest oxide, vanadic acid, VO_3 of Berzelius hence becomes a pentoxide, V_2O_5,. . . the suboxide of Berzelius is a trioxide, V_2O_3, while the terchloride (VCl_3) of Berzelius is an oxychloride with a formula of $VOCl_3$" (Roscoe, 1868, p. 311). He was unable to obtain metallic vanadium by direct reduction of any vanadium compounds that contain oxygen but succeeded in reducing vanadium chloride (free of oxygen) in hydrogen gas, either with or without sodium:

> The reduction can only be effected in platinum boats placed in a porcelain tube, as the metal acts violently on glass and porcelain, and tubes of platinum are porous at a red heat . . . the main points being to guard against diffusion, and to introduce the powdered dichloride into the platinum boat in such a way that it shall not for an instant be exposed to moist air. After all the precautions are taken, the tube is heated to redness, a torrent of hydrochloric acid comes off, and the evolution of this gas continues for 40 to 80 hours. . . After the evolution of any trace of hydrochloric acid has ceased to be perceptible, the tube is allowed to cool, and the boat is found to contain a light whitish grey-colored powder. (Roscoe, 1870, p. 150)

Sir Roscoe was then able to furnish, for the first time, the physical and chemical properties of vanadium:

> Metallic vanadium thus prepared examined under the microscope reflects light powerfully, and is seen to consist of a brilliant shining crystalline metallic mass possessing a bright silver-white luster. Vanadium does not oxidize or even tarnish in the air at ordinary temperature; nor does it absorb oxygen when heated in the ordinary air to 100 °C. It does not decompose water even at 100 °C, and may be moistened with water and dried *in vacuo* without gaining weight. The metal is not fusible or volatile at a bright red heat in hydrogen; the powdered metal thrown into a flame burns with the most brilliant scintillations. Heated quickly in oxygen it burns vividly, forming the pentoxide; but slowly ignited in air, it first glows to form a brown oxide, and then again absorbs oxygen and glows with formation of the black trioxide and blue tetroxide, till it at last attains its maximum degree of oxidation. The specific gravity of metallic vanadium at 15 °C is 5.5. It is not soluble in either hot or cold hydrochloric acid; strong sulfuric acid dissolves it on heating, giving a yellow solution; hydrofluoric acid dissolves it slowly with evolution of hydrogen; nitric acid of all strengths acts violently on the metal, evolving red nitrous fumes and yielding blue solution; fused with sodium hydroxide the metal dissolves with evolution of hydrogen, a vanadate being formed. (Roscoe, 1870, p. 150)

For nearly 60 years, Roscoe's method was used to prepare pure vanadium until Marden and Rich (1927) succeeded in reducing vanadium pentoxide by heating it with calcium and calcium oxide in an iron bomb at 950 °C.

$$V_2O_5 + 5Ca + 5CaCl_2 \rightarrow 2V + 5CaO \cdot CaCl_2$$

After cooling, the resulting mixture is leached with water to remove the calcium salts, and bright steel grey beads of vanadium measuring up to 9 mm in diameter are so obtained.

Up to the end of the nineteenth century, most of the vanadium in commercial use was obtained as a byproduct from silver and other ores of Joachimsthal, Bohemia, vanadiferous slags of ironworks at Creusot, France, vanadate ores of Santa Marta, Estremadura, Spain, and lead ores of Mexico (Mellor, 1952). The first important commercial deposit of vanadium was discovered in 1905 at Minasragra near Cerro de Pasco in the Andes of Peru by engineers of the American Vanadium Company. The deposit is made up of a previously unknown sulfide mineral of vanadium named *patronite* after Antenor Riza Patron, the original discoverer of the deposit. Production from Peruvian deposit promptly filled the growing demand for vanadium by the steel industry (see uses below). Until the early 1920s, the Peru deposit accounted for most of the world's production of this element (Mellor, 1952), as evidenced by the following production figures (in metric tons):

Country/year:	1909	1912	1915	1918	1920	1922
Peru	392	684	804	371	1,110	None
United States	None	300	569	250	462	100

The Peru deposits were supplanted by equally unique sources in Colorado and Utah. Roscoelite, V_2O_5, was first discovered as dark, greenish gray fillings in sandstone in the Placerville district, Colorado. A few kilometers west, mining of the newly discovered mineral carnotite ($K_2O \cdot 2UO_3 \cdot V_2O_5 \cdot 3H_2O$) for its radium began soon after that. Although carnotite was accompanied by roscoelite and other unusual vanadium minerals, vanadium was obtained primarily as a byproduct.

Initial observation of the occurrence of vanadium in soils was made by Beauvallet (1859) and Terreil (1860) in French clays near Paris. Bechi (1879a, 1879b) later found the element in several shales, limestones, and plant ashes. Subsequent studies by Witz and Osmond (1882) and Baskerville (1899) showed that vanadium is a common constituent of many soils, slags, and spring water. Heinz (1911) discovered highly elevated (up to 10%) levels of vanadium in blood of ascidians. The cells (called vanadocyte) are characterized by their unique shape, their greenish color in vivo, their strong reducing properties, and their acidity. The physiological importance of Heinz's observations, later confirmed by Hecht (1918) and Webb (1939), has remained an enigma in the annals of trace metal biochemistry.

Priestley (1876) reported that sodium vanadate was intensely toxic to pigeon, guinea pig, rabbit, cat, and dog. He found that when injected subcutaneously, the lethal dose for rabbits was 9.2–15 mg V_2O_3/kg body weight. Priestley also observed that a slowing and weakening of the heart's action was accompanied by a drop in blood pressure. Further reports on vanadium toxicity during the last century included those by Gamgee and Larmuth (1877), Larmuth (1877), and Dowdeswell (1878). Additional contributions on vanadium toxicity came from Jackson (1911, 1912), who reported that the major effect of vanadium was to induce intense vasoconstriction in the spleen, kidney, and intestine with an increase in blood pressure. Proescher et al. (1917) conducted extensive studies of the toxicity of various vanadium salts to both large and small experimental animals as well as birds and fish, which added much to the growing fund of knowledge on vanadium toxicology.

Vanadium entered into the materia medica of the European culture shortly after it was discovered (Rockwell, 1879). By the 1920s various salts of vanadium were being touted by some French doctors as miracle drugs. For instance, sodium metavanadate in doses of up to 5 mg in 24 h was considered beneficial in the treatment of anemia, tuberculosis, chronic rheumatism, and diabetes (Lyennet et al., 1899; Laran, 1899). Proescher et al. (1917) claimed that intramuscular and intravenous injection of sodium hexavanadate was an effective cure for syphilis.

3. OCCURRENCE

Vanadium is a widely distributed minor element in the solar system, average abundance in the cosmos being 220 cosmic abundance units (c.a.u., $Si = 10^6$) and in solar atmosphere, 300 c.a.u. (Suess and Urey, 1956; Aller, 1961). Average crustal abundance is estimated to be 100 $\mu g/g$, or about 2 times that of copper, 10 times that of lead, and 100 times that of molybdenum (Mason, 1966). Because of the substitutional arrangement, especially for iron, vanadium is more abundant in mafic (high in Fe and Mg) than silicic rocks; typical concentrations in gabbros and norites being 200–300 $\mu g/g$ compared to 5–80 $\mu g/g$ for granites (Table 1). Average concentrations in lunar rock and soil samples generally fall in the range of 20–90 $\mu g/g$ (Table 1) and are comparable to the mean concentration in 863 soil samples collected in various parts of the United States and reported to be 56 $\mu g/g$ (Shacklette et al., 1971).

The levels of vanadium in various geological material are summarized in Table 1. Elevated levels of vanadium have been reported in iron ores (600–4,100 $\mu g/g$), rock phosphate (10–1,000 $\mu g/g$), and superphosphate (50–2,000 $\mu g/g$) (Evans and Landergren, 1978; ATSDR, 1992). Fossil fuels also tend to be enriched in vanadium. Yudovich et al. (1972) estimated the mean concentration of vanadium in hard coals to be 19 $\mu g/g$ (126 $\mu g/g$ in ash) and 10 $\mu g/g$ in brown coal. Average concentrations in crude oil range from 3 $\mu g/g$ (Qatar) to 257 $\mu g/g$ (western Venezuela). Flue-gas deposits from oil-fired furnaces can contain up to 50% of vanadium pentoxide (Grayson, 1983). The role of vanadium in formation of fossil fuel is unknown. Vanadyl porphyrins found in crude oils structurally resemble chlorophyll and hemoglobin and there appears to be a relationship between the enrichment of vanadium in coal and crude oil and the burial and maturation of precursor organic matter into fossil fuels (Hilliard, 1992). Both vanadium and nickel form stable metalloorganic complexes in high-molecular-weight fractions of crude oil. The metal–organic bonds are not broken below a temperature of 300 °C, so that

Table 1 Average Abundance of Vanadium in Geological Material

Material	Abundance/ Concentration (c.a.u.)[a]	Reference
Solar system	630	Goles, 1969
Apollo 11 lunar samples		
Type A	46	Wakita et al., 1970
Type B	75	Wakita et al., 1970
Type C	86	Wakita et al., 1970
Type D	63	Wakita et al., 1970
Apollo 15, soil	110	Laul and Schmitt, 1973
Apollo 16, soil	25	Laul and Schmitt, 1973
Lunar 16, soil	68	Vinogradov, 1973
Lunar 20, soil	47	Laul and Schmitt, 1973
Crustal average	135	Mason, 1966
Basalts	247	Prinz, 1967
Granite	72	Evans and Landergren, 1978
Syenite	110	Evans and Landergren, 1978
Diorite	148	Evans and Landergren, 1978
Sandstone	12	Evans and Landergren, 1978
Deep-sea clay	120	Turekian and Wedepohl, 1961
Bituminous shale, Kupferschiefer (Germany)	1,650	Evans and Landergren, 1978
Bituminous shale, Lower Saxony (Germany)	99	Evans and Landergren, 1978
Bituminous shale, southern England	315	Evans and Landergren, 1978
Bituminous shale, Switzerland	160	Evans and Landergren, 1978
Coal		NAS, 1974
Eastern US	30	
Interior US	34	
Western US	15	
Crude oil		NAS, 1974
US	20	
Western Venezuela	257	
Eastern Venezuela	81	
Iran	75	
Iraq	34	
Kuwait	30	
Qatar	3	
Saudi Arabia	50	

Table 1 (*Continued*)

Material	Abundance/ Concentration (c.a.u.)[a]	Reference
Residual fuel oil		NAS, 1974
United States	30	
Venezuela	156	
Middle East	26	
Libya	0.2	
Angola	2	

[a] Concentrations in $\mu g/g$, unless indicated other wise; c.a.u = cosmic abundance units.

vanadium-to-nickel ratios have been used to fingerprint the sources of crude oil (Hilliard, 1992).

About 80 different minerals containing vanadium have been described (Fleischer, 1987; Clark, 1990). Suites of these minerals are found in four differ- ent geochemical environments, and each milieu is marked by differing crystal chemical behavior by vanadium. The four groups are (1) sulfides; (2) sulfosalts (derived from oxidized sulfide ores) of lead, copper, zinc, and manganese; (3) silicates; and (4) the oxides. The compositions of the principal minerals in each group are shown in Tables 2 and 3. The crystal chemical properties of these minerals are discussed in detailed by Evans and Landergren (1978).

The oxidation of base metal sulfides often results in a group of minerals in which the pentavalent vanadium occurs in structures in a VO_4 isolated tetrahendron (Evans and Landergren, 1978). A number of minerals in this group are isostructural with phosphates and vanadates, including descloizite, vanadinite, and brackebuschite (Table 2). In some muscovite micas, aluminum in the octahedral layer is replaced by V(III) and Fe(II). A typical example is roscoelite from the Placerville district in Colorado, which has been repre- sented by the formula (Evans and Landergren, 1978):

$$(K_{0.75}Ca_{0.02})(V_{1.16}A\ l_{0.72}Fe_{0.09}Mg_{0.08})(Si_{3.18}Al_{0.82})O_{10}(OH)_{1.97}F_{0.03}.$$

Weathering of roscoelite mica leads to vanadium-bearing clays of montmoril- lonite and chlorite types with tetravalent vanadium in the octahedral layer. Further oxidation leads to a breakdown of silicate structure and the release of pentavalent vanadium (Evans and Landergren, 1978).

Minerals that contain oxycations and oxyanions of vanadium are diverse and numerous (Table 3). There are many geochemical reasons for this. (1) Assorted assemblages of vanadium minerals have been found at a number of locations, including Colorado Plateau, Apache Mine, Arizona, and Touissit Mine, Morocco (Evans and Garrels, 1958; Williams, 1990). These deposits have yielded excellent specimens for crystallographic study and provided fertile grounds for assessing the chemical behavior and paragenetic relationships for

Table 2 Mineral Groups That Can Contain Vanadium as an Important Component

Mineral	Composition
Apatite	$A_5(XO_4)_3(F,Cl,OH)$; A = Ba, Ca, Ce, K, Na, Pb, Sr; X = As, P, V (vanadinite)
Autunite	$A(UO_2)_2(XO_4)_2.8 - 12H_2O$; A = Ba, Ca, Cu, Fe^{2+}, Mg, Mn^{2+}; X = As, P, V
Brackebuschite	$A_2B(XO_4)(OH,H_2O)$; A = Ba, Ca, Pb, Sr: B = Al, Cu^{2+}, Fe^{2+}, Mn, Zn; X = As, P, S, V
Crichtonite	$AB_{12}(O,OH)_{38}$; A = Ba, Ca, K, Pb, Sr, Na; B = Cr^{3+}, Fe^{2+}, Fe^{3+}, Mg, Mn^{2+}, Ti, V^{3+}, Zn, Zr
Cryptomelane	AB_8O_{16}; A = Ba, K, Na, Pb; B = Cr^{3+}, Fe^{3+}, Mn^{2+}, Mn^{4+}, Ti, V^{3+}
Descloizite	$PbM(XO_4)(OH)$; M = Cu, Fe^{2+}, Mn^{2+}, Zn; X = As^{5+}, V^{5+}
Epidote	$A_2B_3(SiO_4)_3(OH)$; A = Ca, Ce, Pb, Sr, Y; B = Al, Fe^{3+}, Mn^{3+}, V^{3+}
Hematite	R_2O_3; R = Al, Cr^{3+}, Fe^{3+}, V^{3+}
Spinel	AB_2O_4; A = Divalent metal ion; B = Al, Cr^{3+}, Fe^{3+}, Mg, Mn^{3+}, Ti^{4+}, V^{3+}
Tourmaline	$WX_3Y_6(BO_3)_3Si_6O_{18}(O,OH,F)_4$; W = Ca, K, Na, X = Al, Fe^{3+}, Fe^{2+}, Li, Mg, Mn^{2+}, Y = Al, Cr^{3+}, Fe^{3+}, V^{3+}

vanadium in natural environments. (2) Vanadium oxides and most base metal vanadates are insoluble in water and hence readily precipitated in earth surface environments. (3) Double salts of vanadium, phosphorus, and arsenic oxides with uranyl ions are widespread in certain environments (Table 3). An example is the torbernite group with a formula of $M(II)(UO_2)_2(VO_4)_2 \cdot nH_2O$ or $M(I)_2(UO_2)_2(VO_4) \cdot nH_2O$. Over ten phases in this group have been identified: M(II) is represented by Ca^{2+} (tyuyamunite and metatyuyamunite), Pb^{2+} (curienite), Cu^{2+} (sengierite), Ba^{2+} (francevillite) and Mn^{2+} (fritzscheite), and the M(I) analogues include carnotite (K^+), strelkinite (Na^+), margaritasite, and vanuranylite (Cs^+, K^+, and H^+). (4) Higher oxidation states show an amphoteric tendency to polymerize into complex anionic species, which form insoluble phases with metal ions. A generalized equation for the polymerization reaction is

$$2XO_4^{n-1} + H^+ \rightarrow [O_3X{-}O{-}XO_3]^{2(n-1)} + H_2O$$

Minerals of the polymeric species type include heumulite, $Na_4MgV_{10}O_{28} \cdot 24H_2O$; barnesite, $Na_2V_6O_{16} \cdot 3H_2O$; hewettite, $CaV_6O_{16} \cdot 9H_2O$; hummerite, $KMgV_5O_{14} \cdot 8H_2O$; bariandite $V_2O_4 \cdot 4V_2O_5 \cdot 12H_2O$; grantsite, $Na_4CaV_{12}O_{32} \cdot 8H_2O$; sherwoodite, $Ca_9Al_2V_{28}O_{80} \cdot 56H_2O$; and vanalite, $NaAl_8V_{10}O_{38} \cdot 30H_2O$ (Williams, 1990).

Table 3   Vanadium Minerals

Sulfides
    Patronite, VS_4
    Yushkinite, $V_{1-x}S \cdot n[(Mg, Al)(OH)_2]$
    Sulvanite, Cu_3VS_4
    Arsenovulvanite, $Cu_3(As,V)S_4$
    Colusite, $Cu_{26}V_2(As,Sn,Sb)_6S_{32}$
    Nekrasovite, $Cu_{26}V_2(Sn,As,Sb)_6S_{32}$

Oxides
    Akangite, $Ba(Ti,V,Cr)_8O_{16}$
    Stibivanite, Sb_2VO_5
    Montroseite, $(V^{3+},Fe^{3+})O(OH)$
    Shcherbinaite, V_2O_5
    Paramontroseite, VO_2
    Häggite, $V_2O_2(OH)_3$
    Duttonite, $V^{4+}O(OH)_2$
    Doloresite, $3V_2O_4 \cdot 4H_2O$
    Lenoblite, $V_2O_4 \cdot 2H_2O$
    Navajoite, $V_2O_5 \cdot 3H_2O$
    Vanoxite, $V_4^{4+}V_2^{5+}O_{13} \cdot 8H_2O$
    Bariandite, $V_2O_4 \cdot 4V_2O_5 \cdot 12H_2O$
    Corvusite, $V^{4+}V_6^{5+}O_{17} \cdot nH_2O$
    Simplotite, $CaV_4^{4+}O_9 \cdot 5H_2O$
    Kyzylkumite, $V_2Ti_3O_9$
    Schreyerite, $V_2^{3+}Ti_3O_9$
    Berdesinskiite, $V_2^{3+}TiO_5$
    Tivanite, $V^{3+}TiO_3(OH)$
    Mannardite, $Ba_x(Ti_6V_2^{3+})O_{16}$
    Karelianite, V_2O_3
    Coulsonite, $Fe^{2+}V_2^{3+}O_4$
    Nolanite, $(V^{3+},Fe^{2+},Fe^{3+},Ti)_{10}O_{14}(OH)_2$
    Vuorelainenite, $(Mn^{2+},Fe^{2+})(V^{3+},Cr^{3+})_2O_4$
    Tantite, Ta_2O_5
    Kimrobinsonite, $(Ta,Nb)(OH)_{5-2x}(O,CO_3)_x$, x near 1.2

Silicates
    Cavansite, $Ca(VO)Si_4O_{10} \cdot 4H_2O$
    Pentagonite, $CaVOSi_4O_{10} \cdot 4H_2O$
    Goldmanite, $Ca_3(V, Al, Fe)_2(SiO_4)_3$
    Haradaite, $SrV^{4+}Si_2O_7$
    Saneroite, $Na_2(Mn^{2+},Mn^{3+})_{10}Si_{11}VO_{34}(OH)_4$
    Erlianite, $(Fe^{2+},Mg)_4(Fe^{3+},V^{3+})_2Si_6O_{15}(OH,O)_8$
    Natalyite, $Na(V,Cr)Si_2O_6$
    Roscoelite, $K(V,Al,Mg)_2AlSi_3O_{10}(OH)_2$
    Mukhinite, $Ca_2Al_2V(SiO_4)_3OH$
    Chernykhite, $(Ba,Na)_{1-x}(V,Al,Mg)_2(Si,Al)_4O_{10}(OH)_2$ (x about 0.5)
    Nagashimalite, $Ba_4(V,Ti)_4B_2Si_8O_{27}(O,OH)_2CL$
    Suzukiite, $BaVSi_2O_7$

Table 3 (*Continued*)

Franciscanite, $Mn_6V_{2-x}Si_2(O,OH)_{1.3}$
Kurumsakite, $(Zn,Ni,Cu)_8Al_8V_2Si_5O_{35} \cdot 27H_2O$
Ardennite, $Mn_4(Al,Mg)_6(SiO_4)_2(Si_3O_{10})[(As,V)O_4](OH)_6$
Medaite, $(Mn,Ca)_6(V,As)Si_5O_{18}OH$

Phosphates and arsenates
Sincosite, $CaV_2(PO_4)(OH)_4 \cdot 3H_2O$
Vanadinite, $Pb_5(VO_4)_3Cl$
Kombatite, $Pb_{14}(VO_4)_2O_9Cl_4$
Leningradite, $PbCu_3(VO_4)_2Cl_2$
Cassedanneite, $Pb_5VO_4(CrO_4)_2 \cdot H_2O$
Rankachite, $CaFeV_4W_8O_{36} \cdot 6H_2O$
Tomichite, $(V,Fe)_4Ti_4AsO_{13}OH$

Sulfates and tellurates
Monsmedite, $K_2O \cdot Tl_2O_3 \cdot 8SO_3 \cdot 15H_2O$
Minasragrite, $VOSO_4 \cdot 5H_2O$
Stanleyite, $VO(SO_4) \cdot 6H_2O$
Cheremnykhite, $Pb_3Zn_3TeO_6(VO_4)_2$
Hemloite, $(As,Sb)_2(Ti,V,Fe,Al)_{12}O_{22}OH$

Vanadates of the alkalis and Cu
Metamunirite, β-$NaVO_3$
Munirite, $NaVO_3 \cdot 2H_2O$
Barnesite, $Na_2V_6O_{16} \cdot 3H_2O$
Bannermanite $(Na,K)_xV_x^{4+}V_{6-x}^{5+}O_{15}$
Howardevansite, $NaCuFe_2^{3+}(VO_4)_3$
Huemulite, $Na_4MgV_{10}O_{28} \cdot 24H_2O$
Blossite, α-$Cu_2V_2O_7$
Ziesite, β-$Cu_2V_2O_7$
Stoibertie, $Cu_5V_2O_{10}$
Mcbirneyite, $Cu_3(VO_4)_2$
Fingerite, $Cu_{11}O_2(VO_4)_6$
Volborthite, $Cu_3V_2O_7(OH)_2 \cdot 2H_2O$
Calciovolborthite, $CaCu(VO_4)OH$
Vesignieite, $Cu_3Ba(VO_4)_2(OH)_2$
Lyonsite, $Cu_3Fe_4^{3+}(VO_4)_6$

Vanadates of Mg, Ca, Sr, or Ba
Metarossite, $CaV_2O_6 \cdot 2H_2O$
Rossite, $CaV_2O_6 \cdot 4H_2O$
Metahewettite, $CaV_6O_{16} \cdot 3H_2O$
Hewettite, $CaV_6O_{16} \cdot 9H_2O$
Pintadoite, $Ca_2V_2O_7 \cdot 9H_2O$
Hendersonite, $Ca_2V^{4+}V_8^{5+}O_{24} \cdot 8H_2O$
Fernandinite, $CaV_2^{4+}V_{10}^{5+}O_{30} \cdot 14H_2O$
Pascoite, $Ca_3V_{10}O_{28} \cdot 17H_2O$
Sherwoodite, $Ca_9Al_2V_4^{4+}V_{24}^{5+}O_{80} \cdot 56H_2O$

Table 3 (*Continued*)

Melanovanadite, $Ca_2V_4^{4+}V_6^{5+}O_{25} \cdot nH_2O$
Grantsite, $Na_4Ca(V^{4+}O)_2V_{10}^{5+}O_{30} \cdot 8H_2O$
Hummerite, $KMgV_5O_{14} \cdot 8H_2O$
Metadelrioite, $CaSrV_2O_6(OH)_2$
Delrioite, $CaSrV_2O_6(OH)_2 \cdot 3H_2O$
Straczekite, $(Ca,K,Ba)\ (V^{5+},V^{4+})_8O_{20} \cdot 3H_2O$
Gamagarite, $Ba_2(Fe^{3+},Mn)(VO_4)_2(OH,H2O)$

Vanadates of Al, rare earths, Pb, V, or Bi
Steigerite, $AlV^{5+}O_4 \cdot 3H_2O$
Alvanite, $Al_6(VO_4)_2(OH)_{12} \cdot 5H_2O$
Metaschoderite, $Al_2PO_4VO_4 \cdot 6H_2O$
Schoderite, $Al_2PO_4VO_4 \cdot 8H_2O$
Satpaevite, $Al_{12}VO_2^{4+}V_6^{5+}O_{37} \cdot 30H_2O$
Vanalite, $NaAl_8V_{10}O_{38} \cdot 30H_2O$
Wakefieldite-(Y), YVO_4
Wakefieldite-(Ce), $(Ce,Pb^{2+},Pb^{4+})VO_4$
Chervetite, $Pb_2V_2O_7$
Mottramite, $Pb(Cu,Zn)VO_4OH$
Descloizite, $Pb(Zn,Cu)VO_4OH$
Dreyerite, $BiVO_4$
Pucherite, $BiVO_4$
Clinobisvanite, $BiVO_4$
Schumacherite, $Bi_3[(V,As,P)O_4]_2O(OH)$
Namibite, $CuBi_2VO_6$
Pottsite, $HPbBi(VO_4)_2 \cdot 2H_2O$
Duhamelite, $Cu_4Pb_2Bi(VO_4)_4(OH)_3 \cdot 8H_2O$

Vanadates of U, Mn, Fe, or Ni
Uvanite, $U_2^{6+}V_6^{5+}O_{21} \cdot 15H_2O$
Vanuranylite, $(H_3O)_2(UO_2)_2V_2O_8 \cdot 4H_2O$
Strelkinite, $NaUO_2VO_4 \cdot 3H_2O$
Carnotite, $K_2(UO_2)_2(VO_4)_2 \cdot 3H_2O$
Margaritasite, $(Cs,K,H_3O)_2(UO_2)_2(VO_4)_2 \cdot H_2O$
Sengierite, $Cu_2(UO_2)_2V_2O_8 \cdot 6H_2O$
Metatyuyamunite, $Ca(UO_2)_2(VO_4)_2 \cdot 3\text{–}5H_2O$
Tyuyamunite, $Ca(UO_2)_2V_2O_8 \cdot 5\text{–}8H_2O$
Rauvite, $Ca(UO_2)_2V_{10}O_{28} \cdot 16H_2O$
Francevillite, $(Ba,Pb)(UO_2)_2(VO_4)_2 \cdot 5H_2O$
Metavanuralite, $Al(UO_2)_2(VO_4)_2OH \cdot 8H_2O$
Vanuralite, $Al(UO_2)_2(VO_4)_2OH \cdot 11H_2O$
Curienite, $Pb(UO_2)_2(VO_4)_2 \cdot 5H_2O$
Fritzscheite, $Mn(UO_2)_2(PO_4,VO_4)_2 \cdot 10H_2O$
Palenzonaite, $(Ca,Na)Mn_2V_3O_{12}$
Santafeite, $NaMn_3(Ca,Sr)\ (V,As)_3O_{13} \cdot 4H_2O$
Pyrobelonite, $PbMnVO_4OH$
Brackebuschite, $Pb_2(Mn,Fe^{2+})(VO_4)_2 \cdot H_2O$
Cechite, $Pb(Fe,Mn)VO_4OH$

Table 3 (*Continued*)

Heyite, $Pb_5Fe_2^{2+}(VO_4)_2O_4$
Mounanaite, $PbFe_2^{3+}(VO_4)_2(OH)_2$
Schubnelite, $Fe_{2-x}^{3+}(V^{5+},V^{4+})_2O_4(OH)_4$
Fervanite, $Fe_4^{3+}(VO_4)_4 \cdot 5H_2O$
Kazakhstanite, $Fe_5^{3+}V_3^{4+}V_{12}^{5+}O_{39}(OH)_9 \cdot 9H_2O$
Bokite, $KAl_3Fe_6^{3+}V_6^{4+}{}_{20}^{5+}O_{76} \cdot 30H_2O$
Rusakovite, $(Fe^{3+},Al)_5[(V,P)O_4]_2(OH)_9 \cdot 3H_2O$
Kolovratite, vanadate of Zn and Ni

The effects of oxidation–reduction (Eh) and pH conditions on mineralogy of vanadium have been studied in considerable detail in the uranium–vanadium deposits of the red beds of the Colorado Plateau (Evans and Garrels, 1958; Garrels and Larsen, 1958). In these deposits, the weathering of primary mineral montroseite, VO(OH), leads to the sequence: montroseite → paramontroseite $[VO_2]$ → doloresite $[H_8V_6O_{16}]$ → corvusite $[V_2O_4 \cdot 6V_2O_5 \cdot nH_2O]$ → stanleysite $[VOSO_4 \cdot 6H_2O]$ → hewettite $[CaV_6O_{16} \cdot 9H_2O]$ → carnotite in an acidic environment; the sequence in the alkaline environment is more complicated. Figure 1 shows the equilibrium stability fields and paragenetic sequence for vanadium minerals as a function of Eh and pH.

4. PRODUCTION

There are few ores from which vanadium can be recovered economically as a single product. Most of the vanadium production comes as byproducts and coproducts in the extraction of other elements such as iron, phosphorus, and uranium. Ores from which vanadium can potentially be recovered are found in many parts of the world. The worldwide reserve of recoverable vanadium has been estimated to be 17.4 million tons, and identified resources (include reserves) total about 62 million tons (Table 4). About one-third of the vanadium resources are located in Africa or North America, and about 24% are found in Europe and <4% in both Asia and South America. About 46 million tons of the vanadium resources are contained in various iron ores, 8.5 million tons in crude oil and tar sands, and 7.5 million tons in phosphate rock and phosphatic shale (Morgan, 1980). Vanadium reserves in the United States are estimated to be 115,000 tons and identified resources are estimated to be about 10 million tons (Morgan, 1980).

About 83% of vanadium recently produced from mines comes from vanadiferous magnetite (Fe_2O_3) in South Africa, China, and Russia (Hilliard, 1992). The remaining 17% of worldwide vanadium production (from primary sources) is recovered from the oil industry. During the past 50 years or so, global production of vanadium has grown from about 4,000 tons during the period

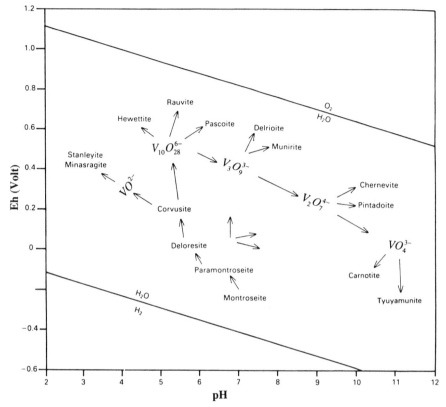

Figure 1. Effects of oxidation–reduction conditions on the formation of secondary vanadium minerals. *Source:* Williams (1990).

1952–1956 to a maximum of slightly over 40,000 tons in 1980 (Table 5); the steepest increase occurred between 1960 and 1980. Outputs during the 1980s and 1990s have fluctuated between 30,000 and 40,000 tons per year (Table 5).

South Africa is the world's leading producer of vanadium and accounts for about 50% of current global output. Other major producing countries include Russia, China, and the United States (Table 5). Prior to about 1985, over 50% of the vanadium mined in the United States was a coproduct with uranium from mineralized sandstone of the Colorado Plateau. More of the V_2O_5 produced in the United States and other countries is now being recovered from refinery residues, fly ash, boiler scales, and spent catalysts; about 2,250 tons of vanadium oxides were recovered from such sources in the United States in 1991 (Hilliard, 1992). Western Australia has large deposits of magnetite ores containing about 0.32% V_2O_5, and the tar sands of Alberta, Canada, represent a huge reservoir of vanadium; these two sources are not currently being tapped. Japan relies heavily on imported vanadium, although some vanadium is recovered from spent catalysts and other secondary sources.

Table 4 Worldwide Vanadium Resources

Region	Reserves	Other	Total
North America	115	20,920	21,035
United States	115	9,920	10,035
Canada	—	11,000	11,000
South America	250	1,500	1,750
Chile	150	100	250
Venezuela	100	1,400	1,500
Europe	8,160	6,680	14,840
Finland	140	—	140
Norway	20	180	200
Sweden	—	1,000	1,000
USSR	8,000	4,000	12,000
East European countries[a]		1,500	1,500
Africa	8,600	12,400	21,000
South Africa	8,600	11,400	20,000
Other		1,000	1,000
Asia	100	1,300	1,400
India	100	300	400
Other		1,000	1,000
Oceania	200	1,800	2,000
Australia	200	1,300	1,500
New Zealand		500	500
World total	17,400	44,600	62,000

Source: Morgan (1980).

[a] Consists of former Communist Block countries

Irrespective of whether vanadium occurs as a primary ore, a coproduct, or in petroleum, the principal starting material for production of all vanadium compounds is V_2O_5. The first stage in processing of vanadiferous ores is the production of oxide concentrate. Crushed ore is mixed with sodium salt and the mixture is roasted at about 850 °C to convert the vanadium to soluble sodium metavanadate. The vanadium chloride formed is extracted by leaching with water and precipitated at pH 3.0 as a red cake (sodium hexavanadate) by adding sulfuric acid. The red cake is fused at 700 °C to yield technical grade (86 weight percent) vanadium pentoxide. The red cake can be purified to 99.8% by various chemical methods (Hilliard, 1992).

Vanadium has been extracted as coproduct with uranium from carnotite by direct leaching of the ore with sulfuric acid. The uranium and vanadium are separated from the pregnant liquor by liquid–liquid extraction techniques (Hilliard, 1992). Recovery from spent catalysts has become an important source of vanadium. The material is first roasted in controlled atmosphere to dissolve the vanadium, and the residue is leached to remove the metal, which is precipitated and purified by various methods. Methods for preparing ferro-vanadium (used for making steel) from the pentoxide are described by Hilli-ard (1992).

Table 5 Worldwide Production of Vanadium

Country	1952–1956	1960	1970	1972	1974	1976	1978	1980	1982	1984	1986	1988	1990
South Africa[a]	593	1,494	7,127	8,650	9,887	11,656	12,885	14,000	12,600	13,798	15,361	16,380	17,106
China	—	—	—	—	—	—	2,200	5,000	5,000	5,000	4,500	4,500	4,500
Former USSR	—	—	3,800	3,720	3,230	8,800	10,500	10,500	10,500	10,500	9,600	9,600	9,000
United States	3,161	4,971	5,319	4,887	4,870	7,376	4,272	6,326	5,611	3,318	2,114	2,950	2,308
Japan	—	—	—	—	—	—	—	710	754	770	843	880	889
Others	270	625	3,250	2,982	2,775	3,377	4,362	3,975	3,700	3,376	—	—	—
Global total	4,024	7,090	19,496	20,239	20,762	31,209	34,219	40,511	38,165	36,762	32,418	34,310	33,803

[a] Including Southwest Africa.

Vanadium is usually traded on the international market in the form of technical grade V_2O_5 and 50–80% FeV. Most of the exports from South Africa and China are in the form of vanadium-bearing iron slag. Aluminum–vanadium master alloys containing fixed ratios of vanadium and aluminum (used in the manufacture of titanium alloys) also represent an important share of the vanadium market.

5.   USES

The relative scarcity, the difficulty of producing pure metal, and the high cost worked against any large-scale commercial applications of vanadium after its discovery. During the 1860s, vanadium salts were used to some extent as catalyst in dyeing aniline black, in the manufacture of vanadium writing ink, and in producing blue, green, brown, and purple colors in glass and pottery (Alexander, 1929). The real impetus for the development of a vanadium industry came with the metallurgical investigations between 1895 and 1900 by French scientists, who had access to large supplies of vanadium in slags of the celebrated Schneider works at Creusot. Although the Firminy Steel Works of France experimented with vanadium in armor plates in 1896, it was the extensive studies by Prof. J. O. Arnold of Sheffield University that revealed its superior alloy properties at the time when automotive and other industries were developing demands for specialty steels that would resist grueling operating conditions and could stand considerable abuse. Henry Ford pioneered the automotive use of vanadium steels in the 1908 Model T. Other early famous applications included the engine of Charles Lindbergh's "Spirit of St. Louis" and the locks of the Panama Canal. By the time of World War I, vanadium was classified as one of the most essential war metals and was practically withdrawn from commercial use.

Following in the heels of crucial metallurgical studies by Arnold and others was the propitious discovery of the first rich deposits of vanadium ore (petronite) in the Peruvian Andes by Riza Patron in 1905. The discovery ensured a plentiful supply of vanadium and this commodity immediately went from being a rare and obscure element to becoming one of the principal metals used in the manufacture of steel alloys.

5.1.   Metallurgical Uses

Early discovery of the hardening of steel by vanadium, owing to grain refinement, stimulated other research on the alloying qualities of the element. It was soon discovered that small quantities of vanadium (<0.2%) imparted pronounced shock and wear resistance to steel, which could be used in such products as crankshafts, pinions, and gears for heavy-duty equipment and machinery. Because of high strength-to-weight ratios, excellent load-bearing characteristics, and improvements in formability of steels containing small amounts of vanadium, such alloys have been favored in the construction of aircrafts, automobiles, and trains. Vanadium forms stable carbides and resists high-temperature abrasion; hence it is used as a component of high-speed tool steels and rotors that operate at high temperature (Morgan, 1980). Over a range of composition and temperatures, vanadium can prevent graphitization or promote uniform size and distribution of carbon flakes and can enhance wear resistance (Morgan, 1980). In describing the use of vanadium in war materials, the Vanadium Corporation of America (1922) contended that

> Vanadium entered into a very wide range of materials including armor for warships, tanks, cars, motor trucks, gunshields and helmets, machine gun barrels and mechanisms, gun mounts, both heavy and light, gun forges, submarine and destroyer engine crank shafts and other forges, valve springs for Diesel oil engines for submarines and for the Liberty and Rolls-Royce airplane engines, forges for the Liberty and Curtiss airplane engines, sheet metal fittings for airplanes and flying boats, motor truck parts and engines, tank engines, tractors and gun tractor mounts, torpedo air flasks, [and] essential parts of nitre plants for production of nitre from atmospheric hydrogen [by the Haber process] (p. 5)

In general, tool, dye, and cutting steels, high-strength structural steel, and wear-resistant cast iron contain 0.01–0.5% vanadium with various amounts of chromium, manganese, molybdenum, and tungsten. Higher vanadium contents are found in steels for special purposes such as in dies, drills, reamers, and finishing tools. Vanadium contents of up to 5% are found in steels in which increased wear resistance is essential; such alloys are found in steam power plants, car and locomotive forges, gears, axles, springs, transmission shafts, rock crushers, dredges, armor plates, and shells (Hudson, 1964).

Vanadium is also a component of many utilitarian nonferrous alloys. Cupro-vanadium (8–12% vanadium) brass and bronze have high strength and resistance to corrosion and have been used to make airplane propeller bushings and boat propellers. Substitution of vanadium for manganese gives bronzes increased tensile strength, and elongation and reduction of area; such bronzes have been used in submarine work. Vanadium–titanium–carbon alloys are temperature-resistant and find application in high-speed aircraft. Chromium–cobalt–tungsten alloys give increased wear-resisting qualities to drills, excavating machinery, and surgical instruments.

Addition of 2.5–40% vanadium has been used to control thermal expansion, electrical resistance, and grain size of aluminum alloys and to strengthen them at high temperatures (Busch and McNulty, 1959). A titanium–aluminum–vanadium alloy, Ti–6A–4V, which is widely used in jet engines, airframes, and other aircraft parts, accounts for a large portion of the titanium-based alloys consumed by industries. Other titanium–aluminum–vanadium alloys such as Ti–10V–2Fe–3A, Ti–15V–3Cr–3Al–3Sn, and Ti–10V–2Fe–3Al have been used in forges of the newer generation of commercial jetliners. Vanadium–columbium alloys have good corrosion resistance and high tensile strength at 1,000 °C and have been used in space vehicles. Vanadium disilicide likewise has been used in the production of high-temperature refractory material (IPCS, 1988).

Metallurgical applications account for about 85–95% of the worldwide vanadium consumption, about 75–85% of this fraction going for steel making (Hilliard, 1992).

5.2. Therapeutic Uses

Vanadium was extensively used in therapeutic agents in France during the early part of this century (Laran, 1899; Lyon, 1899; Lyennet et al., 1899; Manquet, 1899). Alexander (1929) provided the following excellent summary of pharmacological applications for vanadium in the 1920s:

> Vanadium compounds are especially used as a means of supplying the system with needed oxygen, thus compensating in a measure for the deficiency of haemoglobin in anaemias, and assisting the antiseptic and antitoxic reactions of the organism in the acute infections, especially influenza and pneumonia. In gastric catarrhs, they are said to destroy the microbes of fermentation without inhibiting the digestive enzymes. . . . In pulmonary tuberculosis, vanadium compounds are said to increase appetite, improve nutrition, and heighten general resistance. In syphilis, intravenous injection of 5 doses of 3/4 grain (0.05 g) each of sodium orthovanadate at 2-day intervals, associated with inunctions of mercury, lead to symptomatic relief and disappearance of the Wassermann reaction. In chronic rheumatism and rheumatoid arthritis, they have been employed with much the same object, as oxidizers and general stimulants to nutritive process (p. 896)

Topically, vanadic acid was employed as an antiseptic on the skin and on mucous membranes of the bladder, vagina, and uterine canal. A powder containing vanadic acid, bismuth subgallate, and sodium chlorate was used in dusting applications as a replacement for iodoform. A solution of vanadium hypochloride and sodium chlorate was applied to ulcers, superficial abscesses, and infected wound (Laran, 1899; Lyon, 1899; Lyennet et al., 1899; Manquet, 1899). In spite of the many favorable reports claimed by European observers, vanadium lacked official pharmacological recognition in North America and its therapeutic applications were limited.

5.3. Use in Pigments

As a polyvalent transition metal, vanadium compounds show a broad range of hues, from yellow to green and from red to black. Sigmund Lehner (1926) claimed that it was Berzelius who discovered the use of vanadium to make ink by adding ammonium vanadate to a filtered decoction of galls. Good indelible ink was also prepared from hydrogallic acid, powdered gum arabic, and a neutral solution of ammonium vanadate (Alexander, 1929). A widely used black marking ink known as "Jetoline" was made from a mixture of aniline salt solution and sodium chlorate with a small amount of vanadium as a catalyst. Vanadium salts continue to find limited use as catalysts in the manufacture of resinous black printing ink from tar oils and as driers where cobalt and other salts could not be tolerated (Hudson, 1964).

Vanadium salts are used as mordants in the dyeing of cotton and viscose rayon, and, particularly, for fixing aniline black on silk. The art of dyeing by analine black was discovered in the middle of the last century and was done by adding vanadium (as a catalyst) to a mixture of aniline hydrochloride and a soluble source of oxygen such as sodium or potassium chlorate. Ammonium metavanadate has used as a catalyst in the dyeing of leather and fur (NAS, 1974).

Gardner (1921) and Alexander (1929) noted that "linseed oils containing vanadium dryers dry about twice as rapidly as do oils containing an equivalent amount of manganese dryer, and almost five times as fast as do oils containing an equivalent amount of lead dryer" (Gardner, 1921, p. 222). Besides acting as a quick-acting drying agent, the vanadium yielded a tough, uniform film.

Ammonium metavanadate is employed as a colorant or color enhancer in several ceramic glazes, particularly the zircon-vanadium blues. These glazes typically are not used in dinnerware, but in such products as wall and floor tiles (NAS, 1974).

5.4. Use as a Catalyst

Although it is claimed that Berzelius discovered the catalytic properties of vanadium, the first published account is believed to be that of Walter, who in 1895 used the pentoxide as a catalyst in the oxidation of toluene to benzaldehyde and acetic acid, and anthracene to anthraquinone (Alexander, 1929). In 1892 Meyers proposed the use of vanadium pentoxide as a catalyst in making sulfuric acid (Meyers, 1899). Besides being much cheaper, vanadium catalyst is insensitive to many of the platinum "poisons" encountered in sulfuric acid manufacture; it remains one of the primary uses for vanadium. Today, vanadium catalysis is involved in a wide array of chemical syntheses and oxidation reactions (Table 6). Precise data on the quantity of vanadium so consumed are unavailable because of the proprietary nature of some of the applications.

Closely allied to the chemical industry is the polymer industry, in which vanadium compounds are featured extensively as catalysts in copolymerization process, such as in the synthesis of amorphous copolymers derived from ethylene, propylene, and nonconjugated diene (Ritchie, 1968; NAS, 1974). Other vanadium catalysts, such as vanadium oxychloride, have also been used in the polymerization of high-density polyethylene (NAS, 1974). It was estimated that 100–150 tons of vanadium per year was used in the United States in polymer synthesis and processing during the 1970s.

Table 6   Some Catalytic Uses of Vanadium in the Chemical Industry

Production of sulfuric acid
Catalytic combustion of exhaust gases
Manufacture of phthalic anhydride
Manufacture of maleic anhydride
Production of aniline black
Oxidation of cyclohexanol to adipic acid
Oxidation of ethylene to acetaldehyde
Preparation of vinyl acetate from ethylene
Oxidation of anthracene to anthraquinone
Oxidation of toluene or xylene to aromatic acids
Oxidation of furfural to fumaric acid
Oxidation of hydroquinone to quinone
Ammonolysis/oxidation of toluene, xylene, and propylene
Manufacture of cyclohexylamine from cyclohexanol and ammonia
Catalytic synthesis of ethylene–propylene rubber
Polymerization of olefins and dienes, and copolymerization of olefins with dienes
Epoxidizing of cyclohexane
Elimination of disulfide bridges in organic compounds
Hydrorefining of heavy oils

5.5.   Other Uses

It would be impossible to discuss all the commercial applications of vanadium in this report. The additional uses mentioned below represent but a small sample designed to show how deeply vanadium has permeated our industrial culture:

- Vanadium oxides and vanadates have been used in producing glasses of various types and colors, in thermistors and switching elements.
- The pentoxide and other compounds have been used, since 1874, as developers, sensitizers, and coloring agents in photography and cinematography.
- Vanadium hydride can be used as a neutron moderator in atomic reactors;
- Europium-activated yttrium vanadate is used in color television tubes.
- Soluble salts of vanadium have been used in insecticides, fungicides, and "fertilizers";
- Vanadium slags are used in casting shops as molding material.
- Vanadium has been used in window glasses and eyeglasses to curtail the transmission of ultraviolet radiation.
- Vanadium improves the magnetic quality of some material and has been used in construction of magnets; alloys of vanadium–gallium have shown some superconductivity properties.

- Vanadium has extensive applications in the nuclear industry because of its good structural strength and low fission neutron cross section; ductile vanadium metal of high purity has been rolled into sheaths used as cladding material for fuel elements in nuclear reactors
- Rechargeable storage battery is based on a redox flow cell using electrically charged vanadium solution have been built, and a number of vanadium oxides can be used as active components in rechargeable lithium batteries (Skyllas-Kazacos and Grossmith, 1987; Wiesener et al., 1987).

REFERENCES

Alexander, J. (1929). Vanadium and some of its industrial applications. Chem. and Indust. **49**, 871–87; 895–901.

Aller, L. H. (1961). *Abundance of the Elements.* New York, Interscience.

ATSDR. (1992). *Toxicological Profile for Vanadium.* Agency for Toxic Substances and Disease Registry, U.S. Dept. of Health & Human Services, Washington, D.C.

Baskerville, C. (1899). The occurrence of vanadium, chromium and titanium in peats. *J. Am. Chem. Soc.* **21**, 706–710.

Beauvallet, M. P. (1859). Sur la presence du vanadium dans l'argile de Gentilly. *C. R. Acad. Sci. (Paris)* **49**, 301–306.

Bechi, E. (1879a). Sur la composizione delle rocce della minera di Montecatini. *Mem. Accad. Lincei* **3**, 63–70.

Bechi, E. (1879b). Nuove recherche del boro e del vanadio. *Mem. Accad. Lincei* **3**, 401–405.

Beliles, R. P. (1991). Vanadium. In *Patty's Industrial Hygiene and Toxicology.* New York, Wiley, Vol. 2, Part C. pp. 2317–2329.

Busch, P. M., and McNulty, K. W. (1959). Vanadium. In *Bureau of Mines Minerals Yearbook.* U.S. Department of the Interior, Washington, D.C.

Clark, A. M. (1990). *Hey's Mineral Index.* London, Chapman & Hall.

Dowdeswell, G. F. (1878). On the structural changes which are produced in liver under the influence of the salts of vanadium. *J. Physiol. (London),* **1**, 257–260.

Evans, H. T., and Garrels, R. M. (1958). Thermodynamic equilibria of vanadium in aqueous systems applied to the interpretation of the Colorado ore deposits. *Geochim. Cosmochim. Acta* **15**, 131–142.

Evans, H. T., and Landergren, S. (1978). Vanadium. K. H. Wedepohl (Ed.), In *Handbook of Geochemistry* Springer-Verlag, Berlin, Vol. II/2, Section 23.

Fleischer, M. (1987). *Glossary of Mineral Species.* Tucson, Arizona, Mineralogical Record, Inc.

Gamgee, A. E., and Larmuth, L. (1877). On the action of vanadium upon the intrinsic nervous mechanism of the frog's heart. *J. Anat. (London)* **11**, 235–238.

Gardner, H. A. (1921). Vanadium compounds as driers for linseed oil. *J. Indust. Eng. Chem.* **14**, 222–224.

Garrels, R. M., and Larsen, E. S. (1958). Geochemistry and Mineralogy of the Colorado Plateau Uranium Ores. Professional Paper 320. U.S. Geological Survey.

Goles, G. G. (1969). Cosmic abundance. In K. H . Wedephl, (Ed.), *Handbook of Geochemistry.* Springer, Berlin, Vol. 1.

Grayson, M. (1983). Vanadium. In *Kirk-Othmer Encyclopedia of Chemical Technology.* 3d ed. New York, Wiley, Vol. 23, pp. 688–704.

Hecht, S. (1918). The physiology of *Ascidia atra leuseur. Am. J. Physiol.* **45,** 157–162.

Heinz, M. (1911). Untersuchungen uber das Blut der Ascidien. *Hoppe-Seylers Z. Physiol. Chem.* **72,** 494–498.

Hilliard, H. E. (1992). Vanadium. In *Bureau of Mines Minerals Yearbook.* U.S. Department of the Interior, Washington, D.C.

Hudson, T. G. F. (1964). *Vanadium: Toxicological and Biological Significance.* Elsevier, Amsterdam.

IPCS. (1988). Environmental Health Criteria for Vanadium. Report No. 81. International Program on Chemical Safety, World Health Organization, Geneva.

Jackson, D. E. (1911). The pharmacological action of vanadium. *J. Pharmacol. Exp. Ther.* **3,** 477–481.

Jackson, D. E. (1912). The pulmonary action of vanadium together with a study of the peripheral reactions to the metal. *J. Pharmacol. Exp. Ther.* **4,** 1–7.

Laran, M. (1899). Le vanadium et ses composes: Applications therapeutiques. *Presse Med.* **1,** 190–194.

Larmuth, L. (1877). On the poisoning activity of vanadium in ortho-, meta-, and pyrovanadic acid. *J. Anat. (London),* **11,** 251–255.

Laul, J. C. and Schmitt, R. A. (1973). Chemical composition of Luna 20 rocks and soil and Apollo 16 soils. *Geochim. Cosmochim. Acta.* **37,** 927–935.

Lehner, S. (1926). *Ink Manufacture.* London, Scott, Greenwood & Sons.

Lyennet, B., Guinard, L., and Martin, E. (1899). Le metavanadate de soude, son action physiologique. *C. R. Soc. Biol. (Paris)* **51,** 707–711.

Lyon, G. (1899). Le vanadium et ses composes en therapeutiques. *Rev. Ther. Med. Chir.* **65,** 721–725.

Manquet, G. (1899). Les derives du vanadium en therapeutique. *Bull. Med.* **13,** 685–687.

Marden, J. W., and Rich, M. N. (1927). Vanadium. *Indust. Eng. Chem.* **19,** 786–789.

Mason, B. (1966). *Principles of Geochemistry.* 3d ed. Wiley, New York.

Mellor, J. W. (1952). Vanadium In *Comprehensive Treatise on Inorganic and Theoretical Chemistry.* 30, Vol. 9, pp. 714–722.

Meyers, R. (1899). Sulfuric acid catalyst. *Jahrb. Chem.* **9,** 304–308.

Morgan, G. A. (1980). Vanadium. In *Mineral Facts and Problems.* Bureau of Mines Bulletin 671. U.S. Department of the Interior, Washington, D.C.

NAS. (1974). *Medical and Biological Effects of Vanadium.* National Academy of Sciences Press, Washington, D.C.

Priestley, J. (1876). On the physiological action of vanadium. *Phil. Trans., Part* **2,** 495–499.

Prinz, M. (1967). Geochemistry of basaltic rocks: trace elements. In A. Poldervaart and H. H. Hess (Ed), *Basalts* Vol. 1, pp. 271–293.

Proescher, F., Seil, H. A., and Stillians, A. W. (1917). A contribution to the action of vanadium with particular reference to syphilis. *Am. J. Syph.* **1,** 347–349.

Ritchie, P. D. (1968). *Vinyl and Allied Polymers.* Iliffe Books, London, p. 73.

Rockwell, G. J. (1879). Index to the literature on vanadium—1801 to 1877. *Ann. New York Acad. Sci.* **1,** 133–148.

Roscoe, H. E. (1868). On vanadium, one of the trivalent group of elements. *Phil. Mag.* **35,** 307–314

Roscoe, H. E. (1870). Researches on vanadium. Part 2. Phil. Mag. **39,** 146–150.

Sefstrom, N. G. (1831). Sur le vanadium, metal nouveau, trouve dans du fer en barres de Eckersholm, forge que tire sa mine de Taberg dans le Smaland. *Ann. Chim. Phys.* **46,** 105–111.

Shacklette, H. T., Hamilton, J. C., Boerngen, J. G., and Bowlest, J. M. (1971). Elemental composition of surficial materials in the conterminous United States. U.S. Geological Survey, Professional Paper 574-D. U.S. Department of the Interior, Washington, D.C.

Skyllas-Kazacos, M., and Grossmith, F. (1987). Efficient vanadium flow cell. *J. Electrochem. Soc.* **134,** 2950–2953.

Suess, H., and Urey, H. (1956). Abundance of the elements. *Rev. Modern Phys.* **28,** 53–61.

Terreil, M. A. (1860). De la presence du vanadium dans les argiles de Forges-les-Eaux et de Dreux. *C. R. Acad. Sci. (Paris)* **51,** 94–96.

Turekian, K. K. and Wedepohl, (1961). Abundances of elements in principal types of sedimentary rocks. Bull. Geol. Soc. Am. 72: 175–190.

Vanadium Corporation of America. (1922). Vanadium: The master alloy in war and peace. VCA, New York.

Vinogradov, A. P. (1973). Preliminary data on lunar soils collected by the unmanned space craft. *Geochim. Cosmochim. Acta* **37,** 721–727.

Wakita, H., Schmitt, R. A., and Rey, P. (1970). Elemental abundance of major and minor trace elements in Apollo 11 lunar rocks, soil and core samples. *Proc. Apollo 11 Lunar Sci. Conf.,* Vol. 2, pp. 1685–1691.

Webb, D. A. (1939). Observations on the blood of certain ascidians, with special reference to the biochemistry of vanadium. *J. Exp. Biol.* **16,** 499–523.

Weeks, M. E. (1968). *Discovery of the Elements,* 6th Edition. J. Chem. Educ. Incorp., Easton, PA.

Wiesener, K. W., Schneider, D, and Ilis, D. (1987). Vanadium oxides in electrodes for rechargeable lithium cells. *J. Power Sources* **20,** 157–164.

Williams, P. A. (1990). *Oxide Zone Geochemistry.* Ellis Horwood, New York.

Witz, G., and Osmond, F. (1882). Contributions a l'industrie du vanadium. *Bull. Soc. Chim. (Paris)* **38,** 49–52.

Yudovich, Y. E., Korycheva, A. A., Obruchnikov, A. S., and Stepanov, Y. V. (1972). Mean trace-element contents in coal. *Geochim. Cosmochim. Acta* **9,** 1786–1793.

2

EMISSION OF VANADIUM INTO THE ATMOSPHERE

Jerome O. Nriagu

Department of Environmental and Industrial Health, School of Public Health, University of Michigan, Ann Arbor, MI 48109

Nicola Pirrone

CNR Institute for Atmospheric Pollution, c/o DEIS, University of Calabria, 87036 Rende, Italy

Vanadium in the Environment. Part 1: Chemistry and Biochemistry, Edited by Jerome O. Nriagu.
ISBN 0-471-17778-4. © 1998 John Wiley & Sons, Inc.

1. INTRODUCTION

The atmosphere is a medium for the transport and reaction of vanadium and, as such, is important in the environmental cycling of this element. Atmospheric loading information is required in air pollution control programs and in the assessment of exposure and risk associated with airborne vanadium. Characterization of vanadium sources provides the framework for mass balance studies and for understanding the transport and dispersion of vanadium pollution on the global, regional, and local scales. Information on sources of vanadium in the atmosphere is clearly fundamental to our ability to model the biogeochemical cycle of vanadium and the modifications imposed by human activities.

Vanadium in the atmosphere comes from a wide range of natural and anthropogenic sources and pathways. Natural sources include volcanic eruptions, wild forest fires, entrainment of dust particles by wind gusts, and suspension of various biological particles such as pollen, cuticles, waxes, and needles. These sources may involve previously deposited anthropogenic vanadium, so current emission rates from the natural sources (better called natural processes) may be different from those of pretechnological times. In other words, current emissions from natural processes do not provide true background information on the natural atmospheric cycle of vanadium.

Anthropogenic emissions of vanadium can be categorized into area sources (such as mobile sources, agricultural burning, and geothermal power) and point sources, which can further be subdivided into combustion sources (boilers of all types; municipal sludge and hazardous waste incinerators; and wood stoves) and manufacturing sources (ferrous and nonferrous metal mining and smelting, foundries, coke production, chemical and petrochemical production, oil shale retorting, etc.). Previous studies (Nriagu and Pacyna, 1988; U.S. EPA, 1990a; ATSDR, 1992) suggest that combustion sources are the dominant contributor of vanadium to the atmospheric environment, and these are also the focus of this chapter.

2. INVENTORY STRATEGY

Very few inventories of vanadium emission into the atmosphere have been reported. Most of the reported inventories rely on the emission factor-based approach developed for facility-specific observations. This approach requires an emission factor, which is a ratio of the mass of vanadium emitted to a measure of source activity. The emission factors can be derived from emission test data, from engineering analyses based on mass balance techniques, or from extrapolation of information from comparable emission sources. Emission factors typically pertain to average emission across a population of sources within any given source category, implying that reported values are likely to show a wide range.

Table 1   Vanadium Concentrations (μg/g) in Fuel (Oil Shale) and Various Ashes from Two Power Plants

Material	Estonia Power Plant	Baltic Power Plant
Fuel (feed)	22	19
Bottom ash (39%)	33	26
Ash from gas passes (8%)	39	32
Cyclon ash (32%)	38	43
Electrostatic precipitator ash (19%)	62	57
Fly ash, >6 μm (1.5%)	81	61
Fly ash, >6 μm (0.5%)	210	118

Source: Hasanen et al. (1997).

The quality of data on emission of vanadium from individual sources varies widely. In addition to difficulties with the determination of emission factors, uncertainties are also introduced into the estimation of control efficiencies and measures of activity level. Changes in processes over time may also introduce bias into the calculated emission rates. As a consequence, inventories reported by various authors can differ widely. With the present state of knowledge, it is fair to say that emission estimates reported for vanadium from any source category can be in error by a factor of 2 or less.

The emission factors used in this chapter have been compiled from emission studies in Europe, the United States, and Canada (Nriagu and Pacyna, 1988; U.S. EPA, 1990b, Benjey and Coventry, 1992). For oil-burning power plants, the average uncontrolled emission factor adopted was 60 g/ton, the range in reported values being 35–130 g/ton (Nriagu and Pacyna, 1988). For coal combustion in utility and commercial/industrial boilers, the uncontrolled emission factor used was 2.5 g/ton (range, 1.8–3.6 g/ton). The calculations considered the annual combustion of bituminous coal, lignite, and anthracite coal

Table 2   Worldwide Atmospheric Emissions of Vanadium from Natural Sources

Source Category	Annual Emission (tons)	
	Range	Geometric Mean
Continental dusts	1.2–30.0	6.0
Seasalt spray	0.1–7.2	0.85
Volcanoes	0.2–11.0	1.5
Wild forest fires	0.1–3.6	0.60
Biogenic processes	0.1–2.4	0.49
Total	1.6–54.2	9.3

* From Nriagu (1990).

Table 3 Worldwide Industrial Emissions of Vanadium (tons/year) into the Atmosphere, by Source Category, Region, and Year

	1983	1985	1987	1989	1991	1993	1995	2000
North America								
Coal combustion	150	164	168	178	177	176	172	160
Oil combustion	285	296	312	328	318	322	314	292
Miscellaneous	44	46	48	51	50	50	49	45
Total	479	506	528	557	545	548	535	497
Central and South America								
Coal combustion	63	75	78	90	93	92	92	89
Oil combustion	5,459	5,464	6,035	6,135	6,253	6,367	6,465	6,704
Miscellaneous	55	55	61	62	63	65	66	68
Total	5,576	5,594	6,174	6,286	6,409	6,524	6,623	6,861
Western Europe								
Coal combustion	118	126	119	118	119	116	111	100
Oil combustion	392	392	406	416	432	428	412	372
Miscellaneous	51	52	53	53	55	54	52	47
Total	561	569	578	587	606	598	576	519
Eastern Europe and Former USSR								
Coal combustion	2,146	2,179	2,292	2,281	1,698	1,481	1,460	1,397
Oil combustion	21,409	21,382	21,478	20,828	19,277	15,675	15,825	16,170
Miscellaneous	236	236	238	231	210	172	173	176
Total	23,790	23,797	24,008	23,341	21,185	17,327	17,458	17,743

Africa								
Coal combustion	327	355	375	375	442	424	419	401
Oil combustion	3,161	3,399	3,433	3,706	4,008	4,196	4,236	4,328
Miscellaneous	35	38	38	41	44	46	47	47
Total	3,523	3,792	3,846	4,122	4,494	4,667	4,701	4,776
Asia								
Coal combustion	2,448	2,877	3,208	3,563	3,625	3,659	3,681	3,698
Oil combustion	24,265	25,300	27,076	30,742	33,255	36,131	37,198	39,916
Miscellaneous	267	282	303	343	369	398	409	436
Total	26,980	28,459	30,587	34,647	37,249	40,188	41,288	44,050
Oceania								
Coal combustion	11.5	9.7	12.5	14.6	12.8	15.3	15.0	14.1
Oil combustion	12.8	13.7	14.3	15.4	15.8	15.8	15.5	14.5
Miscellaneous	0.2	0.2	0.3	0.3	0.3	0.3	0.3	0.3
Total	24.5	23.6	27.1	30.3	28.8	31.5	30.8	28.9
World Total	60,934	62,741	65,747	69,571	70,516	69,883	71,211	74,475

in power plants as well as in industrial, commercial, and residential facilities. Other source categories considered include oil-burning power plants, production of nonferrous and ferrous metals, and the incineration of sewage sludge (EIA, 1991, 1993; U.S. Bureau of Mines, 1987, 1994). In deriving the annual vanadium emissions, a control efficiency of up to 99% was assumed for developed countries, but no controls were imposed on emitting sources in the developing countries. For the Mediterranean Sea region, a control efficiency of 40–95% was employed for electric utilities and industrial plants, and the commercial and residential facilities were assumed to be operating without any controls.

Combustion of fossil fuels and wastes represent the principal sources of vanadium in the atmosphere. The cycling of vanadium through a power plant can be illustrated by a recent detailed study of two large oil shale-fueled thermal plants in northeastern Estonia, which have a combined electricity output of 3,000 MW. The plants require 22 million tons of oil shale at full capacity, and by the end of 1992 had consumed 595 million tons of oil shale and had generated 264 million tons of ash (Hasanen et al., 1997). The oil shales have a low caloric value (8–10 MJ/kg), high ash content (40–50%), and high sulfur content (1.5%) but low nitrogen content (only 0.1%). Table 1 gives the concentrations of vanadium in the raw fuel and various ashes for the two power plants. From the vanadium balance in the system, an average emission factor of 1.1 g/ton or 0.13 mg/MJ was derived (Hasanen et al., 1997). Of the 440 tons of vanadium that pass through the plant each year, only 30 tons (or 7% of total throughput) are released into the atmosphere. Of the 382 tons that were "captured," 117 tons were in the bottom ash, 28 tons in ash from gas passes, and 128 tons in cyclon ash, and 108 tons remained in ash trapped by electrostatic precipitators. The 28 tons of vanadium in the feet that cannot be accounted for probably reflect inputs into water discharges as well as process losses. The fact that most of the vanadium remains in the ash reflects the low volatility of vanadium and its compounds, contrary to the claim by Hope (1994) that gaseous forms make up 66% of the atmospheric vanadium burden.

Annual worldwide emissions of vanadium were estimated for the period of early 1980 to 1995, and the projection to the year 2000 was made from recent forecasts of fuel and industrial output (coal, oil, nonferrous and ferrous metals) published by the World Bank (EIA, 1996). Inventories for two regions of particular interest are also presented, namely, the Mediterranean Sea countries, which burn a lot of oil, and the Great Lakes region, where there is an accumulation of heavy industries.

3. GLOBAL EMISSIONS

Estimated worldwide emissions of vanadium from natural sources total 1.6–54 tons per year, the geometric mean being 9.3 tons per year (Table 2).

Continental dusts are the dominant natural source of vanadium and can account for up to 54% of the average global emission. Volcanoes can contribute up to 20%, but wild forest fires and biogenic sources account for <5% of the total average natural flux.

Worldwide industrial emissions of vanadium into the atmosphere totaled about 61,000 tons in 1983, compared with 71,000 tons in 1995 (Table 3). These values are in agreement with the geometric mean value of 65,000 tons per year (range, 30–142 tons per year) for the period of 1983/1984 published by Nriagu and Pacyna (1988) and Pacyna et al. (1995). The data in Table 3 show that Asian countries now represent the main source of atmospheric vanadium, accounting for about 56% of the global total in 1995. Eastern Europe and former USSR countries accounted for about 25%, South and Central America (9%) and Africa (7%) also made significant contributions to the atmospheric vanadium burden in 1995 (Table 3). Western Europe and North America together accounted for less than 2% of global vanadium emissions from industrial sources. The data in Table 3 thus point to the fact that the developing countries of Asia, Africa, and South America currently account for most of the anthropogenic emissions of vanadium. In regard to pollution of the atmosphere, the "smoke" of industrial progress has shifted to the developing countries.

There are interesting regional differences in the temporal trends of vanadium emissions. Between 1983 and 1995, the emissions of vanadium increased by 45% (relative to the 1983 value) in Asia, 33% in Africa, and 18% in Central and South America while remaining fairly constant in Europe and North America (Table 3). The 27% decrease in emissions from Eastern Europe and former USSR reflects, presumably, recent deterioration in the economy of the region. Most of the developed countries have imposed stringent controls on industrial emissions of air toxics in their territories. Pollution control measures in the developing countries remain inadequate or unenforced. The high intensities of vanadium emission (Table 3) and the continuing increase in vanadium emission rates in the developing countries must be of concern— these countries are least able to afford any clean-up measures.

A comparison of total global data in Tables 2 and 3 show that the ratios of industrial to natural emissions vary from about 7 to 10. The high ratio is evidence that anthropogenic sources now dominate the inputs of vanadium in the global atmosphere. Previous estimates have reported much lower ratios (<3.0) of industrial to natural emissions (Zoller et al., 1973; ATSDR, 1992; Hope, 1994). It is believed that fluxes from natural sources were overestimated by earlier workers.

4. REGIONAL EMISSIONS

Regional emission inventories contribute to our understanding of the global vanadium cycle. Figure 1 shows the distribution of National Priorities List

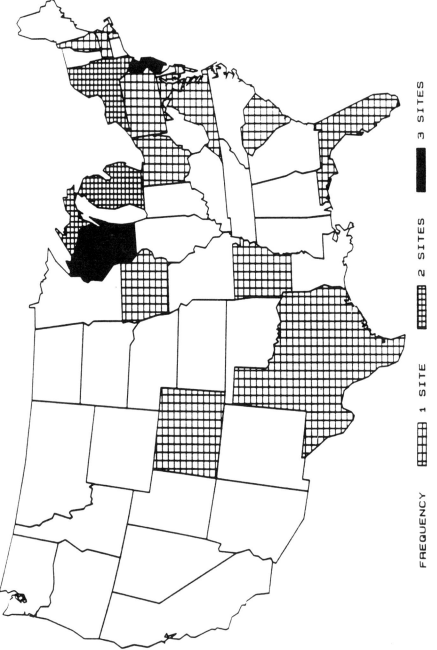

Figure 1. Frequency of Sites on the U.S. National Priorities List with vanadium contamination. *Source:* (ATSDR, 1992).

FREQUENCY ⊞ 1 SITE ⊞ 2 SITES ■ 3 SITES

Table 4 Annual Emissions (ton/year) of Vanadium into the Atmosphere in the Great Lakes Region

	1981	1983	1985	1986	1987	1988	1989	1990	1991	1992	1993	1994	1995
Minnesota	5.8	5.2	5.5	5.4	5.7	6.5	6.6	6.4	6.7	6.7	6.8	6.5	6.4
Wisconsin	6.4	6.5	6.9	7.1	7.1	7.7	7.6	7.5	7.5	7.4	7.5	7.3	7.2
Illinois	16.4	15.7	13.7	14.4	13.8	13.3	12.8	14.3	13.4	12.7	13.0	12.6	12.6
Indiana	16.5	15.4	17.1	16.7	16.8	17.3	18.3	19.1	18.9	18.6	18.8	18.6	18.4
Michigan	11.8	10.3	11.2	11.6	11.8	12.4	11.8	11.2	11.2	10.9	11.1	10.7	10.6
Ohio	21.2	17.4	18.4	18.2	17.9	19.1	19.1	18.5	18.2	18.7	18.8	18.5	18.2
Pennsylvania	26.4	23.3	22.9	23.2	24.3	24.9	25.3	23.1	22.3	22.7	22.8	22.4	22.1
New York	28.8	23.9	23.3	26.6	26.9	29.5	29.3	26.2	23.8	21.6	21.1	20.6	20.4
Ontario	13.0	12.9	11.7	12.7	12.5	12.0	12.5	14.0	11.2	11.1	11.2	10.8	10.6
Total	146.4	130.8	130.9	135.8	136.8	142.7	143.3	140.3	133.2	130.5	131.0	128.0	126.5

Table 5 Annual Emissions (tons/year) of Vanadium into the Atmosphere in the Mediterranean Sea Region

	1983	1984	1985	1986	1987	1988	1989	1990	1991	1992	1993	1994	1995
Albania	70	69	67	61	54	51	43	32	29	19	15	16	13
Algeria	216	238	262	264	265	273	289	308	330	330	345	350	365
Bulgaria	764	759	775	761	742	741	639	533	321	211	190	185	165
Cyprus	43	43	45	47	51	53	64	68	68	68	71	64	62
France	66	64	63	63	62	61	65	64	68	67	64	60	58
Israel	4.9	4.6	4.2	4.3	4.6	5.1	5.3	5.4	5.6	5.9	6	5.8	5.6
Greece	204	211	226	227	246	262	269	290	300	311	315	320	318
Italy	37	35	37	37	39	38	40	39	40	39	38	35	34
Jordan	118	151	151	138	147	138	138	147	140	140	145	142	140
Lebanon	667	636	745	714	745	667	543	512	822	822	850	880	890
Lybia	248	238	253	248	265	281	291	307	329	339	350	365	371
Morocco	236	225	238	250	245	263	295	311	312	312	315	318	325
Spain	38	35	34	34	34	37	40	39	38	45	48	44	45
Syria	322	357	380	424	418	445	459	475	413	413	408	406	398
Tunisia	138	134	129	143	145	143	150	148	169	169	175	177	182
Turkey	746	761	805	877	1,006	985	971	1,070	1,105	1,153	1,185	1,195	1,201
Egypt	829	895	948	937	974	963	977	1,025	1,005	1,067	1,105	1,145	1,176
Jugoslavia	658	812	829	772	795	797	790	804	633	414	380	330	285
Total	5,403	5,668	5,992	5,999	6,238	6,206	6,069	6,177	6,127	5,924	6,005	6,038	6,034

(NPL) sites with vanadium contamination in the United States. The figure is derived from the Toxic Release Inventory (TRI) database and as such has a number of severe limitations, including the fact that some key sources of vanadium in the environment, such as coal and residual burning facilities, are not even represented. Nevertheless, the figure does point to the Great Lakes basin as being the area with highest releases of vanadium into the environment. It is therefore of interest to evaluate the emissions of vanadium into the atmosphere in this region (Table 4). In 1995 the leading states in regard to vanadium emission are Pennsylvania (17%), New York (16%), Indiana (15%), and Michigan (14%). Only 8% of emission in the basin comes from Canada (Ontario). Emissions from the Great Lakes states and the Province of Ontario represent about 24% (relative to 1992 flux) of total industrial emissions in North America. Between 1981 and 1995, emissions of vanadium by industries in the Great Lakes basin declined by about 14% (Table 4), in large measure because of the pollution control programs that have been developed to protect the Great Lakes ecosystems.

Unlike in the Great Lakes basin, emissions in the Mediterranean Basin (MB) have continued to rise, the increase between 1983 and 1995 being estimated to be 12% (Table 5). The leading vanadium emitters in the MB are Turkey (20%), Egypt (19%), and Lebanon (15%); these three countries together account for over 50% of the total vanadium emission in the region in 1995. It should be noted that the total emission in the MB in 1995 exceeds that for Europe by nearly a factor of 12. The fraction of the vanadium released into the atmosphere in the MB that eventually gets exported to other parts of Europe is unknown. The big disparity in emission intensities speaks for a more coordinated approach to air quality management in Europe and the Middle East.

Once emitted into the atmosphere, the vanadium is subject to transport, dispersion, and physical/chemical/photochemical reactions, and it ultimately gets scavenged and removed by wet and dry deposition. The distance from actual emissions to removal depends on the physical and chemical properties of the vanadium-bearing aerosols, the effective height of the release, meteorological conditions, and the physiography of the region. These issues are considered in greater detail in Chapter 3.

REFERENCES

ATSDR. (1992). *Toxicological Profile for Vanadium.* U.S. Department of Health and Human Services, Agency for Toxic Substances and Disease Registry, Washington, DC.

Benjey, W. G., and Coventry, D. H. (1992). Geographic distribution and source type analysis of toxic metal emissions. In *Proceedings of the Intl. Symposium on Measurement of Toxics and Related Pollutants. Durham, NC, May 3–8, 1992.*

EIA. (1991). *International Energy Annual Report.* Energy Information Administration, Washington, DC.

EIA (1993). *International Energy Annual Report.* Energy Information Administration, Washington, DC.

EIA. (1996). *International Energy Annual Report.* Energy Information Administration, Washington, DC.

Hasanen, E., Aunela-Tapola, L., Kinnunen, V., Larjava, K., Mehtonen, A., Salmikangas, T., Leskela, J., and Loosaar, J. (1997). Emission factors and annual emissions of bulk and trace elements from oil shale fueled power plants. *Sci. Total Environ.* **198,** 1–12.

Hope, B. K. (1994). A global biogeochemical budget for vanadium. *Sci. Total Environ.* **141,** 1–10.

Nriagu, J. O. (1990). Global metal pollution: poisoning the biosphere. *Environment* **32,** 7–11, 28–33.

Nriagu, J. O. and Pacyna, J. M. (1988). Quantitative assessment of worldwide contamination of air, water and soils by trace metals. *Nature* **333,** 134–139.

Pacyna, J. M., Scholtz, M. T., and Li, Y. F. (1995). Global budget of trace metal sources. *Environ. Rev.* **3,** 145–159.

U.S. Bureau of Mines. (1987). *Minerals Yearbook: Metals and Minerals.* U.S. Government Printing Office, Washington, DC.

U.S. Bureau of Mines. (1994). *Minerals Yearbook: Metals and Minerals.* U.S. Government Printing Office, Washington, DC.

U.S. EPA. (1990a). Toxic Air Pollutant Emission Factors: A Compilation for Selected Air Toxic Compounds and Sources. Report No. 450/2-90-011. Emission Inventory Branch, US Environmental Protection Agency, Washington, DC.

U.S. EPA. (1990b). Factor Information Retrieval (FIRE) System, Version 3.0. Report No. 94-298-130-23-06. Emission Inventory Branch, U.S. Environmental Protection Agency, Washington, DC.

Zoller, W. H., Gordon, G. E., Gladney, E. S. and Jones, A. G. (1973). The sources and distribution of vanadium in the atmosphere. In E. L. Kothny, (Ed.), Trace Elements in the Environment *Advances in Chemistry Series,* Vol. 123. American Chemical Society, Washington, DC, pp. 31–47.

3

VANADIUM IN THE ATMOSPHERE

Y. Mamane

Technion–Israel Institute of Technology, Haifa 32000, Israel

Nicola Pirrone

CNR Institute for Atmospheric Pollution, c/o: DEIS University of Calabria, 87036 Rende, Italy

Vanadium in the Environment. Part 1: Chemistry and Biochemistry, Edited by Jerome O. Nriagu.
ISBN 0-471-17778-4. © 1998 John Wiley & Sons, Inc.

1. INTRODUCTION

Vanadium occurs commonly but not uniformly in the earth's crust, not as a free metal but rather as vanadates of copper, zinc, lead, iron, and other elements. The average concentration in earth's crust is from 100 to 150 mg/kg, which makes vanadium as abundant as Ni, Zn, and Pb. Crude oil and, to a lesser extent, coal from certain regions contain high levels of vanadium, which are partly emitted into the air during combustion. Lesser quantities of vanadium are emitted into the atmosphere by high-temperature metallurgical processes (van Zinderen Bakker and Jaworski, 1980; NAS, 1974, and references therein). In Canada fuel oil combustion contributed 94.1% (in 1972) to the emissions of vanadium into the atmosphere, coal and coke contributed 4.5%, and metallurgical and other processes contributed 1.2% (van Zinderen Bakker and Jaworski, 1980).

The global anthropogenic emission of vanadium into the atmosphere has been estimated to be $2.1 \cdot 10^5$ tonnes/year, three times larger than the natural emissions (Galloway et al., 1982). Other estimates showed that anthropogenic emissions of particulate-bound vanadium ($0.9 \cdot 10^5$ tonnes/year) were not greatly in excess of natural atmospheric inputs from continental ($0.7 \cdot 10^5$ tonnes/year) or volcanogenic dusts ($0.1 \cdot 10^5$ tonnes/year) and accounted for about 53% of total atmospheric loading. Anthropogenic vanadium accounted for only ~5% of the dissolved vanadium reaching the oceans, and it might increase seawater vanadium concentrations on a global basis by 1% (Hope, 1994). Combustion of heavy fuels, especially in oil-fired power plants, refineries, and industrial boilers, are the most dominant source of anthropogenic emissions of vanadium into the atmosphere.

Residual fuel oils manufactured from U.S. crude oils contain 25–50 ppm of vanadium, whereas Venezuelan residual oils have 200–300 ppm, Middle Eastern have 10–20 ppm, and North African have 50–90 ppm. Bouhamra and Jassem (1994) reported slightly higher values for Kuwait crude oils, 26–45 ppm. During crude oil refining virtually all vanadium remains in the heavier oil fractions (NAS, 1974). Vanadium is highly enriched relative to other elements in heavy fuel oils owing to the presence of vanadium porphyrins (Zoller et al., 1973), and for this reason vanadium is extensively used as a marker of emissions from fuel oil combustion (Gordon, 1988; Ganor et al., 1988; Juichang et al., 1995; Divita et al, 1996, and references therein).

During combustion, the organovanadium complexes in residual oils are oxidized and transformed into various compounds, including vanadium pentoxide, tetroxide, trioxide, and dioxide; they are emitted as fly ash into the atmosphere. It has been suggested that vanadium (as well as other elements) volatilizes at high temperatures and condenses uniformly on surfaces of entrained fly ash particles as the temperature falls beyond the combustion zone (Linton et al., 1976).

2. CONCENTRATIONS IN URBAN, RURAL, AND REMOTE AREAS

Measurements of trace elements associated with particulate matter in the atmosphere have been reported over the last 25 years for a large number of locations throughout the world. Vanadium is one of the metals found both in the fine (<2.5 μm) and coarse (2.5–10 μm) fractions. Schroeder et al. (1987) provided a concise review of published literature pertaining to sampling and analysis of ambient measurements of toxic trace elements including vanadium. Table 1 shows the concentration range of vanadium based on the information compiled by Schroeder et al. (1987). Most of the studies reported therein were from the 1970s, and only a few were from the early 1980s. The range of concentrations is from one to four orders of magnitude, a range too large to be used in any assessment.

In ambient air when vanadium is accompanied by nickel, and the correlation between the two elements is high, crude oil combustion is the source for both elements. Figure 1 is a plot of nickel versus vanadium for fine particles (d. <2.5 μm) collected in an industrial urban site using crude oil as the major source of energy. Twenty-four-hour samples were collected for one month in Haifa. The linear correlation coefficient is 0.88 (Mamane et al., 1996). High correlation coefficients between nickel and vanadium were also reported downwind of the Kuwaiti oil fires (Madany and Raveendran, 1992).

In the last 15 years, new information has emerged from ambient measurements of the mass and chemical composition of fine (d. >2.5 μm) and coarse (2.5 < d. < 10 μm) particles in several urban and rural locations (Gordon, 1988; Chow et al., 1994). Some of these studies reported also the elemental concentrations of the particulate emissions of the major sources in the area. The combined ambient and source fingerprint data were used in receptor modeling studies to apportion the contribution of the various sources. The following survey of vanadium concentrations in the atmosphere is based on

Table 1   Concentration Range of Vanadium Associated with Particulate Matter in the Atmosphere (ng/m³)

Location	V Concentration, ng/m³
Remote	0.001–14
Rural	2.7–97
Urban	
Canada	10–130
United States	0.4–1460
Europe	11–73
Other	1.7–180

Source: Schroeder et al. (1987).

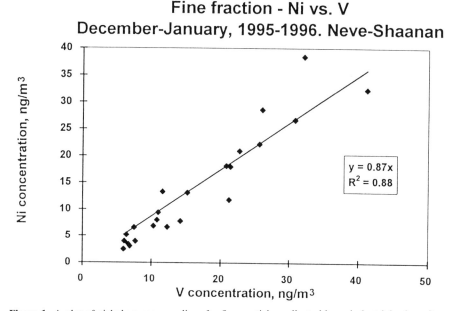

Figure 1. A plot of nickel versus vanadium for fine particles collected in an industrial urban site. Heavy oil is used as a source of energy for power plants and a refinery. A high correlation coefficient points to crude and heavy oil combustion as the source for vanadium. *Source:* Mamane et al. (1996).

figures from the literature. They are based on different methods of sampling and analysis. Different analytical methods do not produce quite comparable results at very low concentrations of vanadium. The samples are often taken over a short period of time (few days to few weeks). The samplers are quite different, from fine-particles (<2.5 μm), low-volume sampler to total suspended particulate (d. < ~30 μm), high-volume sampler. The data included in Tables 2–7 below should be regarded as examples of vanadium concentrations encountered in the atmosphere; their representativeness cannot always be guaranteed. Tables 2–4 list vanadium concentrations in urban, rural, and remote areas, providing some information on the particle size distribution.

It is difficult to draw generalized conclusions from the data presented in Tables 2–4, as the values vary from less than 1 to over 1,000 ng/m³. Nevertheless, the following points can be made.

1. Vanadium concentrations vary widely with location from 1 to 15 ng/m³ in urban locations in the West Coast and Midwest of the United States and up to 10–40 ng/m³ in the urban sites of the East Coast. Urban locations are sites in a city or elsewhere directly influenced by local anthropogenic emissions.

2. There are indications that current vanadium concentrations in urban locations are lower than reported concentrations in the sixties and seventies.

Table 2 Ambient Concentrations of Vanadium in Urban Areas

Location	Land Use[a]	Conc. (ng/m³)	Size (μg)	Time	Analytical Method	No. of Samples	Reference
Beijing	(dust storm)	5.3–88		4/1993	INAA		Zhou et al. (1996)
Haifa	Residential	16.2 ± 9.6	<2.5 μm	12/95–1/96	XRF	30	Mamane et al. (1996)
		9.0 ± 1.7	2.5 < d. < 10 μm				Ibid.
Burbank, CA	Ind	5.2–6.2	<10	6–12/87	XRF	40	Chow et al. (1994)
Los Angeles, CA	Res	5.2–9.2	<10	6–12/87	XRF	17	Ibid.
		5.5–8	<2.5	6–12/87	XRF	17	Ibid.
Hawthorne, CA	Ind	5.8–12.9	<10	6–12/87	XRF	17	Ibid.
		6.1–13.2	<2.5	6–12/87	XRF	17	Ibid.
Anaheim, CA	Res/Com	5.1–11.6	<10	6–12/87	XRF	17	Ibid.
		5.6–7.9	<2.5	6–9/87	XRF	17	Ibid.
Long Beach, CA	Ind	6.8–12.5	<10	6–9/87	XRF	17	Ibid.
		6.8–8.9	<2.5	6–9/87	XRF	17	Ibid.
Rubidoux, CA	Res/Ind	8.4	<10	6–9/87	XRF	11	Ibid.
		5.4	<2.5	6–9/87	XRF	11	Ibid.
Azusa, CA	Res/Com	7.5	<10	6–9/87	XRF	11	Ibid.
		5.7	<2.5	6–9/87	XRF	11	Ibid.
Claremont, CA	Res/Com	6	<10	6–9/87	XRF	11	Ibid.
		5.3	<2.5	6–9/87	XRF	11	Ibid.
Riverside, CA	Res/Com	9.9	<10	11–12/87	XRF	6	Ibid.
		5.6	<2.5	11–12/87	XRF	6	Ibid.
Chicago, IL	Ind./Com	0.7 ± 1.8	<2.5	6–7/1991	XRF	38	Keeler (1994)
	Ind./Com	1.9 ± 2.2	2.5–10	6–7/1991	XRF	38	Ibid.
	5 km off-shore from Chicago	0.5 ± 1.4	<2.5	6–7/1991	XRF	18	Ibid.
		0.7 ± 1.1	2.5–10	6–7/1991	XRF	18	Ibid.

41

Table 2 (*Continued*)

Location	Land Use[a]	Conc. (ng/m³)	Size (μg)	Time	Analytical Method	No. of Samples	Reference
Beijing	(dust storm)	350	Bulk	26/4/1990	PIXE		Zhang et al. (1993)
Beijing	(non-dust-storm)	210	Bulk	4/1990	PIXE	3	Ibid.
Chicago, IL	Ind.	3 ± 2.7	<2.5	9/85–6/88	INAA	50–100	Sweet and Vermette (1993)
St. Louis, IL	Ind	3.7 ± 2.7	2.5–10	9/85–6/88	INAA	50–100	Ibid.
		3.0 ± 2.4	<2.5	9/85–6/88	INAA	50–100	Ibid.
		3.0 ± 2.0	2.5–10	9/85–6/88	INAA	50–100	Ibid.
Marseille	Res/Com and Ind	8–38	TSP	1982–1991	AAS		Grimaldi et al. (1993)
Birmingham (UK)	Ind	17.2	2.1–10	6–8/1992	PIXE		Harrison et al. (1993)
		47.2	<2.1				Ibid.
S.E. Chicago, IL		3.0 ± 2.7	<2.5	1985–1988	XRF-INAA		Vermette et al. (1991)
		3.7 ± 2.5	2.5–10		XRF-INAA		Ibid.
Reading, PA		6.75		7–11/1982	XRF		Lioy et al. (1989)
Salamanca (Spain)	City Center	n.d.	n.a.	5/1978–5/1982	XRF	800	Fidalgo et al. (1988)
		10	n.a.	5/1978–5/1982	XRF	800	Ibid.
Camden NJ	Ind./Com	19 ± 4	<2.5	7–8/1982	XRF	50	Dzubay et al. (1988)
		13 ± 0.6	<2.5		INAA	50	Ibid.
		13 ± 1.4	<2.5		INAA	50	Ibid.
Camden NJ		7 ± 4	2.5–10		XRF	50	Ibid.

Location		<2.5	2.5–15		1979–1981	XRF		Thurston and Spengler (1985)
Boston				22.1	1979–1981	XRF		Thurston and Spengler (1985)
Albany, NY	Ind	3.34	n.a.	37	17–22/7/1976	AAS	6	Ibid.
University of Maryland Campus		n.a.		20	6–8/1976	INAA	130	Husain et al. (1983)
Prince George Gen. Hospital, Cheverly, MD				22		INAA	130	Kowalczyk et al. (1982)
District of Columbia Gen. Hospital				34		INAA	130	Ibid.
Public Health Center, Alexandria, VA				34		INAA	130	Ibid.
WRC-TV studio				21		INAA	130	Ibid.
West end library				37		INAA	130	Ibid.
Am. Chem. Soc. Building				30		INAA	130	Ibid.
Nuclear Reactor Building, Gaithersburg, MD				9.1		INAA	130	Ibid.
Beltsville, MD				15		INAA	130	Ibid.
Hammond High School, Alexandria, VA				29		INAA	130	Ibid.

Table 2 (*Continued*)

Location	Land Use[a]	Conc. (ng/m³)	Size (μg)	Time	Analytical Method	No. of Samples	Reference
N.Y. Medical Center	Res/Com	874		1969	AAS		Ibid.
		68.9		1972	AAS		Ibid.
		86		1973	AAS		Ibid.
		72.6		1974	AAS		Ibid.
		38.8		1975	AAS		Ibid.
N.Y.–Bronx		1,230		1968	AAS		Ibid.
		795		1969	AAS		Ibid.
		53		1972	AAS		Ibid.
		80		1973	AAS		Ibid.
N.Y.–Queens		34		1974	AAS		Ibid.
		19.4		1975	AAS		Ibid.
San Francisco Bay Area (different stations)							
Benicia	Res	4.9	n.a.	13–17/7/1971	NAA	n.a.	Martens et al. (1973)
Berkeley, Lawrence Radiation Lab	Res	6.9		12–16/7/1971	NAA		Ibid.
Firemont	Res	7.6		12–16/7/1971	NAA		Ibid.
Livermore	Res	7.5		12–16/7/1971	NAA		Ibid.
Pittsburgh	Ind	12		12–16/7/1971	NAA	n.a.	Martens et al. (1973)
Richmond	Res	5		12–16/7/1971	NAA		Ibid.
Redwood City	Ind.	3.3		12–16/7/1971	NAA		Ibid.
San Francisco		2.7		12–16/7/1971	NAA		Ibid.
San Jose	Res/Com	6.7		12–16/7/1971	NAA		Ibid.

Osaka, Japan		n.a.			n.a.	Sugimae and Hasegawa (1973)
Toyonaka (city office)	50	n.a.	6/1970–3/1971	Emission spectrography	n.a.	Ibid.
Suita (city office)	120		6–10/1969			Ibid.
Toyonaka (fire department)	240		6/1970–3/1971			Ibid.
Osaka (Environmental Pollution Control Center)	210		6/1970–3/1971			Ibid.
Higashiosaka (city office)	120		6/1970–3/1971			Ibid.
Yao (city office)	120		6/1970–3/1971			Ibid.
Sakai City (Institute of Hygiene)	550		6–10/1969			Ibid.
Sakai (Kanaoka Elementary School)	110		6/1970–3/1971			Ibid.
Sakai (Hamadera Union High School)	200		6/1970–3/1971			Ibid.
Takaishi (city office)	100		6/1970–3/1971			Ibid.
Kishiwada (city office)	60		6/1970–3/1971			Ibid.

[a] Ind, industrial; Res, residential; Com, commercial.

Table 3 Ambient Concentrations of Vanadium in Rural Areas

Place	Type	Conc. (ng/m³)	Size (μg)	Time	Analytical Method	No. of Samples	Reference
Mediterranean coast line (Israel)		8.6 (0.1–35)	<2.5	5–6,1993	XRF	10	Gertler (1994)
San Nicholas Island, CA		4.1	<10	6–9/87	XRF	11	Chow et al. (1994)
Point Petre, Ontario	Rural	4.9	<2.5	6–9/87	XRF	11	Ibid.
		2.54	TSP	1989	NAA		Hoff and Brice (1994)
		1.78	TSP	1990	NAA		Ibid.
		1.47	TSP	1991	NAA		Ibid.
		1.33	TSP	1992	NAA		Ibid.
Bondville, IL		0.8 ± 0.4	<1–2.5	9/85–6/88	INAA	50–100	Sweet and Vermette (1993)
		1.2 ± 1.3	2.5–10	9/85–6/88	INAA	50–100	Ibid.
China							
Shapuotou (desert area)	(dust storm)	420	Bulk	10/4/1990	PIXE		Zhang et al. (1993)
40 km N. of Xian	(dust storm)	210	Bulk	6/4/1990	PIXE		Ibid.
40 km N. of Xian	(dust storm)	500	Bulk	11/4/1990	PIXE		Ibid.
Shapuotou	(non-dust-storm)	160	Bulk	4/1990	PIXE	6	Ibid.
40 km N. of Xian	(non-dust-storm)	110	Bulk	3/1990–2/1991	PIXE	15	Ibid.

Location		Concentration	Particle size	Date	Method	N	Reference
South Haven, MI		0.4 ± 1.7	<2.5	6–7/1991	XRF	32	Keeler (1994)
Kankakee, IL		0.7 ± 2.5	2.5–10	6–7/1991	XRF	32	Ibid.
		0.5 ± 1.2	<2.5	6–7/1991	XRF	34	Ibid.
		1.0 ± 2.3	2.5–10	6–7/1991	XRF	34	Ibid.
Champaign, IL		0.8 ± 0.4	<2.5	9/1985–9/1987	XRF-NAA	104	Vermette et al. (1991)
		1.2 ± 1.4	2.5–10				Ibid.
Arad, Israel (isolated desert town)	Res	9–20	d. < 30 μm	5/1987	ICP	11	Ganor and Foner (1989)
Shenandoah Valley, VA		1.54 ± 1.04	<2.5	7–8/1980	INAA	32	Tuncel et al. (1985)
Holland, NY		~2		7/1976	AAS	6	Husain et al. (1983)
Schoharie, NY		3.1		7/1976	AAS	6	Ibid.
High Point, NJ		6.2		7/1976	AAS	5	Ibid.
Whiteface MT., NY		2.1		7/1976	AAS		Ibid.
Colstrip, MT		1.4 ± 1.1		1975	INAA-XRF	47	Crecelius et al. (1980)
Puerto Rico							Martens et al. (1973)
Coastline	(navy base)	0.62–1.7		23–28/11/1971	NAA	n.a.	Ibid.
Pico del Este		1–2.7		22–30/11/71	NAA	n.a.	Ibid.
Puerto Rico							Martens et al. (1973)
Coastline navy base	Off shore	0.62		12–16/7/1971	NAA	83 m^3	Martens et al. (1973)
Coastline navy base	Off shore	1.7		25–28/11/1971	NAA	125 m^3	Ibid.

Table 4 Ambient Concentrations of Vanadium in Remote Areas

Place	Type	Conc. (ng/m³)	Size (µg)	Time	Analytical Method	No. of Samples	Reference
Arctic Ocean	Ocean subset	0.029	<2	8–10/1991	INAA-PIXE	78	Maenhaut et al. (1996)
		0.043	<10			78	Ibid.
	Pack ice	<0.01	<2	8–10/1991	INAA-PIXE	78	Ibid.
		0.043	<10		INAA-PIXE	78	Ibid.
	Sea level	0.041	<2.5	Summer, 1984–86–87	INAA-PIXE	58	Ibid.
Canadian high Arctic		0.1–2.0		1983–1988	ICP	180	Landsberger et al. (1992)
Ny Alesund (Norwegian Arctic)		1.99	<2.5	Winter, 1983	INAA-PIXE	14	Maenhaut and Vitols (1989)
		0.42	<2.5	Winter, 1984	INAA-PIXE	18	Ibid.
		0.022	<2.5	Summer, 1984	INAA-PIXE	13	Ibid.
Vardo (Norwegian Arctic)		0.47	<2.5	Winter, 1986	INAA-PIXE	14	Ibid.
		1.96	<2.5	Winter, 1983	INAA-PIXE	13	Ibid.
		1.61	<2.5	Winter, 1984	INAA-PIXE	14	Ibid.
		0.26	<2.5	Summer, 1984	INAA-PIXE	12	Ibid.
Bermuda		1.2		3–11/1974	INAA	78	Chen and Duce, (1983)
North Pacific	Sea level	<0.09–0.21				15	Duce and Hoffman (1976)
North Atlantic (near Iceland)	Sea level	0.06–0.09					Ibid.
Azores	Sea level	0.14–0.9					Ibid.

Zoller et al. (1973) suggested that the U.S. cities could clearly be separated into two distinct groups: Group A (Los Angeles, San Francisco, Chicago), characterized by vanadium concentrations of 5–20 ng/m³, and Group B (Boston, New York), in which overall vanadium concentrations ranged between 150 and 1,400 ng/m³. Kleinman et al. (1980) reported vanadium concentrations of 20–90 ng/m³ in New York City (1972–1975), but also some extreme levels of 795, 874, and 1,230 ng/m³ (in the years 1968–1969). In another review by van Zinderen Bakker and Jaworski (1980), it had been suggested that industrialized urban centers with a high consumption of residual fuel oil might have maximum levels of up to 10,000 ng/m³, and an average range of 500–2,000 ng/m³. Sugimae and Hasegawa (1973) reported vanadium levels in Japan of 50–550 ng/m³ in various urban locations in the years 1969–1971. For Israel, Donagi et al. (1979) and Foner and Ganor (1986) reported fairly high vanadium concentrations in Tel-Aviv: 130–160 ng/m³ (using high-volume sampler). High values were recently reported in China: 110–210 ng/m³ in urban sites (Zhang et al., 1993). The high vanadium levels may represent the use of less refined crude oil in uncontrolled power plants or refineries.

3. Recent data from rural sites show vanadium concentrations ranging from 0.3 to 5 ng/m³, although most sites showed a typical value of around 1 ng/m³. In the seventies van Zinderen Bakker and Jaworski (1980) had suggested levels of 1–40 ng/m³ for rural sites, and as high as 65 ng/m³. Those levels are higher than the recent vanadium data in Table 3. Rural sites represent a regional background not directly influenced by local anthropogenic emissions.

4. In arid and semiarid zones a major fraction of the suspended particulate matter is of mineral origin. Since vanadium is found also in soil and rocks, dust may contain vanadium of natural origin. Ganor (1975) reported that vanadium concentrations during dust storms in Israel might reach levels of 50 ng/m³. Zhang et al. (1993), in a study to determine the flux of trace elements during dust and non-dust storm periods in China, measured vanadium concentrations of 210–500 ng/m³ during the April dust storms (1990). During the non-dust season (dry and wet seasons), the vanadium level was only 110–210 ng/m³. But even in the "clear" season soil minerals dominated the particles collected at these sites.

5. In remote locations, such as the Arctic Ocean, vanadium concentrations ranged from less than 0.01 to 0.04 ng/m³ (summers of 1984–1991). Earlier data from the Norwegian Arctic (1983–1986) indicated that summer values (0.022 – 0.26 ng/m³) are almost an order of magnitude lower than the winter levels (0.42 – 1.99 ng/m³) (Landsberger et al., 1992). During the winter season, the use of energy is higher, and the meteorological conditions (increased stability of the air mass) in the northern hemisphere tends to cause high concentrations of pollutants in the lower atmosphere. These, in turn, are transported to the remote sites. Remote sites are defined as any area of lowest concentration.

3.   VANADIUM CONCENTRATIONS IN EPISODES

High vanadium concentrations were found during periods of high emission rates and unique meteorological conditions. Three cases will be reported here: (a) high vanadium concentrations during the winter season along the northeastern urban regions of the United States and Canada; (b) high vanadium concentrations of natural origin during dust storms; and (c) high vanadium levels during the burning of the Kuwaiti oil fires.

The average concentration of atmospheric vanadium over urban areas of the United States in recent years ranges from less than 1 to more than 40 ng/m^3. The higher concentrations from combustion of vanadium-rich fuel oil were found along the eastern seaboard. In remote areas concentrations are often less than 2 ng/m^3. The highest concentration of atmospheric vanadium ever recorded (in the order of 10,000 ng/m^3) was measured in New York City, where the average was about 460 ng/m^3 (van Zinderen Bakker and Jaworski, 1980). The average fall and winter levels in eastern seaboard cities in the United States were twice those of the average spring and summer values. Peak vanadium concentrations in the air of Canadian cities, where winter temperatures are much lower, may well reach levels of 10,000–20,000 ng/m^3 if vanadium-rich fuel oils are used (van Zinderen Bakker and Jaworski, 1980). This is due to the use of high-vanadium fuel oil for heating purposes (NAS, 1974).

High vanadium concentrations, from 210 to 500 ng/m^3, were reported by Zhang et al. (1993) during dust storms in Northern China. Vanadium is found in rocks in the order of 20–300 μg/g. Basalts and phosphorites appear to have the highest vanadium levels. Vanadium is also found in soils, from 20 to 250 μg/g, depending on its content in the parent rock. Zhang et al. (1993) reported that soil-derived elements (Al, Si, K, Ca, Ti, and Fe) accounted for $\geq$95% of the total mass attributable to the 17 elements determined by PIXE analysis. For example, in the Shapuotou dust storm sample aluminum reached 82 μg/m^3. This is equivalent to a mineral aerosol concentration of about 1,000 μg/m^3. If vanadium content in soil is assumed to be around 100–200 μg/g, then soil-derived particles during the dust storm may contribute 100–200 ng/m^3 of natural vanadium. Zhang et al. (1993), however, reported a measured value of 460 ng/m^3 vanadium in Shapuotou. Zhang et al. (1993) have shown by using different approaches that most of the mass collected is of soil origin, and thus it is likely that most of the vanadium is of natural sources. Therefore, dust storm and arid zones may be associated with elevated nonanthropogenic vanadium levels.

Other new information was related to the Kuwaiti oil fires from March 1991 to July 1992. Several studies reported smoke and atmospheric fallout in Kuwait and in neighboring countries (Madany and Raveendran, 1992; Al-Arfaj and Alam, 1993; Stevens et al., 1993; Sadiq and Mian, 1994).

Table 5 summarizes vanadium data collected at Dhahran during the Kuwaiti oil fires. Dhahran is located 400 km south of Kuwait, and the regional northerly

Table 5   Summary of V Concentrations Collected at Dhahran, Saudi Arabia, during the Kuwaiti Oil Fires

Month	Particle Sample	Vanadium Concentrations (ng/m^3)		
		Average	Minimum	Maximum
March 91	TSP	24	1.8	78
	PM-10	—		
Apr 91	TSP	32	—	68
	PM-10	23	20	26
May 91	TSP	55	7.3	160
	PM-10	25	7.0	89
June 91	TSP	72	27	138
	PM-10	26	8.6	48
July 91	TSP	61	34	112
	PM-10	30	14	53
Aug 91	TSP	32	16	71
	PM-10	13	9.6	1,170
Sept 91	TSP	26	9.3	49
	PM-10	11	—	28
Oct 91	TSP	24	13	41
	PM-10	16	9.1	29
Nov 91	TSP	12	8.6	16
	PM-10	9	6.7	27
Dec 91	TSP	11	5.2	20
	PM-10	5	1.7	12
Jan 92	TSP	13	0.5	27
	PM-10	7	2.4	15
Feb 92	TSP	25	7.9	53
	PM-10	17	2.3	53
Mar 92	TSP	14	7.8	36
	PM-10	12	2.9	40
Apr 92	TSP	20	13	25
	PM-10	13	9.0	38
May 92	TSP	24	16	33
	PM-10	18	13	24
June 92	TSP	49	34	79
	PM-10	32	25	46
July 92	TSP	24	19	29
	PM-10	17	12	20

Source: Sadiq and Mian, 1994.

high winds, blowing from April to July, transported large quantities of smoke and particulates to Saudi Arabia. The maximum 24-h vanadium concentration ranged from about 10 to 90 ng/m^3 (for PM-10 samples), not including an extreme event, in which vanadium concentration reached 1,166 ng/m^3. Figure 2 shows the monthly vanadium concentrations in Saudi Arabia from March 1991 to July 1992, for total suspended particulate (TSP, d. < ~30 μm) and for PM-10 (d $\leq$ 10 μm). In this unique case vanadium is also associated with particles larger than 10 μm. vanadium monthly concentrations in Dhahran, Saudi Arabia, are quite high for northerly winds and reached levels of 20–30 ng/m^3 (Sadiq and Mian, 1994).

The monthly mean concentration of vanadium in TSP is higher than PM-10 during the northerly dry "Shamal" winds in Saudi Arabia. Dust storms are more frequent during the Shamal season. Sadiq and Mian (1994) have normalized the vanadium data to particulate weight. They showed that in fact the normalized concentration (in micrograms vanadium per gram aerosol) in the TSPs were lower than those found in PM-10 samples. Here, too, dust storms contributed to elevated vanadium concentrations in the local atmosphere.

A similar study was carried out for 1 week in Bahrain during the burning of oil fields in Kuwait. Bahrain is located 320 km south-southeast from Kuwait. The 24-h vanadium concentration ranged from 11 to 42 ng/m^3. As in Saudi Arabia, a high and significant correlation between nickel and vanadium points to the burning of crude oil in the Kuwait oil fields. No other source of vanadium could be identified (Madany and Reveendran, 1992).

Both studies reported similar vanadium concentration of around 10–90 ng/m^3 for 24-h samples. Those concentrations were measured 300–400 km south of the sources in Kuwait. Near the source, say 30–40 km from the oil fires, average concentrations (PM-10, 24-h average) could have been 10 times higher, 100–900 ng/m^3 vanadium, if concentration is inversely proportional to distance. Peak concentrations were probably even higher.

4. SIZE DISTRIBUTION OF VANADIUM-CONTAINING PARTICLES

Earlier studies (Duce and Hoffman, 1976, and references therein) indicated that the mass median diameter of vanadium, in both urban and remote areas, was approximately 1 μm, with a range of 0.4–5.0 μm. The small particle distribution of vanadium would be expected for an element with a high-

---

Figure 2. A plot of the monthly vanadium concentrations in Saudi Arabia from March 1991 to July 1992, for total suspended particulate (TSP, d. < ~30 μm) and for PM-10. In this unique case vanadium is also associated with particles larger than 10 μm. Vanadium monthly concentrations in Dhahran, Saudi Arabia, are quite high for northerly winds and reached average levels of 20–30 ng/m^3. *Source:* Sadiq and Mian (1994).

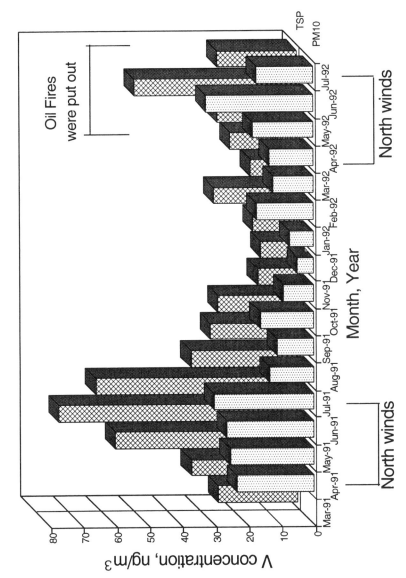

Monthly Average V Concentrations Downwind of the Kuwaiti Oil Fires

temperature combustion source, such as the burning of fossil fuels. Recently Divita et al. (1996, and references therein) studied the size spectra of vanadium-containing aerosol in Washington, D.C., using an eight-stage impactor with cut-off diameters from 0.09 to 15 μm. They found that the vanadium size spectra were most often unimodal, with modal diameters of 0.22–0.4 μm. These diameters are smaller than reported earlier, since most oil-fired power plants in the United States are often efficiently controlled. Therefore, only fine vanadium-particles, or submicrometer particles that are not efficiently captured by control devices, are emitted into the atmosphere.

Recent data, summarized in Tables 2 and 3, indicate that most vanadium is found in the fine fraction (<2.5 μm). These data are based on sampling ambient particles into two size fractions with a dichotomous sampler: d. < 2.5 μm, and 2.5 μm < d. < 10 μm. This sampler does not provide information on larger particles; that may be of importance where mineral particles dominate the atmosphere aerosol concentration. Larger particles (d. > 10 μm) are of importance near the source, where large vanadium-containing particles (such as cenospheres) may be found in the atmosphere or are deposited to the ground.

5. DEPOSITION OF ATMOSPHERIC VANADIUM

5.1. Wet and Dry Deposition

An increase in metal emissions is followed by an increase in atmospheric metal concentrations and in metal deposition (Pirrone et al., 1995b, 1996a,b). There are three ways to assess whether human activity or natural emission processes control the current concentrations of metals in precipitation: (1) by comparing the actual metal emission rates of human sources and natural sources (mobilization factor—MF); (2) by comparing the ratios of metal concentrations in the atmosphere to the ratios of metal concentrations in the natural sources (enrichment factor—EF); and (3) by determining historical trends of metal concentrations in atmospheric deposition (Galloway et al., 1982; Pirrone et al., 1995b, 1997).

For vanadium the mobilization factor was estimated as 3.2, with natural and anthropogenic emissions of 0.65 and 2.1 $\times$ 10^5 tonnes per year, respectively. The enrichment factor was around 2–20 for the South Pole and North Atlantic Westerlies, respectively (Galloway et al., 1982; Duce and Hoffman, 1976).

The third technique uses historical records of atmospheric deposition preserved in glaciers, soil, or lake sediments. These sediments are taken from lakes in a relatively steady state, undisturbed by human agency, and, preferably, with no inlets or outlets. Galloway and Likens (1979), reporting data on 47 metals from a core from Woodhull Lake, New York, found that Ag, Au, Cd, Cr, Cu, Pb, Sb, V, and Zn showed increases in the order Sb > Pb > Au > Cd > Ag > Cu > Zn > V. The increase began about 30 years ago. They attribute the enrichment to increase in the rate of atmospheric deposition. Vesely et al. (1993) documented the long-term effects of atmospheric deposition of pollutants associated with the industrial revolution, by analyzing metal concentrations in sediments of Certovo Lake, Czech Republic. They found that the accumulation of anthropogenic vanadium was insignificant till the early 1970s. In the middle 1980s the deposition of anthropogenic vanadium is substantial, around 0.27 μm/cm^2 · year, or 0.027 kg/hectare · year, probably owing to increased consumption of crude oil in Europe. Munch (1993) studied the concentration profiles of several metals in forest soil beside a relatively busy road. In the soil lying directly at the roadside edge, vanadium contamination was up to 51 mg/kg, in comparison with background concentration of 27 mg/kg.

Galloway et al. (1982, and references therein) reported the ranges of toxic metal concentrations in wet deposition. More data are listed in Table 6.

	Range (μg/L)	Median (μg/L)
Urban	16–68	42
Rural	0.13–23	9
Remote	0.016–0.3	0.16

Galloway et al. also presented selected data for deposition rates of toxic metals, including vanadium. However, the quality of the data vary and they are not strictly comparable. The deposition rates of vanadium for dry, wet-only, and bulk (dry and wet) were quite similar for urban and rural locations: <0.25, <4.8, and 8.3–12.4 kg/hectare · year. Others reported significantly lower values for urban and rural locations, 0.04 kg/hectare · year (Galloway et al., 1982).

Church et al. (1984) reported vanadium monthly deposition to the western Atlantic Delaware coast and to Bermuda, respectively, at about 69 and 12 μg/m^2 (or 0.008 and 0.0014 kg/hectare · year). Church et al. suggested that sea salt aerosol may contribute to metal deposition near ocean sites.

Similar results were obtained by Atteia (1994), who studied precipitation chemistry in the western part of Switzerland far from any important industrial center. The vanadium deposition rate was around 0.003–0.007 kg/hectare · year.

Ganor et al. (1988) reported dry fallout of vanadium in Tel-Aviv in the vicinity of a 480-MW oil-fired power plant: 4 kg/hectare · year in the immediate vicinity of the power plant, and 0.04–0.4 kg/hectare · year 1–10 km from the power plant. Foner et al. (1989) reported slightly higher values (0.4–1.6 kg/hectare · year) for dry fallout in the city of Ashdod near a 1,200-MW oil-fired power plant. Bacskai et al. (1994) measured vanadium and nickel in dustfall in a residential site near an oil power plant and a major refinery outside of Budapest. They reported a vanadium deposition rate of 14.2 kg/hectare · year

Table 6   Vanadium Concentrations Measured in Rain, Snow, and Ice Samples

Sample	Type	Location	Time	Concentration (μg/L)	Reference
Bulk		Western Switzerland	Jan 1990–Nov 1991	0.12–0.65 (range) 0.45 (mean)	Atteia (1994)
Remote	Wet	Middle Atlantic coast (Lewes, DE)	Aug 1982–May 1983	0.2–1.16 (range) 0.67 (average)	Church et al. (1984)
Remote	Wet	Midwestern Atlantic (Bermuda)	Oct 1982–May 1983	0.049–0.111 (range) 0.096 (average)	Church et al. (1984)
Urban	(rain)	New York	n.a.	68	Galloway et al. (1982) and references therein
Urban	(rain)	Seattle		46	Ibid.
Urban	(rain)	New York		16	Ibid.
Rural	(rain)	Argonne		23	Ibid.
Rural	(rain)	Beaverton, OR		23	Ibid.
Rural	(rain)	Livermore, CA		13	Ibid.
Rural	(rain)	Netherlands		4.7	Ibid.
Rural	(rain)	Puerto Rico		1.1	Ibid.
Rural	(snow)	Alaska		0.13	Ibid.
Rural	Bulk	Swansea, UK		13	Ibid.
Rural	Bulk	England		8.9	Ibid.
Rural	Bulk	Wraimres, UK		4.1	Ibid.
Remote	Ice/snow	Northern Norway		0.31	Ibid.
Remote	Ice	Greenland		0.022	Ibid.
Remote	Ice	Greenland		0.016	Ibid.

near (200 m) the power plant to 0.3 kg/hectare · year at 3,000 m from the plant.

An estimate of dry deposition (D) rate could be obtained from multiplying ambient concentration (C) by deposition velocities (V). Akeredolu et al. (1994) suggested a deposition velocity of 0.4 cm/s for vanadium, in agreement with the reported values in the literature of 0.2–1.4 cm/s (Duce and Hoffman, 1976; Slinn et al., 1978; Milford and Davidson, 1985; Pacyna et al, 1989). Galloway et al. (1982), on the basis of actual measurements of dry and total metal deposits, suggested that the dry fraction is between 0.3 and 0.6. Akeredolu et al. (1994) estimated that 470 tonnes per year of vanadium, 1.7% of total emissions in Europe and northern Asia, is transported toward the Arctic, and 102 tonnes per year is deposited there. This is equivalent to a deposition rate of 0.00005 kg/hectare · year if the vanadium is uniformly deposited in the Arctic cap area ($2.1 \cdot 10^{13}$ m^2).

Based on ambient concentration data in Table 4, dry deposition rates of vanadium may be calculated for the Arctic region as follows:

$$D = V \cdot C$$

where C is the summer concentration in the Arctic, 0.02–0.3 ng/m^3; V is the deposition velocity, 0.4 m/s; and D is calculated deposition rate in the Arctic, 2.5×10^{-5} to 3.75×10^{-4} kg/hectare · year.

Based on concentration data in the Arctic, and assuming that dry equals wet deposition, then the total deposition rates for the summer are about 0.00005–0.0007 kg/hectare · year, and for the winter it is less than 0.005 kg/hectare · year (vanadium winter concentrations are often less than 2 ng/m^3; see Figure 4 of Landsberger et al., 1992).

Similar results (within an order of magnitude) were obtained by Duce and Hoffman (1976) for deposition of vanadium to open ocean regions. They used three different approaches to obtain three similar estimates, all between 3×10^{-4} and 5×10^{-4} kg/hectare · year.

Atmospheric dry and wet deposition are complex phenomena that depend on the physics and chemistry of the particles deposited (particle size, solubility), on the meteorological conditions (wind, atmospheric stability, precipitation), and on topography (Pirrone et al., 1995a). Therefore, deposition rates vary from site to site and data are not always comparable.

The following rates are proposed for vanadium deposition rates (total deposition).

- Urban or other sites affected by strong sources: 0.1–10 kg/hectare · year.
- Regional sites (or urban) not affected by local sources: 0.01–0.1 kg/hectare · year.
- Remote sites: 0.001 or less to 0.01 kg/hectare · year.

5.2   Deposition of Acid Smuts

Acid smuts are agglomerates, or flakes, of acid particles and fly ash generated by oil-fired utility installations. Smuts have a size range of 0.5–5 mm, and tend to deposit near the source (power plants, industrial boilers), causing local problems. It has been shown that high H_2SO_4 emissions, with concomitant acid smut and plume capacity, has been associated with combustion of high-sulfur- and high-vanadium-containing fuel (Dietz and Wieser, 1983). Sources of acid smut include:

1. Hollow cenospheres, which have a light inorganic structure coated with unburned carbon and are formed by incomplete combustion.
2. Acid flakes, resulting from the deposition of fine particulate and sulfuric acid on colder duct and stack surfaces after the air heating stage in power plant (Jaworowski and Goldberg, 1979). As the atmospheric temperature and pressure change, the deposited layer flakes off in relatively large pieces and is carried out by the flue gas. These flakes are highly acidic and can cause spotting and corrosion on metallic or painted surfaces.
3. Acidic metal sulfate corrosion products from boiler surfaces running at low loads.

The formation of acid smut is associated with the conversion of SO_2 to SO_3. This reaction depends on boiler temperature, excess air, residence time, and the availability of oxidation catalysts such as vanadium, iron, and nickel oxides (Jaworowski and Goldberg, 1979).

Figure 3 shows photomicrographs of cenospheres found in acid smuts collected near oil-fired boilers—a power plant (top) and an industrial boiler (bottom). X-ray elemental analysis of these particles show the presence of vanadium and sulfur and occasionally a small amount of iron. Figure 4 shows a spectrum of an oil-fired cenosphere, with a very large peak of sulfur and vanadium-suggesting the presence of vanadium sulfates, or vanadium oxides and sulfuric acid. Thus, acid smuts deposition is accompanied by metal corrosion and metal deposition including vanadium.

In general mechanical collectors are the preferred method for avoiding emission of acid smuts. Other methods include optimization of fuel oil and burner efficiency, and a complete understanding of the temperature profile of the system from the air heater to the stack outlet. Chemical additives (MgO, NH_3, etc.) for improved fuel burning to minimize oxidation of SO_2 to SO_3 and neutralized condensed acid significantly contribute to acid smut control (Davies et al., 1981).

The deposition of vanadium associated with acid smut in the neighborhood of oil-fired power plants has been reported by Ganor et al. (1988), Foner et al. (1989), and Mamane and Melamed (1991, 1994, 1995).

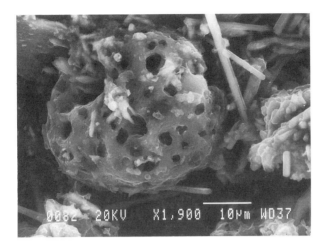

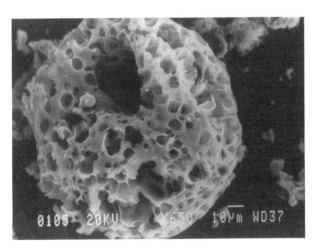

Figure 3. Scanning electron photomicrographs of cenospheres found within acid smuts in the vicinity of an oil-fired power plant (top) and near an oil-fired industrial boiler (bottom).

Figure 5 shows the dry deposition rates of vanadium in the vicinity of an uncontrolled 480-MW oil power plant during the summers of 1990 and 1993. Samples were collected on a 400-cm^2 flat painted surface for 1 month during the dry season. Vanadium deposition rates ranged from 2–11 kg/hectare · year near the plant in 1990 to 0.3–1.4 kg/hectare·year in 1993. A few hundred meters (500–1,500 m) from the power plant stack, the deposition rate decreased rapidly with distance. Improvements in fuel combustion and the use of magnesium additives resulted in a better control of acid smut.

In conclusion, the dry deposition rates of vanadium in the vicinity of an uncontrolled oil power plant may reach levels of 1–10 kg/hectare · year,

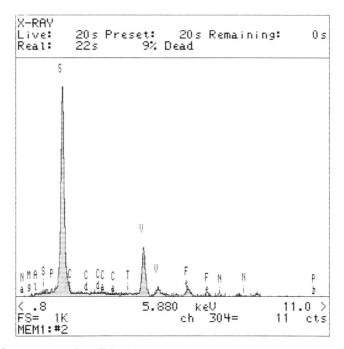

Figure 4. X-ray spectrum of an oil-fired power plant cenosphere containing a large peak of sulfur and vanadium. Elements below atomic number 11 (<Na) are not detected in this analysis. Most of the cenospheres contain sulfur and vanadium or iron, as well as some mineral elements (Al, Si, K, Ca).

one to two orders of magnitude higher than deposition rates in regional locations.

6.   VANADIUM CONCENTRATION IN FLY ASH

Crude oil has an average vanadium content of 50 mg/kg, with a range of 0.6–1,400 mg/kg (van Zinderen Bakker and Jaworski, 1980). The vanadium is present in the oil in the form of an organometallic porphyrin compound (Zoller et al., 1973). During refining and distillation, the vanadium remains in the residual oils because of its low volatility, and as a result it becomes more concentrated than in the original crude (NAS, 1974). The average concentration of vanadium in the Venezuelan nos. 5 and 6 residual oils is 870 mg/kg, compared with 112 mg/kg for the crude (Zoller et al., 1973). Residual oils are mainly used for electric power generation. The bulk of the Venezuelan oil imported into Canada and the United States is used on the East Coast (van Zinderen Bakker and Jaworski, 1980).

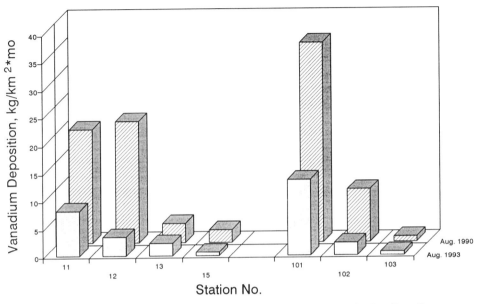

Figure 5. Deposition of vanadium (in kg/km$_2$·month) in the vicinity of the Reading oil power plant in Tel-Aviv, during the summer months of 1990 and 1993. Sites 11, 12, and 101 are near the plant. Sites 13, 15, 102, and 103 are several hundred meters from the stack.

During combustion, most of the vanadium in residual heating oils is released into the atmosphere in the form of V_2O_5 as part of the fly ash particulates (EPA, 1977). Sixty percent of the fly ash particles produced during combustion of oil in furnaces have a diameter of less than 10 μm, that is, they are of respirable size and can be deposited in the lungs. This occurs because the vanadium compounds are present in the vapor state in the combustion zone, and as the gas cools off, fine particulate fumes are formed that are emitted as submicrometer particles into the atmosphere (Sugimae and Hasegawa, 1973).

The vanadium content of some fly ash (see Table 7) has been determined to be between 1.5 and 7.0% (15,000–70,000 μg/g) (Tullar and Suffet, 1975; Bacci et al., 1983; Mamane et al., 1986). Oil bottom ash may contain a higher percentage (12.5%) in comparison with oil fly ash (1.4%) (Bache et al., 1991). In the Philadelphia Aerosol Study fly ash particles emitted from controlled oil power plants (with a cyclone or electrostatic precipitator) were mostly in the fine fraction (<2.5 μm). Based on XRF elemental analysis, it was possible to reconstruct the compounds emitted as fine particles: sulfates, ≈40%; carbon-

Table 7 Vanadium Concentrations in Fly Ash

Location	Position	Size (µg)	Concentration	Analytical Method	Reference
Boiler (high-S bituminous coal)	ESP inlet	>8.2	230 µg/g	ICP-AES	England et al. (1994)
	ESP inlet	4.1–8.2	290 µg/g	ICP-AES	Ibid.
	ESP inlet	<4.1	370 µg/g	ICP-AES	Ibid.
	Stack	>7	60 µg/g	ICP-AES	Ibid.
	Stack	3–7	40 µg/g	ICP-AES	Ibid.
	Stack	<3	20 µg/g	ICP-AES	Ibid.
Barcelona (waste incinerator)	Electrostatic ash filters		40 µg/g (range 10–60)	INAA-ICP-AAS	Fernandez et al. (1992)
Coal boiler (bituminous coal)	Fly ash	0.01–10	131 µg/g	PIXE	Kauppinen and Pakkanen (1990)
	Bottom ash		110 µg/g		
	Coal		41 µg/g		
	In stack:				
	Fine mode	d. < 0.1	3.4 mg/g		Ibid.
	Coarse mode	0.1 < d. < 10	0.7 mg/g		Ibid.
Coal-fired power plant		<2.5	1 mg/g (VO_2^+)	XRF, mass balance	Mamane et al. (1986)
Oil-fired power plant	Cyclone		16 mg/g (VO_2^+)		Ibid.
Oil-fired power plant	ESP inlet		23 mg/g (VO_2^+)		Ibid.
Coal utility boiler (subbituminous coal)	Upstream of particulate control device	d. < 10.5 d. < 0.1	91 µg/g ≈10% of the total mass	XRF-INAA	Markowski and Filby (1985)
320-MW oil-fired power plant (Italy)	In stack: 130 MW	TSP	5.0 ± 0.5% (50 mg/g)	AAS + PIXE	Bacci et al. (1983)
	280 MW	TSP	1.0 ± 0.5% (10 mg/g)		

62

Location	Sample	Size/fraction	Concentration	Method	Reference
Morgantown, WV	Stack ash	$1.9 < d. < 4.1$	190 μg/g	SSMS	Weissman et al. (1983)
		$1 < d. < 2$	180 μg/g		
		$0.7 < d. < 1.1$	130 μg/g		
Coal-fired power plant (U.S.)	Airborne (center of the stack)		0.02–0.05%	XRD FT-IR	Henry and Knapp (1980)
Oil-fired power plant (U.S.)	Airborne (center of the stack)		1.1–12.85%[a]	XRD FT-IR	Henry and Knapp (1980)
Steel plant	Breathing zone of workers		<5 μg/m^3 max 44 μg/m^3		EPA (1977)
Coal-fired (Illinois)	Inlet of ESP	5 μm (mmd)	225 μg/g	AAS	Lee et al. (1975)
	Outlet of ESP	1.6 μm (mmd)	172 μg/g		
Steel mill (Pennsylvania)	Inlet	4.8 μm (mmd)	392 μg/g		
	Outlet		Nondetectable		
Coal-fired power plant (southern Indiana coal)	Fly ash retained in plant	>74	150 μg/g	AAS	Davison et al. (1974)
	"	44–74	260 μg/g		
	"	>40	250 μg/g		
	"	30–40	190 μg/g		
	"	20–30	340 μg/g		
	"	10–20	320 μg/g		
	"	5–10	330 μg/g		
	"	<5	320 μg/g		
	Stack, airborne fraction	>11.3	150 μg/g	AAS	
	"	7.3–11.3	240 μg/g		
	"	4.7–7.3	420 μg/g		
	"	3.3–4.7	230 μg/g		
	"	2.06–3.3	310 μg/g		
	"	1.06–2.06	480μg/g		

[a] Most of the vanadium was found in the water-soluble phase of the fly ash, and was associated with sulfates. V_2O_5 was identified in the water-insoluble fraction.

rich material, ≃35%; minerals, ≃10%; vanadium compounds, ≃1.5%; and the rest were other metal oxides. Based on individual analysis of several hundred particles by scanning electron microscope coupled with X-ray analysis, the following observations can be made (Mamane et al., 1996).

1. The size distribution of morphological and chemical properties are trimodal: Particles larger than 3 μm (actual diameter) were often carbonaceous and rich in sulfur particles with a "spongy" or "lacy" appearance; particles of 0.5–3 μm were mineral spheres or crystalline; particles smaller than 0.5 μm were mostly sulfates.
2. Vanadium or nickel were detectable in about 50–60% of the "mineral" and the submicrometer sulfate particles. This conclusion is based on individual analysis of coarse and fine particles collected in the stack.

Figure 6 (top) is a photomicrograph of ≃2-μm oil fly ash particles collected in the stack of a power plant in Philadelphia. The particles are spherical, as in coal-fired power plants, but some are crystalline and others are carbonaceous. Smaller particles are mainly sulfates. Figure 6 (bottom) is an X-ray spectrum of the crystalline particle on the right. The particle contains organic material (high background in the spectrum), a high peak of V, Ca, S, and some Si and Fe.

Coal in the United States has an average vanadium content of 25 mg/kg (Bertine and Goldberg, 1971; Zoller et al., 1973; NAS, 1974). Upon combustion, the trace elements present in the coal are transferred to the bottom ash, fly ash, or gases emitted. In the more common pulverized coal boilers, as much as 90% of the total ash produced is in the form of fly ash (Klein et al., 1975). Vanadium is only partially incorporated into the boiler slag. A fraction of the fly ash and particulates is discharged into the atmosphere through the stack. The pathways of various trace elements through a cyclone-fed, coal-fired power plant have been investigated by Klein et al. (1975), who observed the following concentrations for vanadium: 28.5 μg/g in coal, 260 μg/g in slag, 440 μg/g in precipitator intake ducts, and 1,180 μg/g in precipitator output. Mamane et al. (1986) found similar vanadium concentrations of 900 μg/g in the fine fly ash (<2.5 μm) and 700 μg/g in the coarse fraction (2.5 μm < d < 10 μm) emitted from an ESP-controlled coal-fired power plant. This again indicates that the vanadium is associated with the smaller particulates, which are likely to pass through the electrostatic precipitators. Coal fly ash has been found to have a vanadium content of 50–5,000 μg/g (Natusch et al., 1975; van Zinderen Bakker and Jaworski, 1980).

Fine fly ash particles emitted from a coal-fired power plant in Philadelphia were also investigated. There, the sulfates contributed only 15% to the total mass, while minerals contributed 82%, and the rest were carbonaceous material and other elements (Mamane et al., 1986, and unpublished data from that study). Natusch et al. (1975) determined that the vanadium concentration

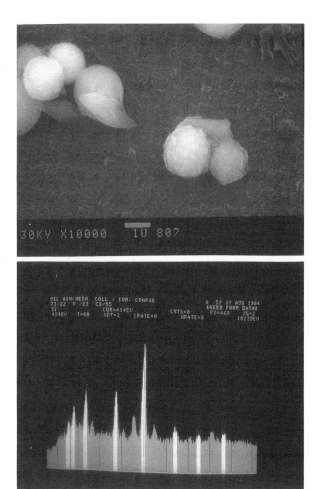

Figure 6. Top. A scanning electron photomicrograph of micrometer-size oil fly ash particles, collected in a stack of an oil power plant in Philadelphia. The particles are spherical and some are crystalline. **Bottom.** An X-ray spectrum of the crystalline particle. It shows in the middle a very high peak of vanadium, some Ca and S, and small amount of Si and Fe. Smaller oil fly ash particles are mainly sulfates.

of the particulates they examined was twice as high at the particle surface (760 μg/g) as at an arbitrary depth of 50 nm (380 μg/g). Thus, when these particles are deposited in the respiratory tract, the respiratory tissues are exposed to higher vanadium levels than indicated by bulk chemical analyses.

Figure 7 is a photomicrograph of coal fly ash particles and their typical X-ray spectrum. Here 95% of particles are smooth spheres of the following mineral content: Al, Si, K, Ca, Ti, and Fe. Some sulfur is associated with the coal fly ash particles, apparently on their surfaces.

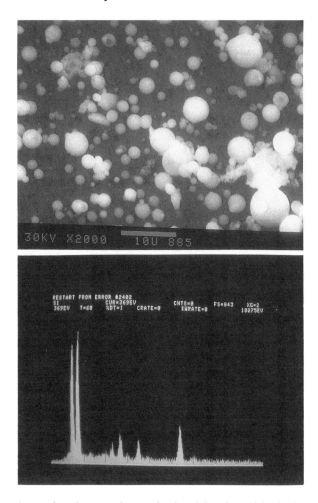

Figure 7. Top. A scanning electron micrograph of coal fly ash particles in the coarse fraction. Particles are smooth spheres. **Bottom.** A typical X-ray spectrum showing the mineral elements of the fly ash (from left to right the peaks are: Al, Si, K, Ca, Ti, and Fe). The low background suggests that no carbonaceous material is associated with these fly ash particles.

Table 7 lists vanadium concentrations in fly ash of oil and coal power plants. Although the data vary quite significantly the following observations could be made:

1. Vanadium in coal fly ash ranges from 100 to 1,000 μg/g, and in oil from 10 to 50 mg/g, two orders of magnitude difference.
2. The emission of vanadium into the atmosphere is controlled by the vanadium content of the fuel (coal, crude oil versus residuals, etc.), by the combustion process (what portion of the ash formed is deposited in

the boiler or in the plant), and by the type of control device installed. Oil-fired power plants are often not controlled for particulate emissions. The desulfurization process to control gaseous sulfur emissions also reduces the vanadium levels; there seems to be an almost one-to-one relationship between the vanadium removal and the degree of desulfurization (EPA, 1977).

3. The variability of vanadium content in fly ash would severely affect global estimates of vanadium emissions to the atmosphere.

ACKNOWLEDGMENTS

The authors wish to acknowledge the assistance of Ms. Elvira Nucaro, of the University of Calabria, and of Ms. Hadas Mamane and Mr. Eugene Lokshin, of Technion, Haifa, in collecting vanadium concentration data.

REFERENCES

Akeredolu, F. A., Barrie, L. A., Olson, M. R., Oikawa, K. K., Pacyna, J. M., and Keeler, G. L. (1994). The flux of anthropogenic trace metals into the Arctic from the mid-latitudes in 1979/80. *Atmos. Environ.* **28,** 1557–1572.

Al-Arfaj, A. A., and Alam, I. A. (1993). Chemical characterization of sediments from the Gulf area after the 1991 oil spill. *Mar. Pollut. Bull.* **27,** 97–101.

Atteia, O. (1994). Major and trace elements in precipitation on Western Switzerland. *Atmos. Environ.* **28,** 3617–3624.

Bacci, P., Del Monte, M., Longhetto, A., Piano, A., Prodi, F., Redaelli, P., Sabbioni, C., and Ventura, A. (1983). Characterization of the particulate emission by a large oil fuel fired power plant. *J. Aerosol Sci.* **14,** 557–572.

Bache, C. A., Rutzke, M., and Lisk, D. J. (1991). Absorption of vanadium, nickel, aluminum and molybdenum by Swiss chard grown on soil amended with oil fly ash or bottom ash. *J. Food Safety* **12,** 79–85.

Bacskai, G., Vaskovi, B., Vagvolgyi, G., Rudnai, P., Kertesz, M., and Gulyas, A. (1994). Case study: Vanadium and nickel contents of dustfall in Szazhalombatta. *Int. J. Environ. Health Res.* **4,** 113–114.

Bertine, K. K., and Goldberg, E. D. (1971). Fossil fuel combustion and the major sedimentary cycle. Science **173,** 233–235.

Bouhamra, W. S., and Jassem, F. M. (1994) . Estimation of the mass transfer coefficients and diffusion coefficients of Ni and V in Kuwait crude oils. *Environ. Technol.,* **15,** 645–656.

Chen, L., and Duce, R. A. (1983). The sources of sulfate, vanadium and mineral matter in aerosol particles over Bermuda. *Atmos. Environ.* **17** (10), 2055–2064.

Chow, J. C., Watson, J. G., Fujite, E. M., Lu, Z., and Lawson, D. R. (1994). Temporal and spatial variations of $PM_{2.5}$ and PM_{10} aerosol in the Southern California air quality study. *Atmos. Environ.* **28** (12), 2061–2080.

Church, T. M., Tramontano, J. M., Scudlark, J. R., Jickells, T. D., Tokos, J. J., and Knap, A. H. (1984). The wet deposition of trace metals to the western Atlantic Ocean at the mid-Atlantic coast and on Bermuda. *Atmos. Environ.* **18** (12), 2657–2664.

Crecelius, A., Lepel, A., Laul, J. C., Rancitelli, L. A., and McKeever, R. L. (1980). Background air particulate chemistry near Colstrip, Montana. *Environ. Sci. Technol.* **14,** 422–428.

Davies, I., Laxton, J. W., and Owers, M. J. (1981). The use of magnesium-based flue gas additives in minimizing smut emissions from oil-fired boilers. *J. Inst. Energy* **54,** 21–30.

Davison, R. L., Natusch, F. S., Wallace, J. R., and Evans, C. A. (1974). Trace elements in fly ash. *Environ. Sci. Technol.* **8** (13), 1107–1113.

Dietz, R. N., and Wieser, R. F. (1983). Sulfate formation in oil-fired power plant plume. Vol. 1, Report, EPRI-EA-3231, EPRI, California.

Divita, F., Ondov, J. M., and Suarez, A. E. (1996). Size spectra and atmospheric growth of V-containing aerosol in Washington, D.C. *Aerosol Sci. Technol.* **25,** 256–273.

Donagi, A. E., Ganor, E., Shenhar, A., and Cember, A. (1979). Some metallic trace elements in the atmospheric aerosols of the Tel-Aviv area. *J. Air Pollut. Control Assoc.* **29,** 53–54.

Duce, R. A., and Hoffman, G. L. (1976). Atmospheric vanadium transport to the ocean. *Atmos. Environ.* **10,** 986–996.

Dzubay, T. G., Stevens, R., Gordon, G. E., Olmez, I., and Sheffleld, A. E. (1988). A composite receptor method applied to Philadelphia aerosol. *Environ. Sci. Technol.* **22,** 46–52.

England, G. C., McGrath, P., Riseq, R. G., and Hansell, D. (1994). Trace metal emissions from coal fired power plants: Chromium speciation and effect of particle size. In *Air Pollution Measurement Methods and Monitoring Studies, 87th Annual Meeting Proceedings,* Vol. B, 94-WA73.05, Cincinnati, Ohio.

EPA (U.S. Environmental Protection Agency, 1977). Scientific and technical assessment report on vanadium. U.S. EPA, HERL, EPA-600/6-77-002, Research Triangle Park, NC.

Fernandez, M. A., Martinez, L., Segarra, M., Garcia, J. C., and Esplell, F. (1992). Behavior of heavy metals in the combustion gases of urban waste incinerators. *Environ. Sci. Technol.* **26** (5), 1040–1047.

Fidalgo, M. R., Mateos, J., and Garmendia, J. (1988) The origin of some of the elements contained in the aerosols of Salamnca (Spain). *Atmos. Environ.* **22,** 1495–1498.

Foner, H. A., and Ganor, E. (1986). Elemental and mineralogical composition of total suspended matter in Tel-Aviv, Israel. In Environmental Quality and Ecosystem Stability, Vol. IIIB. Bar Ilan University Press, Ramat-Gan, Israel.

Foner, H. A., Lahav, D., Zohar, Y., and Goldberger, Z. (1989). A study of vanadium and nickel contamination in the Ashdod area: Preliminary results. The Fourth International Conference of Environmental Quality and Ecosystems Stability, Israel. Jerusalem, June 4–8, 1989.

Galloway, J. N., and Likens, G. E. (1979). Atmospheric enhancement of metal deposition in Adirondack Lake sediment. *Limnol. Oceanogr.* **24,** 427–433.

Galloway, J. N., Thornton, J. D., Norton, S. A., Volchok H. L., and MacLean, R.A.N. (1982). Trace metals in atmospheric deposition: A review and assessment. Atmos. Environ. **16,** 1677–1700.

Ganor, E. (1975). Atmospheric Dust in Israel, Ph.D. thesis (in Hebrew). The Hebrew University, Jerusalem, Israel.

Ganor, E., Altshuller, S., Foner, H. A. Brenner S., and Gabbay, J. (1988). *Vanadium* and nickel in dustfall as indicators of power plant pollution. *Water Air Soil Pollut.,* **42,** 241–252.

Ganor, E. S., and Foner, H. A. (1989). Composition of some urban atmospheric aerosols in Israel. The Fourth International Conference of Environmental Quality and Ecosystems Stability. Israel. Jerusalem, June 4–8, 1989.

Gertler, A. W. (1994). A preliminary apportionment of the sources of fine particles impacting on the Israeli coast. *Israel J. Chem.* **34,** 425–433.

Gordon, G. E. (1988). Receptor models. *Environ. Sci. Technol.* **22,** 1132–1142.

Grimaldi, F., Bascou H., Viala A., Muls E., Esberard N., and Casabianca S. (1993). Studies of the Marseilles-Provence Committee of APPA on atmospheric metallic pollutants. *Pollut. Atmos.,* Juillet-September, pp. 98–103.

Harrison, R. M., Luhana, L., and Smith, D. J. T. (1993). Urban air pollution by heavy metals: Sources and trends. In Allan and J. D. Nriagu (Eds.), *Heavy Metals in the Environment,* Vol. 1, pp. 81–84.

Henry, W. H., and Knapp, K. T. (1980). Compound forms of fossil fuel fly ash emissions. *Environ. Sci. Technol.* **14,** 451–456.

Hoff, R. M., and Brice, K. A. (1994). Atmospheric dry deposition of PAHs and trace metals to Lake Ontario and Lake Huron. Paper 94-RA110.04, 87th AWMA Annual Meeting and Exhibition. Cincinnati, Ohio, June 19–24.

Hope, B. K., (1994). A global biogeochemical budget for Vanadium. *Sci. Total Environ.* **141,** 1–10.

Husain, L., Webber, J., and Canelli, E. (1983). Erasure of midwestern Mn/V signature in an area of high vanadium concentration. *J. Air Pollut. Control Assoc.* **33** (12), 1185–1188.

Jaworowski, R. J., and Goldberg, H. J. (1979). Chemical control of acid smut from oil fired boilers. *Combustion* **51,** 39–44.

Juichang, R., Freedman, B., Coles, C., Zwicker, B., Holzbecker, J., and Chatt, A. (1995). *Vanadium* contamination of lichens and tree foliage in the vicinity of three oil-fired power plants in Eastern Canada. *J. Air Waste Manage. Assoc.* **45,** 461–464.

Kauppinen, E. I., and Pakkanen, T. A. (1990). Coal combustion aerosols: A field study. *Environ. Sci. Technol.* **24** (12), 1811–1817.

Keeler, G. J. (1994). Lake Michigan Urban Air Toxics Study, US-EPA Interim Report.

Klein, D. H., Andrew, A. W., Carter, J. A., Emergy, J. F., Feldman, C., Fulkerson, W., Lyon, N. S., Ogle, J. C., Talmi, Y., van Hook, R. I., and Bolton, N. (1975). Pathways of thirty-seven trace elements through coal fired power plants. *Environ. Sci. Technol.* **9,** 973–979.

Kleinman, M. T., Pasternack, B. S., Eisenbud, M., and Kneip, T. J. (1980). Identifying and estimating the relative importance of airborne particulates. *Environ. Sci. Technol.* **14,** 62–65.

Kowalczyk, G. S., Gordon, G. E., and Rhelngrover, S. (1982) Identification of atmospheric particulate sources in Washington, D.C., using chemical element balances. *Environ. Sci. Technol.* **16,** 79–90.

Landsberger, S., Vermette, V. G., Stuenkel, D., Hopke, P. K., and Barrie, L. A. (1992). Elemental source signatures of aerosols from the Canadian high Arctic. *Atmos. Pollut.* **75,** 181–187.

Lee, R. E., Crist, H., Riley, A. E., and Macleod, K. E. (1975). Concentration and size of trace metals emissions from a power plant, a steel plant, and a cotton gin. Environ. Sci. Technol. **9,** 643–647.

Linton, R. W., Loh, A., Natusch, D. F. S., Evans, C. A., and Williams, P. (1976). Surface predominance of trace elements in airborne particles. *Science* **191,** 852–854.

Lioy, P. J., Zelenka, M. P., Cheng, M., Reiss, N. M., and Wilson, W. (1989) The effect of sampling duration on the ability to resolve source types using factor analysis. *Atmos. Environ.* **23** (1), 239–254.

Madany, I. M., and Raveendran, E. (1992). Polyciclic aromatic hydrocarbons, nickel and *vanadium* in air particulate matter in Bahrain during the burning of oil fields in Kuwait. *Sci. Tot. Environ.* **116,** 281–289.

Maenhaut W., Duncastel G., Leck, K., Nilsson, E., and Heintzenberg, J. (1996). Multi-elemental composition and sources of the high Arctic atmospheric aerosol during summer and autumn. *Tellus* **48B** (2), 300–321.

Maenhaut, W., and Vitols, V. (1989). Trace element composition and origin of the atmospheric aerosol in the Norwegian Arctic. *Atmos. Environ.* **23** (11), 2551–2569.

Mamane, Y., Lokshin, E., and Melamed, E. (1996). *Apportionment of Aerosols, Including Sulfates, in the Haifa Urban Area.* Interim Report, Technion, Haifa 32000, Israel.

Mamane, Y., and Melamed E. (1991, 1994, 1995). *Chemical Analysis and Microscopy of Acid Smuts in the Vicinity of the Reading Power Plant (Tel-Aviv).* Annual Research Reports, Technion, Haifa, Israel.

Mamane, Y., Miller, J. L., and Dzubay, T. G. (1986). Characterization of individual fly ash particles emitted from coal- and oil-fired power plants. *Atmos. Environ.,* **20** (11), 2125–2135.

Markowski, G. R., and Filby, R. (1985). Trace element concentration as a function of particle size in fly ash from a pulverized coal utility boiler. *Environ. Sci. Technol.* **19** (9), 796–804.

Martens, C. S., Wesolowski, J. J., Kaifer, R., John, W., and Harriss, R. C. (1973). Source of vanadium in Puerto Rican and San Francisco Bay area aerosol. *Environ. Sci. Technol.* **7,** 817–820.

Milford, J. B., and Davidson, C. I. (1985). The size of particulate trace elements in the atmosphere—A Review. *J. Air Pollut. Control Assoc.* **35,** 1249–1260.

Munch, D. (1993). Concentration profiles of arsenic, cadmium, chromium, cooper, lead, mercury, nickel, zinc, vanadium and polynuclear aromatic hydrocarbons (PAH) in forest soil beside an urban road. *Sci. Tot. Environ.* **138,** 47–55.

NAS (National Academy of Sciences). (1974). *Vanadium.* NAS Committee on Biological Effect of Atmospheric Pollutants, 117 pp., Washington, D.C.

Natusch, D. F. S., Bauer, C. F., Matusiewica, H., Evans, C. A., Bakker, J., Loh, A. H., Linton, R. W., and Hopke, P. K. (1975). Characterization of trace elements in fly ash. *Proceedings of the International Conference on Heavy Metals in the Environment,* Oct. 27–31, Toronto, Vol. 2 (2), 553–575.

Pacyna, J. M., Bartonova, A., Cornille P., and Mainhaut, W. (1989). Modeling of long-range transport of trace elements: A case study. *Atmos. Environ.* **23,** 107–114.

Pirrone, N., Allegrini, I., Keeler, G. J., Nriagu, J. O., Rossmann, R., and Robbins, J. A. (1997). Historical atmospheric mercury emissions and depositions in North America compared to mercury accumulations in sedimentary records. *Atmos. Environ.* (In press).

Pirrone, N., Keeler, G. J., and Holsen, T. M. (1995a). Dry deposition of trace elements over Lake Michigan: A hybrid receptor-deposition modeling approach. *Environ. Sci. Technol.* **29,** 2112–2122.

Pirrone, N., Keeler, G. J., and Nriagu, J. O. (1996a). Regional differences in worldwide emissions of mercury to the atmosphere. *Atmos. Environ.* **30,** 2981–2987.

Pirrone, N., Keeler, G. J., Nriagu, J. O., and Warner, P. O. (1996b). Historical trends of airborne trace metals in Detroit from 1971 to 1992. *Water, Air Soil Pollut.* **88,** 154–165.

Pirrone, N., Keeler, G. J., and Warner, P. O. (1995b). Trends of ambient concentrations and deposition fluxes of particulate trace metals in Detroit from 1982 to 1992. *Sci. Total Environ.* **162,** 43–61.

Sadiq, M., and Mian, A. A. (1994). Nickel and *vanadium* in air particulates at Dhahran (Saudi Arabia) during and after the Kuwait oil fires. *Atmos. Environ.* **28,** 2249–2253.

Schroeder, W. H., Dobson, M., Kane, D. M., and Johnson, N. D. (1987). Toxic trace elements associated with airborne particulate matter: A Review. *J. Air Pollut. Control Assoc.* **37,** 1267–1285.

Slinn, W. G. N., Hasse, L., Hicks, B. B., Hogan, A. W., Lal, D., Liss, P. S., Munnich, K. O., Sehmel, G. A., and Vittori O. (1978). Some aspects of the transfer of atmospheric trace constituents past the air–sea interface. *Atmos. Environ.* **12,** 2055–2087.

Stevens, R., Pinto, J., Mamane, Y., Ondov, J., Abdulraheem, M., Al-Majed, N., Sadek, M., Cofer, W., Ellenson, W., and Kellogg, R. (1993). Chemical and physical properties of emissions from Kuwaiti oil fires. *Water Sci. Technol.* **27,** 223–233.

Sugimae, A., and Hasegawa, T. (1973). Vanadium concentrations in atmosphere. *Environ. Sci. Technol.* **7,** 444–448.

Sweet, C. W., and Vermette, S. J. (1993). Source of toxic trace elements in urban air in Illinois. *Environ. Sci. Technol.* **27,** 2502–2510.

Thurston, G. D., and Spengler, J. D. (1985). A quantitative assessment of source contributions to inhalable particulate matter pollution in metropolitan Boston. *Atmos. Environ.* **19** (1), 9–25.

Tullar, I. V., and Suffet, I. H. (1975). The fate of vanadium in an urban air shed: The lower Delaware River Valley. *J. Air Pollut. Control Assoc.* **25**, 282–286.

van Zinderen Bakker, E. M., and Jaworski, J. F. (1980). *Effects of Vanadium in the Canadian Environment.* Report NRCC No. 18132, 94 pp. National Research Council of Canada, Ottawa.

Tuncel, S. G., Olmez, I., Parrington, J. R., and Gordon, G. E. (1985). Composition of fine particle regional sulfate component in Shenandoah Valley. *Environ. Sci. Technol.* **19**, 529–537.

Vermette, S. J., Williams A. L., and Landsberger, S. (1991). *PM$_{10}$ source apportionment using local surface dust profiles: Examples from Chicago.* Specialty Conference on PM-10, Air Waste Management Association, Pittsburgh, PA.

Vesely, J., Almquist-Jacobson, H., Miller, L. M., Norton, S., Appleby, P., Dixit, A. S., and Smol, J. P. (1993). The history and impact of air pollution at Certovo Lake, Southwestern Czech Republic. *Paleolimnol.* **8**, 211–231.

Weissman, H., Carpenter, R. L., and Newton, G. (1983). Respirable aerosol from fluidized bed coal combustion. 3. Elemental composition of fly ash. *Environ. Sci. Technol.* **17** (2), 65–71.

Zhang, X., Arimoto, R., An, Z., Chen, T., Zhang, G., Zhu, G., and Wang, X. (1993). Atmospheric trace elements over source regions for Chinese dust: Concentrations, sources and atmospheric deposition on the Loess Plateau. *Atmos. Environ.* **27A** (13), 2051–2067.

Zhou, M., et al. (1996). *Atmos. Res.* **40**, 19–31.

Zoller, W. H., Gordon, G. E., Gladney, E. S. and Jones, A. G. (1973). The sources and distribution of *vanadium* in the atmosphere. In E. L. Kothny (Ed.), *Trace Elements in the Environment.* Advances in Chemistry, Series 123, American Chemical Society, Washington, DC.

4

CHEMISTRY OF RELEVANCE TO VANADIUM IN THE ENVIRONMENT

Debbie C. Crans, Sean S. Amin, and Anastasios D. Keramidas

Department of Chemistry, Colorado State University, Fort Collins, CO 80523

Vanadium in the Environment. Part 1: Chemistry and Biochemistry, Edited by Jerome O. Nriagu.
ISBN 0-471-17778-4.   © 1998 John Wiley & Sons, Inc.

1. INTRODUCTION

The chemistry of vanadium of relevance to the environment encompasses solid-state chemistry (vanadium minerals) (Rehder, 1991; Crans, 1995), solution chemistry (aqueous vanadium chemistry) (Chasteen, 1981; Pope and Müller, 1991; Crans, 1994a), and gas phase chemistry (airborne vanadium particles) of vanadium compounds. Vanadium, as an early transition metal, exhibits extraordinarily rich chemistry; it can readily convert between oxidation states under mild conditions and thus spans an unsurpassed range of reactions under environmental conditions. Its affinity for oxygen allows it to form stable vanadium–oxide cations and anions such as VO^{2+}, VO_2^+, $H_2VO_4^-$, and HVO_4^-, which are exceedingly important to the environmental chemistry of vanadium. The various protonated forms of the vanadium(V)–oxide anion, vanadate is a well-known analog of phosphate and its different protonated states. The analogy exists with respect to pK_a values, electronic and structural properties presumably accounting for the many reports of vanadate acting as a phosphate analog in biological systems. Vanadium compounds are most abundant naturally in the form of minerals and bioaccumulated material in the solid state and as vanadate in most surface waters. This chapter outlines the basic environmental chemistry of the element named after the Nordic love and beauty goddess, Vanadis (Weeks and Leicester, 1968).

The trace element vanadium has a somewhat controversial history, in part because of its complex chemistry and in part because of the effects various compounds can induce in biological systems (Shaver et al., 1995; Posner et al., 1994). Indeed, vanadium compounds are renowned for their potent insulin action both in vitro and in vivo. In spite of countless reports (Shechter, 1990) of positive studies, a significant group of the scientific community pointing to the toxic effects of these compounds, opposes use of these active compounds in mammals and humans. In humans vanadium poisoning can manifest itself in a number of symptoms, ranging from minor symptoms such as tremors of the hands and eye irritation to appearance of a green tongue and ultimately to death (Lewis, 1959). Although other mammals, such as cats, show related symptoms, unfortunately mice do not. The greenish coloration of the tongue is a curious effect that has been observed in humans exposed to high levels of V_2O_5 and disappears in 2–3 days once the patient no longer continues to be exposed to the vanadium. The chemical explanation for the green tongue is most likely the formation of vanadium(III) and vanadium(IV) complexes in the mouth by the action of bacteria and ptyalin (Wyers, 1946). Since the salts are not simply removed by cleaning, it is not just a question of formation of the salts on the surface of the tongue.

Thus, the scientific community is divided into two groups—one opposing (Domingo, 1996) and another in favor of pharmaceutical application of vanadium compounds (Posner et al., 1994; Shechter, 1990; Ide, 1995). There is no doubt that reconciliation of the two camps will be beneficial not only for the area of study to progress at a more rapid pace, but for mankind to benefit from the positive information many of the vanadium compounds provide. The problem, of course, is compounded by complex vanadium chemistry, which is usually a result of the decomposition of the compounds under investigation (a fact that has escaped many of those working with these compounds) (Crans et al., 1995). At this time little is known about which vanadium compounds are active, since redox and hydrolytic reactions preclude studies of one species in the absence of others, and few studies are conducted in a manner such that an analysis of the effects of each species can be carried out (Crans et al., 1994b).

Few areas in the world naturally contain high levels of vanadium, one of these being the Rocky Mountain Plateau region. Uravan is a small town located in this region in southwestern Colorado. The town's name derives from the uranium and vanadium mining that began in the early 1900s in this area (Fischer, 1968). The ore minerals in this region are typically low-valent oxides and silicates of uranium and vanadium. The abundance of vanadium is actually greater than of uranium in the Morrison region of Colorado, in contrast to most other regions that mine uranium. Since vanadium is a by-product in uranium mining, by virtue of the extensive mining that has taken place in the Uravan area significant discharge of this trace metal into the environment has occurred over the years. Mining companies and the U.S. Bureau of Mines have specifically addressed environmental problems in this region through various programs (Froisland et al., 1982).

This review will contain a section concerning the fundamental chemistry of vanadium and a section describing common and interesting vanadium-containing minerals, and will discuss the vanadium-content in petroleums, systems supporting bioaccumulation of vanadium, and processes currently in use in the modern world. Throughout this chapter we will focus on how to describe and refer to vanadium chemistry in each area affecting the environment.

2. VANADIUM CHEMISTRY

2.1. General Comments Relating to Vanadium Chemistry

Vanadium can exist in a multitude of different oxidation states from -2 to $+5$, although the forms naturally found in the environment are 3, 4, and 5. Vanadium chemistry is extremely rich in part because of the flexibility of this element to take on many different coordination numbers and coordination geometries and in part because of the extensive redox chemistry readily available to this element even under the mild conditions commonly observed in the environment. In addition, vanadium(IV) and -(V) can undergo complex hydration reactions that lead to a multitude of compounds, many of which have so far been characterized only in the solution phase. On top of these complexities, vanadium compounds, particularly in oxidation states 3, 4, and 5, are involved in protonation equilibria. However, when considered appropriately, this rich chemistry offers the researcher a wide range of geometric and spectroscopic tools to study vanadium.

2.2. Solid-State Vanadium Chemistry

The coordination number and geometry of vanadium minerals and other compounds found in the environment is dependent upon the oxidation state of vanadium and the type of coordinating ligand. Vanadium(III) complexes can have up to five different types of coordination sphere, being matched in this respect only by oxidation state 5. Here, coordination numbers range from three to eight. Vanadium(IV) has four different types of coordination sphere, including four, five, six, and eight; interestingly the seven coordination sphere is not observed for this oxidation state (Vilas Boas and Costa Pessoa, 1987).

The geometry in vanadium complexes is tied to coordination number; some of them are illustrated in Figure 1: tetrahedral(4), trigonal bipyramidal(5), square pyramidal(5), octahedral(6), and pentagonal bipyramidal(7). In vanadium(III), -(IV), and (V) complexes, the five and six coordination geometries are the most common. Most vanadium complexes in oxidation states 4 and 5 contain at least one oxo group, and many vanadium(V) complexes contain two oxo groups (cis dioxo) (Holloway and Melnik, 1985). As the oxidation state of vanadium increases, so does the preference of vanadium for harder bases such as oxygen. The predominance of vanadium(IV) and -(V) oxide minerals, as observed in Table 1, reflects not only the favorable match between vanadium and oxide, but also the stability of these materials.

2.3. Aqueous Solution Vanadium Chemistry

Dissolution of vanadium compounds in aqueous solution, by leaching of vanadium-containing rocks, for example, will rapidly convert all vanadium compounds to vanadium(V) in the form of vanadate. Figure 2 shows a speciation diagram in which the oxidation state and major species are indicated based on pH and reduction potential. From Figure 2 it is clear that oxidation state 5 is the most prevalent under most environmental conditions, although

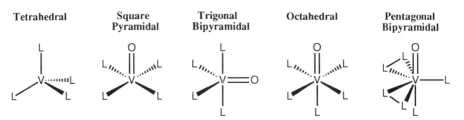

Figure 1. Idealized geometries of vanadium complexes in a number of different coordination spheres. L, Ligand.

Table 1 Formulas, Oxidation States, Location, and Characteristics of Selected Vanadium-Containing Minerals

Mineral	Formula	Oxidation State	Location[a]	Characteristic
Cuprodescloizite	$Pb(Cu)(VO_4)(OH)$	5	USA	Main V mineral
Descloizite	$Pb(Zn, Cu)(VO_4)(OH)$	5	USA; Namibia	Main V mineral
Mottramite	$Pb(Cu, Zn)(VO_4)(OH)$	5	USA; England	Main V mineral
Patronite	VS_4 (?)	?	Peru	Main V mineral
Roscoelite	$K(V, Al, Mg)_2AlSi_3O_{10}(OH)_2$	3	USA; Australia	Main V mineral
Vanadinite	$Pb_5(VO_4)_3Cl$	5	USA; Mexico	Main V mineral
Carnotite	$K_2(UO_2)_2(VO_4)_2 \cdot 3H_2O$	5	USA; Zaire	Main & U-V mineral
Metatyuyamunite	$Ca(UO_2)_2(VO_4)_2 \cdot 3H_2O$	5	USA; USSR	U-V mineral
Tyuyamunite	$Ca(UO_2)_2V_2O_8 \cdot 5-8H_2O$	5	USA; USSR	U-V mineral
Corvusite	$V_2V_{12}O_{34} \cdot nH_2O$	4/5	USA; USSR	Mixed valence
Fernandinite	$CaV_2V_{10}O_{30} \cdot 14H_2O$	4/5	Peru; USA	Mixed valence
Franciscanite	$Mn_6V_2Si_2(O,OH)_{14}$	3/5	USA	Mixed valence
Haggite	$V_2O_2(OH)_3$	3/5	USA; Pakistan	Mixed valence
Minasragrite	$VO(SO_4) \cdot 5H_2O$	4	Peru	
Munirite	$NaVO_3 \cdot 2H_2O$	5	Pakistan	

Source: Brief condensation of data in Roberts et al. (1990).

[a] Principal location of occurrence.

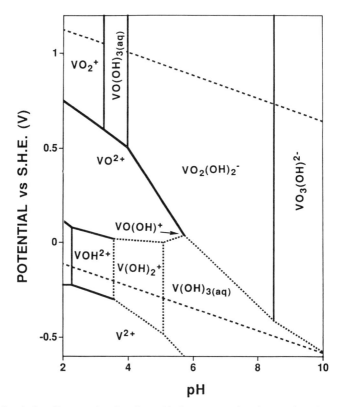

Figure 2. Speciation diagram showing the oxidation state and major species of vanadium as a function of pH and reduction potential (versus standard hydrogen electrode). At boundary lines species in adjacent regions are present in equal concentrations. Boundaries indicated by short dashed lined are less certain than those indicated by bold lines. The upper and lower longer dashed lines indicate the upper and lower limits of the stability of water. Adapted from Baes and Mesmer (1976).

vanadium in oxidation state 4 also exists under some conditions. However, the dissolution and leaching of vanadium(III)- and -(IV)-containing minerals will also result in oxidation of the vanadium.

2.3.1. Vanadium(V)

Vanadium(V) is the most mobile form of vanadium in the pH ranges commonly found in surface waters. Vanadium(V) undergoes hydrolytic reactions in water, generating many oligomeric anionic species including vanadate in water. Vanadate, an analog of phosphate, also forms a wide range of condensed forms, each of which exists in various protonation states, and at low pH vanadium(V) complexed to ligands exists as the cation VO_2^+, which bears no resemblance to any form of phosphate. In the literature the term *vanadate* is generally used loosely to describe the colorless solutions that are generated from metavanadate, orthovanadate, or vanadium pentoxide. The species in these solutions are sensitive to the conditions briefly described here; however, the complex speciation in these solutions is often not recognized, and the term *vanadate* is often used synonymously for the pH-dependent mixture of $H_2VO_4^-$, HVO_4^{2-}, and VO_4^{3-}. Since vanadium(V) is a d^0 metal ion, vanadium-51 is an NMR active nucleus. This method is responsible for much of the information currently available to describe this element in oxidation state 5. The ^{51}V NMR chemical shift is very sensitive to nuclearity and protonation state of the vanadium species (Heath and Howarth, 1981). This is one reason why a significant amount of work in the area of aqueous speciation of vanadium oligomers and their protonated forms have been carried out.

The equilibrium of colorless vanadium(V) species in solution is exceedingly sensitive to vanadium concentration, pH, ionic strength, and other solution components including potential (Heath and Howarth, 1981; Pettersson et al., 1985). As can be seen from Table 2, the monomer, commonly referred to as vanadate, has three different protonation states. At neutral pH and most environmental conditions the monomer will exist mainly as $H_2VO_4^-$. Vanadate monomer is in a number of equilibria with vanadate dimer ($H_3V_2O_7^-$), the cyclic tetramer ($V_4O_{12}^{4-}$), and cyclic pentamer ($V_5O_{15}^-$) (Crans et al., 1990a). In addition to these major species, minor species are observed, some of which are limited to specific concentration and pH conditions. Minor complexes include linear trimer and tetrameres as well as pentamers and other less defined species (Howarth, 1990). The colorless vanadium(V) species in solution exchange on a millisecond time scale so that equilibria involving oligomers will establish immediately in the environment under investigation (Crans et al., 1990). Under acidic conditions (pH < 6) the yellow/orange decamer forms. Although decamer is thermodynamically unstable at neutral or basic pH, the slow hydrolysis of this oligomer at pH > 7 allows it to persist for up to 2 days (Pope, 1983). Figure 3a shows the ^{51}V NMR spectrum of a solution of 0.1 mM NH_4VO_3 in water at pH 7.0; the major resonance is due to vanadate monomer and the additional ~8% vanadium species is vanadate dimer. However, a corresponding spectrum of a solution containing 2.5 mM NH_4VO_3 at pH 7.0 (Figure 3b) shows that monomer, dimer, and tetramer are present in solution. At millimolar vanadium concentrations significant concentrations of oligomers are present; vanadate monomer is no longer the major species in solution. A 25-fold dilution of this solution will generate a solution with speciation identical to that shown in Figure 3a. These spectra thus demonstrate the effect of vanadium concentration and rapid equilibration of vanadium species in solution.

Low vanadium concentrations are typically found in most natural environments, where detectable freshwater concentrations range from 2 to 300 μg/L, with an average concentration of 40 μg/L (300 nM) (Kopp and Kroner, 1968) for rivers in the United States. This is considerably higher than the seawater concentration (Pacific), which is 1.6 μg/L (35 nM) (Rehder, 1991). In

Table 2 Formulas, pK_a Values, and Chemical Shifts of the Major Oxovanadates in Aqueous Solution

Formula	pK$_a$	Chemical Shift ^{51}V NMR (ppm)	Color and Structure	References
VO$_4^{3-}$	—	−541.2	Colorless; linear	(Heath and Howarth, 1981)
HVO$_4^{2-}$	~12	−538.8	Colorless; linear	(Heath and Howarth, 1981)
H$_2$VO$_4^{-}$	7.1	−560.4	Colorless; linear	(Heath and Howarth, 1981)
V$_2$O$_7^{4-}$	—	−561.0	Colorless; linear	(Heath and Howarth, 1981)
HV$_2$O$_7^{3-}$	8.9	−563.5	Colorless; linear	(Heath and Howarth, 1981)
H$_2$V$_2$O$_7^{2-}$	7.2	−572.7	Colorless; linear	(Heath and Howarth, 1981)
V$_4$O$_{13}^{6-}$	—	} ~ −564 to ~ −572	Colorless; cyclic	(Pettersson et al., 1985)
HV$_4$O$_{13}^{5-}$	8.3		Colorless; cyclic	(Pettersson et al., 1985)
V$_4$O$_{12}^{4-}$	—	−574.9	Colorless; cyclic	(Pettersson et al., 1985)
V$_5$O$_{15}^{5-}$	—	−582.7	Colorless; cyclic	(Pettersson et al., 1985)
V$_{10}$O$_{28}^{6-}$	—	−423, −497, −514	Yellow/orange; cluster	(Pettersson et al., 1985)
HV$_{10}$O$_{28}^{5-}$	5.7	−424, −500, −516	Yellow/orange; cluster	(Pettersson et al., 1985)
H$_2$V$_{10}$O$_{28}^{4-}$	3.6	−425, −506, −524	Yellow/orange; cluster	(Pettersson et al., 1985)
H$_3$V$_{10}$O$_{28}^{3-}$	1.6	−427, −515, −534	Yellow/orange; cluster	(Pettersson et al., 1985)

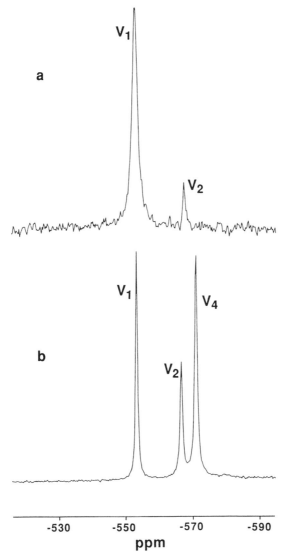

Figure 3. ^{51}V NMR spectra at 78.9 MHz of ammonium metavanadate a. 0.1 mM, at pH 7. b. 2.5 mM, at pH 7.0.

both environments the mononuclear mono ($H_2VO_4^-$) and dianionic (HVO_4^{2-}) species predominate. However, in most higher than micromolar concentration studies of and with vanadium(V) the concentration of vanadium is sufficiently high so that oligomeric species of vanadium(V) form. It has recently been recognized that these oligomeric species represent non-innocent bystanders both with respect to chemical reactivity and biological properties since the oligomeric oligoanions have both different chemical and biological properties (Crans et al., 1995; Crans, 1994a; Crans et al., 1990b). *Chemical and biological studies of vanadium(V) with higher than micromolar concentrations of vanadate must include considerations of the oligomeric species that are commonly present in model systems for environmental studies.* As recently recognized these oligomerization equilibria can be used to the researchers advantage, since they maintain the monomeric vanadium concentration within a small concentration range (Crans, 1994a).

2.3.2.   Vanadium(IV)

Vanadium(IV) is readily formed from vanadium(V) in the presence of any reducing components in aqueous solution. Vanadium(IV) is probably the principal oxidation state at which the trace level of vanadium is found in living humans, including mammals. Mononuclear vanadium(IV) complexes contain a free electron and thus are observable by electron paramagnetic resonance (EPR) spectroscopy, which allows deciphering of the equatorial ligand environment around the vanadium atom (Chasteen, 1981). EPR spectroscopy has been instrumental in the identification of different heteroatoms in porphyrin and nonporphyrin complexes and other naturally occurring sources of vanadium (Dickson et al., 1972). In aqueous solution the addition of $VOSO_4$ gives rise to the hydrated form of VO^{2+}, presumably $VO(H_2O)_5^{2+}$, at acidic pH (below pH 3). As the pH approaches 7, the hydrated form of VO^{2+} undergoes a series of hydrolytic reactions. Such solutions generate in a complex equilibrium a number of species that are all EPR-silent at neutral pH. The exact structure and stoichiometry of these species are complicated by the fact that no experimental method is currently available to characterize them. Some species are present below detection limits of available methods, and other species dimerize or polymerize to generate EPR-silent species.

$$VO^{2+}_{(aq)} + H_2O \rightleftharpoons VOOH^+_{(aq)} + H^+ \qquad (1)$$

$$2VO^{2+}_{(aq)} + 2H_2O \rightleftharpoons (VOOH)_2^{2+}_{(aq)} + 2H^+ \qquad (2)$$

$$VO^{2+}_{(aq)} + 2OH^- \rightleftharpoons VO(OH)_{2(s)} \qquad (3)$$

$$VO(OH)_{2(s)} + OH^- \rightleftharpoons VO(OH)_3^-_{(aq)} \qquad (4)$$

Despite these obvious experimental limitations a reasonable speciation scheme has been outlined for the physiological pH range (Chasteen, 1981). Above pH 4, equations 1 and 2 illustrate the formation of the $VOOH^+_{(aq)}$ species and its dimer. The former species is present at 10^{-7} M or less, depending on pH, and thus cannot be detected by EPR. In the dimer the electrons are paired and so the species is EPR-silent. At neutral pH the insoluble hydroxide $VO(OH)_2$ forms as described in equation 3 and, given its low precipitation product, the amount of species in solution is undetectable by EPR. Above pH 11 the vanadium species observed by EPR spectroscopy is consistent with

$VO(OH)_3^-$. The equilibrium between $VO(OH)_2$ and $VO(OH)_3^-$ (eq 4) is only slowly established, as dissolution of the insoluble $VO(OH)_2$ is slowly reversible or incomplete (Chasteen, 1981).

2.3.3   Vanadium(III)

In nature vanadium(III) exists in very reducing environments or is complexed to organic ligands; vanadium(III) containing minerals are immediately oxidized upon leaching from soil. The major form of vanadium in ascidians is in oxidation state 3 as described below. In general the information available on vanadium(III) speciation is limited owing to the restricted information obtainable with UV-Vis spectroscopy, potentiometry, and electrochemical methods. In aqueous solution vanadium(III) undergoes complex hydrolytic chemistry and several species are well established, including $[V(H_2O)_6]^{3+}$, $[V(OH)(H_2O)_5]^{2+}$, $[V(OH)_2(H_2O)_4]^+$ and $[V_2(\mu_2-O)(H_2O)_{10}]^{4+}$ (Meier,1995). In analogy with vanadium(V) aqueous solutions contain trimers and tetramers (Meier, 1995; Cotton et al., 1986). Measurements made on the magnetic susceptibilities of aqueous solutions containing hydrolysis products correlated with the formation of a homodimer, such as $[V(OH)_2V]^{4+}$, with hydroxide bridges rather than a species with an oxo bridge $[V-O-V]^{4+}$ (Pajdowski and Jezowka-Trzebiatowska, 1966). In contrast to the aqueous forms of vanadium(IV) and -(V), aqueous vanadium(III) species do not contain oxo groups.

2.4. Chemistry of Airborne Vanadium

Because vanadium is a nonvolatile metal, most of the airborne chemistry of vanadium involves particular forms of vanadium in the air (Keeler and Pirrone, 1996). The airborne vanadium settles in the soil and sediments of lakes, rivers, and oceans. As a consequence, measuring the levels of vanadium in sediments is indicative of the levels of airborne vanadium (Klein, 1975). For example, studies have been undertaken of the Great Lakes region to establish the sources of 28 elements, including vanadium, present in Lake Michigan. For vanadium the aerosol input was estimated to be 120 tons per year (Klein, 1975). Vanadium from the burning of fossil fuels (coal and oil) is emitted as oxides. Vanadium oxides form from oxidation state 2–5. Well-characterized oxides include VO, V_2O_3, VO_2, and V_2O_5. A range of oxides between these simple oxides that contain fractional stoichiometries of vanadium and oxygen have also been identified (Robin and Day, 1967). V_2O_5 undergoes hydrolytic reactions in water, generating vanadate solutions; these reactions are faster at basic pH than at neutral or acidic pH. "Intact" V_2O_5 is slowly and sparingly soluble in water, and the oxides have similar solubility properties. Thus the chemistry of airborne vanadium involves reactions of these particles with water, generating various forms of vanadic acid.

Natural sources of airborne vanadium include volcanic action, marine aerosols, and continental dust. Airborne vanadium concentrations in unpopulated areas such as the South Pole are low ($0.001–0.002$ ng/m^3) (Zoller et al., 1974).

Along the eastern seaboard of the United States concentrations in metropolitan areas can exceed 100 ng/m^3 (Waters, 1977). The elevated concentrations are a result of fossil fuel burning, primarily the burning of oil and coal. Thus, there is no doubt that natural airborne vanadium is very small in comparison with industrial sources of airborne vanadium. Few of the particles emitted by combustion consist of simple oxides; as a matter of fact a number of other metal ions, including sodium, nickel, and iron, can be significant constituents (National Academy of Sciences, 1970). No generalizations can be made concerning the size of these particles generated during combustion.

3.   VANADIUM-CONTAINING ROCKS

Vanadium exists in a variety of minerals in nature, but not as the free metal (Roberts et al., 1990). The average concentration of vanadium in the earth's crust is 150 ppm (Vinogradov, 1959). This classifies vanadium as a rare element, though it is more prevalent than copper, lead, or zinc.

Some rocks are found to have much higher concentrations of vanadium than others. For instance, the concentrations in igneous rock and shale are 135 and 130 ppm, respectively (Schroeder, 1970), while the concentrations in sandstone and limestone are far lower, at 20 ppm. Selected vanadium-containing minerals are listed in Table 1. The main vanadium-containing minerals include carnotite, cuprodescloizite, descloizite, mottramite, patronite, roscoelite, and vanadinite. (The minerals cuprodescoizite, descloizite, and mottramite all have the same chemical formula.) In these minerals the vanadium exists in oxidation states from 3–5; in the higher oxidation states the minerals are generally more soluble, and the vanadium(V) minerals are easily leached from soils into water. The vanadium oxides (carnotite, cuprodescloizite, descloizite, mottramite, and vanadinite) are mostly vanadium(V) minerals that contain cations such as K, Pb, Ca, Zn, and Cu. These oxides make up the bulk of the vanadium-containing minerals, the remaining two minerals in this series being roscoelite and patronite. The vanadium in the last two minerals is not in oxidation state 5. Roscoelite is in oxidation state 3. Patronite is not a homogeneous substance and the exact chemical composition is not known. A tentative formula of VS_4 has been given to this mineral (Palache et al., 1944). Two minor vanadium–oxide minerals are metatyuyamunite and tyuyamunite; these two and carnotite are uranium–vanadium minerals. Metatyuyamunite and tyuyamunite are particularly abundant in the Uravan mineral belt of Colorado (Fischer, 1968), where extensive mining has taken place for both vanadium and uranium.

Table 1 also lists four mixed-valence minerals, two each of vanadium (III/IV) (franciscanite; haggite) and vanadium(IV/V) (corvusite; fernandinite). Mixed-valence vanadium compounds that have been studied include vanadium oxides prepared from a mixture of V_2O_3 and V_2O_5 (Hoscheck and Klemm, 1939). The nature of mixed-valence species is difficult to define (Robin and

Day, 1967). As a matter of fact both corvusite and fernandinite are often observed with different ratios of the vanadium oxidation states. Finally, vanadyl sulfate and sodium metavanadate are commonly used in laboratory studies of all types, and it is interesting that these compounds actually occur naturally as minerals.

4. VANADIUM-CONTAINING PETROLEUM

The vanadium content of oils can vary dramatically, Venezuelan oils typically having the highest concentrations of vanadium (Table 3) (Biggs et al., 1985; Fish and Komlenic, 1984). The oxidation state of vanadium in oils is almost exclusively +4. The VO^{2+} ion exists in oils as either porphyrin or nonporphyrin complexes. An example of a vanadium porphyrin and nonporphyrin complex is illustrated in Figure 4a and 4b. Both types of complexes have five-coordinate square pyramidal geometries with an apically bound oxo group. The porphyrin found in oils can account for up to 50% of the vanadium, leaving the remaining as nonporphyrin vanadium complexes. Thus, an appreciable percentage of vanadium is complexed to nonporphyrin ligands. Porphyrin complexes of vanadium have been characterized by mass spectroscopy, UV-Vis spectroscopy, and EPR spectroscopy (Yen, 1975). The nonporphyrin complexes are poorly defined in the literature, presumably reflecting the greater structural differences among these types of vanadium complexes in various types of oils. Limited characterization of nonporphyrin complexes has been reported: low molecular weight (<400); extraction from all molecular weight ranges in crudes; and physical properties dissimilar to those of vanadium porphyrin complexes (Fish and Komlenic, 1984).

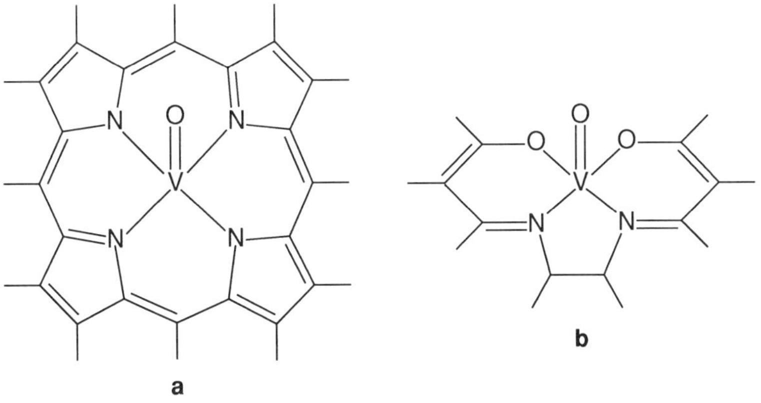

Figure 4. General structures for vanadium porphyrin (a) and vanadium nonporphyrin complexes (b).

Table 3 Vanadium Content of Selected Crude Oils, Asphaltenes, and Oil Shales and Their Geographical Location

Oil	Type	V Content (ppm)	Geographical Location	References
Boscan	Crude	1,180	Venezuela	(Biggs et al., 1985)
Athabasca	Asphaltene	640	Canada	(Kotlyar et al., 1988)
Cerro Negro	Crude	550	Venezuela	(Fish and Komlenic, 1984)
Morichal	Crude	282	Venezuela	(Biggs et al., 1985)
Maya	Crude	243	Mexico	(Biggs et al., 1985)
Nowruz	Crude	140	Iran	(Sadiq and Zaidi, 1984)
Beta	Crude	134	USA	(Biggs et al., 1985)
Arabian Heavy	Crude	55	Middle East	(Biggs et al., 1985)
Wilmington	Crude	49	USA	(Fish and Komlenic, 1984)
Prudhoe Bay	Crude	19	USA	(Fish and Komlenic, 1984)
Stuart	Oil shale	0.4–1.0	Australia	(Patterson et al., 1983)
Rundle	Oil shale	<0.3	Australia	(Patterson et al., 1983)
Moslavina	Crude	0.02–2.5	Yugoslavia	(Ugarkovic and Premerl, 1987)

By taking into consideration such factors as abundance of the metal ion, environment, stability of the metal–nitrogen bond, and lability of bridge positions of metal porphyrin complexes, it becomes apparent why vanadium porphyrin complexes are so abundant. (Quirke, 1987). The vanadium porphyrin complexes are among the most stable vanadium complexes known (Walker et al., 1975). Our search revealed no numerical formation constants, but quantitative information regarding the stability of these complexes is available (see below). An empirical relationship has been suggested for the stability of the metal–nitrogen bond in a porphyrin metal complex based on the conditions required for demetallation for the various types of complexes that exist in petroleum (Buchler, 1975). Five classes of stability, ranging from stability class I, where there is incomplete demetallation after 2 h at 100% H_2SO_4, to stability class V, where complete demetallation occurs in 1 : 1 $H_2O–CH_2Cl_2$ mixture. On the basis of this relationship, vanadium porphyrin complexes are categorized in stability class I, whereas nickel porphyrin complexes are in stability class II. Other metals, such as zinc and magnesium, are in classes III and IV, respectively. Despite the fact that vanadium is considered a trace element, the high stability of the vanadium porphyrin complexes has allowed these complexes to prevail over other metal ion porphyrin complexes.

During the distillation of crude oil the vanadium porphyrin, given its low volatility, accumulates in the residue. When this residue is used as fuel oil, it may contain high levels of vanadium, depending on the vanadium content of the original fuel oil used. In Table 3 crude oils, asphaltenes, and oil shales from various geographical locations are listed with their vanadium content (Biggs et al., 1985; Fish and Komlenic, 1984; Kotlyar et al., 1988; Sadiq and Zaidi, 1984; Patterson et al., 1983; Ugarkovic and Premerl, 1987). From Table 3 it is clear that Boscan, Athabasca, and Cerro Negro oil contain much higher vanadium levels than other types of oil. Thus, fuels of these types prepared by concentration of the residue are likely to produce significant levels of airborne vanadium upon burning. The distillates, on the other hand, contain very low vanadium levels and the combustion of such fuels regardless of oil source will not generate high levels of airborne vanadium particles. In addition to vanadium content, sulfur content in oils is instrumental to application of the fuels. Sulfur-rich heavy fuel oils must have their sulfur content reduced, and these treatments also reduce the vanadium content of these oils significantly (Radford and Rigg, 1970). Because stricter regulations are being adopted to reduce sulfur emissions into the atmosphere (reduction of acid rain), the current levels of airborne vanadium emission are likely to follow. Although the vanadium levels in airborne, aqueous, and solid phases may be under observation, vanadium levels are generally not believed to be a problem in most environments, although wide ranges in projected emissions have been reported based on available data (Lee, 1983). Exceptions such as those described in the Uravan region have been documented.

5. VANADIUM IN BIOACCUMULATORS AND OTHER SYSTEMS OF RELEVANCE TO THE ENVIRONMENT

Vanadium, a trace element, is considered by some to be an essential element for humans and several mammals. Several laboratories have had difficulties repeating the original studies that supported claims of essentiality, and the question is currently not resolved (Nielsen and Uthus, 1990; Mackey et al., 1996). The reader is referred to a detailed discussion elsewhere in this volume of these aspects of vanadium biology and chemistry. In contrast, the roles of vanadium as a cofactor for haloperoxidases (Vilter, 1995) and nitrogenases (Eady, 1995; Bayer, 1995), and as a metal ion in the natural product amavadine (Bayer, 1995; Michibata, 1993), have been identified. Thus the existence and specific association of vanadium in several seaweed, bacteria, lichen, and fungi is indisputable. Although vanadium plays an obvious role in these enzymes, it still remains a trace element in these organisms.

Haloperoxidases include a wide range of enzymes that catalyze the oxidation of halides in the presence of hydrogen peroxides; some of these enzymes contain a prosthetic group while others have been found to contain one mole of vanadium per mole of enzyme (Vilter, 1995). A recent 2.1-Å-resolution X-ray structure has been solved of the azide complex of a chloroperoxidase from the fungus *Curvularia inaequalis* (Messerschmidt and Wever, 1996). In the crystal isolated from the 2 mM azide mother liquor, the vanadium center has a trigonal bipyramidal coordination geometry with three nonprotein oxygen ligands in the equatorial plane. One axial position is occupied by an OH group and the other by a nitrogen atom of histidine 495.

The environmentally important nitrogenases catalyze the conversion of N_2 to NH_3 and are responsible for the cycling of N_2 from the atmosphere to the soil (≈ 108 tons per year) (Eady, 1995). Since 1985 the existence of Mo-independent route(s) to N_2 fixation has been generally accepted, two Mo-independent systems now are known, one based on a vanadium nitrogenase and the other based on an iron nitrogenase (Eady, 1995). All known nitrogenase enzymes are multisubunit enzymes with complex structures involving polynuclear metal ion centers. The VFe protein from two related species of *Azotobacter* are hexamers ($\alpha_2\beta_2\delta_2$) with a molecular weight about 250,000, and they contain from 0.7 to 2 vanadium atoms and from 9 to 19 iron atoms (Robson et al., 1986; Hales Case, et al., 1986a; Hales, Langosch et al., 1986b). Regardless of the role of the vanadium in these proteins, there is no doubt that sulfur–vanadium cluster chemistry (Simonnet-Jégat et al., 1996; Reynolds et al., 1995; Klich et al., 1996) will be important to understanding the structure and catalytic function of the vanadium atom. However, given that the protein reduces N_2 to NH_3, it is very possible that vanadium–nitrogen chemistry (Gailus et al., 1994; Song et al., 1994; Rehder et al., 1992) will contribute to the catalytic process in this enzyme. Model reactions on both systems are currently active areas within vanadium chemistry (see elsewhere in this volume) (Simonnet-Jégat et al., 1996; Reynolds et al., 1995; Klich et al., 1996; Gailus et al., 1994; Song et al., 1994; Rehder et al., 1992).

Vanadium is essential to several species of ascidians, which are capable of efficiently accumulating vanadium from the ocean waters. The greater than millionfold concentration mechanism of these organisms is not well understood, nor has the role of the vanadium in these organisms been defined, even though many suggestions have been made (Smith et al., 1995; Ishii and Nakai, 1995). Although most work in this area has been carried out with tunicates, a new vanadium accumulator, the fan worm *Pseudopotamilla occelata* has recently been identified (Ishii and Nakai, 1995). Significant progress in characterization of the tunicates has been achieved, and it is now known that vanadium is present in oxidation state 3 and/or 4, depending on tunicate species and cell type. In addition, when vanadium(III) and vanadium(IV) are exposed to blood plasma and oxygen, vanadium(V) will form (Andersen and Swinehart, 1991). Tunicates contain high levels of tunichromes, which are electron-rich catechols and pyrogallois and viable candidates for the native reducing and complexing agents of vanadium. The tunichrome complextion and redox chemistry of vanadium are exceedingly pH- and oxygen-sensitive. Recent preparation and isolation of tunichromes, combined with the spectroscopic studies of both the whole tunicates and the fan worm, have led to a better appreciation of the role of ligands in these organisms (Smith et al., 1995; Ishii and Nakai, 1995; Frank et al., 1994).

Amavadin (Fig. 5) is a natural product found in mushrooms, specifically in various *Amanita* species (Bayer, 1995). The vanadium levels vary from 36 to 250 mg/kg dry weight. The structure for amavadin was initially mischaracterized. Extensive characterization of model compounds (Carrando et al., 1988) and additional spectroscopic studies have now revealed the correct structure. The complex formed between vanadium(IV) and two equivalents of *N*-hydroxyliminoacetic acid forms a compound containing eight coordinate vanadium in a dodecahedral geometry (Fig. 5) (Carrando et al., 1988). Thus, this

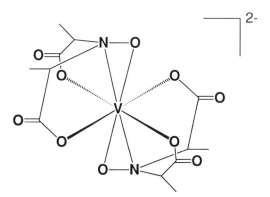

Figure 5. Dodecahedral geometry of vanadium(IV) amavadin.

complex not only is an example of a non-oxo vanadium(IV) compound, but it is eight-coordinate and furthermore is among the most stable vanadium(IV) complexes known (Bayer et al., 1987).

6. APPLICATIONS OF VANADIUM IN THE MODERN WORLD

Vanadium has numerous uses in industrial processes, such as the processing of steels, chemical production, polymer synthesis, ceramics, and electronics (National Academy of Sciences, 1970). The steel industry accounts for 80–85% of the world's vanadium consumption and is by far the greatest user of any form of vanadium (Hilliard, 1992). As a catalyst vanadium is used in the contact process for the production of sulfuric acid. About 40 million tons of sulfuric acid is generated annually. Additionally, a variety of common organic precursors are prepared by the use of vanadium catalysts.

Steels alloyed with vanadium show increased strength, which explains the widespread metallurgical use of vanadium as an alloying component. Negligible loss of vanadium as fumes or airborne dust is observed in a normal steel-smelting operation. However, increasing proportions of scrap steel are now recycled by use of a basic-oxygen furnace (National Academy of Sciences, 1970). This method refines steel with a high-pressure jet of oxygen, resulting in large fume clouds. These fumes are mostly metallic oxides, iron oxides being the major constituents. Vanadium emissions have been estimated at 0.02% of the particulate matter discharged (Davis, 1971). However, the recycling of steel has increased the levels of vanadium in the scrap steel, and it likely has increased the vanadium contribution to the large oxide fume clouds.

A key step in the production of sulfuric acid is the catalytic conversion of sulfur dioxide to sulfur trioxide (Chang, 1988). The catalyst for this reaction is vanadium pentoxide on alumina. Although the scale of sulfuric acid production is large, the alumina vanadium oxide catalysts have long lifetimes and high efficiency (National Academy of Sciences, 1970). The environmental impact of this process on the production of airborne vanadium remains small in a global sense. The United States is already processing spent catalyst in place of simply discarding such waste. The high vanadium content in this oxide mixture makes it fairly attractive to process it into other vanadium products.

Several organic precursors are prepared by the use of vanadium catalysts. Such chemicals include phthalic anhydride, maleic anhydride, aniline black, adipic acid, acetaldehyde, anthraquinone, fumaric acid, quinone, *m*-xylene, *p*-xylene, propylene, ethylene, cyclohexylamine, and aromatic acids. Specialty chemicals, high-purity metals, and intermetallics are prepared with vanadium trichloride in stereospecific catalyst systems (National Academy of Sciences, 1970). Also, vanadium compounds are used as mordants in the dyeing and printing of cotton and silk.

7. SUMMARY

As a transition metal vanadium exists in the environment in oxidation states 3, 4, and 5. In oxidation states III and IV vanadium acts as a cation, whereas in oxidation state 5 it reacts both as a cation and as an anion analog of phosphate. Both oxidation states 4 and 5 form strong vanadium oxide ions, which are responsible for most of the chemistry occurring in the environment. V_2O_5, $NaVO_3$, and $NaVO_4$, upon dissolution in water, all generate solutions containing vanadate, that is, vanadium in oxidation state 5. $VOSO_4$ and V_2O_4 both generate solutions of vanadium(IV). The speciation in these solutions is primarily dependent on vanadium concentration. For example, in this review we show in a ^{51}V NMR spectrum that a 0.1 mM solution of NH_4VO_3 at pH 7.0 contains only 92% $H_2VO_4^-$ and 8% $H_2V_2O_7^{2-}$, whereas a 2.5 mM solution of NH_4VO_3 at pH 7.0 contains 0.84 mM $H_2VO_4^-$, 0.53 mM $H_3V_2O_7^-$, and 1.13 mM $V_4O_{12}^{4-}$. At naturally occurring vanadium concentrations the vanadium species is monomeric; however, laboratory studies conducted at higher concentrations must consider the other species present, as the oligomeric species are not innocent bystanders in reactions or to biological and environmental responses.

Vanadium(V) is the most mobile form of vanadium entering rivers, lakes, and oceans through leaching; however, when minerals containing other oxidation states dissolve, the vanadium is rapidly oxidized to form vanadium(V). Vanadium accumulates in biological systems mostly in oxidation states 3 and 4, given the fact that such environments are mostly reducing. Given their high stability, the vanadium(IV) porphyrin complexes are prevalent in many oils. Other interesting aspects of this metal ion include the fact that vanadium is a cofactor in haloperoxidases and nitrogenases. In addition, a vanadium-containing natural product, amavadine, exists in mushrooms. Amavadine is a rare example of a naturally existing 'bare' (non-oxo) vanadium complex and is unique not only in its structure and chemistry but also in its extraordinary high stability.

In view of the ease by which vanadium species interconvert under the mild conditions of the environment, any studies with vanadium compounds, be they chemical, biological, toxicological, or environmental, must consider the underlying chemistry involved. Not only does vanadium chemistry impose limitations on experimental studies, but the properties of these species also provide additional tools that can be used to great advantage when studying this fascinating element.

ACKNOWLEDGMENTS

We thank the National Institutes of Health for funding this work.

REFERENCES

Andersen, D. H., and Swinehart, J. H. (1991). The distribution of vanadium and sulfur in the blood cells, and the nature of vanadium in the blood cells and plasma of the ascidian, *Ascidia ceratodes. Comp. Biochem. Physiol.* **99A,** 585–592.

Baes, C. F., and Mesmer, R. E. (1976). *The Hydrolysis of Cations.* Wiley-Interscience, New York, pp. 197–210.

Bayer, E. (1995). Amavadin, the vanadium compound in amanitae. In H. Sigel and A. Sigel (Eds.), *Metals Ions in Biological Systems,* Vol. 31. Marcel Dekker, New York, pp. 407–422.

Bayer, E., Koch, E., and Anderegg, G. (1987). Amavadin, an example for selective binding of vanadium in nature: Studies of its complexation chemistry and a new structural proposal. *Angew. Chem. Int. Ed. Engl.* **26,** 545–546.

Biggs, W. R., Fetzer, J. C., Brown, R. J., and Reynolds, J. G. (1985). Characterization of vanadium compounds in selected crudes. I. Porphyrin and non-porphyrin separation. *Liquid Fuels Technol.* **3,** 397–421.

Buchler, J. W. (1975). *Porphyrins and Metalloporphyrins.* Elsevier, Amsterdam, p. 157.

Carrando, A. A. F. d. C. T., Duarte, M. T. L. S., Pessoa, J. C., Silva, J. A. L., da Silva, J. J. R. F., Vaz, M. C. T. A., and Vilas-Boas, L. F. (1988). Bis-(*N,* hydroxyiminodiacetate)vanadate(IV), a synthetic model of "amavadin." *J. Chem. Soc. Chem. Commun.,* pp. 1158–1159.

Chang, R. (1988). Stoichiometry: The arithmetic of chemistry. *Chemistry.* Random House, New York, pp. 83–134.

Chasteen, N. D. (1981). Vanadyl(IV) EPR spin probes inorganic and biochemical aspects. *Biological Magnetic Resonance.* Plenum Press, New York, pp. 53–119.

Cotton, F. A., Extine, M. W., Falvello, L. R., Beck Lewis, D. B., Lewis, G. E., Murillo, C. A., Schwotzer, W., Thomas, M., and Troup, J. M. (1986). Four compounds containing oxo-centered trivanadium cores surrounded by six μ, η^2-carboxylato groups. *Inorg. Chem.* **25,** 3505–3512.

Crans, D. C. (1994a). Aqueous chemistry of labile oxovanadates: Relevance to biological studies. *Comments Inorg. Chem.* **16,** 1–33.

Crans, D. C. (1994b). Enzyme interactions with labile oxovanadates and other poly oxometalates. *Commun. Inorg. Chem.* **16,** 35–76.

Crans, D. C. (1995). Interaction of vanadates with biogenic ligands. In H. Sigel and A. Sigel (Eds.), *Metal Ions in Biological Systems.* Marcel Dekker, New York, pp. 147–209.

Crans, D. C., Mahroof-Tahir, M., Shin, P. K., and Keramidas, A. D. (1995). Vanadium chemistry and biochemistry of relevance for use of vanadium compounds as antidiabetic agents. *Mol. Cell. Biochem.* **153,** 17–24.

Crans, D. C., Rithner, C. D., and Theisen, L. A. (1990a). Application of time-resolved ^{51}V 2-D NMR for quantitation studies of kinetic exchange between vanadate oligomers. *J. Am. Chem. Soc.* **112,** 2901–2908.

Crans, D. C., Willging, E. M., and Butler, S. K. (1990b). Vanadate tetramer as the inhibiting species in enzyme reactions in vitro and in vivo. *J. Am. Chem. Soc.* **112,** 427–432.

Davis, W. E., and associates. (1971). *National Inventory of Sources and Emissions. Arsenic, Beryllium, Manganese, Mercury and Vanadium.* Section V: *Vanadium.* Report for Environmental Protection Agency. W. E. Davis and Associates, Leawood, Kansas, p. 53.

Dickson, F. E., Kunesh, C. J., McGinnis, E. L., and Petrakis, L. (1972). Use of electron spin resonance to characterize the vanadium(IV)–sulfur species in petroleum. *Anal. Chem.* **44,** 978–981.

Domingo, J. L. (1996). Vanadium: A review of the reproductive and developmental toxicity. *Reprod. Toxicol.* **10,** 175–182.

Eady, R. R. (1995). Vanadium nitrogenases of *Azotobacter.* In H. Sigel and A. Sigel (Eds.), *Metal Ions in Biological Systems.* Marcel Dekker, New York, pp. 363–406.

Fischer, R. P. (1968). The uranium and vanadium deposits of the Colorado Plateau region. In J. D. Ridge (Ed.), *Ore Deposits of the United States, 1933–1967.* American Institute of Mining, Metallurgical, and Petroleum Engineers, New York, pp. 736–746.

Fish, R. H., and Komlenic, J. J. (1984). Molecular characterization and profile identifications of vanadyl compounds in heavy crude petroleums by liquid chromatography/graphite furnace atomic absorption spectrometry. *Anal. Chem.* **56**, 510–517.

Frank, P., Hedman, B., Carlson, R. M. K., and Hodgson, K. O. (1994). Interaction of vanadium and sulfate in blood cells from the tunicate *Ascidia ceratodes:* Observations using X-ray absorption edge structure and EPR spectroscopies. *Inorg. Chem.* **33**, 3794–3803.

Froisland, L. J., Placek, P. L., and Shirts, M. B. (1982). *Restoration of Surface Vegetation on Uranium Wastes at Uravan, Colorado.* BuMines, Rhode Island.

Gailus, H., Woitha, C., and Rehder, D. (1994). Dinitrogenvanadates(-I): Synthesis, reactions and conditions for their stability. *J. Chem. Soc. Dalton Trans.* **23**, 3471–3477.

Hales, B. J., Case, E. E., Morningstar, J. E., Dzeda, M. F., and Mauterer, L. A. (1986a). Isolation of a new vanadium-containing nitrogenase from *Azotobacter vinelandii. Biochemistry* **25,** 7251–7255.

Hales, B. J., Langosch, D. J., and Case, E. E. (1986b). Isolation and characterization of a second nitrogenase Fe-protein from *Azotobacter vinelandii. J. Biol. Chem.* **261**, 15310–15306.

Heath, E., and Howarth, O. W. (1981). Vanadium-51 and oxygen-17 nuclear magnetic resonance study of vanadate(V) equilibria and kinetics. *J. Chem. Soc. Dalton Trans.,* pp. 1105–1110.

Hilliard, H. E. (1992). *Vanadium.* U.S. Department of the Interior, Washington, DC.

Holloway, C. E., and Melnik, M. (1985). Vanadium coordination compounds: Classification and analysis of crystallographic and structural data. *Rev. Inorg. Chem.* **7**, 75–159.

Hoschek, E., and Klemm, W. (1939). Weitere Beiträge zur Kenntnis der Vanadinoxyde. *Z. Anorg. Allgem. Chem.* **242,** 63–69.

Howarth, O. W. (1990). Vanadium-51 NMR. *Prog. NMR Spectrosc.* **22**, 453–483.

Ide, C. W. (1995). A survey of exposure to and urinary excretion of vanadium and response to heat stress in boiler cleaners working in Scotland. *Int. J. Environ. Health Res.* **5**, 269–280.

Ishii, T., and Nakai, I. (1995). Biochemical significance of vanadium in a polychaete worm. In H. Sigel and A. Sigel (Eds.), *Metal Ions in Biological Systems,* Vol. 31. Marcel Dekker, New York, pp. 491–510.

Keeler, G. J., and Pirrone, N. (1996). Atmospheric transport and deposition of trace elements to Lake Erie from urban areas. *Water Sci. Technol.* **33**, 259–265.

Klein, D. H. (1975). Fluxes, residence times, and sources of some elements to Lake Michigan. *Water, Air, Soil Pollut.* **4**, 3–8.

Klich, P. R., Daniher, A. T., Challen, P. R., McConville, D. B., and Youngs, W. J. (1996). Vanadium(IV) Complexes with Mixed O, S Donor Ligands. Syntheses, structures and properties of the anions tris(2-mercapto-4-methylphenolato)vanadate(IV) and bis(2-mercaptophenolato)oxovanadate(IV). *Inorg. Chem.* **35**, 347–356.

Kopp, J. F., and Kroner, R. C. (1968). *Trace Metals in Water of the United States.* Federal Water Pollution Control Administration, Cincinnati, Ohio.

Kotlyar, L. S., Ripmeester, J. A., Sparks, B. D., and Woods, J. (1988). Comparative study of organic matter derived from Utah and Athabasca oil sands. *Fuel* **67**, 1529–1535.

Lee, K. (1983). Vanadium in the aquatic ecosystem. In J. O. Nriagu (Ed.), *Aquatic Toxicology.* John Wiley & Sons, New York, pp. 155–187.

Lewis, C. E. (1959). The biological effects of vanadium. II. The signs and symptoms of occupational vanadium exposure. *Am. Med. Assoc. Arch. Ind. Health* **19**, 497–503.

Mackey, E. A., Becker, P. R., Demiralp, R., Greenberg, R. R., Koster, B. J., and Wise, S. A. (1996). Bioaccumulation of vanadium and other trace metals in livers of Alaskan cetaceans and pinnipeds. *Arch. Environ. Contam. Toxicol.* **30**, 503–512.

Meier, R. (1995). Solution properties of vanadium(III) with regard to biological systems. In H. Sigel and A. Sigel (Eds.), *Metal Ions in Biological Systems,* Vol. 31. Marcel Dekker, New York, pp. 45–88.

Messerschmidt, A., and Wever, R. (1996). X-ray structure of a vanadium-containing enzyme: Chloroperoxidase from the fungus *Curvularia inaequalis. Proc. Natl. Acad. Sci. USA* **93,** 392–396.

Michibata, H. (1993). The mechanism of accumulation of high levels of vanadium by ascidians from seawater: Biophysical approaches to a remarkable phenomenon. *Adv. Biophys.* **29,** 105–133.

National Academy of Sciences. (1970). *Trends in the use of Vanadium. A Report of the National Materials Advisory Board.* Publication NMAB-267. Clearinghouse for Federal Scientific and Technical Information, Springfield, VA, p. 46.

Nielsen, F. H., and Uthus, E. O. (1990). The essentiality and metabolism of vanadium. In N. D. Chasteen (Ed.), *Vanadium in Biological Systems: Physiology and Biochemistry.* Boston, Kluwer Academic Publishers, pp. 51–62.

Pajdowski, L., and Jezowka-Trzebiatowska, J. (1966). Magnetochemical study of the hydrolsis of vanadium(III) ion. *J. Inorg. Nucl. Chem.* **28,** 443–446.

Palache, C., Berman, H., and Frondel, C. (1944). *The System of Mineralogy of James Dwight Dana and Edward Salisbury Dana.* 7th ed. Vol. 1, John Wiley and Sons, New York, p. 347.

Patterson, J. H., Dale, L. S., and Chapman, J. F. (1983). Partitioning of trace elements during the retorting of Australian oil shales. *Fuel* **67,** 1353–1356.

Pettersson, L., Andersson, I., and Hedman, B. (1985). Multicomponent polyanions. 37. A potentiometric and ^{51}V-NMR study of equilibria in the H^+-HVO_4^{2-} system in 3.0 M-Na(ClO$_4$) medium covering the range $1 \leq -1g[H^+] \leq 10$. *Chem. Scripta* **25,** 309–317.

Pope, M. T. (1983). *Heteropoly and Isopoly Oxometalates.* Springer-Verlag, New York, p. 180.

Pope, M. T., and Müller, A. (1991). Polyoxometalate chemistry: An old field with new dimensions in several disciplines. *Angew. Chem. Int. Ed. Eng.* **30,** 34–48.

Posner, B. I., Faure, R., Burgess, J. W., Bevan, A. P., Lachance, D., Zhang-Sun, G., Fantus, I. G., Ng, J. B., Hall, D. A., Soo Lum, B., and Shaver, A. (1994). Peroxovanadium compounds. A new class of potent phosphotyrosine phosphatase inhibitors which are insulin mimetics. *J. Biol. Chem.* **269,** 4596–4604.

Quirke, J. M. E. (1987). Rationalization for the predominance of nickel and vanadium porphyrins in the geosphere. In R. H. Filby and J. F. Braanthaver (Eds.), *Metal Complexes in Fossil Fuels. Geochemistry, Characterization, and Processing.* American Chemical Society, Washington, DC, pp. 74–83.

Radford, H. D., and Rigg, R. C. (1970). New way to desulfurize resids. *Hydrocarbon Process.* **49,** 187–191.

Rehder, D. (1991). The bioinorganic chemistry of vanadium. *Angew. Chem. Int. Ed. Eng.* **30,** 148–167.

Rehder, D., Woitha, C., Priebsch, W., and Gailus, H. (1992). *trans*-[Na(thf)][V(N$_2$)$_2$ (Ph$_2$PCH$_2$CH$_2$PPh$_2$)$_2$]: Structural characterization of a dinitrogenvanadium complex, a functional model for vanadiumnitrogenase. *J. Chem. Soc. Chem. Commun.* **4,** 364–365.

Reynolds, J. G., Sendlinger, S. C., Murray, A. M., Huffman, J. C., and Christou, G. (1995). Synthesis and characterization of vanadium(II, III, IV) complexes of pyridine-2-thiolate. *Inorg. Chem.* **34,** 5745–5752.

Roberts, W. L., Campbell, T. J., and Rapp, G. R. (1990). *Encyclopedia of Minerals.* 2nd ed. Van Nostrand Reinhold Company, New York.

Robin, M. B., and Day, P. (1967). Mixed valence chemistry—A survey and classification. In H. J. Emeléus and A. G. Sharpe (Eds.), *Advances in Inorganic Chemistry and Radiochemistry.* Academic Press, New York, pp. 247–422.

Robson, R. L., Eady, R. R., Richardson, T. H., Miller, R. W., Hawkins, M., and Postgate, J. R. (1986). The alternative nitrogenase of *Azotobacter chroococcum* is a vanadium enzyme. *Nature* **322,** 388–390.

Sadiq, M., and Zaidi, T. H. (1984). Vanadium and nickel content of Nowruz spill tar flakes on the Saudi Arabian coastline and their probable environmental impact. *Bull. Environ. Contam. Toxicol.* **32,** 635–639.

Schroeder, H. A. (1970). *Air Quality Monographs.* Monograph 70-13. *Vanadium.* American Petroleum Institute, Washington, DC, p. 32.

Shaver, A., Hall, D. A., Ng, J. B., Lebuis, A.-M., Hynes, R. C., and Posner, B. I. (1995). Bisperoxovanadium compounds: Synthesis and reactivity of some insulin mimetic complexes. *Inorg. Chim. Acta* **229,** 253–260.

Shechter, Y. (1990). Insulin-mimetic effects of vanadate. Possible implications for future treatment of diabetes. *Diabetes* **39,** 1–5.

Simonnet-Jégat, C., Delalande, S., Halut, S., Marg, B., and Sécheresse, F. (1996). Synthesis and characterization of a novel oxo-disulfidotetravadate(V) anion with a bridging tetrasulfido ligand $[\{V(O)(S_2)_2\}_2)(\mu^2 - S_4) \{V(O)(S_2)_2\}_2]^{6-}$. *Chem. Commun.,* pp. 423–424.

Smith, D. F., Stults, N. L., and Mercer, W. D. (1995). Bioluminescent immunoassays using streptavidin and biotin conjugates of recombinant aequorin. *Am. Biotechnol. Lab.* pp. 17–18.

Song, J.-I., Berno, P., and Gambarotta, S. (1994). Dinitrogen fixation, ligand dehydrogenation, and cyclometalation in the chemistry of vanadium(III) amides. *J. Am. Chem. Soc.* **116,** 6927–6928.

Ugarkovic, D., and Premerl, D. (1987). Trace-metal distribution in Moslavina basin crude oil and oil products. *Fuel* **66,** 1431–1435.

Vilas-Boas, L. V., and Costa Pessoa, J. (1987). Vanadium. In G. Wilkinson, R. D. Gillard, and J. A. McCleverty (Eds.), *Comprehensive Coordination Chemistry. The Synthesis, Reactions, Properties and Applications of Coordination Compounds.* Pergamon Press, New York, pp. 453–583.

Vilter, H. (1995). Vanadium-dependent haloperoxidases. In H. Sigel and A. Sigel (Eds.), *Metal Ions in Biological Systems,* Vol. 31. Marcel Dekker, New York, pp. 325–362.

Vinogradov, A. P. (1959). *The Geochemistry of Rare and Dispersed Elements in Soil.* 2nd ed. Consultants Bureau, New York, p. 209.

Walker, F. A., Hui, E., and Walker, J. M. (1975). Electronic effects in transition metal porphyrins. I. The reaction of piperidine with a series of para- and meta-substituted nickel(II) and vanadium(IV) tetraphenylporphyrins. *J. Am. Chem. Soc.* **97,** 2390–2397.

Waters, M. D. (1977). Toxicology of vanadium. In R. A. Goyer and M. A. Mehlman (Eds.), *Advances in Modern Toxicology.* John Wiley & Sons, New York, pp. 147–189.

Weeks, M. E., and Leicester, H. M. (1968). *Discovery of the Elements.* 7th ed. Chemical Education Publishing, Easton, PA, p. 351.

Wyers, H. (1946). Some toxic effects of vanadium pentoxide. *Br. J. Ind. Med.* **3,** 177–182.

Yen, T. F. (1975). Vanadium and its bonding in petroleum. *The Role of Trace Metals in Petroleum.* Ann Arbor Science Publishers, Ann Arbor, MI, pp. 167–181.

Zoller, W. H., Gladney, E. S., and Duce, R. A. (1974). Atmospheric concentration and sources of trace metals at the South Pole. *Science* **183,** 198–200.

5

HIGH VANADIUM CONTENT IN MT. FUJI GROUNDWATER AND ITS RELEVANCE TO THE ANCIENT BIOSPHERE

Tatsuo Hamada

Department of Animal Science, Tokyo University of Agriculture, Sakuragaoka 1-1-1, Setagaya, Tokyo 156, Japan

Vanadium in the Environment. Part 1: Chemistry and Biochemistry, Edited by Jerome O. Nriagu.
ISBN 0-471-17778-4. © 1998 John Wiley & Sons, Inc.

1. INTRODUCTION

Most continental waters show a vanadium concentration of less than 3 μg/L (Bradford et al., 1968; Nojiri et al., 1985; Sandler et al., 1988; Söremark, 1967; Sugawara et al., 1956; Sugiyama, 1989). However, a certain underground water in California shows the highest concentration of 20.5 μg/L (Linstedt and Kruger, 1970). Vanadium concentration in the Colorado River Basin varies between 0.2 and 49.2 μg/L and higher concentration values are found in the vicinity of uranium–vanadium milling operations (Linstedt and Kruger, 1969). In some waters from Wyoming, Idaho, Utah, and Colorado, vanadium concentrations of 2.0–9.0 μg/L are found (Parker et al., 1978). Drinking waters in Cleveland show a mean vanadium concentration of 5 μg/L with a maximum value of 100 μg/L (Strain et al., 1982). Spring waters in the eastern region of Mt. Fuji contain 24.5–81.5 μg/L of vanadium, and estuary waters at the mouths of the Fuji and Kano Rivers contain 11–18 μg/L of vanadium (Okabe and Morinaga, 1968; Okabe et al., 1981). Vanadium concentration in a tapwater sample in Kanagawa prefecture, under the influence of Mt. Fuji groundwater, also shows the highest value of 22.6 μg/L among 21 cities across Japan and the United States (Tsukamoto et al., 1990). Mt. Fuji groundwater has been used for drinking and other industrial purposes. I examined the vanadium concentration of Mt. Fuji groundwater as well as that of some organisms at the southwestern foot of Mt. Fuji.

In shales of marine origin there is a tendency for vanadium to be enriched in deposits of high organic content. A relationship between vanadium and organic carbon is evident by their strong correlation in marine oil shales and black shales (Breit and Wanty, 1991). However, vanadium concentration is not significantly correlated with bitumen content or organic carbon content (Lewan and Maynard, 1982), and samples of various nonmarine and marine shales with equal amounts of vanadium may differ by a factor of 3 in their amount of organic carbon (Tourtelot, 1964). High levels of vanadium and nickel are associated with petroleum, and these metals are incorporated into organic-rich shales and complexed to porphyrins, which are degradation products of chlorophyll (Lewan and Maynard, 1982). As shown by Breit (1992), factors that control the vanadium contents of fossil fuels are poorly understood. In this chapter I will speculate on the origin of the variable amounts of vanadium in fossil fuels and propose a hypothetical theory on the anomalous distribution of vanadium in ancient sediments. Furthermore, I will speculate on the effect of vanadium enrichment on the decline of the dinosaurs.

2. MT. FUJI AND ITS VANADIUM CONTENT

Mt. Fuji is the most typical stratovolcano in Japan. It has gentle slopes on all sides, rising to a height of 3,776 m above sealevel with a base diameter of 50 km, its southern slope reaching the shores of Suruga Bay. Mt. Fuji, a rather young volcano, consists of three volcanoes, Komitake, Older Fuji (Ko-Fuji) and Younger Fuji (Shin-Fuji). Komitake is believed to be hidden in the northeastern flank of the present Mt. Fuji. Older Fuji started its activity about 80,000 years ago and ceased activity 11,000 years ago, forming the largest bulk, a bulk equal to 250 km^3 of dense rocks, which is equivalent to the thickest accumulation of mud flows derived from pyroclastic flows and the debris of avalanche deposits (Miyaji et al., 1992; Tsuya, 1971). Younger Fuji, at nearly the same position as Older Fuji, has exhibited various types of eruptions during the succeeding 11,000 years, about 83% of the total amount was erupted as lava flows, the last eruption occurring in 1707 A.D. The lava flows of Younger Fuji are equal to 43 km^3 of dense rock and cover Older Fuji from the summit side (Miyaji et al., 1992; Tsuya, 1971). More than 99% of Mt. Fuji is made up of island-arc tholeiitic basalts with $SiO_2 < 52.5\%$ (Miyashiro, 1975; Togashi et al., 1991). Mt. Fuji belongs to the Izu-Mariana arc. From the viewpoint of plate tectonics the northern tip of the oceanic Philippine Sea plate is colliding with the continental Eurasian plate at the northern margin of the Izu Peninsula in central Japan. The Izu Borderland, which includes Mt. Fuji, is a convergent plate boundary on land that combines the Suruga Trough with the Sagami Trough. Intense Quaternary crustal movements, such as active faulting and rapid rifting or subsiding, occur along this island plate boundary (Yamazaki, 1992). There are no rivers or springs more than 1,000 m above sea level on Mt. Fuji. All the rain and snow falling there sinks into the mountain basement, and after having passed through cracks and holes present in the lava, the water emerges again as springs and rivers.

In Table 1 the means and standard deviations are calculated from several trace element contents of Japanese and other countries' rock reference samples presented by Yoshida et al. (1992). In the GSJ rock references (Ando et al., 1989), the SiO_2 contents of basalts (4 samples), andesites (3 samples), and granitic rocks (4 samples) are 51.0–53.2, 56.2–64.1 and 67.1–77.0%, respec-

Table 1   Comparison of Trace Element Contents (ppm) in Basalts, Andesites, and Granites[a]

	Basalts ($N = 10$)	Andesites ($N = 4$)	Granites ($N = 9$)
V	305 ± 40(148–578)*	132 ± 14(105–172)[†]	29 ± 8(3–73)[†]
Ni	141 ± 31(13–267)*	49 ± 32(2–142)	6 ± 1(2–13)[†]
V/Ni	9.2 ± 4.7(0.6–41.3)	16.4 ± 12.1(0.9–52.5)	4.2 ± 0.7(1.4–7.2)
Zn	103 ± 9(66–160)	78 ± 7(63–91)	81 ± 22(13–235)
Mo	4.6 ± 3.3(0.5–34.0)	1.7 ± 0.5(0.5–2.7)	1.4 ± 0.5(0.2–5.2)
U	1.1 ± 0.3(0.025–2.5)*	1.5 ± 0.4(0.34–2.4)*	6.6 ± 1.8(2.0–18.0)[†]
Cr	268 ± 54(16–469)*	138 ± 110(7–465)	18 ± 6(4–65)[†]
Cu	106 ± 21(19–227)*	44 ± 6(29–60)[†]	10 ± 3(0.4–33)[†]

[a] Each value shown as mean ± *SD* (min–max).
*,[†] There are significant differences between the values denoted by superscripts * and † ($P < 0.05$).

tively. The vanadium contents of basalts are significantly higher than those of andesites or granites, but the nickel contents of basalts show a large variation, so that between basalts and andesites there is no significant difference. The ratios of vanadium to nickel are not significantly different among basalts, andesites, and granites.

Ninety samples of Mt. Fuji basalts show a mean vanadium content of 348 ± 77 (SD) ppm (Arculus et al., 1991). Twenty-three samples of Deccan flood tholeiitic basalts erupted at the Cretaceous–Tertiary boundary show a mean vanadium content of 340 ± 61 ppm (Mahoney, 1988). Fifty-one samples of Columbia River flood basalt of the mid-Miocene show a mean vanadium content of 311 ± 88 ppm (Hooper, 1988). Among these vanadium contents a significant difference is obtained between Mt. Fuji and Columbia River samples $(P < 0.05)$. The SiO_2 contents of Mt. Fuji, Deccan basalts, and Columbia basalts are 50.6 ± 1.2, 50.7 ± 1.2, and $51.8 \pm 3.7\%$, respectively, and their nickel contents are 39 ± 20, 102 ± 52, and 53 ± 45 ppm, respectively. Takahashi et al. (1991) have compared the chemical composition of the rocks from Older to Younger Fuji, and their mean vanadium content (374 ± 66 ppm) is relatively higher than that stated above. The vanadium contents of Mt. Fuji and Oshima basalts in the Izu-Mariana arc are higher than other types of alkali basalts or andesites around Mt. Fuji (Katsura, 1956).

During the fractional crystallization of a basic magma, the overall order of the elements should display the following sequence: initially $Ni > V > Co$, then $V > Ni > Co$, and finally $V > Co > Ni$; and in the tholeiitic series the general trend during differentiation appears to be that of enrichment of vanadium at the expense of nickel (Ishikawa, 1968). Magnetite is a clay mineral related to vanadium enrichment, and magnetite fractionation controls vanadium abundance in the Tongan volcanic islands (Ewart et al., 1973). The positive correlation found between vanadium and TiO_2 in the Lesser Antilles volcanic island arc suggests a Ti-magnetite fractionation control that results in the trend of basalts to andesites and then to dacites (Brown et al., 1977).

3.   VANADIUM IN MT. FUJI GROUNDWATER

3.1.   Survey in the Mt. Fuji Area

Mt. Fuji has high levels of precipitation (an annual average of over 2,500 mm). According to Sugawara et al. (1980), waters in Mt. Fuji area can be classified into three chemical types. The first is meteoric precipitation; the concentrations of its major elements, such as Na, K, Mg, Ca, Cl, S, and Si, are lower than the average in Japan by 0.45% (Si) to 15% (Cl). The second is groundwaters; the fallen precipitate is easily sucked into the pervious volcanic ejecta to emerge at lower levels as springwater, which is greatly enriched with major elements such as K, Mg, Ca, and Si. This reflects their supply from ground material, and concentrations of Cd, Hg, Pb, and Fe decrease as they are taken up by the ground matrix. The third type is the five lake waters, and the concentrations of the tested elements are intermediate. The lake waters apparently are fed partly (75%) from the groundwaters and partly (25%) from the direct supply of meteoric water and surface runoff. There are no hot springs outcrops or anomalous geothermal activity. The temperature of the groundwaters in this mountainous area is in the range of 13–17 °C.

In order to know the vanadium concentration of Mt. Fuji groundwater, samples were taken from the locations shown in Figure 1. Vanadium concentration was determined by using a high-resolution inductively coupled plasma-mass spectrometry (ICP-MS) (Tsumura and Yamasaki, 1992; Yamasaki and Tsumura, 1992) after filtering the samples through 0.45-μm cellulose acetate filters. These water samples were classified with respect to their origins into four categories: deep-well water, fall or spring water, swamp water, and river water. As shown in Table 2, deep-well water had the highest vanadium concentrations, at 89.2–146.9 μg/L. According to the geological maps (Tsuya, 1988), the waters of deep wells (F, G) belong to the groundwaters in Older Fuji. Fall or spring water showed the second-highest vanadium concentrations, at 43.4–82.7 μg/L; these waters originated from the boundary between the overlying beds of porous, cracked lava of Younger Fuji and the underlying mud flow deposits of Older Fuji that had become less permeable to water. Swamp water contained various vanadium concentrations of 2.5–16.4 μg/L, in which the Fumoto sample (A) showed the lowest concentration, at 2.5 μg/L, since only this water came from the neighboring Miocene andesite volcanics. River waters showed vanadium concentrations of 17.7–48.8 μg/L. Of these waters E, K, and M samples came from an area originally covered by Mt. Fuji volcanic rocks, whereas the waters of L and O were mixed with those of a river coming from a different area.

The vanadium concentrations in Mt. Fuji waters showed a tendency to become higher at places nearer the summit and deeper in the ground. According to Okabe and Morinaga (1968), vanadium concentrations at Shiraito Fall (I in Fig. 1) and Wakutama Pond (N in Fig. 1) are 84.5 and 76.6 μg/L, respectively; in the estuary of the Fuji River 20–35% of the total vanadium concentration (11–18 μg/L) is in a suspended-matter fraction, and in the surface water about 2 km off the mouth of the Fuji River this value falls to 2% of the total concentration (2 μg/L). In nonmarine waters of Japan no higher vanadium concentrations than those found in Mt. Fuji have been reported except for hot springs, in which higher vanadium concentrations of 51–330 μg/L have been reported (Kuroda, 1942). In the thermal springwaters of the Big Horn Basin of Wyoming, vanadium concentrations of 4–8 μg/L have been shown, and the mud samples of that spring contain 0.1–0.2% vanadium (Egemeier, 1981).

The concentrations of several minerals such as V, Zn, Na, K, Ca, and Mg were compared between deep wells and other locations in the Mt. Fuji area (Table 3). The vanadium and zinc concentrations of deep-well samples (F, G, H) were significantly higher than those of spring and swamp samples (B, C,

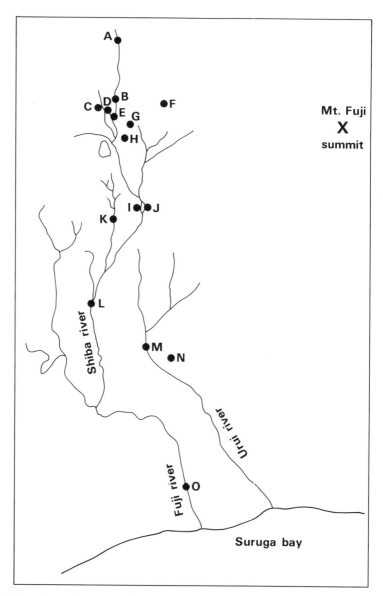

Figure 1. Locations of water sampling in the Mt. Fuji area. The sites A–O correspond to the analytical values in Table 2.

Table 2   Vanadium Concentrations (μg/L) of Fresh Waters in Mt. Fuji Area

Kinds of Water	Vanadium Concentration	Sampling Location on Map
Deep well	89.2	119-m-deep well (H)
Deep well	102.7	150-m-deep well (F)
Deep well	146.9	120-m-deep well (G)
Fall	54.4	Shiraito Fall (I)
Fall	54.0	Komadome Fall (J)
Spring	82.7	Wakutama Pond (N)
Spring	43.4	Trout Research Station (B)
Swamp	14.8	Uematsu (C)
Swamp	16.4	Nishikawa (D)
Swamp	2.5	Fumoto (A)
River	48.6	Shiba River (E)
River	48.8	Shiba River (K)
River	17.7	Shiba River (L)
River	38.2	Urui River (M)
River	21.0	Fuji River (O)

D, N) ($P < 0.05$); there is a tendency for vanadium and zinc solubilization to occur simultaneously. In addition, such minerals as Al, Ni and Sc in deep-well (G) and spring (N, B) water samples ($N = 3$) were analyzed by ICP-MS. Ranges of water Al, Ni, and Sc concentrations were 64 (N) through 148 (G), 5 (G,B) through 21 (N), and 5 (G) through 12 (B) ng/L, respectively. Compared with a 100-m-deep well sample from Tsuchiura City in Ibaraki prefecture (Al: 2,000, Ni: 120, Sc: 20, V: 395 ng/L), Mt. Fuji groundwater does not seem to leach nickel and aluminum to any significant extent.

Table 3   Comparison of Mineral Concentrations in Freshwater Samples of Deep Wells (H, F, G) and Those of Springs and Swamps (N, B, C, D)[a]

	Deep-well Samples ($N = 3$)	Other Samples ($N = 4$)
V (μg/L)	112.9 (89.2–146.9)*	39.3 (14.8–82.7)*
Zn (μg/L)	55 (36–93)*	13 (10–16)*
Na (mg/L)	13.2 (5.5–25.0)	4.7 (3.5–6.3)
K (mg/L)	1.5 (0.9–2.2)	1.2 (0.8–1.9)
Ca (mg/L)	10.0 (5.8–15.8)	7.8 (6.6–10.5)
Mg (mg/L)	2.9 (1.7–5.1)	2.0 (1.3–3.4)

[a] Each value shown as mean (min–max).
* There are significant difference between the values denoted by asterisks ($P < 0.05$).

3.2. Variation of Vanadium Concentration in Waters

The vanadium concentration in Mt. Fuji springwaters is positively correlated with the pH ($r = 0.734$) and negatively correlated with the discharge rate ($r = -0.987$) (Okabe et al., 1981). In 63 rivers and estuaries, dissolved vanadium concentrations are <0.026 to 2.05 μg/L; the vanadium concentration tends to increase with increasing river pH, which indicates that there is a greater weathering rate in the basins of alkaline rivers (Shiller and Boyle, 1987). According to Ikeda (1982), the pH of Mt. Fuji groundwaters ranges from 7.0 to 8.5, and the average pH in Younger Fuji (7.15) is lower than that in Older Fuji (7.84). The Cl^-, SO_4^{2-}, and SiO_2 concentrations in Mt. Fuji groundwaters are 5.6–5.9, 13.7–13.9, and 34.4–41.6 mg/L, respectively. In the groundwaters of Younger and Older Fuji the average HCO_3^- concentrations are 63.8 and 87.4 mg/L, respectively. In an in vitro leaching experiment using rocks from Mt. Fuji, the HCO_3^- concentration increased linearly as the immersion period increased from 20 to 500 days (Ikeda, 1982). Sediments of lakes located in volcanic regions in Japan are reported to contain higher amounts of vanadium than those in nonvolcanic regions (Sugawara et al., 1956).

Minerals that are leached by Mt. Fuji groundwater are not in volcanic ashes but in lava flows (Suzuki et al., 1984). In Mt. Fuji's case porous lava flows that contain extremely high amounts of vanadium can be leached by slightly alkaline and oxygen-containing water. Ochiai and Kawasaki (1970) have shown that it takes from several years to more than 30 years for Mt. Fuji groundwaters to come out again in springs and falls. Some of the waters confined in the deeper layers of Older Fuji remain for much longer periods of time. Vanadium is present in pentavalent anionic and tetravalent cationic forms as well as in a neutral complexed form. Relative abundances of pentavalent, tetravalent, and complexed vanadium in unpolluted, meteoric water of the inner Alps are 41%, 59%, and 0.7%, respectively (Orvini et al., 1979). In Mt. Fuji groundwater dissolved organic matter derived from plants and microbes may not be abundant, but the carbonate complex of vanadyl is supposed to exist, which leads to an increase in its stability (Wanty and Goldhaber, 1992).

The concentration of vanadium in surface seawater is generally in the range of 1–3 μg/L (Burton, 1966). The concentration of vanadium in the ocean deviates slightly from conservative behavior by exhibiting 10% depletion in surface waters, where concentration changes correlate with phosphate (Collier, 1984). Pentavalent anionic vanadium species is regarded to be predominant in oxic seawater. However, according to one report (Sugimura et al., 1978), more than 80% of vanadium dissolved in seawater is present in organic forms regardless of the depth. In many of the basalts, plagioclase is the phase most resistant to weathering and alteration, followed in order of decreasing resistance by pyroxenes and finally olivine (Engel and Engel, 1963). Experimentally, vanadium in oceanic basalts cannot be hydrothermally leached (Humphris and Thompson, 1978), and little vanadium can be solubilized from basaltic rocks even under extreme conditions of low pH (Seyfried and Mottl, 1982).

In contrast to the vanadium concentration of surface seawater, the interstitial waters show exceedingly high concentrations of vanadium. The interstitial waters from the anoxic sediment (40–50 cm depth) of the Gulf of California (Brumsack and Gieskes, 1983), an oxic sediment (a few centimeters depth) of the Guatemala Basin (Emerson and Huested, 1991), and a metalliferous sediment (0–0.5 cm depth) in the eastern equatorial Pacific (Heggie et al., 1986) contain vanadium at 221–330, more than 70, and 15–24 μg/L, respectively. In these waters there are abundant complexes of vanadyl and organic matter. In the groundwater of pampas in Argentina an extremely high average vanadium concentration of 281 μg/L has been reported, with a maximum concentration of 1,715 μg/L (Nicolli et al, 1989). This water also contains toxic levels of arsenic and fluorine.

4. VANADIUM IN ORGANISMS

4.1. Survey in the Mt. Fuji Area

I compared the vanadium contents of freshwater plants (*Ceratophyllum demersum, Batrahium nipponicum, Lemna paucicostata, Elodea nuttallii, Oenanthe javanioca, Nasturtium officinale, Cardamine scutata, Blyxa japonica, Hepaticae* sp.) in Inokashira Pond (B in Fig. 1; water vanadium concentration: 43.4 μg/L) with contents of plants (*Nasturtium officinale, Eichhornia crassipes, Lectuca scariola, Vesicularia dubyana, Ceratopteris thalictroides, Ludwigia repens, Rotala indica, Higrophila polysperma, Myriophyllum* sp., *Echinodorus bleheri, Hygrophila stricta, Cardamine lyrata, Limnophila sessiliflora, Vallisneria spiralis, Varisneria gigantea, Camomba caroliniana, Egeria densa*) in Utsunomiya Pond and Tsuchiura Market for water plants (water vanadium concentrations: 0.72 and 0.40 μg/L, respectively). The vanadium contents in these samples were analyzed by flameless atomic absorption spectrophotometry, as shown previously (Hamada, 1994b). The mean vanadium content of water plants in Inokashira Pond was 21.8 $\pm$ 11.3 (*SD*) with a range of 5.6–43.7 μg/g dry weight, whereas that in Utsunomiya Pond and Tsuchiura Market was 0.79 $\pm$ 0.52, with a range of 0.22–1.91 μg/L dry weight. Significant difference is noted between them ($P < 0.01$). I found the green alga (*Cladophoraceae* sp.) in Inokashira Pond to contain an exceedingly high amount of vanadium, at 118–168 μg/g dry weight. Vanadium is an essential element for the green alga *Scenedesmus obliquus* (Arnon and Wessel, 1953). At low concentrations of vanadium (<1 mg/L), primary production and growth of diatoms (*Bacillariophyceae*) is suppressed, whereas some of the green and blue-green algae appear to be stimulated (Lee, 1983). Compared with other water plants living in a low-vanadium environment, water plants in the Mt. Fuji area show a greater enrichment in vanadium, which indicates a positive relationship between vanadium concentration of waters and that of water plants or algae.

Farming and breeding of rainbow trout have been practiced in the Mt. Fuji area for more than 60 years. Two ponds in Mt. Fuji were selected to catch rainbow trout (*Oncorhynchus mykiss*). One was Inokashira Pond (B in Fig. 1) and the other was Wakutama Pond (N in Fig. 1). As a control, rainbow trout were caught from Utsunomiya Pond of the Trout Breeding Station of Tochigi Prefecture in Utsunomiya City. This pond utilized the groundwater originated from an upper stream of the Kinu River in Nikkou, and the vanadium concentration was 0.72 μg/L. The vanadium levels of all trout tissues except that of muscles tended to be elevated as freshwater vanadium concentration increased from 0.72 μg/L (Utsunomiya) to 43.4 (Inokashira) or 82.7 μg/L (Wakutama). In Utsunomiya trout (average body weight 884 g), Inokashira trout (average body weight 707 g), and Wakutama trout (average body weight 1,988 g), the bone vanadium contents were 0.87, 4.77, and 17.2 μg/g dry weight, respectively, and the kidney vanadium contents were 0.43, 2.38, and 4.63 μg/g dry weight, respectively (unpublished result). The muscle vanadium contents in these trout were not different (0.016–0.024 μg/g dry weight). All the trout in the three ponds were apparently in good condition irrespective of varying vanadium intake.

In Suruga Bay, 1,200 fish species have been found, in comparison with the total 2,500 species found near the Japanese archipelago. More than 250 kinds of new species have been found in the bay. Ancient-type marine organisms such as *Chlamydoselachus anguineus* (frill shark), *Misukurina owstoni* (goblin shark), *Diaphus watasei* (sardine), *Sergestes lucens* (shrimp), and *Phylum mollusca* (mollusks) were found in this bay (Hoshino, 1962). Some of them are known as living fossils. Away from the mouth of the Fuji river is the best spawning ground for *S. lucens,* which can live only in this bay (Konagaya et al., 1984; Omori et al., 1988). It is said among fishermen that in years of greater rainfall *S. lucens* are caught in larger quantitics.

According to a survey conducted in 1980, 14–19 billion *S. lucens* were estimated to spawn in the bay, but their numbers have been gradually decreasing. The yearly *S. lucens* commercial catch amounts to about 3,000 tons. Diatoms such as *Skeletonema costatum* and *Nitzchia seriata* and dinoflagellates such as *Ceratium, Peridinium, Gynodinium, Prorocentrum,* and *Nactiluca* are very enriched off shore (Konagaya et al., 1984). These plankton are food for *S. lucens.* Then *S. lucens* become the prey of fish such as *Mictophida.* In turn *Mictophida* become the food of larger fish such as *Trichiurus lepturus* (cutlassfish). *S. lucens* are distributed over depths of 200–300 m. The mixing of the Black Current, the Kurile Current, river water, and the exceedingly deep seawater on the steep continental slope around a deep submarine canyon adjacent to the mouth of the river is favorable to the shrimp's ecology (Omori et al., 1988). However, this topographic consideration does not explain why *S. lucens* gather off shore of the Fuji River in high numbers. The vanadium content of *S. lucens* was 0.56 μg/g dry weight, with the cranial portion containing 8.3 times as much vanadium as the caudal portion (unpublished result).

A large influx of vanadium into the Suruga Bay through the Fuji River might be responsible for establishing the chemical environment of offshore waters suitable for the reproduction of *S. lucens*. Vanadium might become an essential trace mineral or a growth promoter in the above-stated food chain. Vanadium-containing enzymes such as bromo- and iodoperoxidases are found in some brown and red algae and in some bacteria; vanadium may serve as a cofactor in nitrogenase (da Silva and Williams, 1991; Rehder, 1992). Vanadium might have been used in nitrogen fixation or CO_2 reduction in photosynthetic processes in ancient-type organisms (da Silva and Williams, 1991).

4.2. Vanadium Accumulation by Organisms

Zooplankton (radiolarians) collected in Monterey Bay, California, show a maximum vanadium concentration of 41.5 μg/g dry weight (Martin and Knauer, 1973). As skeletal remains of radiolarians are often found in deep-sea sediments, much of the material in the microplankton samples may have been ready for transport to the sea floor. Vanadium concentrations of mixed plankton samples from both the East Pacific and near-shore areas off southern California range from 2.57 to 101.6 (average 25.0) μg/g dry weight, and most are likely to be concentrated by the Sr-rich species (e.g., radiolarian) (Knauss and Ku, 1983). Vanadium incorporation into foraminiferal calcite is proposed to be proportional to seawater vanadium concentration, which might be used as an index for changes in vanadium concentration of ancient seawater (Hasting et al., 1990). From the numerous works summarized for algae (Lee, 1983), it appears that vanadium concentrations of <100 to 500 μg/L often reveal positive effects (e.g., stimulation of biomass production and chlorophyll synthesis), whereas inhibitory effects and various LC_{50} values are distinctly in the milligram per liter range.

According to Bertrand (1950), the vanadium content in 62 plants varies from 0.152 to 4.2 μg/g (dry aerial part) and from 0.100 to 12.14 μg/g (dry root part). Lower forms of plants contain more vanadium than seed-producing plants. Several species of the Amanita family sequester vanadium, the fly agaric (*Amanita muscaria*) showing an average concentration of 112 μg/g (Bertrand, 1950) and a maximal concentration of 325 μg/g dry matter (Rehder, 1992). A natural product, amavadine, which contains vanadium, has been isolated from it. Of the 82 species of marine algae in Japan, *Portieria homemanni* (Rhodophyceae) shows the highest vanadium content, at 204 μg/g in dry matter (Asakawa, 1992). Leaves of an aquatic plant (*Pontedaria cordate*) contain 80 μg/g of vanadium in dry weight (Cowgill, 1973). Five species of water plants such as *Potamogeton* from the Ukraine show vanadium contents of 7.2–57 μg/g dry matter (Hutchinson, 1975). Plants growing in areas of selenium toxicity may be high in vanadium; there are vanadium accumulator plants (legumes) such as *Astragarus confertiflorus* and *Astragarus preussii;* and vanadium absorption by plants may be deterred by the presence of calcium in the soil (Cannon, 1963). Various species of the genus *Homalium* from New Caledonia are hyperaccumulators of nickel (Brooks et al., 1977). Lichens such as *Parmelia physodes,* near oil-fired power plants in Sweden, can accumulate 637–790 μg/g (dry matter) of vanadium (Juichang et al., 1995).

The average vanadium contents of invertebrates and vertebrates are 1.2 and 0.1 μg/g dry matter, respectively (Bertrand, 1950). However, ascidians, a phylogenic ancestor of vertebrates, can accumulate vanadium from seawater against a 10^6- to 10^7-fold concentration gradient and store it as trivalent or tetravalent vanadium (Biggs and Swinehart, 1976; Rehder, 1992). The purpose of accumulating vanadium is not known, but it may be for protection or, alternatively, to serve as an electron and proton sink (da Silva and Williams, 1991). There is a large variation in vanadium content among species (0–7,000 μg/g whole dry matter) (Biggs and Swinehart, 1976; Dingley et al., 1981). Older ascidian families tend to accumulate more vanadium (Hawkins et al., 1983). Edible ascidians (*Holocynthia roretzi*) taken from the sea off the Tohoku district of Japan showed vanadium contents in soft tissues and outer hard tissues (tunica) at 1.36 and 20.48 μg/g dry matter, respectively (unpublished results). This species is not a vanadium accumulator, but more vanadium is retained in the tunica portion by a surface sorption process.

The presence of considerable amounts of vanadium suggests that at least some ascidian species participate in the concentration of this element in black shales (Bonham, 1956; Vinogradov, 1953). In some areas the vanadium concentration of seawater might have been elevated in ancient times when ascidians flourished, and they must have adapted to such conditions because of their immobility. Consequently, they have acquired the capacity to store vanadium as inert vanadium(III) or vanadium(IV) with sulfate or other agents in special cells called vanadocytes. In the near-shore environment water column anoxicity may be a direct result of organic enrichment in the euphotic zone. In the euphotic zone there is a zone with a very low O_2 concentration (Didyk et al., 1978). Ascidians can survive in it by means of vanadocytes, which serve as an electron and proton sink (Smith, 1989). According to Glikson et al. (1985), the vanadium in the Toolebuc oil shale could have been concentrated by organisms such as tunicates. Abundant tunicate fossils have been observed in vanadium-rich coal in China (Zhang, 1987).

There are other vanadium accumulators. A holothurian named *Sticopus möbii,* collected at the Tortugas, contains vanadium equal to 1,230 μg/g dry weight (Phillips, 1918). The fan worm (*Pseudopotamilla occelata*) taken near the shore of the Sanriku coast of Japan contains vanadium at 510 μg/g whole dry matter, the bipinnate radiole part showing a high content of vanadium (0.55% of dry weight) (Ishii et al., 1994). After the death of plants the lignin is transformed into insoluble humic acids, which are able to concentrate uranium and other cations from very dilute solutions in natural waters (Szalay, 1964; Szalay and Szilagyi, 1967). Hydroxy-containing compounds such as carbohydrates and glycols play important roles in reduction and complexation of vanadium (Bandwar and Rao, 1995), and chitosan shows a strong capacity for adsorption of vanadyl (Jansson-Charrier et al., 1996). Direct surface sorption processes are important for vanadium bioaccumulation by mollusks, crustaceans, and echinoderms.

Bertrand (1950) has speculated on the possibility that in ancient geological periods there were some plants with a vanadium content approximately 20 times higher (maximally) than the norm, which would explain how some fossil fuels (coals) contain in their ash 10–20 times as much vanadium as that found in most others. As shown above, there are many organisms that can retain vanadium up to a hundred times greater concentration than those (1–5 μg/g dry matter) which are found in most plants and animals as an uppermost value. Vanadium might have been used more commonly in the earlier stages of evolution and replaced by more effective metals later on (Rehder, 1992). A vanadium-containing enzyme, bromoperoxidase, is found in brown and red algae. A role in the host defense system has been suggested for this enzyme (Wever et al., 1991). Vanadium in waters becomes a growth stimulator of algae at lower concentrations but a toxic substance at higher concentrations (Lee, 1983). Vanadium can be retained well by plankton and algae, but in higher animals and plants the uptake of vanadium is restricted by many devices. In higher mammals such as ruminants, vanadium absorption through the gastrointestinal tract appears to be insignificant. Anaerobic fermentation in the rumen may decrease vanadium availability for the ruminant. The vanadium contents of calf kidneys and milk samples from a dairy farm of the Mt. Fuji area (H in Fig. 1) were not different from those of other districts far away from Mt. Fuji (unpublished results).

5. RIDDLE OF VANADIUM ENRICHMENT IN FOSSIL FUELS

The concentrations of Zn, Cu, Pb, Bi, Cd, Ni, Co, Ag, Cr, Mo, W, and V in seawater are thousands of times less than the calculated amounts of these elements supplied to the sea during all of geological time, and vanadium and nickel are usually the two that show the greatest enrichment in organic sediments (Krauskopf, 1956). A notable characteristic of the most vanadium-rich shales is that they were derived from deposits composed mostly of marine-derived organic matter (Tourtelot, 1964). Vanadium retained by organisms was postulated as a source of the metal enrichment in fossil fuels (Bertrand, 1950; Vinogradov, 1953), but the concentrations of trace metals in organisms may not be enough to account for the observed enrichment of trace metals in black shales (Wedepohl, 1971). The metal enrichment in organic-rich sediments is due to chemical precipitation and to reactions with dead organic remains (Nicholls and Loring, 1962; Holland, 1979). Earlier ideas (Holland, 1979; Lewan and Maynard, 1982) tended to favor a nearly constant oceanwide flux of metals from seawater to sediments. The explanation of the origin of organic-rich sediments and rocks invokes deposition from normal seawater under conditions of anoxia (Brumsack, 1980).

In the Late Jurassic and Cretaceous Atlantic ocean, high productivity of organic matter may have been caused by a fluvial nutrient supply and by intensified upwelling (Stein et al., 1986). Sporadic temporal and spacial increases in primary production constitute a more tenable explanation for the occurrence of Cretaceous black shales (Pedersen and Calvert, 1990). There is strong environmental evidence that very low or zero bottom water O_2 concentrations are necessary to promote formation of measurable authigenic vanadium enrichment (Emerson and Huested, 1991). In episodes of anoxia or very low bottom-water oxygen concentrations associated with increased productivity, more efficient release of reactive phosphorus from sediments occurs (a positive feedback), which may result in a higher primary productivity if the regenerated phosphorus is transferred to the ocean surface (Calvert et al., 1996). Adsorption and possibly subsequent incorporation of vanadyl as a solid solution is probably inorganic sink for vanadyl in reducing sediments (Wehrli and Stumm, 1989). Hydrogen sulfide can further reduce vanadyl to trivalent vanadium in sediments (Wanty and Goldhaber, 1992).

Hodgson (1954) has proposed that it is perhaps more reasonable to believe that a distinct change may have occurred in the sea as a result of the changing surface of the terrestrial earth during the post-Cambrian period, and Le Riche (1959) has speculated that conditions were almost certainly very different in the Liassic sea from those prevailing at the present, since it would appear that anaerobic conditions have been more common in the past geological time than at present. In contrast, others (Brumsack and Gieskes, 1983; Holland, 1979; Wanty and Goldhaber, 1992) have proposed that the trace metals in sea waters have always existed in the same proportions. However, metal enrichment of the organic fraction in normal seawater requires geochemical enrichment factors as great as tens of millions of times; a plausible syngenetic hypothesis requires some prior concentration of the metals in seawater (Vine and Tourtelot, 1970). If the vanadium contents of ancient seawaters are regarded as being the same as in present seas, sorption of vanadium from a standing seawater column is not enough to explain the exceedingly high vanadium enrichment in the vanadiferous zone of the Phosphoria Formation of Wyoming and Idaho (McKelvey et al., 1986). Breit (1992) also has contended that to concentrate the vanadium level in a layer of sediment to 1,000 ppm in such shallow, restricted basins as the Health Formation and the Toolebuc Formation, the above seawater columns must have contained higher abundances of vanadium than the present vanadium content.

Vanadium is preferentially concentrated in organic-rich sediments in anoxic basins or upwelling areas where bottom-water oxygen content is low. According to the analysis of modern sediments from the Gulf of Paria by Hirst (1962), the ratio of nondetrital vanadium to organic carbon content has an inverse relationship with the increase of organic carbon content, which might be caused by a limited supply of vanadium to water columns. The role of chemical scavenging by particles is given more emphasis as a general mechanism for removal of trace elements from seawater, and such a process is expected to lead to significant regional variation in rates of scavenging (Thomson et al., 1984). The Cretaceous La Luna Formation of the Maracaibo Basin in western Venezuela is extensive, and the source beds in it are believed to have generated the major part of the oil produced in the Maracaibo area (Didyk et al., 1978). The range of vanadium abundance in petroleum in Venezuela is from 0.3 to 1,400 ppm, but the factors that controlled such an anomalous distribution are still poorly understood (Breit, 1992).

6. PALEOGEOGRAPHICAL TREND IN VANADIUM DISTRIBUTION

Since the late Paleozoic (about 300 Ma) Earth has been subjected to numerous volcanic episodes consisting of huge outpourings of basaltic lavas onto continental surfaces (Cox, 1988). Especially in the Mesozoic and Early Tertiary era, vigorous activities of basaltic volcanoes have been recorded (Vogt, 1972). In the Cretaceous with the return of moist climates, terrestrial productivity rebounded to levels well above those of the Paleozoic, and subsiding landscapes became massive burial sites of organic carbon (Robinson, 1990). In the western United States, Upper Cretaceous igneous rocks older than about 80 million years are found mainly in California and western Nevada; uppermost Cretaceous (Laramide) igneous rocks (70–65 Ma) occur in north-central Nevada and as far east as the southern Rocky Mountains (Lipman et al., 1971). The Laramide uplift in the central Rocky Mountains that occurred near 70 Ma was produced by the dynamics of moving plates and possible complex interactions between North America and one or more Pacific plates to the west as in major changes of plate motions, probably on a global scale (Coney, 1976). Predominantly andesitic volcanism in the continental interior (Montana, Wyoming, and Idaho) terminated about 40 Ma; and predominantly calc–alkalic intermediate volcanism terminated in the southwestern United States at about the end of the Oligocene, but it continued in Miocene and Pliocene time in parts of western Nevada and eastern California and through the Quaternary in the Cascade Range (Lipman et al., 1971). There are anomalous enrichments of trace minerals (Cr, Mo, Se, Ni, U, V) in a Cretaceous rock field in North America, and several indicator plants for such elements are known (Kubota, 1980). During the Late Cretaceous and early Paleocene a large outpouring of different kinds of magma took place along the western and southwestern margins of the Fort Union Formation basin of the United States. Bentonitic beds in this formation indicate contemporaneous volcanism (Zubovic et al., 1961).

Such countries as Venezuela, the United States (California), Angola, and Canada show relatively higher vanadium and nickel contents in crude oils, while China (northwest China) and New Zealand show lower contents (Frankenberger et al., 1994). In the Upper Devonian to Upper Cretaceous crude oils from western Canada, the concentrations of vanadium and nickel varied with sulfur content but varied inversely with A.P.I. gravity (Hodgson, 1954). There is a positive correlation between the vanadium and nickel contents of crude oils in the whole world, but in New Zealand crude oils there is no such correlation; instead, vanadium is highly significantly correlated with aluminum ($r = 0.747$) (Frankenberger et al., 1994). New Zealand oils were derived from coal sources, and higher expulsion temperatures of coals might decompose organometallic species. Not only the source rocks but also the redox potential, hydrogen ion activity, and sulfide activity of depositional environments can affect the proportionality of vanadium to nickel in crude oils (Lewan, 1984).

According to an analysis by Patterson et al. (1986) on trace element distribution in oil shale from Julia Creek, Australia, vanadium is associated with either kerogen (organic matter) or mixed-layer mica–montmorillonite clay mineral, while nickel is associated with either kerogen or pyrite fraction. Kettle-bottom coal and coalified trees at the Bull Mountain coal field show exceedingly high enrichment of vanadium (250–283 ppm) compared with the associated coal beds (9.3 ppm), suggesting that vanadium accumulation by plants must be involved in addition to a postdepositional process of vanadium enrichment (Zubovic et al., 1961). In these samples not only vanadium but also other metals such as Cr, Ga, and Ge are very highly concentrated. Vanadium and nickel concentrations in bitumens extracted from a variety of organic sedimentary rock types of different geological ages and geographical areas range from <0.2 to 4,760 ppm and <7 to 1,240 ppm, respectively, and vanadium concentrations show a polymodal frequency distribution, while nickel concentrations show a nearly normal frequency distribution (Lewan and Maynard, 1982).

Paleogeographical trends are recognized in variations of vanadium abundance. For example, in the Esopus Formation of New York there appears to be a rise in vanadium concentration with increased height in the stratigraphic section (Fenner and Hagner, 1967). In Lower Pennsylvanian oils from the Seminole area, Oklahoma, the concentrations of vanadium and nickel are greatest near the ancient shoreline and decrease basinward (Bonham, 1956). In the coal beds of the Tongue River section of the Fort Union Formation in Montana and North Dakota, the eroding source rocks are assumed to be in the west and southwest part, and only subbituminous coal samples of the Bull Mountain field show exceedingly high vanadium enrichment (Zubovic et al., 1961). The marked gradient concentrations of cationic elements from the source rocks are present in certain peat deposits, which demonstrates the efficiency with which peat components can trap a variety of cations from solution as well as filter out mineral grains (Given and Miller, 1987). Some geographical trends of vanadium enrichment appear to exist in Iraqi oil fields (Al-Shahristani and Al-Atyia, 1972), crude oils of Canada (Hodgson, 1954), and Italian oils and asphalts (Colombo and Sironi, 1961). According to Premovic et al. (1986) and McKelvey et al. (1986), volcaniclastic materials and volcanic ashes are the most likely sources of vanadium in ancient marine sedimentary rocks. In deep sea sediments the highest vanadium concentrations are found to be associated with basic volcanics of the Hawaiian Islands (Goldberg and Arrhenius, 1958) or the submarine active ridge of the Indian Ocean (Boström and Fisher, 1971). Vanadium is more enriched in the volcaniclastic sediments of the Tonga–Kermadec Ridge of basaltic volcanism than in the sediments around the White Island of andesitic volcanism (Hodkinson et al., 1986).

Zibermintz (1935; cited by Breit, 1992) has attributed the high vanadium content of coals in the Ural Mountains to waters transporting vanadium that was weathered from mafic massifs containing vanadiferous magnetite deposits. As a source of such vanadium enrichment, basaltic volcanoes containing exceedingly high amounts of vanadium must have been present. The dissolved vanadium in the groundwaters would contribute to the vanadium enrichment in the organisms such as plankton and algae. The vanadium contents of these organisms would increase as water vanadium concentration increases as shown above (Section 4.1). An anomalous sedimentary vanadium enrichment pattern may be caused by an anomalous distribution of basaltic volcanoes. Mt. Fuji has been discharging a considerable amount of vanadium into Suruga Bay for about 80,000 years. However, since Suruga Bay has an extraordinary depth (about 2,500 m), connecting to the Nankai Trough, it is not a suitable place to form a vanadium-rich sediment such as the shelf type of metal-rich black shale. Suruga Bay's sediments show rather low levels (55–98 ppm) of vanadium enrichment (Sato and Okabe, 1978). In ancient times of vigorous volcanic activity, the vanadium flux from basaltic volcanoes into a warm, shallow sea might have been greater in scale and duration than is characteristic of the present Mt. Fuji.

There are geographical trends of vanadium enrichment that could be related to volcanic activity. Maximum vanadium contents are 2,700, 1,150, 9,900, 740, 2,800 and 5,000 ppm, respectively, in the Serpiano marl (carbonate sediment) in the Swiss Alpine complex (Premovic et al., 1986), La Luna shaly limestone of the Maracaibo basin in northwest Venezuela (Premovic et al., 1986), Mecca Quarry shale (coal) of the Phosphoria Formation in Wyoming (McKelvey et al., 1986), Cretaceous offshore marine shale of the western interior of North America (Tourtelot, 1964), Julia Creek oil shale of the Toolebuc Formation in Australia (Patterson et al., 1986), and black shale of the Kupfershiefer Formation in Germany (Wedepohl, 1964). In these sediments volcanic evidence such as volcanic ash layer, subaerial and submarine volcanism, and clay minerals of volcanic origins have been found.

7. POSSIBLE EFFECT OF VANADIUM ON DINOSAURS

Vanadium may be a nutritionally essential element for higher animals (French and Jones, 1993). Trace elements of vanadium and copper, for example, act as essential nutrients with optimum intakes but act as toxic substances in excess (Hamada, 1995). Pentavalent vanadium compounds in food and drinking water are reduced to tetravalent vanadyl in the body, and intracellularly vanadyl remains as complexes with other agents such as transferrin, ferritin, or phosphate compounds (ATP, etc.). Free vanadyl produces hydroxyl radical and hemolyzes vitamin E-deficient erythrocytes (Hamada, 1994a). Vanadate is an inhibitor of ATPase (Cantley et al., 1977). The accumulation of vanadium in kidneys affects kidney function (Phillips et al., 1983). In laboratory animals such as rats, mice, and hamsters, there is a linear equation; $y = 34.87 x + 129.8$ ($r^2 = 0.90$) between food vanadium content (x, mg/kg) and kidney vanadium content (y, ng/g dry matter) (Hamada, 1992). When aqueous solutions containing VCl_3, $VOSO_4$, or $NaVO_3$ are administered into the air sacs of 14-day fertilized chick eggs and then incubated for a further 5 days, the growth rate of chick embryos decreases dose-dependently and the death rate increases dose-dependently irrespective of different vanadium compounds administered (Hamada, 1994b). The concentration of vanadium in eggs to cause 50% embryonic death is 0.6–0.7 mg/kg.

The effects of vanadium concentration in drinking waters on human health are not precisely known. Strain et al. (1982) have stated that higher intakes of vanadium from drinking water (up to 100 μg/L) may be an important factor in the lower death rates for all causes associated with the Great Plains states of the United States and regional differences in New York State. However, there are some vanadium-related problems. In the hospitals of northeast Thailand there are many patients with renal tubular acidosis (Nilwarangkur et al., 1990), and high intake of vanadium from drinking water has been suspected as a cause (Dafnis et al., 1992). Some uremic patients who received dialysis treatment in Kanagawa prefecture show exceedingly high plasma vanadium content resulting from the increased intake of vanadium from drinking water (Tsukamoto et al., 1990). According to Tajiri and Hirasawa (1985), the blood vanadium level of chronic hemodialyzed patients in a hospital of Niigata City is significantly higher than that of controls (2.27 vs. 0.24 μg/L). Vanadium is selectively retained by the kidneys, and it is suspected that patients who suffer from kidney malfunctions might accumulate more vanadium in their bodies. It is not known whether such patients would absorb vanadium more efficiently. Considering the various pharmacological effects of vanadium on diabetes, psychosis, and cancer, epidemiological surveys are important in areas that exhibit exceedingly high vanadium concentration in drinking water.

The Cretaceous era is an unusual period in several respects: the proliferation of epicontinental marine sediments due to a high sea level, including the invasion of marine depositional environments into the western interior of North America, and the influx of volcanic material into the Cretaceous marine environment (Nadeau and Reynolds, 1981). The Late Cretaceous was a time of enormous volcanic activity all around the Pacific; this volcanism is related to an immense increase in mantle plume activity for a relatively short geological time interval and the mantle plume activity probably occurred simultaneously at several localities throughout the world, including the Deccan Plateau (Officer and Drake, 1983, 1985). It has been speculated that the cause of the iridium abundance anomaly occurring in a thin zone at the planktonic Cretaceous–Tertiary (K-T) boundary in marine sediments was the volatile iridium compound (IrF_6) in volcanic gases rather than being of extraterrestrial origin (Zoller et al., 1983). At the Raton Basin of New Mexico and Colorado, the K-T boundary bed's vanadium contents are 110–187 ppm, in contrast to the 10–67 ppm of beds above or below the K-T boundary, suggesting the influence of basic volcanism (Gilmore et al., 1984). Large-scale extinctions of marine plankton (Bramlette, 1965) and of dinosaurs occurred at the K-T transition period. The Deccan flood basalt coincided with the volcanic activities of the K-T boundary (Baksi, 1990). Rocks of the Deccan flood basalt contain as much vanadium as those of Mt. Fuji (Section 2 above).

Cloud (1959) and Vogt (1972) have supported the trace element hypothesis as the cause of global extinctions in geological time. According to Cloud (1959) the faunal community, and particularly marine or marsh-land organisms, are sensitive to variations in trace metal concentrations in their aquatic milieu, and the great herbivorous dinosaurs could have become extinct because general chemical changes either killed the swamp flora on which they fed or reached concentrations in food or water in amounts lethal to the dinosaurs. Although Cloud (1959) considered a toxic effect of copper as primary, vanadium is a more likely element because of its more abundant distribution in fossil fuels.

If tremendously large amounts of vanadium were emitted from basaltic volcanoes into surrounding environments, even in higher animals excessive accumulation of vanadium into renal and reproductive systems would occur through the absorption from the gastrointestinal tract and lungs, which would surpass the animals' abilities to retain vanadium as inert vanadyl–chelate complexes. Then vanadium would have become a radical-forming, toxic element. Thus, dinosaurs might have suffered from renal disorders caused by a considerable vanadium accumulation in the kidneys. If vanadium was transferred into fertilized eggs of dinosaurs to reach a concentration of 0.6–0.7 mg/kg, even if the embryos had been able to survive, newborns could not have lived much longer owing to significant growth retardation.

8. CONCLUDING REMARKS

In order to explain the anomaly of vanadium enrichment in sediments it is necessary to consider the presence of basaltic volcanoes like Mt. Fuji around the sedimentary basin. Vanadium can be leached from basaltic rocks into groundwaters in oxic and slightly alkaline conditions. So, volcanic ashes located in seafloors (Premovic et al., 1986) and submarine volcanism (Wedepohl, 1971) are unlikely sources because of rather anoxic ambient conditions and because the accumulation had taken place in restricted, shallow basins, apparently isolated from hydrothermal vents. Basaltic volcanoes of high vanadium content must have been present on the land and their basaltic lava flows were stripped of their vanadium by oxygenated groundwater, which then ran into the sea. For example, in the coal province of the Northern Great Plains of the United States, vanadium and other trace elements were leached from the Bull Mountain field and were further conveyed from the west to the east by the Tongue River (Zubovic et al., 1961). Kapo (1978, cited by Breit, 1992) suggested that the high vanadium content in some Venezuelan oil and coal is due to a source in surrounding uplands, and Wedepohl (1964) has proposed that metals enriched in the bituminous shale of the Kupferschiefer in Germany originated from the dissolution of iron oxides in the red bed units surrounding the sea. My hypothesis is well in accord with these circumstances.

If the concentrations of vanadium in nonmarine and marine waters increased, plankton, algae, lichens, crustaceans, ascidians, some legumes, and microbes would accumulate more vanadium into their bodies by metabolism and sorption processes, as shown in Section 4. Brongersma-Sanders (1968, cited by Vine and Tourtelot, 1970) suggested that trace metals are concentrated by living plankton in aerated surface waters and the dead bodies are carried by subsurface countercurrents to accumulate in great masses. She further suggested that the trade winds provide the constant wind direction necessary to produce the countercurrents and upwelling adjacent to land. The vanadium content of most present organisms may reflect a relatively low vanadium environment. However, in the ancient biosphere of high-vanadium load, the vanadium contents in organisms, as shown above, would increase to more significant extents than the levels found in present organisms. Therefore, the contribution of organisms to vanadium enrichment must be reevaluated.

ACKNOWLEDGMENTS

I thank Mrs. E. Nirasawa, of the National Institute of Animal Industry, for technical assistance; Dr. A. Tsumura, of the National Institute of Agro-Environmental Sciences, for analyses of water samples by ICP-MS; Mr. T. Motizuki, Mr. Y. Okitsu, and Mr. Y. Tsuchiya, of the Animal Husbandry Experimental Station of Shizuoka Prefecture, for water and water plant samples in the Mt. Fuji area; Mr. Y. Kanamori, of Sengen Shrine at Fujinomiya City, and Mr. N. Fukutomi, of the Trout Breeding Station of Tochigi Prefecture, for trout samples; and Dr. S. Togashi, of the Geological Survey of Japan, for geological advice on Mt. Fuji.

REFERENCES

Al-Shahristani, H., and Al-Atyia, M. J. (1972). Vertical migration of oil in Iraqi oil fields: Evidence based on vanadium and nickel concentrations. *Geochim. Cosmochim. Acta* **36,** 929–938.

Ando, A., Kamioka, H., Terashima, S., and Itoh, S. (1989). 1988 values for GSJ rock reference samples, Igneous rock series. *Geochem. J.* **23**, 143–148.

Arculus, R. J., Gust, D. A., and Kushiro, I. (1991). Fuji and Hakone. *Natl. Geogr. Res. Explor.* **7**, 276–309.

Arnon, D. I., and Wessel, G. (1953). Vanadium as an essential element for green plants. *Nature* **172**, 1039–1040.

Asakawa, A. (1992). Survey of vanadium-accumulating abilities of sea algae (Japanese). *Summary Rep. Annu. Meet. Jpn. Soc. Sci. Fisheries,* p. 179.

Baksi, A. K. (1990). Timing and duration of Mesozoic-Tertiary flood-basalt volcanism. *Eos Trans. Am. Geophys. Union* **71**, 1835–1840.

Bandwar, R. P., and Rao, C. P. (1995). Relative reducing abilities in vitro of some hydroxy-containing compounds, including monosaccharides, towards vanadium(V) and molybdenum(VI). *Carbohydrate Res.* **277**, 197–207.

Bertrand, D. (1950). Survey of contemporary knowledge of biogeochemistry. 2. The biogeochemistry of vanadium. *Bull. Am. Mus. Nat. Hist.* **94**, 403–456.

Biggs, W. R., and Swinehart, J. H. (1976). Vanadium in selected biological systems. In H. Sigel (Ed.), *Metal Ions in Biological Systems Vol. 6, Biological Action of Metal Ions.* Marcel Dekker, New York, pp. 141–196.

Bonham, L. C. (1956). Geochemical investigation of crude oils. *Bull. Am. Assoc. Pet. Geol.* **40**, 897–908.

Boström, K., and Fisher, D. E. (1971). Volcanogenic uranium, vanadium and iron in Indian Ocean sediments. *Earth Planet. Sci. Lett.* **11**, 95–98.

Bradford, G. R., Bair, F. L., and Hunsaker, V. (1968). Trace and major element content of 170 High Sierra Lakes in California. *Limnol. Oceanogr.* **13**, 526–530.

Bramlette, M. N. (1965). Massive extinctions in biota at the end of Mesozoic time. *Science* **148**, 1696–1699.

Breit, G. N. (1992). Vanadium—Resources in fossil fuels. In J. D. Young, Jr., and J. Hammerstrom, (Eds.), *Contributions to Commodity Geology Research. U.S. Geol. Surv. Bull.,* 1877-K, pp. 1–8.

Breit, G. N., and Wanty, R. B. (1991). Vanadium accumulation in carbonaceous rocks: A review of geochemical controls during deposition and diagenesis. *Chem. Geol.* **91**, 83–97.

Brooks, R. R., Lee, J., Reeves, R. D., and Jaffre, T. (1977). Detection of nickeliferous rocks by analysis of Herbarium specimens of indicator plants. *J. Geochem. Explor.* **7**, 49–57.

Brown, G. M., Holland, J. G., Sigurdsson, H., Tomblin, J. F., and Arculus, R. J. (1977). Geochemistry of the Lesser Antilles volacanic island arc. *Geochim. Cosmochim. Acta* **41**, 785–801.

Brumsack, H. J. (1980). Geochemistry of Cretaceous black shales from the Atlantic Ocean. *Chem. Geol.* **31**, 1–25.

Brumsack, H. J., and Gieskes, J. M. (1983). Interstitial water trace-metal chemistry of laminated sediments from the Gulf of California, Mexico. *Mar. Chem.* **14**, 89–106.

Burton, J. D. (1966). Some problems concerning the marine geochemistry of vanadium. *Nature* **212**, 976–978.

Calvert, S. E., Bustin, R. M., and Ingall, E. D. (1996). Influence of water column anoxia and sediment supply on the burial and preservation of organic carbon in marine shales. *Geochim. Cosmochim. Acta* **60**, 1577–1593.

Cannon, H. L. (1963). The biogeochemistry of vanadium. *Soil Sci.* **96**, 196–204.

Cantley, Jr., L. C., Josephson, L., Warner, R., Yanagisawa, M., Lechene, C., and Guidotti, G. (1977). Vanadate is a potent (Na, K)-ATPase inhibitor found in ATP derived from muscle. *J. Biol. Chem.* **252**, 7421–7423.

Cloud, Jr., P. E. (1959). Paleoecology—Retrospect and prospect. *J. Paleon.* **33**, 926–962.

Collier, R. W. (1984). Particulate and dissolved vanadium in the North Pacific Ocean. *Nature* **309**, 441–444.

Colombo, U., and Sironi, G. (1961). Geochemical analysis of Italian oils and asphalts. *Geochim. Cosmochim. Acta* **25,** 24–51.

Coney, P. J. (1976). Plate tectonics and the Laramide Orogeny. In L.A. Woodward and S.A. Northrop (Eds.), *Tectonics and Mineral Resources of Southwestern North America.* New Mexico Geol. Soc., Special Publ., No. 6, pp. 5–10.

Cowgill, U. M. (1973). The determination of all detectable elements in the aquatic plants of Linsley pond and Cedar Lake (North Bradford, Connecticut) by X-ray emission and optical emission spectroscopy. *Appl. Spectrosc.* **27,** 5–9.

Cox, K. G. (1988). Gradual volcanic catastrophes? *Nature* **333,** 802.

Dafnis, E., Spohn, M., Lonis, B., Kurtzman, N. A., and Sabatini, S. (1992). Vanadate causes hypokalemic distal renal tubular acidosis. *Am. J. Physiol.* **262,** F449–F453.

Da Silva, J. J. R. F., and Williams, R. J. P. (1991). *The Biological Chemistry of the Elements: The Inorganic Chemistry of Life.* Clarendon Press, Oxford, pp. 411–435.

Didyk, B. M., Simoneit, B. R., Brassell, S. C., and Eglinton, G. (1978). Organic geochemical indicators of paleoenvironmental conditions of sedimentation. *Nature* **272,** 216–222.

Dingley, A. L., Kustin, K., Macara, I. G., and McLeod, G. C. (1981). Accumulation of vanadium by tunicate blood cells occurs via a specific anion transport system. *Biochim. Biophys. Acta* **649,** 493–502.

Egemeier, S. J. (1981). Cavern development by thermal waters. *Natl. Speleol. Soc. Bull.* **43,** 31–51.

Emerson, S. R., and Huested, S. S. (1991). Ocean anoxia and the concentrations of molybdenum and vanadium in seawater. *Mar. Chem.* **34,** 177–196.

Engel, C. G., and Engel, A. E. J. (1963). Basalts dredged from the Northeastern Pacific Ocean. *Science* **140,** 1321–1324.

Ewart, A., Bryan, W. B., and Gill, J. B. (1973). Mineralogy and geochemistry of the younger volcanic islands of Tonga, S. W. Pacific. *J. Petrol.* **14,** 429–465.

Fenner, P., and Hagner, A. F. (1967). Correlation of variations in trace elements and mineralogy of the Esopus Formation, Kingston, New York. *Geochim. Cosmochim. Acta* **31,** 237–261.

Frankenberger, A., Brooks, R. R., Alvarez, H. V., Collen, J. D., Filby, R. H., and Fitzgerald, S. L. (1994). Classification of some New Zealand crude oils and condensates by means of their trace element contents. *Appl. Geochem.,* **9,** 65–71.

French, R. J., and Jones, P. J. H. (1993). Role of vanadium in nutrition: Metabolism, essentiality and dietary considerations. *Life Sci.* **52,** 339–346.

Gilmore, J. S., Knight, J. D., Orth, C. J., Pillmore, C. L., and Tschudy, R. H. (1984). Trace element patterns at a non-marine Cretaceous-Tertiary boundary. *Nature* **307,** 224–228.

Given, P. H., and Miller, R. N. (1987). The association of major, minor and trace inorganic elements with lignites. III. Trace elements in four lignites and general discussion of all data from this study. *Geochim. Cosmochim. Acta* **51,** 1843–1853.

Glikson, M., Chappell, B. W., Freeman, R. S., and Webber, E. (1985). Trace elements in oil shales, their source and organic association with particular reference to Australian deposits. *Chem. Geol.,* **53,** 155–174.

Goldberg, E. D., and Arrhenius, G. O. S. (1958). Chemistry of Pacific pelagic sediments. *Geochim. Cosmochim. Acta* **13,** 153–212.

Hamada, T. (1992). Problems of trace metals (Japanese). In *Renal Nutrition.* Special Issue of *Kidney and Dialysis,* Vol. 33, Tokyo Igakusha, Tokyo, pp. 474–479.

Hamada, T. (1994a). Vanadium induced hemolysis of vitamin E deficient erythrocytes in Hepes buffer. *Experientia* **50,** 49–53.

Hamada, T. (1994b). A new experimental system of using fertile chick eggs to evaluate vanadium absorption and antidotal effectiveness to prevent vanadium uptake. *J. Nutr. Biochem.* **5,** 382–388.

Hamada, T. (1995). Antioxidant and prooxidant roles of copper in Tween 20-induced hemolysis of hamster and pig erythrocytes containing marginal vitamin E. *Experientia* **51,** 572–575.

Hasting, D., Emerson, S., Mix, A., and Nelson, B. (1990). Vanadium incorporation into planktonic foraminifera as a tracer for the extent of anoxic bottom water (abstract). *Eos Trans. Am. Geophys. Union* **11,** 1351.

Hawkins, C. J., Kott, P., Parry, D. L., and Swinehart, J. H. (1983). Vanadium content and oxidation state related to ascidian phylogeny. *Comp. Biochem. Physiol.* **76B,** 555–558.

Heggie, D., Kahn, D., and Fischer, K. (1986). Trace metals in metalliferous sediments, MANOP site M: Interfacial pore water profiles. *Earth Planet. Sci. Lett.* **80,** 106–116.

Hirst, D. M. (1962). The geochemistry of modern sediments from the Gulf of Paria—II. The location and distribution of trace elements. *Geochim. Cosmochim. Acta* **26,** 1147–1187.

Hodgson, G. W. (1954). Vanadium, nickel, and iron trace metals in crude oils of western Canada. *Bull. Am. Assoc. Pet. Geol.* **38,** 2537–2554.

Hodkinson, R., Cronan, D. S., Glasby, G. P., and Moorby, S. A. (1986). Geochemistry of marine sediments from the Lau Basin, Havre Trough, and Tonga-Kermadec Ridge. *N. Z. J. Geol. Geophys.* **29,** 335–344.

Holland, H. D. (1979). Metals in black shales—A reassessment. *Econ. Geol.* **74,** 1676–1680.

Hooper, P. R. (1988). The Columbia River Basalt. In J. D. Macdougall (Ed.), *Continental Flood Basalts.* Kluwer Academic Publishers, Dordrecht, pp. 1–33.

Hoshino, M. (1962). History of Suruga Bay (Japanese). In Marine Mus. Tokai Univ. (Ed.), *Nature of Suruga Bay.* Shizuoka Education Publ., Shizuoka-shi, pp. 14–23.

Humphris, S. E., and Thompson, G. (1978). Trace element mobility during hydrothermal alteration of oceanic basalts. *Geochim. Cosmochim. Acta* **42,** 127–136.

Hutchinson, G. E. (1975). *A Treatise on Limnology.* Vol. III: *Limnological Botany.* John Wiley & Sons, New York, pp. 325–327.

Ikeda, K. (1982). A study of chemical characteristics of ground water in Fuji area (Japanese). *J. Groundwater Hydrol.* **24,** 77–93.

Ishii, T., Otake, T., Okoshi, K., Nakahara, M., and Nakamura, R. (1994). Intracellular localization of vanadium in the fan worm *Pseudopotamilla occelata. Mar. Biol.* **121,** 143–151.

Ishikawa, H. (1968). Some aspects of geochemical trends and fields of the ratios of vanadium, nickel and cobalt. *Geochim. Cosmochim. Acta* **32,** 913–917.

Jansson-Charrier, M., Guibal, E., Roussy, J., Delanghe, B., and Cloirec, P. L. (1996). Vanadium(IV) sorption by chitosan: Kinetics and equilibrium. *Water Res.* **30,** 465–475.

Juichang, R., Freedman, B., Coles, C., Zwicker, B., Holzbecker, J., and Chatt, A. (1995). Vanadium contamination of lichens and tree foliage in the vicinity of three oil-fired plants in eastern Canada. *J. Air Waste Manage. Assoc.* **45,** 461–464.

Katsura, T. (1956). Geochemical studies of Japanese volcanoes (No. 33). Vanadium content of volcanic rocks in Fuji Volcanic Belt (Japanese). *Nippon Kagaku Zasshi* **77,** 358–363.

Knauss, K., and Ku, T-L. (1983). The elemental composition and decay-series radionuclide content of plankton from the East Pacific. *Chem. Geol.* **39,** 125–145.

Konagaya, T., Nakamura, Y., and Tsukui, F. (1984). *Report on Propagation of S. lucens* (Japanese). Shizuoka Prefecture, pp. 1–106.

Krauskopf, K. B. (1956). Factors controlling the concentrations of thirteen rare metals in seawater. *Geochim. Cosmochim. Acta* **9,** 1–32B.

Kubota, J. (1980). Regional distribution of trace element problems in North America. In B. E. Davies, (Ed.), *Applied Soil Trace Elements.* John Wiley & Sons, New York, pp. 441–466.

Kuroda, K., (1942). Vanadin-, Chrom- und Molybdängehalt einiger Mineralquellen Japans. *Bull. Chem. Soc. Jpn.* **17,** 213–215.

Lee, K. (1983). Vanadium in the aquatic ecosystem. In J. Nriagu, (Ed.), *Aquatic Toxicology.* John Wiley & Sons, New York, Vol. 13, pp. 155–187.

Le Riche, H. H. (1959). The distribution of certain trace elements in the Lower Lias of southern England. *Geochim. Cosmochim. Acta* **16**, 101–122.

Lewan, M. D. (1984). Factors controlling the proportionality of vanadium to nickel in crude oils. *Geochim. Cosmochim. Acta* **48**, 2231–2238.

Lewan, M. D. and Maynard, J. B. (1982). Factors controlling enrichment of vanadium and nickel in the bitumen of organic sedimentary rocks. *Geochim. Cosmochim. Acta* **46**, 2547–2560.

Linstedt, K. D., and Kruger, P. (1969). Vanadium concentrations in Colorado river basin waters. *J. Am. Water Works Assoc.* **61**, 85–88.

Linstedt, K. D., and Kruger, P. (1970). Determination of vanadium in natural waters by neutron activation analysis. *Anal. Chem.* **42**, 113–115.

Lipman, P. W., Prostka, H. J., and Christiansen, R. L. (1971). Evolving subduction zones in the western United States, as interpreted from igneous rocks. *Science* **174**, 821–825.

Mahoney, J. J. (1988). Deccan traps. In J. D. Macdougall (Ed.), *Continental Flood Basalts*. Kluwer Academic Publications, Dordrecht, pp. 151–194.

Martin, J. H., and Knauer, G. A. (1973). The elemental composition of plankton. *Geochim. Cosmochim. Acta* **37**, 1639–1653.

McKelvey, V. E., Strobell Jr., J. D., and Slaughter, A. L. (1986). The vanadiferous zone of the Phosphoria Formation in western Wyoming and southeastern Idaho. *U.S. Geol. Survey, Prof. Pap.*, **1465**, 1–27.

Miyaji, N., Endo, K., Togashi, S., and Uesugi, Y. (1992). Tephrochronological history of Mt. Fuji. In *29th IGC Field Trip Guide Book, Volcanoes and Geothermal Fields of Japan*. Geological Survey of Japan, Tsukuba, Vol. 4, pp. 75–109.

Miyashiro, A. (1975). Origin of the troodos and other ophiolites: A reply to Hynes. *Earth Planet. Sci. Lett.* **25**, 217–222.

Nadeau, P. H., and Reynolds Jr., R. C. (1981). Volcanic components in pelitic sediments. *Nature* **294**, 72–74.

Nicholls, G. D., and Loring, D. H. (1962). The geochemistry of some British carboniferous sediments. *Geochim. Cosmochim. Acta* **26**, 181–223.

Nicolli, H. B., Suriano, J. M., Peral, M. A. G., Ferpozzi, L. H., and Baleani, O. A. (1989). Groundwater contamination with arsenic and other trace elements in an area of the pampa province of Cordoba, Argentina. *Environ. Geol. Water Sci.* **14**, 3–16.

Nilwarangkur, S., Nimmannit, S., Chaovakul, V., Susaengrat, W., Ong-Aj-Yooth, S., Vasuvattakul, S., Pidetcha, P., and Malasit, P. (1990). Endemic primary distal renal tubular acidosis in Thailand. *Q. J. Med., New Ser.* **74**, 289–301.

Nojiri, Y., Kawai, T., Otsuki, A., and Fuwa, K. (1985). Simultaneous multielement determination of trace metals in lake waters by ICP emission spectrometry with preconcentration and their background levels in Japan. *Water Res.* **19**, 503–509.

Ochiai, T., and Kawasaki, H. (1970). Behavior of groundwater flowing in lava beds (Japanese). *Bull. Agr. Eng. Res. Sta. Jpn.* **8**, 67–83.

Officer, C. B., and Drake, C. L. (1983). The Cretaceous-Tertiary transition. *Science* **219**, 1383–1390.

Officer, C. B., and Drake, C. L. (1985). Terminal Cretaceous environmental events. *Science* **227**, 1161–1167.

Okabe, S., and Morinaga, T. (1968). Vanadium and molybdenum in the river and estuary waters which pour into the Suruga Bay, Japan (Japanese). *Nippon Kagaku Zasshi* **89**, 284–287.

Okabe, S., Shibasaki, M., Oikawa, T., Kawaguchi, Y., and Nihongi, H. (1981). Geochemical studies of spring and lakewaters on and around Mt. Fuji (1) (Japanese). *J. Fac. Mar. Sci. Technol. Tokai Univ.* **14**, 81–105.

Omori, M., Ukishima, Y., and Muranaka, F. (1988). New record of occurrence of *Sergia lucens* (Hansen) (Crustacea, Sergetidae) off Tung-kang, Taiwan, with special reference to phylogeny and distribution of the species (Japanese). *J. Oceanogr. Soc. Jpn.* **44**, 261–267.

Orvini, E., Lodola, L., Sabbioni, E., Pietra, R., and Goetz, L. (1979). Determination of the chemical forms of dissolved vanadium in freshwater as determined by ^{48}V radiotracer experiments and neutron activation analysis. *Sci. Total Environ.* **13**, 195–207.

Parker, R. D. R., Sharma, R. P., and Miller, G. W. (1978). Vanadium in plants, soils and water in the Rocky Mountain region and its relationship to industrial operations. *Trace Subst. Environ. Health* **12**, 340–350.

Patterson, J. H., Ramsden, A. R., Dale, L. S., and Fardy, J. J. (1986). Geochemistry and mineralogical residence of trace elements in oil shales from Juria Creek, Queensland, Australia. *Chem. Geol.* **55**, 1–16.

Pedersen, T. F., and Calvert, S. E. (1990). Anoxia vs. productivity: What controls the formation of organic-carbon-rich sediments and sedimentary rocks? *Am. Assoc. Pet. Geol. Bull.* **74**, 454–466.

Phillips, A. (1918). A possible source of vanadium in sedimentary rocks. *Am. J. Sci.* **46**, 473–475.

Phillips, T. D., Nechay, B. R., and Heidelbaugh, N. D. (1983). Vanadium: Chemistry and the kidney. *Fed. Proc.* **42**, 2969–2973.

Premovic, P. I., Pavlovic, M. S., and Pavlovic, N. Z. (1986). Vanadium in ancient sedimentary rocks of marine origin. *Geochim. Cosmochim. Acta* **50**, 1923–1931.

Rehder, D. (1992). Structure and function of vanadium compounds in living organisms. *BioMetals* **5**, 3–12.

Robinson, J. M. (1990). Lignin, land plants, and fungi: Biological evolution affecting Phanerozoic oxygen balance. *Geology* **15**, 607–610.

Sandler, A., Brenner, I. B., and Halicz, L. (1988). Trace element distribution in waters of the northern catchment area of Lake Kinneret, Northern Israel. *Environ. Geol. Water Sci.* **11**, 35–44.

Sato, Y., and Okabe, S. (1978). Vanadium in sea waters and deposits from Tokyo Bay, Suruga Bay and Harima Nada (Japanese). *Mem. Tokai Univ. Oceanogr. Dep.* **11**, 1–19.

Seyfried Jr., W. E., and Mottl, M. J. (1982). Hydrothermal alteration of basalt by seawater under seawater-dominated conditions. *Geochim. Cosmochim. Acta* **46**, 985–1002.

Shiller, A. M., and Boyle, E. A. (1987). Dissolved vanadium in rivers and estuaries. *Earth Planet Sci. Lett.* **86**, 214–224.

Smith, M. J. (1989). Vanadium biochemistry: The unknown role of vanadium-containing cells in ascidians (sea squirts). *Experientia* **45**, 452–457.

Söremark, R. (1967). Vanadium in some biological specimens. *J. Nutr.* **92**, 183–190.

Stein, R., Rullkötter, J., and Welte, D. H. (1986). Accumulation of organic-rich sediments in the Late Jurassic and Cretaceous Atlantic Ocean—A synthesis. *Chem. Geol.,* **56**, 1–32.

Strain, W. H., Varnes, A. W., Drenski, T. L., Paxton, C. A., and McKinney, B. M. (1982). Vanadium content of drinking water. *Trace Subst. Environ. Health* **16**, 331–337.

Sugawara, K., Naito, H., and Yamada, S. (1956). Geochemistry of vanadium in natural waters. *J. Earth Sci. Nagoya Univ.* **4**, 44–61.

Sugawara, K., Yoshiwara, T., Yanagi, K., and Ambe, M. (1980). A new chemical approach to the study of waters in the Mt. Fuji environs. Abstr. *21st Assembly of Societas Internationalis Limnologiae (Kyoto),* 1–7, pp. 166.

Sugimura, Y., Suzuki, Y., and Miyake, Y. (1978). Chemical forms of minor metallic elements in the ocean. *J. Oceanogr. Soc. Jpn.* **34**, 93–96.

Sugiyama, M. (1989). Seasonal variation of vanadium concentration in Lake Biwa, Japan. *Geochem. J.* **23**, 111–116.

Suzuki, T., Hirano, T., and Oki, Y. (1984). Water-chemistry of Fuji five lakes (Japanese). *Bull. Hot Springs Res. Inst. Kanagawa Prefecture* **15**, 107–114.

Szalay, A. (1964). Cation exchange properties of humic acids and their importance in the geochemical enrichment of UO_2^{++} and other cations. *Geochim. Cosmochim. Acta* **28**, 1605–1614.

Szalay, A., and Szilagyi, M. (1967). The association of vanadium with humic acids. *Geochim. Cosmochim. Acta* **31**, 1–6.

Tajiri, M., and Hirasawa, Y. (1985). Vanadium retention and its possible toxic effects on membrane transports in uremic patients (Japanese). *Jpn. J. Nephrol.* **27**, 499–504.

Takahashi, M., Hasegawa, Y., Tsukui, M., and Nemoto, Y. (1991). Evolution of the magma-plumbing system beneath Fuji Volcano: From the viewpoint of whole-rock chemistry (Japanese). *Kazan (Bull. Volcano. Soc. Jpn.),* **36**, 281–296.

Thomson, J., Carpenter, M. S. N., Colley, S., Wilson, T. R. S., Elderfield, H., and Kennedy, H. (1984). Metal accumulation rates in northwest Atlantic pelagic sediments. *Geochim. Cosmochim. Acta* **48**, 1935–1948.

Togashi, S., Miyaji, N., and Yamazaki, H. (1991). Fractional crystallization in a large tholeiitic magma chamber during the early stage of the Younger Fuji Volcano, Japan (Japanese). *Kazan (Bull. Volcano. Soc. Jpn.)* **36**, 269–280.

Tourtelot, H. A. (1964). Minor-element composition and organic carbon content of marine and nonmarine shales of Late Cretaceous age in the western interior of the United States. *Geochim. Cosmochim. Acta* **28**, 1579–1604.

Tsukamoto, Y., Saka, S., Kumano, K., Iwanami, S., Ishida, O., and Marumo, F. (1990). Abnormal accumulation of vanadium in patients on chronic hemodialysis. *Nephron* **56**, 368–373.

Tsumura, A., and Yamasaki, S. (1992). Direct determination of rare earth elements and actinoids in fresh water by double-focussing and high resolution ICP-MS. *Radioisotopes* **41**, 185–192.

Tsuya, H. (1971). A report on topography and geology of Mt. Fuji (Japanese). In *Mt. Fuji Research Report.* Fuji Express, Tokyo, pp. 1–109.

Tsuya, H. (1988). *Geological Map of Mt. Fuji.* 2nd ed. Geological Survey of Japan, Tsukuba-shi.

Vine, J. D., and Tourtelot, E. B. (1970). Geochemistry of black shale deposits—A summary report. *Econ. Geol.* **65**, 253–272.

Vinogradov, A. P. (1953). *The Elementary Chemical Composition of Marine Organisms.* Memoir II. Sears Foundation for Marine Research, Yale University, New Haven, pp. 418–429.

Vogt, P. R. (1972). Evidence for global synchronism in mantle plume convection, and possible significance for geology. *Nature* **240**, 338–342.

Wanty, R. B., and Goldhaber, M. B. (1992). Thermodynamics and kinetics of reactions involving vanadium in natural systems: Accumulation of vanadium in sedimentary rocks. *Geochim. Cosmochim. Acta* **56**, 1471–1483.

Wedepohl, K. H. (1964). Untersuchungen am Kupferschiefer in Nordwestdeutschland: Ein Beitrag zur Deutung der Genese bituminöser Sedimente. *Geochim. Cosmochim. Acta* **28**, 305–364.

Wedepohl. K. H. (1971). Environmental influences on the chemical compositions of shales and clays. In L. H. Ahrens, F. Press, S. K. Runcorn, and H. C. Urey (Eds.), *Physics and Chemistry of the Earth.* Pergamon Press, Oxford, Vol. 8, pp. 307–333.

Wehrli, B., and Stumm, W. (1989). Vanadyl in natural waters; adsorption and hydrolysis promote oxygenation. *Geochim. Cosmochim. Acta* **53**, 69–77.

Wever, R., Tromp, M. G. M., Krenn, B. E., Marjani, A., and Tol, M. V. (1991). Brominating activity of the seaweed *Ascophyllum nodosum:* Impact on the biosphere. *Environ. Sci. Technol.* **25**, 446–449.

Yamasaki, S., and Tsumura, A. (1992). Determination of ultra-trace levels of elements in water by high resolution ICP-MS with an ultrasonic nebulizer. *Water Sci. Technol.* **25**, 205–212.

Yamazaki, H. (1992). Tectonics of a plate collision along the northern margin of Izu Peninsula, central Japan. *Bull. Geol. Survey Jpn.* **43**, 603–657.

Yoshida, T., Yamasaki, S., and Tsumura, A. (1992). Determination of trace and ultra-trace elements in 32 international geostandards by ICP-MS. *J. Min. Pet. Econ. Geol.* **87,** 107–122.

Zhang, A. (1987). Fossil appendicularians in the early Cambrian. *Sci. Sin.* **30,** 888–896.

Zoller, W. H., Parrington, J. R., and Kotra, J. M. P. (1983). Iridium enrichment in airborne particles from Kilauea Volcano: January 1983. *Science* **222,** 1118–1121.

Zubovic, P., Stadnichenko, T., and Sheffey, N. B. (1961). Geochemistry of minor elements in coals of the Northern Great Plains coal province. *U.S. Geol. Survey Bull.* **1117-A,** 1–57.

6

WATER QUALITY CRITERIA FOR VANADIUM WITH REFERENCE TO IMPACT STUDIES ON THE FRESHWATER TELEOST *Nuria denricus* (HAMILTON)

S.A. Abbasi

Centre for Pollution Control and Energy Technology, Pondicherry University, Kalabet, Pondicherry–605 014, India

Vanadium in the Environment. Part 1: Chemistry and Biochemistry, Edited by Jerome O. Nriagu.
ISBN 0-471-17778-4.   © 1998 John Wiley & Sons, Inc.

1. INTRODUCTION

Vanadium is an ubiquitous element and is more widely dispersed than such essential ones as zinc, copper, molybdenum, and cobalt (Soni, 1990). It has a mean geochemical abundance of 150 ppm and is present in rocks and soils at levels ranging from 5 to 250 ppm and from 20 to 500 ppm, respectively (Thornton, 1983; Adriano, 1980). Vanadium is therefore more readily accessible to most plants and animals. However, although the role of vanadium as an essential micronutrient has still to be established, there is little doubt that this element can be toxic to biological systems. There is increasing evidence of the damage caused to plants and animals through exposure to vanadium (Morrell et al., 1986; Abbasi and Soni, 1987). There is also evidence of a significant and increasing release of vanadium into the biosphere through industrial emmissions and the burning of fossil fuels (Abbasi et al, 1987). Vanadium levels in unpolluted freshwaters are of the order 2×10^{-4} ppm (Abaychi & Dou Abul, 1985; Borg, 1987), but abnormally high levels of this element have been reported in waters, plants, and animals from sites affected by industrial pollution (Kovacs et al., 1985; Dissanayake and Weerassoriya, 1986). The reported levels of vanadium in polluted natural waters (average 1.5, maximum 7.4 ppm) (Nojiri et al., 1985; Borg, 1987) are far higher than the concentration of 10^{-7} M (5.1 ppb) that is thought to be physiologically significant or the concentration of 0.5 ppm that has been found to be toxic to fish (Soni, 1990). The present study was therefore undertaken with a threefold objective: to determine the relative toxicity of vanadium; to review the existing irrigation water and effluent discharge standards for the metal; and to evaluate the safe level of the metal. The last-named objective is essentially to help in evolution of a realistic water quality criterion for safe environmental management of vanadium.

2. MATERIALS AND METHODS

2.1. Dilution Water

Filtered well water was used in the bioassays. The quality of the test water was determined periodically according to standard methods. Only slight variation in water quality was observed during the course of the experiment (Table 1). The well water presumably did not contain vanadium, as the solvent extraction–spectrophotometry and atomic absorption spectrometry methods (Abbasi, 1987; Rand et al., 1989) (sensitivity 0.001 mole/L^{-1}) failed to detect the presence of this metal.

2.2. Metal Solution

Stock solutions of vanadium were prepared by dissolving reagent-grade ammonium metavanadate in distilled water.

Table 1   Characteristics of the Water Used in the Bioassays

Parameter	Range
Electrical conductivity (μS/cm)	30–32
pH	6.2–6.5
Alkalinity (mg/L)	3.5–6.5
Total hardness (mg/L)	4.5–5.5
Calcium hardness (mg/L)	2.0–3.0
Sulfate (mg/L)	Absent (<1.0)
Chlorides (mg/L)	11.2
Nitrate (mg/L)	Absent (<0.05)
Nitrite (mg/L)	Absent (<0.05)
Iron (mg/L)	0.31
Bicarbonate (mg/L)	5.8
Total dissolved solids (mg/L)	86.1
Appearance	Clear

2.3. Bioassays

Healthy, adult *Nuria denricus* (average length 5.1 cm, average weight 498 mg) were collected from a freshwater pond and were acclimated for 2 weeks. No mortality was observed during acclimation. Subsequently, batches of 40 organisms were randomly picked and released into glass aquaria containing 40 L of water. The aquaria, in duplicate, were treated with metal vanadium concentrations varying from 0 to 20 mg/L by adding measured volumes of stock solution plus dilution water to achieve the desired overall concentrations in the aquaria. The rest of the bioassay was carried out in accordance with standard methods (Rand et al., 1989). Each test aquarium was observed continuously for the first day (when mortality in higher concentrations was high) and subsequently four times a day. The organisms were provided with an adequate supply of food and air throughout the period of acclimation and metal exposure. The median lethal level (LC_{50}) values were calculated by a computer-aided analysis of mortality data as detailed earlier (Abbasi and Soni, 1984, 1986, 1987; Abbasi et al., 1988).

3. RESULTS

The LC_{50} values of vanadium and its lower and upper limits at 95% confidence levels for *N. denricus* are presented in Table 2. The estimated maximum allowable toxicant concentration (MATC) and safe concentration (SC) of the metal with existing water quality standards for effluent disposal and irrigation set by the Indian Standards Institution (ISI, 1981, 1982), and the United States Department of Interior (USDI, 1968) are presented in Table 3.

Table 2   Median Lethal Concentration (LC_{50}) of Vanadium for *N. denricus*

No. of hours	LC_{50} (mg/L)	95% Confidence Level	
		Lower	Upper
24	13.30	12.23	14.24
96	2.60	2.08	3.13
288	1.0	0.79	1.26
336	0.85	0.69	1.03
456	0.80	0.67	0.97
480	0.78	0.66	0.92

4. DISCUSSION

Vanadium (above 0.5 ppm) induced marked changes in the swimming and balancing patterns of the fishes. The stressed fish exhibited erratic and rapid twisting movements, increased frequency of surface breathing, and vertical swimming, compared with the smooth and predominantly horizontal swimming of fishes in the control aquaria. A marked effect on feeding was observed above 1.0 mg/L vanadium. Fishes exposed to 0–1/mg/L vanadium concentrations consumed food completely, whereas in 5–20 mg/L vanadium food consumption was lower than normal: a clear trend towards loss in food consumption with increase in metal concentrations. The stress gradually led to sluggishness of the animals, and death.

A comparison of LC_{50} levels for vanadium for exposures of 96–480 h (Tables 2 and 3) with the allowable limit for discharge of this metal in irrigation waters for long-term and short-term use (USDI 1968; ISI 1981) indicates that the prevailing standards are approximately 3.9–12.9 times higher. The freshwater teleost *N. denricus* reaches paddy fields along irrigation canals. In addition many important edible catfishes such as *Channa* sp, *Clarias batrachus,* and *Heteropneustus fossilis* also dwell in the paddy fields, and during rains frogs spawn in these confined waters. The present disparity between safe

Table 3   Comparison of Estimated Safe Levels of Vanadium for *Nuria denricus* and Water Quality Standards

Findings/standards	Vanadium (mg/L)
1. Maximum allowable toxicant concentration (MATC)	Less than 0.5
2. Safe concentration (SC)	0.015
3. Effluent discharge standards (ISI, 1982)	Not available
4. Irrigation Water Standards	
1. Long-term use (USDI, 1968; ISI, 1982)	10.0
2. Short-term use (USDI, 1968; ISI, 1981)	10.0

concentrations and preveiling irrigation water standards for vanadium indicates that these levels may adversely affect the normal health and population levels of aquatic animals dwelling in the irrigation channels and paddy fields.

4.1 Significance of the Findings in Selecting Methods of Treatment of Vanadium

The most common methods of vanadium removal from effluent are coagulation, precipitation, ion exchange, distillation, and electrodyalysis (Gillies, 1978). These methods can maximally remove 97% of vanadium from the effluents. The average concentration of vanadium in mining and industrial effluent is 34–112 mg/L (Jayasheela et al., 1984). After treatment by the above-mentioned methods the effluent may still contain 1–3.5 mg/L of vanadium, which is several times higher than estimated safe levels of vanadium. This indicates that for vanadium more efficient methods of removal from industrial and mining effluents need to be developed.

REFERENCES

Abaychi, J. K., and Dou Abul, A. A. Z. (1985). Trace metals in Shat-al-Arab river, Iraq. *Water Res.* **19,** 457–462.

Abbasi, S. A. (1987). Trace analysis of vanadium in environment as its ternary complex. *Anal. Lett.* **20,** 1347–1362.

Abbasi, S. A., Nipaney, P. C., and Soni, R. (1988). Studies on environmental management of mercury(II), chromium(VI) and zinc(II) with respect to the impact on some arthropods and protozoans—Toxicity of zinc(II). *Int. J. Environ. Studies* **32,** 181–187.

Abbasi, S. A., and Soni R. (1984). Lower-than-permissible toxicity levels of chromium(VI). *Environ. Pollut.* **A35,** 75–82.

Abbasi, S. A., and Soni, R. (1986). An examination of environmentally safe levels of cadmium(II), zinc(II), and lead(II) with reference to channel fish *Nuria denricus. Environ. Pollut.* **40A,** 37–51.

Abbasi, S. A., and Soni, R. (1987). Water quality criteria for the management of chromium and vanadium in irrigation waters and industrial effluents with reference to impact studies on the freshwater teleost *Nuria denricus.* In M. W. H. Chan, R. W. M. Hoare, P. R. Holmes, R. J. S. Law, and S. B. Reeds (Eds.), *Pollution in the Urban Environment.* Elsveier, London and New York, 1987, pp. 566–574.

Abbasi, S. A., Soni, R., and Kunhahmed, T. (1987). *Effect of Heavy Metals on the Growth of Black Gram,* Cicer aeretenium. Technical Report TR-21, Centre for Water Resources Development and Management, 1987, pp. 1–27.

Adriano, D. C. (1980). *Trace Elements in the Aquatic Environment.* Springer-Verlag, New York.

Borg, H. (1987). Trace metals and water chemistry of forest lakes in Northern Sweden. *Water Res.* **21,** 65–72.

Dissanayake, C. B., and Weerassoriya, S. V. R. (1986). The environmental chemistry of Mahaweli River, Sri Lanka. *Int. J. Environ. Studies,* **28,** 207–229.

Gillies, M. T. (1978). *Drinking Water Detoxification.* Noyes Data Corporation, New Jersey, p. 325.

ISI (Indian Standards Institution). (1981). *Draft Indian Standard Tolerance Limits for Industrial Effluents Used for Irrigation Purposes.* IS: 3307. Personal correspondence with ISI, 1981.

ISI (Indian Standards Institution), (1982). *Tolerance Limits for Industrial Effluents. Part I. General Limits.* IS-2490 (I) 1982. ISI, New Delhi.

Jayasheela, H. M., Itnal, M. P., Tenginakai, S. G., and Puuranik, S. C. (1984). Mechanisms controlling the chemistry of groundwaters of the Gadag Schist Belt, Karnataka. In K. A. and K. I. Vasu (Eds.), *Proceedings of Symposium on the Environmental Impact of Mining and Processing of Minerals.* Department of Metallurgy, Indian Institute of Science, Bangalore.

Kovacs, M., Nyari, I., and Toth, L. (1985). The concentration of microelements in the aquatic weeds of Lake Balaton. *Symp. Biol. Hung.* **29,** 67–81.

Morrell, B. G., Lepp, N. W. and Phipps, D. A. (1986). *Environ. Geochem. Health* **8,** 14.

Nojiri, Y., Kawai, T., Otsuki, A., and Fuwa K. (1985). Simultaneous multielemental determination of trace elements in lake waters by ICP emmission spectrometry with preconcentration and their background levels in Japan. *Water Res.* **19,** 503–509.

Rand, M. C., Greenberg, A. R., and Taras, M. J. (1989). *Standard Methods for the Examination of Water and Wastewater.* 17th APHA, Washington, DC.

Soni, R. (1990). *Studies on the Environmental Management of Heavy Metals and Pesticides with respect to Their Toxicity Towards Aquatic Organisms.* PhD thesis, University of Calicut, p. 680.

Soni, R., and Abbasi S. A. (1980) *Chromium in Biological Systems—An Indexed Bibliography.* Centre for Water Resources Development and Management, Calicut.

Thornton, I. (1983). *Applied Environmental Geochemistry.* Academic Press, London.

USDI (United States Department of Interior), (1968). *Characteristics and Pollution Problems of Irrigation Return Flow.* Federal Pollution Control Administration, U.S. GDO, Washington DC, p. 47.

7

SPECTROSCOPIC METHODS FOR THE CHARACTERIZATION OF VANADIUM COMPLEXES

Giovanni Micera and Daniele Sanna

Dipartimento di Chimica, Università di Sassari, Via Vienna 2, 07100 Sassari, Italy

Vanadium in the Environment. Part 1: Chemistry and Biochemistry, Edited by Jerome O. Nriagu. ISBN 0-471-17778-4. © 1998 John Wiley & Sons, Inc.

1. INTRODUCTION

Metal ions occur in a variety of forms in the environment. Some of them can be regarded as true coordination compounds formed by interaction with organic compounds and inorganic anions present in the aqueous phase or with complex biomolecules often exhibiting very specific binding properties. Colloidal particles are also very active in complex formation processes and can involve natural organic polymers and inorganic components, for examples, clays and (oxo)hydroxides. All these interactions provide means through which "free" metal ions are transformed into rather stable forms. The properties (solubility, transport, biological effects, etc.) of the formed complexes may be vastly different from those of the "free" metal ions themselves. On the other hand, complex formation can affect the reactivity and the structure of the bound ligands.

A major challenge of environmental chemistry is to establish the distribution of metal ions among the large number of available ligands, to assess the status and chemical forms of trace elements, and to evaluate the thermodynamics and kinetics of processes. Concepts and investigation techniques typical of metal coordination chemistry are currently applied in these studies. In this context, the spectral properties of the complexes may be useful in identifying the oxidation state of the metal, assigning structures to metal sites, characterizing ligands as well as their coordination mode, conformation, or configuration, measuring any influences of pH, and so forth. This is a rather difficult task and often leads to only tentative assignments even when very sophisticated and expensive techniques are involved. Indeed, spectroscopic methods would measure only the effects of the complex formation, which may be rather complicated and difficult to interpret, especially when the species involve metal sites of irregular symmetry and ligands adopting various conformations. However, an advantage of the spectroscopic approach derives from the fact that it is readily applicable to noncrystalline samples and to aqueous solutions.

Very often a single technique alone is not sufficient for an unequivocal conclusion. In order to arrive at a definitive description of the structure, it may be necessary to combine information from several techniques. Distinctive spectroscopic data may be gathered for model systems. By establishing relationships between the structure and the spectral features of well-defined molecules, it may be possible to extract precise and detailed data from rather complicated natural and biological systems. In any case, in the absence of single-crystal X-ray analysis, complementary equilibrium studies, performed by potentiometric or voltamperometric techniques, are strongly recommended for a reliable assignment of complex structure even with simple ligands.

Vanadium exists in several oxidation states. Some of them, namely +3, +4, and +5, can yield stable complex species of environmental and biochemical significance. A wide range of interactions is available to the element, including redox reactions, incorporation into metalloenzymes, formation of oxo and hydroxo species, and so forth. A survey of the application of spectral techniques that are useful in the characterization of vanadium complexes is presented in this short review. Space limitations do not allow this presentation to be complete in any sense.

2. VANADIUM(III) COMPLEXES

The chemistry of vanadium(III) has been attracting much attention, especially in relation to the intracellular status of vanadium in certain species of marine organisms known as tunicates. Tunicates accumulate vanadium in unusually large quantities from sea water, where it is found predominantly in the pentavalent form, via some type of reductive process. The element is stored in the blood cells (vanadocytes) primarily in the +3 oxidation state. Lysates of blood cells rapidly turn bright red, yielding the so-called Henze's solution.

The d^2 configuration of V^{III} makes this ion difficult to study by electron paramagnetic resonance (EPR) conventional techniques. Most spectroscopic studies on this oxidation state involve electronic absorption and vibrational spectroscopic methods.

Six-coordinate octahedral compounds of V^{III} usually exhibit two electronic absorption bands corresponding to transitions from $^3T_{1g}(F)$ (in term diagram) to $^3T_{2g}$, and $^3T_{1g}(P)$ (see Scheme 1a). The third spin-allowed transition to $^3A_{2g}$ is expected at rather high energy owing to the large value of the crystal field strength for a trivalent metal ion. Indeed, the aquaion shows distinct bands at 560 nm and 390 nm (Lever, 1984).

A few tetrahedral complexes are known. For such a geometry transitions from the $^3A_2$ ground state to the $^3T_2$, $^3T_1(F)$, and $^3T_1(P)$ transitions are predicted (see Scheme 1b). The first band, split into three components because of the deviation from the regular geometry, is observed in the range 1,200–1,000 nm for $[VCl_4]^-$. The band at 670 nm contains the transitions to the $^3T_1(F)$ level (Lever, 1984).

The coordination number 7 was recently found in V(III) chelates of aminopolycarboxylic acids (Robles et al., 1993). A broad absorption between 700

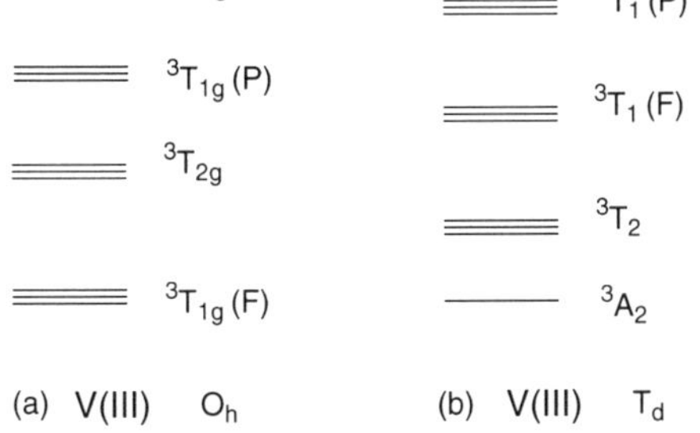

(a) V(III) O_h (b) V(III) T_d

SCHEME 1

and 800 nm is the typical feature of this coordination number (Meier et al., 1995).

Oxo- and hydroxocomplexes of V^{III} occupy an important place in the bioinorganic chemistry of vanadium(III) because of their resemblance to vanadocyte hemolysates. In fact, it is assumed that the so-called Henze's solution contains a (μ-oxo)divanadium(III) dimer that perhaps includes tunichrome (a reducing blood pigment) as a ligand. Studies of simple (μ-oxo) divanadium(III) complexes have concentrated on monobridged (μ-oxo) species, in which the V–O–V bridging angle falls in the range 165–180°, and tribridged species, in which the oxo bridge is supported by two cobridging carboxylate or phosphate groups, resulting in a V–O–V angle that falls in the range 130–145°.

A dinuclear species arising from the hydrolysis of hexaaquavanadium(III) ion above pH 1.5 exhibits an intense absorption band at 415–425 nm that resembles that of the vanadocyte hemolysates (Boeri and Ehrenberg, 1954). The absorption was previously attributed to the charge-transfer (CT) transition of a dinuclear vanadium(III) complex with a bis(μ-hydroxo) bridge (Pajdowski and Jezowska-Trzebiatowska, 1966). Oxo- or hydroxo-bridged V^{III} dimers can be satisfactorily characterized by vibrational spectroscopy, in particular by resonance Raman (RR) spectroscopy. This technique could make it possible to assign the vibrational modes proper of the bridge and to prove the nature of the intense electronic transitions. In fact, by exciting at wavelengths close to the CT transition, a strong enhancement of Raman scattering is expected for vibrational motions that are part of the chromophore.

A Raman resonance study was carried out on the dimer formed by the hydrolysis of hexaaquavanadium(III) using the excitation corresponding to the CT absorption band (Kanamori et al., 1991). The spectral features at 1,533 and 774 cm^{-1} were very similar to those measured on the μ-oxo dimer $[V_2O(ttha)]^{2-}$ (ttha = triethylenetetraaminehexaacetate), which also exhibited an intense electronic absorption at 450 nm. These observations clearly indicated that the hydrolytic dimer has a μ-oxo rather than a bis(μ-hydroxo) bridge. The bridging mode was definitively determined from the Raman spectra of the hydrolytic dimer formed in a $1:1$ $H_2{}^{16}O/H_2{}^{18}O$ mixture. For a bis(μ-hydroxo) bridge three isotopic species, $V({}^{16}OH)_2V$, $V({}^{16}OH)({}^{18}OH)V$, and $V({}^{18}OH)_2V$, would be expected in a $1:2:1$ ratio. In contrast, only two isotopic species, $V–{}^{16}O–V$ and $V–{}^{18}O–V$, in a $1:1$ ratio should be formed. Since only bands attributable to two isotopically pure species were observed, it was concluded that the hydrolytic dimer is a μ-oxo species.

Resonance Raman spectroscopy is able to detect even subtle changes in the V–O–V bridge angle of V^{III}–O–V^{III} dimers. A crystal structure analysis revealed that the complex $[V_2O(L\text{-his})_4]\cdot2H_2O$ exhibits a V–O–V angle of 153.9°, which is unusually small for an unsupported oxo bridge connecting two vanadium(III) ions (Kanamori et al., 1993). A careful reexamination of the structure (Czernuszewicz et al., 1994) indicates that the rather acute angle is due to two intramolecular hydrogen bonds between histidine ligands, which effectively assist the oxo bridge. The resonance Raman study of the complex in the solid state reveals bands at 721 and 1,437 cm^{-1}, assigned to ν_{as} and $2\nu_{as}$ (the fundamental vibration and the first overtone of the asymmetric stretch) of the V–O–V fragment, respectively. The bands shift to 734 and 1,454 cm^{-1} in H_2O (699 and 1,387 cm^{-1} in $H_2{}^{18}O$). These differences are consistent with an opening of the V–O–V angle caused by a breaking of the hydrogen bonding contacts in solution. In such a way, a linear structure is available to the V–O–V fragment, as expected for an unsupported V(III)–O–V(III) dimer. Theoretical calculations confirm the results and predict ν_{as} modes at 735 and 699 cm^{-1} for the ^{16}O- and ^{18}O-isotopomers, respectively, in nearly perfect agreement with the experimental data.

Dinuclear species with the (μ-oxo)bis(μ-carboxylato) or (μ-oxo)bis(μ-phosphato) core contain a bent V–O–V bridge of C_{2v} symmetry. Electronic and magnetostructural correlations dependent on the V–O–V angle were obtained (Knopp and Wieghardt, 1991; Bond al., 1995), which could be useful in assigning structures to complex biosystems containing the V–O–V core. Very intense electronic absorption maxima, arising from ligand-to-metal charge transitions at 660–710, 440–460, and in some cases 260–275 nm (all with $\varepsilon > 2,000$ M^{-1} cm^{-1}), dominate the UV-visible spectra of these complexes, which are of a deep green color. In some instances, it is possible to protonate the oxo ligand, forming a μ-hydroxo-bridged species. A dramatic color change from green to red occurs because the protonated species exhibit in the visible region only moderately intense maxima attributable to the d–d transitions of a V^{III} ion in an approximately octahedral environment. Therefore, the bleaching of the oxo $\rightarrow$ V^{III} CT bands in the visible region indicates well the protonation process, so that the analysis of the absorbance versus [H$^+$] allows calculation of the equilibrium constant for the reaction.

A bent M–O–M array of C_{2v} symmetry will give rise to three normal modes of vibration: symmetric (ν_s), asymmetric (ν_{as}), and bending (δ) vibrations belonging to species A$_1$, B$_2$, and A$_1$, respectively. These modes, which are both IR- and Raman-active, were examined by resonance Raman scattering with variable-wavelength excitation for the bent V–O–V bridge of (μ-oxo)bis(μ-carboxylato)- or (μ-oxo)bis(μ-phosphato)divanadium(III) complexes (Bond et al., 1995). All three vibrational modes of the bent bridge can be observed in RR spectra excited with red and violet radiation. The excitation profile of the ν_s(V–O–V) mode tracks in particular the red absorption. This proves unambiguously that most of the intensity of the strong visible absorptions is due to O $\rightarrow$ V^{III} CT transitions. In [V$_2$O(O$_2$CCH$_3$)$_2$(HB(pz)$_3$)$_2$], where HB(pz)$_3$ is hydridotris(pyrazolyl)borate, the most intense Raman band at 536 cm^{-1}, which shifts to 520 cm^{-1} upon ^{18}O isotopic substitution, is assigned to the symmetric V–O–V stretch (Bond et al., 1995). The first and second overtone vibrations of this stretching mode are also detected: $2\nu_s$ at 1,070 cm^{-1} and $3\nu_s$ at 1,606 cm^{-1}. The asymmetric mode is located at 685 cm^{-1} (fundamental) and 1,346 cm^{-1} (first overtone) on the basis of the ^{16}O $\rightarrow$ ^{18}O downshift (32 and 64 cm^{-1} for ν_s and $2\nu_s$, respectively). The substitution of acetato with diphenylphosphato in the dinuclear complex leads to a larger V–O–V angle. The pattern of vibrational frequencies and isotopic shifts (lower frequency values and smaller ^{18}O shift for ν_s but higher frequency and larger ^{18}O shift for ν_{as}) in the phosphate complex is consistent with the larger V–O–V angle observed by crystal structure determination.

3. VANADIUM(IV) COMPLEXES

3.1. Oxovanadium(IV) Complexes

3.1.1. Electronic and CD Spectroscopy

$V^{IV}O$ species exhibit very specific electronic absorption spectra that are quite distinct from those of other vanadium(IV) complexes. The bonding with the oxo ligand has a profound influence on the electronic properties of the ion and determines a peculiar sequence of energy levels. The ground state is d_{xy} and three potential intra-d-shell transitions are expected. The highest possible symmetry around the metal ion would be C_{4v}. The order established by Ballhausen and Gray (1962) for $[VO(H_2O)_5]^{2+}$ is $3d_{xy} < 3d_{xz} = 3d_{yz} < 3d_{x^2-y^2} << 3d_{z^2}$ (or $b_2 < e < b_1 < a_1$) (see Scheme 2a). The absorptions are observed at 760 and 625 nm for the $d_{xy} \rightarrow d_{xz,yz}$ (band I) and $d_{xy} \rightarrow d_{x^2-y^2}$ (band II) transitions, respectively (Bridgland et al., 1965). The transition to $3d_{z^2}$ (band III) moves to higher energy values (350–400 nm) and may be masked by charge-transfer absorptions (Selbin, 1966; Lever, 1984).

The relative energy of the b_1 ($d_{x^2-y^2}$) and e ($d_{xz,\,dyz}$) levels is mainly determined by a balance between the σ donor strength of equatorial ligands and the V–O π bonding. For $[VOCl_4]^{2-}$ bands I and II are located at 730 and 860 nm, respectively, which is the reverse of the sequence assumed for $[VO(H_2O)_5]^{2+}$ (Collison et al., 1980). Therefore, the molecular orbital approach is indispensable for treating the electronic structure of the oxovanadium(IV) complexes (Deeth, 1991).

A criterion for distinguishing five- from six-coordinate complexes, for instance $[VO(NCS)_4]^{2-}$ from $[VO(NCS)_4(H_2O)]^{2-}$, was proposed by Collison et al. (1980). For the five-coordinate complexes all of the expected d–d-type transitions occur above 330 nm. For the six-coordinate complexes only bands

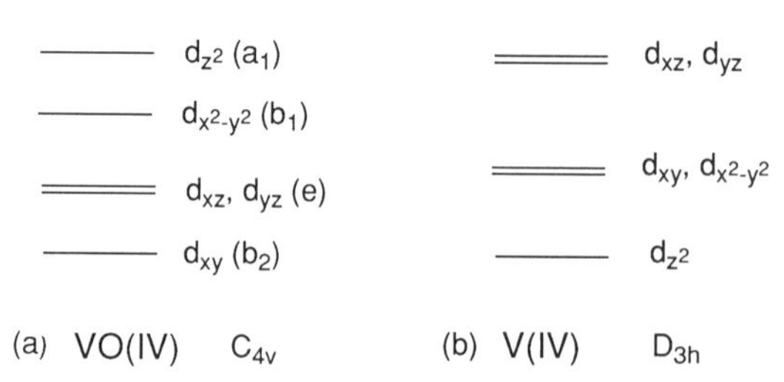

——— d_{z^2} (a$_1$)	══ d_{xz}, d_{yz}
——— $d_{x^2-y^2}$ (b$_1$)	
══ d_{xz}, d_{yz} (e)	══ d_{xy}, $d_{x^2-y^2}$
——— d_{xy} (b$_2$)	——— d_{z^2}
(a) VO(IV) C_{4v}	(b) V(IV) D_{3h}

SCHEME 2

I and II are observed above 330 nm, and band III is shifted to higher energy owing to the presence of the axial donor.

A distortion from C_{4v} symmetry to C_{2v} usually occurs, for example, in chelate complexes, and an additional $d-d$ transition arises from the splitting of e orbitals. For the *trans*-$[VOCl_2(H_2O)_2]$ C_{2v} chromophore, the level sequence $3d_{xy} < 3d_{xz} < 3d_{yz} \sim 3d_{x^2-y^2} << 3d_{z^2}$ was found, by locating the z, x, and y principal axes along the $V=O$ and nearly along the $V-OH_2$ and $V-Cl$ bonds, respectively (Azuma and Ozawa, 1994). In this case, the electron donation from the p_z orbitals of the chloro ligands to the $3d_{yz}$ orbital may elevate the $3d_{yz}$ level from that with no π donation. Since the aqua ligand has a smaller π-donating ability than the chloro ligand, the $3d_{yz} > 3d_{xz}$ sequence is possible. Therefore, three $d-d$ transitions, corresponding to $3d_{xy} \rightarrow 3d_{xz}$, $3d_{xy} \rightarrow 3d_{yz}$, and $3d_{xy} \rightarrow 3d_{x^2-y^2}$, are observed.

Four absorption bands are presented by $V^{IV}O$ complexes with very low symmetry (Chasteen et al., 1969; Micera et al., 1993). In particular, the bis-chelated complexes of α-hydroxycarboxylates with $2(COO^-, O^-)$ donor set have a trans coordination at the metal ion. A distortion toward the trigonal bipyramid was proposed so as to determine the ordering $3d_{xy} < 3d_{x^2-y^2}$ *(or $3d_{yz}$)* $< 3d_{yz}$ (or $3d_{x^2-y^2}$) $< 3d_{xz} < 3d_{z^2}$, (Chasteen et al., 1969). In a series of $V^{IV}O$ α-hydroxycarboxylates it was found that, irrespective of the ligand, three absorption bands exhibited almost the same energy values (ca. 410, 530, and 600 nm). The fourth absorption, while falling around 800–810 nm in the complexes of glycolic, lactic, and mandelic acids, was shifted to higher wavelength (ca. 850 nm) in the complexes of 2-hydroxyisobutyric, 2-hydroxy-2-methylbutyric, 2-ethyl-2-hydroxybutyric, and benzilic acids. The shift can be taken as a measure of the distortion of the coordination geometry from the square pyramid toward the trigonal bipyramid because of the substitutions at the α-C atom of the ligands (Micera et al., 1993).

The absorption spectra of dinuclear $V^{IV}O$ species formed in alkaline solutions by the D (or L) and racemic forms of the dihydroxydicarboxylic ligand tartaric acid (see below) are strikingly different (Chasteen et al., 1969). The DL complex exhibits a three-band spectrum in contrast to the four bands displayed by the DD (or LL) isomer. The finding indicates clearly that the (COO^-, O^-)-bidentate ligands are arranged trans in the DD (or LL) and cis in the DL isomer.

Many natural molecules like amino acids, peptides, or sugars are optically active species. Circular dichroism (CD) spectroscopy proves useful in the study of their $V^{IV}O$ complexes. An example is given by the complexes of lactobionic acid (LH, **1**) (Kozlowski et al., 1991), which coordinates $V^{IV}O$ through the flexible gluconic moiety. The complexation in aqueous solution starts at rather low pH (<3) with formation of $[VOL_2]$ (**2**), in which the carboxylate function acts as an anchor site toward the metal ion. The stepwise deprotonation and metal-coordination of one and two vicinal hydroxyl groups yields $[VOL_2H_{-1}]^-$ (**3**) and $[VOL_2H_{-2}]^{2-}$ (**4**), which adopt the $(COO^-, OH; COO^-, O^-)$ and $2(COO^-, O^-)$ binding modes, respectively. The formation of

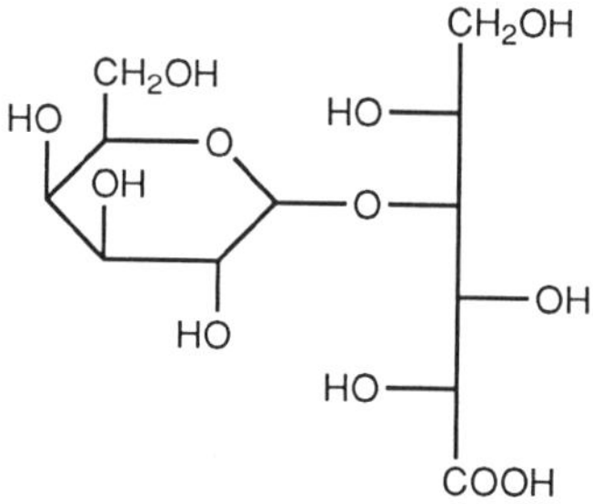

Figure 1.

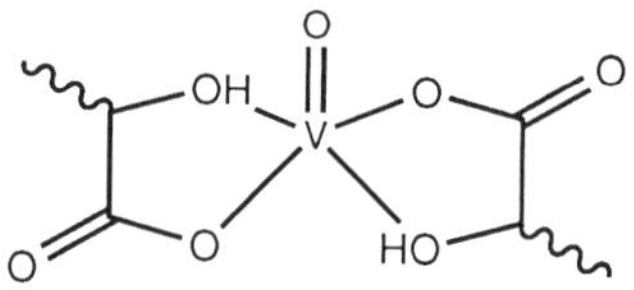

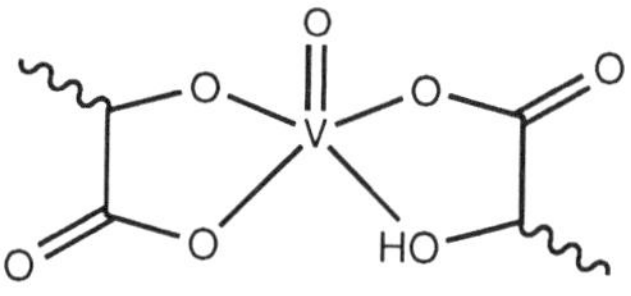

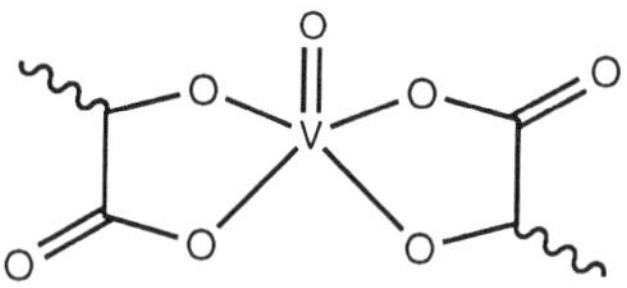

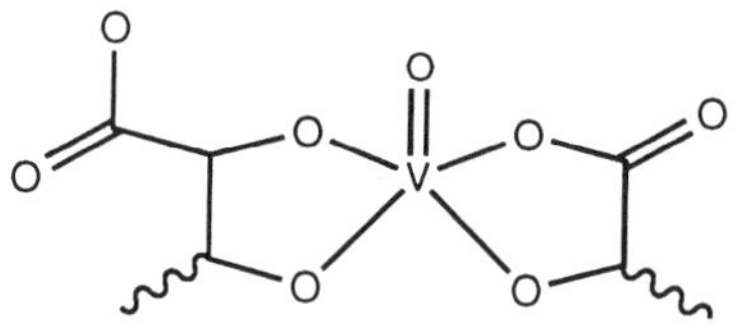

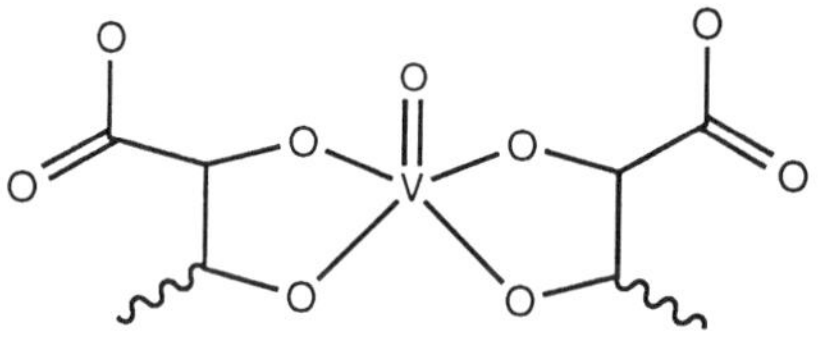

$[VOL_2H_{-3}]^{3-}$ (**5**) and $[VOL_2H_{-4}]^{3-}$ (**6**) above pH 8 suggests that the metal can induce the deprotonation of two adjacent hydroxyls in each molecule and yield the $(COO^-, O^-; O^-, O^-)$ and $2(O^-, O^-)$ donor sets in **5** and **6**, respectively. CD effects observed for the d–d absorptions (Table 1) rule out hydroxo binding and substantiate that lactobionic acid is bound to vanadium. The distinct increase of Cotton effects between pH 8 and 11 indicates only the variation of the binding sites.

3.1.2. EPR Spectroscopy

Vanadyl ion exhibits an orbitally nondegenerate ground state for which sharp and simple EPR spectra are expected. Typical eight-line patterns are observed due to ^{51}V hyperfine (hf) interaction ($I = 7/2$ and natural abundance = 99.75%). In C_{4v} symmetry, the ground state is predominantly d_{xy} and the following spectral features are predicted: $g_\| < g_\perp < 2.0023$ and $A_\| > A_\perp$, where the subscripts $\|$ and $\perp$ refer to directions that normally are parallel and perpendicular to the V=O bond, respectively. In complexes of chelated ligands, the lowering of symmetry to C_2 or C_{2v} and the mixing of d_{z^2} and $d_{x^2-y^2}$ in the ground state do not significantly affect this spectral trend (Hitchman et al., 1969).

EPR parameters are sensitive to the equatorial donors, and empirical relationships were derived for a large variety of square pyramidal and six-

Table 1 Characteristic Absorption and CD Spectra for VO(IV) Complexes of Lactobionic Acid[a]

Species	Absorptions, λ [nm], (ε)	CD, λ [nm], ($\Delta\varepsilon$)	Donor Set
$[VOL_2]$	780(21)	730(+0.02)	$2(COO^-, OH)$
	527(7)		
$[VOL_2H_{-1}]^-$	770(19)	700(+0.2)	$(COO^-, OH)(COO^-, O^-)$
	540(15)	565(−0.001)	
	395(14)	530(+0.01)	
$[VOL_2H_{-2}]^{2-}$	760(16)	780(−0.01)	$2(COO^-, O^-)$
	622(19)	640(+0.06)	
	520(20)	520(+0.025)	
	392(26)	400(+0.03)	
$[VOL_2H_{-3}]^{3-}$	782(15)	780(−0.01)	$(COO^-, O^-)(O^-, O^-)$
	625(16)	610(+0.06)	
	531(16)	560(+0.01)	
	407(24)	400(+0.03)	
$[VOL_2H_{-4}]^{4-}$	717(24)	765(−0.27)	$2(O^-, O^-)$
	720(8)	737(−0.20)	
	425(19)	615(+0.30)	
		416(+0.10)	

[a] ε and $\Delta\varepsilon$ values are expressed in M^{-1} cm^{-1}.

coordinate complexes. With increase in the equatorial donor strength, there is a trend to decrease in the isotropic metal coupling constant, $A_o = (2A_\perp + A_\parallel)/3$ and to increase in the average g factor ($g_o = 2g_\perp + g_\parallel)/3$, which are measured in fluid solution. The same trend is valid for the $g_\perp$, $g_\parallel$, $A_\perp$, and $A_\parallel$ parameters measured in anisotropic spectra, for example, those recorded on frozen solutions or in solid-state samples. These parameters often cluster in domains that depend on the identity of equatorial donor atoms, for example, O_4, O_2N_2, O_2S_2, N_4, and S_4 and are useful in assigning the coordination set. The coupling constants, mainly A_o and $A_\parallel$, may be predicted to a reasonable approximation by use of additive contributions for equatorial donors (L_i), that is, $A = (A_{L1} + A_{L2} + A_{L3} + A_{L4})/4$ (Chasteen, 1981; Cornman et al., 1995).

EPR spectroscopy can contribute to the study of low-molecular-weight complexes that are part of or are closely related to systems of biological or environmental interest. For instance, mono- and bis-chelated $V^{IV}O$ species of low-molecular-weight o-catecholic ligands formed in plant roots upon adsorption of vanadium were identified and distinguished from the VO^{2+} ions bound to macromolecules (Micera and Dessì, 1989). At room temperature, the former exhibited isotropic spectra consisting of eight lines, which is characteristic of $V^{IV}O$ species tumbling rapidly in fluid solution. The latter were distinguished by anisotropic spectra that were indicative of the immobilization of the paramagnetic ion. Also polygalacturonic acid, which is well suited to mimicking the extracellular metal adsorption sites of plant roots, forms inner-sphere complexes with $V^{IV}O$ through the direct coordination via the carboxylate groups (Deiana et al., 1980). The $V^{IV}O$ adsorption by amorphous oxides yielding ions bound to the oxygen donors of the matrix was also reported (Gessa et al., 1981). The intracellular reduction of vanadium(V), leading to formation of $V^{IV}O$ glutathione complexes, was followed by EPR spectroscopy in adipocytes (Degani et al., 1981).

EPR spectroscopy substantiated that a minor part of vanadium (less than 5% in vanadocytes exists in the +4 form and is tumbling too rapidly to be coordinated to a macromolecule (Frank et al., 1986). The g and A values were consistent with the aquaion of $V^{IV}O$ and the line widths of the resonances allowed to estimate the acidity (pH ca. 1.8) of the intracellular environment of *Ascidia ceratodes*. However, the spectra of the blood cells of other ascidians were different, suggesting a different blood chemistry (Sakurai et al., 1987).

The hf interaction of the unpaired electron with magnetic nuclei of ligands is also known as superhyperfine (shf) interaction. The shf splittings of EPR spectra may be useful in assigning the type and number of coordinated ligands. ^{31}P ($I = 1/2$) shf structure was observed in the EPR spectra of $V^{IV}O$ complexes formed by phosphorus-containing ligands, e.g., dithiophosphinates and dithiophosphates, carrying coordination units involving four-membered chelated structures (Miller and McClung, 1973; Mukherjee and Shastri, 1989). The magnitude of this splitting was assumed as characteristic of the bidentate chelating behavior of these ligands.

[31]P splittings of similar magnitude were observed on a $V^{IV}O$ orthophosphate complex formed in aqueous solution above pH 8 (Alberico and Micera, 1994). The intensity ratio of $1:2:1$ of the shf components indicated two orthophosphate ligands bound to the metal. The results were therefore assumed as an indication of a bidentate chelation of orthophosphate to $V^{IV}O$. Above pH 11, a species showed a well-resolved interaction with only one ^{31}P nucleus. Therefore, a hydrolytic complex in which the metal ion is chelated by only a phosphate group was assumed.

The ^{31}P shf structure in the EPR spectra of adenosine-5'-diphosphate (ADP)-VO^{2+} system was useful in assigning the metal binding sites of the nucleotide. A quintet pattern ($1:4:6:4:1$) due to the coupling of the unpaired electron with four equivalent ^{31}P nuclei was detected on the $M_I(^{51}V) = -3/2$ perpendicular absorption (Mustafi et al., 1992; Alberico et al., 1995). The magnetic equivalence of the four phosphorus atoms with respect to the metal center can be achieved only if two diphosphate moieties chelates the metal ion in a $[VOL_2]^{4-}$ species (L is the molecule fully deprotonated at the diphosphate chain). In the complex $[VOL_2H_{-2}]^{6-}$, formed in basic solution, the quintet pattern was gradually replaced by a triplet with an intensity ratio 1:2:1. This indicates that two equivalent ^{31}P nuclei are interacting with the metal ion, which agrees with the hypothesis that one of the two nucleotides binds the ion through the polyphosphate moiety, while the other is chelated through the deprotonated hydroxyls of the ribose moiety. The transfer of the metal ion to two ribose moieties in $[VOL_2H_{-4}]^{8-}$ removed the ^{31}P shf coupling.

In the oxovanadium(IV)–guanosine-5'-monophosphate (GMP) system a monomer $\rightleftarrows$ dimer equilibrium, arising from the intermolecular binding of phosphate groups, was substantiated (Micera et al., 1996; Jiang and Makinen, 1995). Indeed, two $(O^-,O^-)(O^-,O^-)$-coordinated monomeric complex species (**7**) formed by the ribose moieties can interact with each other through intramolecular axial bonds of phosphate to vanadium (**8**). While the ^{51}V EPR parameters were hardly affected by the dimer formation process (ruling out any gross changes in the equatorial donor set), the shf splitting ($1:1$ doublets) attributed to ^{31}P clearly indicated the dimetallic arrangement.

EPR spectroscopy detects oligo- or polynuclear $V^{IV}O$ complexes and provides structural information, for example, on the metal–metal distance. Hydroxo-bridged dinuclear species are EPR-silent because of the strong magnetic coupling between the metal ions. In some instances, polymetallic complexes are formed by ligands bridging adjacent vanadium centers. Such species often involve cyclic structures in which equivalent metal ions are weakly magnetically coupled because of the relatively large intermetallic distance. The best-characterized systems involving dinuclear species are $V^{IV}O$-tartrates (Belford et al., 1969; Kiss et al., 1995). Several studies have demonstrated the dinuclear interaction (the V-V distance is around 4 Å) and its stereoselectivity with the D, L, and DL isomeric forms of the ligand (H_2L). The different EPR properties arise from the relative magnitude of J, the electron–electron exchange interaction term, and A, the vanadium hf interaction. The complex

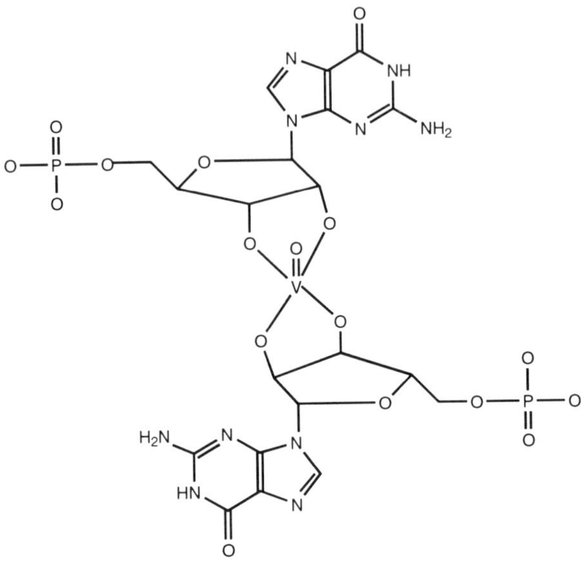

7

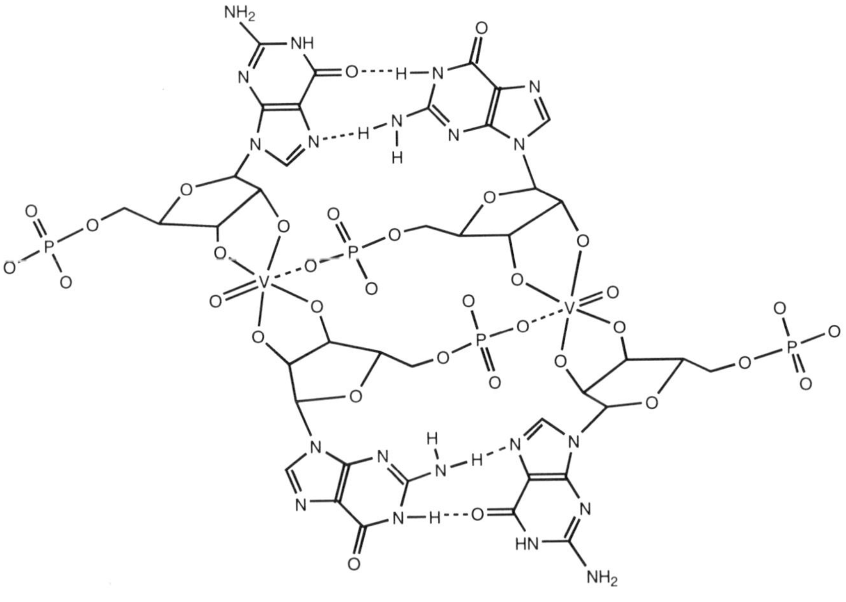

8

Figure 2.

$[(VO)_2L_2H_{-1}]^-$ (**9**) is detected at room temperature with either D- or L-tartaric acid below pH $\sim$ 4, as indicated mainly by two sets of satellite lines flanking the resonances of the aquaion. The species $[(VO)_2L_2H_{-2}]^{2-}$ (**10**) is distinguished by an EPR isotropic pattern consisting of 14 or 15 major lines irregularly spaced and accompanied by satellites, in accordance with J $\sim$ 5A and 10A, for **9** and **10**, respectively (Dunhill and Symons, 1968; James and Luckhurst, 1970). A

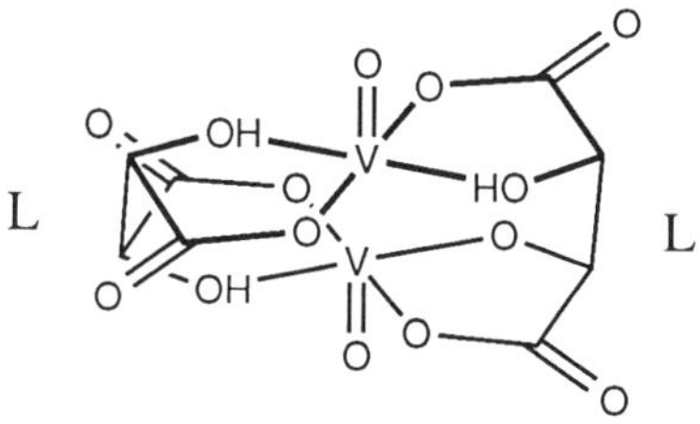

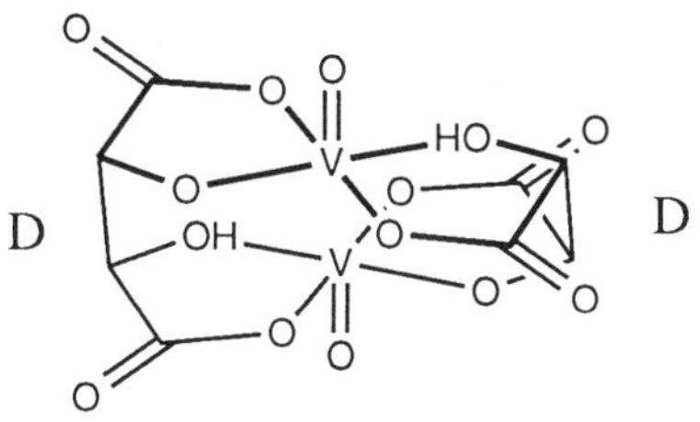

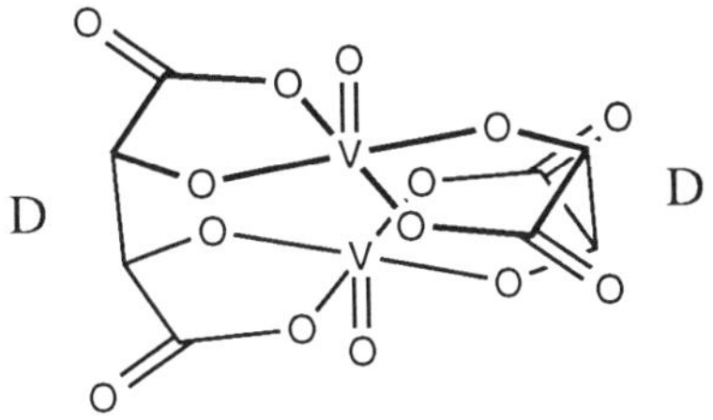

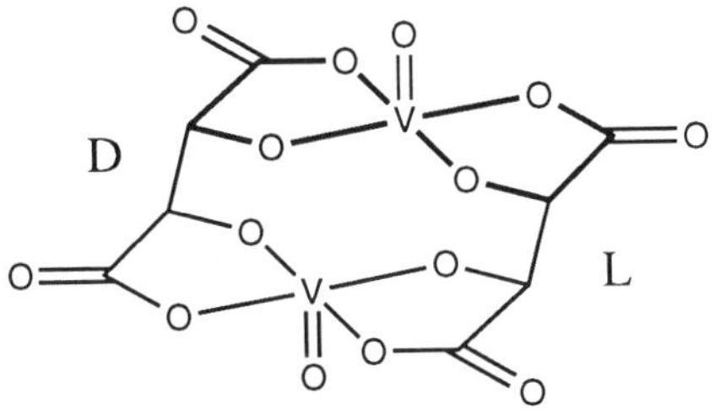

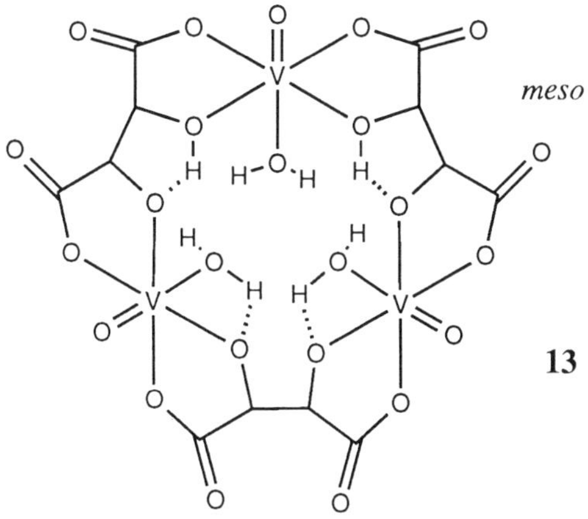

Figure 3.

major dimer $[(VO)_2L_2H_{-4}]^{4-}$ (**11**) is formed by two fully deprotonated ligands. In this case, the condition $J \gg h\nu$ (ν is the microwave frequency value) causes 15 equispaced lines to be expected in the isotropic EPR spectrum, although only the pattern at lower fields is resolved (Belford et al., 1969). In **11** the metal ion has a trans coordination geometry as in the solid complex characterized by X-ray diffraction (Forrest and Prout, 1967). Anisotropic spectra indicate the triplet state for both species **10** and **11,** through two sets of 15 equispaced lines, separated by $2D,$ where D is the zero-field splitting value (Table 2). DL-tartaric acid yields a dimer (**12**) with cis coordination geometry, as proven by X-ray crystallography (Garcia-Jaca et al., 1993, 1994). EPR spectra indicate the different symmetry of the complexes.

A different $V^{IV}O$ binding mode has been observed for *meso*-tartaric acid (Kiss et al., 1995). A resolved isotropic EPR pattern consisting of 22 almost equispaced lines is distinguished between pH 4 and 6, with a ^{51}V hf coupling constant that is nearly one-third that typical of a mononuclear $V^{IV}O$ species. These features are distinctive of cyclic trimers formed by three magnetically coupled equivalent nuclei. Speciation curves indicate $[(VO)_3L_3H_{-4}]^{4-}$, which has four of six alcoholic-OH groups deprotonated and involved in coordination (**13**). Some fluxional behavior or an extensive hydrogen bonding network within the polynuclear cage could make the coordination of the three metal ions equivalent at room temperature.

In the $V^{IV}O$–citric acid (H_3L) system, titration curves suggested the formation of a species with $[(VOLH_{-1})_n]^{2n-}$ stoichiometry in the pH range 5–9 (Kiss et al., 1995). No EPR signal was observed at room temperature, indicating magnetically coupled $V^{IV}O$ centers. Dimetallic features were substantiated by the anisotropic spectra recorded on frozen solutions (Dunhill and Smith, 1968). Therefore, the species was assigned a $[(VO)_2L_2H_{-2}]^{4-}$ composition. The zero-field splitting value measured by EPR (Table 2) indicates a metal–metal distance shorter than in the case of the tartrates. By assuming an axially symmetric dimer and a purely dipolar interaction, a value of ca. 3.3 Å was calculated, which is in full agreement with that measured for **14** by single-crystal X-ray diffraction (Zhou et al., 1995).

A dimeric species resembling those formed by tartrate ligands has been found, in aqueous solution, with the carboxylic sugar D-galacturonic acid (**15**) (Branca et al., 1990c). In equimolar solution, at a metal concentration as high as 10^{-2} M and pH 10–11, EPR spectral features distinctive of a dimetallic species are observed. The EPR analysis indicates a distance around 5 Å between the coupled centers in an axially symmetric dimer. The only way to accommodate the metal ions at such a distance and to fit the spectroscopic data is to assume that the sugar acts as a tetradentate ligand, coordinating through the $[COO^-, O(4)^-]$ and $[O(3)^-, O(2)^-]$ donor sets and adopting an open-chain structure (**16**).

A $V^{IV}O$ dinuclear arrangement was also found with pyrazole-3,5-dicarboxylic acid (H_2L). The four donors (two nitrogens and two carboxylate oxygens) in the pyrazole ring make the ligand suitable to bind two V^{IV} centers.

Table 2 EPR and Electronic Absorption Parameters of Oxovanadium(IV) Oligonuclear Complexes of Tartaric and Citric Acids

Ligand	Species	$g_\parallel$	$A_\parallel$, 10^{-4} cm^{-1}	D, 10^{-4} cm^{-1}	g_o	A_o, 10^{-4} cm^{-1}	Absorption maxima,[a] nm
L-Tartrate	$[(VO)_2L_2H_{-2}]^{2-}$	1.964	73	190			545(13), 770(19)
	$[(VO)_2L_2H_{-4}]^{4-}$	1.957	76	336	1.978	36	400(35), 530(20), 580(16), >850
DL-Tartrate	$[(VO)_2L_2H_{-2}]^{2-}$	1.967	74	186			545(13), 765(20)
	$[(VO)_2L_2H_{-4}]^{4-}$	1.953	72	341	1.966	40	420(46), 535(23), 730(18)
meso-Tartrate	$[(VO)_3L_3H_{-4}]^{4-}$				1.966	31	400(16), 550(12), 730(15)
Citrate	$[(VO)_2L_2H_{-2}]^{4-}$	1.953	77	666			560(27), 620sh

[a] Meaured at the maximum extent of formation of the species; extinction coefficients, in M^{-1} cm^{-1} units, in parentheses.

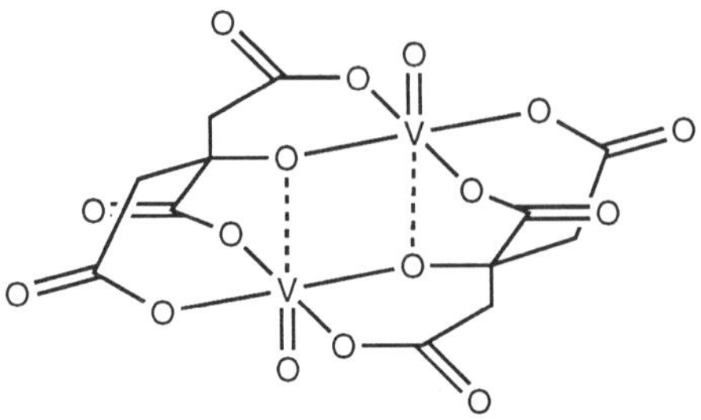

14

Figure 4.

Upon deprotonation of the pyrrolic site a $[(VO)_2L_2H_{-2}]^{2-}$ species (**17**) is formed (Sanna et al., in press). Here the donor arrangement permits all eight equatorial donors to be approximately coplanar. EPR spectra clearly indicate this feature and measure an intermetallic distance that is in quite good agreement with that revealed by X-ray diffraction on $(Bu_4N)_2[(VO)_2 (3,5,-dcp)_2]$ $(3,5-dcpH = LH_{-1})$ (Hahn et al., 1992).

The diphosphate anion $P_2O_7^{4-}$ (L) forms a trimeric species $[(VO)_3L_3]^{6-}$ in the pH range 5–9 (**18**). A cyclic structure, with the ligand adopting a chelating and bridging mode through the four unshared oxygen atoms, is attributed to the species (Parker et al., 1970; Hasegawa et al., 1969; Hasegawa; 1971; Buglyò et al., 1995b). The isotropic EPR spectrum consists of 22 equally spaced hf lines with intensity ratios close to those predicted for the interaction of one unpaired electron with three equivalent vanadium nuclei. Only weak hf components are resolved in the anisotropic EPR spectra, with a splitting value that is one-third of that expected for the mononuclear complexes. Similar species are formed by methylenediphosphonic and 1-hydroxyethylidenediphosphonic ligands (Buglyò et al., 1996).

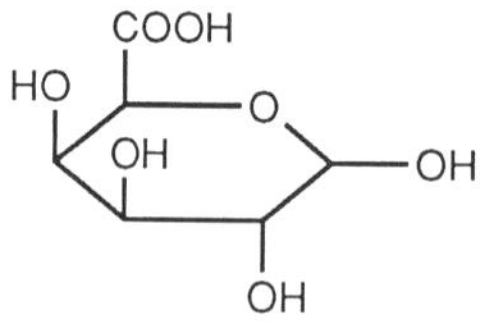

15

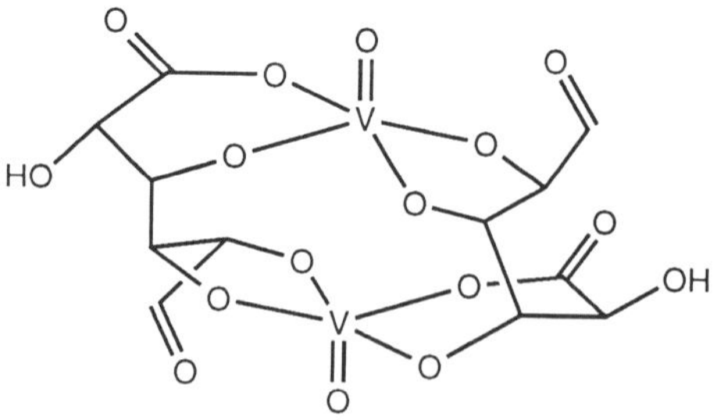

16

Figure 5.

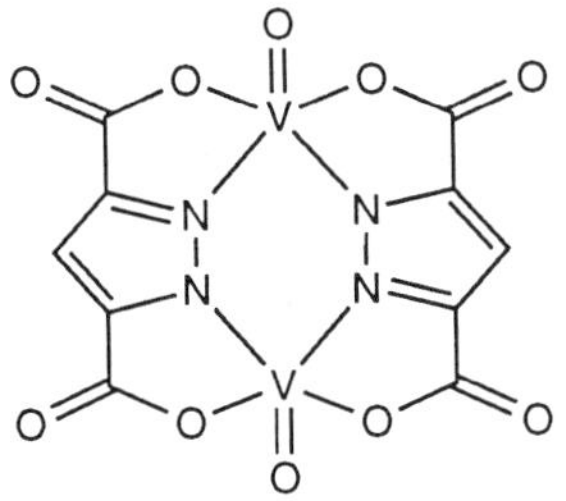

17

Figure 6.

Aromatic ligands provided with carboxylic and phenolic donors, which serve as versatile ligands to chelate toxic and nutrient metal ions in living systems and in soil, can yield cyclic oligonuclear complexes. Well suited for such arrangements are 2,3-dihydroxyterephthalic (2,3-dhtp, H_2L) and 2,3,4-trihydroxybenzoic (2,3,4-thb) acids, which have four adjacent donors on the aromatic ring (Kiss et al., 1994). In the oxovanadium(IV)–2,3-dhtp system, a 22-line isotropic EPR pattern is distinctive of the cyclic trimer $[(VO)_3L_3H_{-6}]^{6-}$ (**19**). With the ligand 2,3,4-thb, an oligomeric complex $[(VOLH_{-2})_n]^{2n-}$ is formed by LH_{-2}, the fully deprotonated ligand. In this species all four donor atoms are coordinated to the metal ions. The isotropic ^{51}V hyperfine coupling constant (22×10^{-4} cm^{-1}) indicates a cyclic tetramer (**20**). Frozen solution spectra show almost all of the 29 ^{51}V parallel resonances expected in the case of four coupled oxovanadium(IV) ions in equivalent environments.

With imidazole-4,5-dicarboxylic acid the high stability of the complex, involving a bridging mode for the ligand, promotes the deprotonation of the pyrrole ring at pH values as low as 4. The one-atom spacing between the two chelating sets of the ligand hinders the closure of a di- or trinuclear ring, and a larger

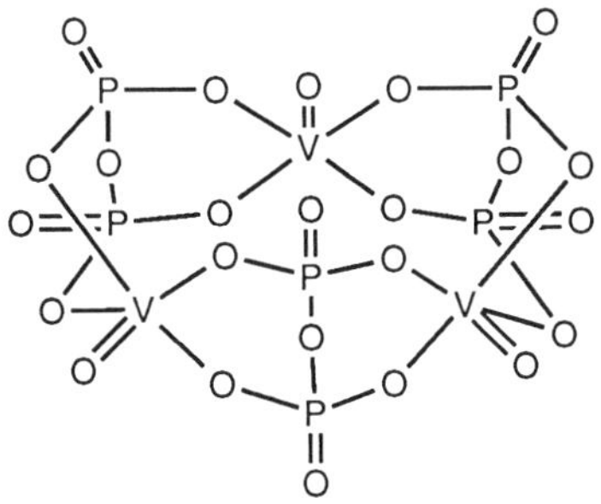

18

Figure 7.

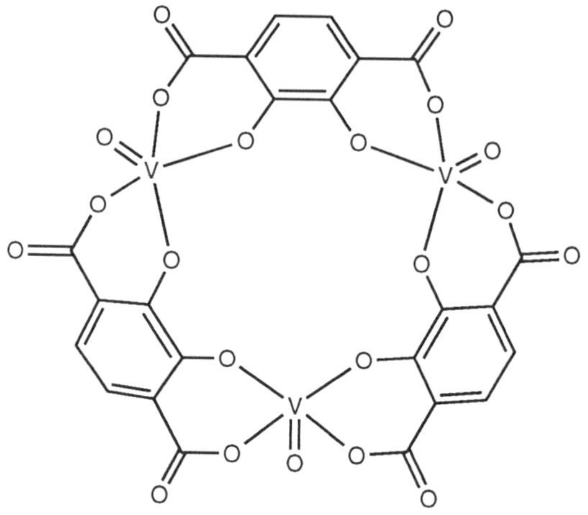

19

Figure 8.

tetrametallic arrangement is observed in the cyclic structure **21** (Sanna et al., in press).

3.1.3   *ENDOR Spectroscopy*

ENDOR spectroscopy can resolve even weak magnetic interactions within a distance of 5–6 Å, making it possible to regain the hyperfine or superhyperfine

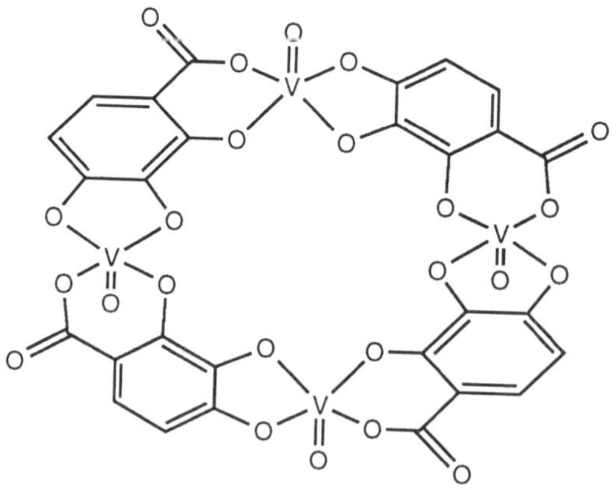

20

Figure 9.

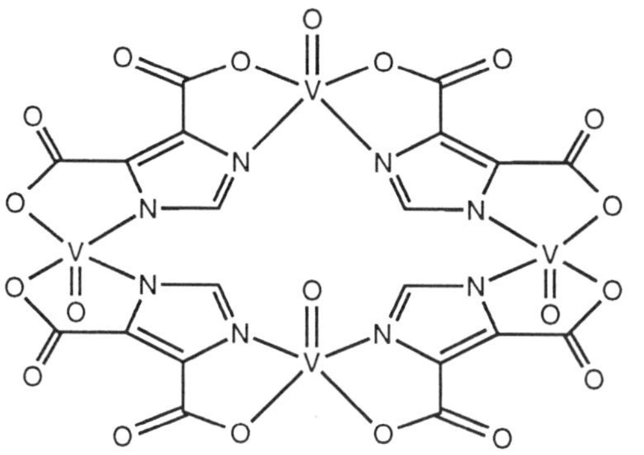

21

Figure 10.

details that are missing in the EPR spectrum. According to theory, each class of equivalent nuclei interacting with unpaired electrons contributes a set of ENDOR resonances. For a system with a single unpaired electron, this set consists of a pair of transitions at $\nu\pm = |\nu_0 \pm A/2|$, where ν_0 is the free nuclear Larmor frequency and A is the same hf constant as that expected in the EPR spectra. Therefore, the interacting nuclides may be distinguished by the proper frequency ranges and each group of equivalent nuclei contributes only two lines (for nuclei with $I \geq 1$ quadrupole splitting is also expected). Hf coupling constants A are made up of two main contributions. The isotropic contribution A_i, mainly due to Fermi-contact and pseudocontact interaction, does not depend on the molecular orientation with respect to the magnetic field. The anisotropic contribution A_a depends on the molecular orientation and in most cases, for example, with d^1 ions and distant interacting nuclei, can be approximated by the point-dipole equation:

$$A_a = gg_n\beta\beta_n(3\cos^2\alpha - 1)/hr^3 \tag{1}$$

where g and g_n are the electron and nuclear g-factors, β and β_n are the Bohr and nuclear magnetons, h is the Planck constant, r is the nucleus-electron distance, and α is the angle between the nucleus–electron axis and the magnetic field direction. Hence, an analysis of the hf interaction requires knowledge of the molecular orientation and this is possible in single-crystal measurements. However, one can obtain crystal-like information for randomly oriented molecules (powders or frozen solutions) provided that g or A anisotropy is large. In such a case, the EPR spectrum identifies well-defined molecular directions (the principal axes of g- and A-tensors) along which the hf interaction can be measured by ENDOR.

$V^{IV}O$ is a suitable probe for ENDOR spectroscopy. ^1H, ^{14}N, and ^{15}N ENDOR studies assign the binding sites in $V^{IV}O$ complexes of ligands of low molecular weight, for example, imidazoles and sugars (Mulks et al., 1982; Branca et al., 1989).

In $V^{IV}O$ complexes with strictly axial geometry like those of *o*-catecholic and salicylic ligands, the ^1H ENDOR analysis may be quite straightforward. The ligands are almost coplanar with the metal, and the proton–vanadium distance is large enough to make the interaction purely dipolar. Therefore the ^1H couplings measured along the principal directions of the $A(V)$-tensor can be used to locate the protons and evaluate their distance from metal in terms of the point-dipole model (eq 1). For instance, in the ENDOR spectra of bis(catecholato)oxovanadate(IV) (Branca et al., 1990a) the aromatic protons adjacent to the donor groups exhibit shf couplings with magnitudes and an orientation dependence consistent with the X-ray crystallographic data. If the ligand is salicylic acid, the protons near the donor group are not equivalent and yield distinct couplings that permit accurate estimates of the distances (see Table 3).

3.2   Non-oxo Vanadium(IV) Complexes

There are only a relatively few examples of straight V^{IV} complexes. Nevertheless, vanadium(IV) complexes of rather high stability may be formed with ligands able to displace, partly or fully, oxygen(s) from the stable VO^{2+} oxocation. Catechols and hydroxamic acids, which are particularly effective in stabilizing highly charged metal ions in aqueous solution, belong to this class of chelators (Cooper et al., 1982; Branca et al., 1990d; Batinic-Haberle et al.,

Table 3 ^1H ENDOR Data for the Oxovanadium(IV) Complexes of *o*-Catecholic and Salycilic Ligands^a

	^1H ENDOR
Complex	A_H
22	1.80
23	1.80
24	1.70
25	1.70, 0.95
26	0.95
27	1.70
28	1.70, 0.95

^{a} Coupling constants in MHz. Measurements at 110 K. A_H values are the largest splittings for the aromatic protons adjacent to donor groups, taken from perpendicular ^1H ENDOR spectra in D_2O.

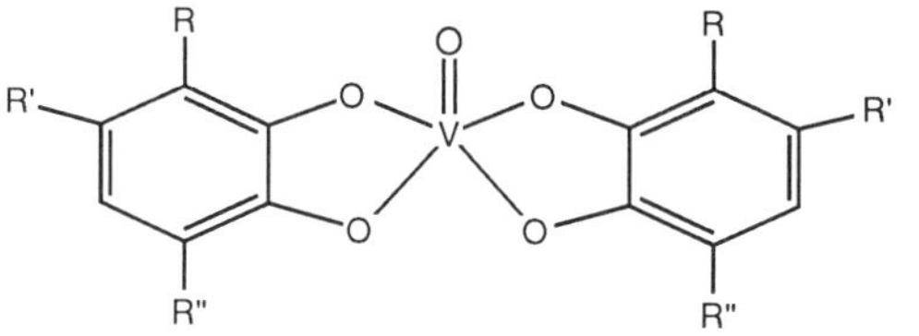

22: R = R" = H; R' = H
23: R = R" = H; R' = CO_2^-
24: R = CO_2^-; R' = H; R" = H

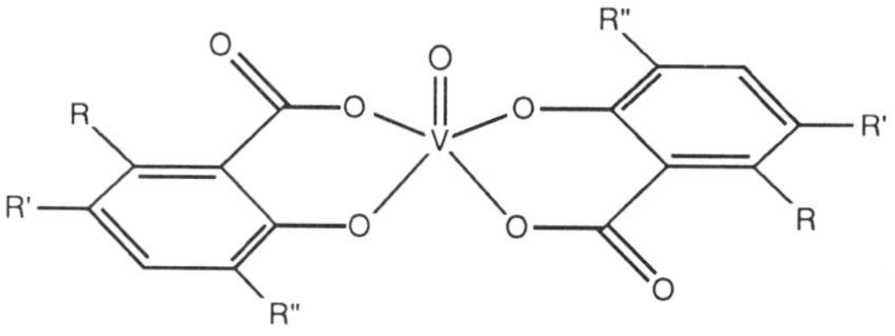

25: R = H; R' = H; R" = H
26: R = H; R' = H; R" = OH
27: R = OH; R' = H; R" = H

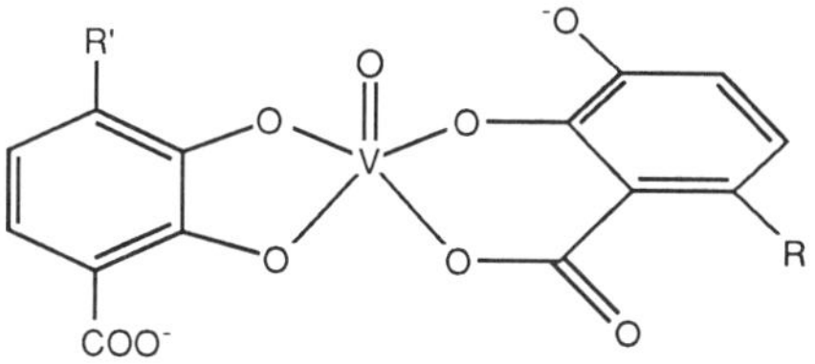

28: R = H; R' = H

Figure 11.

1991; Dessì et al., 1992; Buglyò et al., 1993, 1995a). Also tannic compounds, which are important organic components of soil and exhibit the o-catecholic type of coordination, are effective in yielding stable non-oxo complexes of vanadium(IV), where the bare metal ion is tris-chelated by three o-diphenolate donor sets (Branca et al., 1990b). Non-oxo vanadium(IV) complexes with β-diketonato or o-catecholato ligands are intensely colored species because of the strong O → V^{IV} charge transfers, which dominate the electron absorption spectra in the visible region and obscure the $d \rightarrow d$ absorptions (Von Dreele and Fay, 1972; Cooper et al., 1982; Hambley et al., 1987). The oxo-ligand displacement by an o-catecholic ligand involves the reaction of a ligand molecule (H_2L), having the o-diphenolic site undissociated, with the bis-chelated complex of $V^{IV}O$ (VOL_2) and the formation of a water molecule:

$$VOL_2 + H_2L \rightleftharpoons VL_3 + H_2O \qquad (2)$$

The optical properties of the tris-chelated species can be used for the determination of the equilibrium constant of the reaction. For instance, Buglyò and Kiss (1991) reported the spectrophotometric study of $V^{IV}O$-tiron (1,2-dihydroxybenzene-3,5-disulphonate) system. The non-oxo V^{IV} complex formation with the linear trihydroxamic siderophore deferoxamine B was studied by monitoring an absorption at 385 nm (Buglyò et al., 1995a). Ligands with O,S or SS donor sets are also effective in the displacement of the vanadyl(IV) oxygen. The non-oxo vanadium(IV) tris-complexes of 2-mercaptophenol (mp) and 2-mercapto-4-methylphenol (mmp) were reported (Klich et al., 1996). Like their o-catecholic analogues, the complexes exhibited four very intense bands in the visible region. For the mmp complex a red shift relative to the mp analog was documented, presumably due to the electron-releasing properties of the 4-methyl group. Also non-oxo $V(SS)_3$ or $V(ONS)_2$ complexes show intense CT bands (Dutton et al., 1988; Welch et al., 1988).

Hexa-coordinate non-oxo vanadium(IV)-chelated complexes exhibit a severe distortion from the octahedral geometry. With five-membered chelators, like o-catecholic, mercaptophenolic, or hydroxamic ligands, the geometry is close to the trigonal-prismatic, and there is a switch of the ground state to d_{z^2} (Scheme 2b). In this case, the theory requires $g_x \sim g_y < g_z \sim 2.0023$ and $A_z < A_y \sim A_x$ (Desideri et al., 1978) compared with the spectral features expected for the ground state $d_{xy} : g_\parallel < g_\perp < 2.0023$ and $A_\parallel > A_\perp$. Therefore, EPR can be used as a diagnostic tool to distinguish oxo- from non-oxo vanadium(IV)-chelated complexes formed in solution. In Table 4 the EPR parameters measured for o-catecholate and similar complexes of V(IV) are compared.

3.3. Vibrational Spectroscopy for V^{IV} or $V^{IV}O$ Complexes

Vibrational spectroscopy may be applied to the study of oxo and non-oxo complexes of V^{IV}. The IR-active mode distinctive of oxovanadium(IV) is the $V=O$ stretch, for which a range of 900–1,035 cm^{-1} is reported (Selbin, 1966; Collison et al., 1980). However, its frequency is affected by a number of factors, including electron donation from ligand atoms in the basal plane, solid-state effects, and the involvement of oxygen in hydrogen bonding or in additional interaction with metal ions. In oxo-bridged chain polymers, wave number values as low as 850–875 cm^{-1} are measured because of the decreased $V=O$ bond order due to donation of π electrons from a $V=O$ unit to the next vanadium atom (Cornman et al., 1995). The $V=O$ stretch is often difficult to distinguish from ligand vibrations. A way to solve the problem is to exchange the ^{16}O for an ^{18}O atom, as was done for the complex bis[3-hydroxy-3-methylglutarato oxovanadate(IV)] (Castro et al., 1995). The exchange resulted in the replacement of the absorption at 976 cm^{-1} by a band at 935 cm^{-1}. Accordingly, a shift of 42 cm^{-1} is predicted by Hooke's law.

V-Cl stretches were used to follow cis/trans equilibria in di-chloro complexes of V^{IV}. In the solution spectra of $[V(dpm)_2Cl_2]$ (dpm = t-C_4H_9-CO-CH-CO-CH-t-$C_4H_9^-$) a band at 385-cm^{-1} gains intensity, with increasing solvent dielectric constant, at the expense of an absorption at 362 cm^{-1} (Von Dreele

Table 4 EPR Parameters for $V^{IV}O$ and V^{IV} Complexes

Complex[a]	Donor Set	g_z	g_x	g_y	A_z, 10^{-4} cm^{-1}	A_x, 10^{-4} cm^{-1}	A_y, 10^{-4} cm^{-1}	Reference
[VO(mp)]$^{2-}$	VO(OS)$_2$	1.975	2.007	1.999	150	40	40	Klich et al. 1996
[VO(cat)]$^{2-}$	VO(OO)$_2$	1.947	1.981	1.981	154	50	50	Branca et al. 1990d
[VO(mnt)]$^{2-}$	VO(SS)$_2$	1.973	1.986	1.986	135	43	43	Atherton and Winscom, 1973
[VO(tsalen)]	VO(N$_2$S$_2$)	1.978	1.986	1.986	148	51	51	Dutton et al. 1988
[VO(salen)]	VO(N$_2$O$_2$)	1.951	1.985	1.985	159	59	59	Jezierski and Raynor 1981
[V(mmp)]$^{2-}$	V(OS)$_3$	2.002	1.980	1.958	17	75	115	Klich et al. 1996
[V(cat)]$^{2-}$	V(OO)$_3$	1.991	1.937	1.937	14	107	107	Branca et al. 1990d
[V(mnt)]$^{2-}$	V(SS)$_3$	2.000	1.974	1.974	9	91	91	Kwik and Stiefel, 1973
[V(bha)$_3$]$^{2-}$	V(OO)$_3$	1.98	1.955	1.955	13	100	100	Dessì et al. 1992

[a] cat, catecholato(2−); mnt, 1,2-dicyanoethylene-1,2-dithiolato(2−); mp, 2-mercaptophenolato(2−); mmp, 2-mercapto-4-methylphenolato(2−); bha, benzohydroxamato(2−); salen, N,N'-ethylenebis(salicylideneiminato)(2−); tsalen N,N'-ethylenebis(tiosalicylideneiminato)(2−).

and Fay, 1972). Only the latter band is present in the spectrum of the solid. The study assigns the 385-cm^{-1} band to the cis isomer and the 362-cm^{-1} band to the trans isomer.

4.   VANADIUM(V) COMPLEXES

4.1.   Introduction to Vanadium(V)

Vanadium in the +5 oxidation state is involved in biological and environmental systems. Although V^V is easily reduced by organic matter, the reduced V^{III} and V^{IV} ions may be reoxidized in aerobic conditions. Indeed, vanadium(V) is present in sea water and in intracellular compartments. In most of its bioreactions vanadate becomes a part of a large biomolecule (Rehder, 1988). In this respect, the biological responses induced by vanadate(V) are noteworthy, including the inhibitory effects on the Na,K-ATPase and on enzymes having a phosphoprotein intermediate in their catalytic cycle, and the stimulation of a number of enzymes like adenylate cyclase. Therapeutic effects have been proven for V^V compounds, for example, the insulin-mimetic activity. Vanadium(V) has a versatile coordination geometry with coordination numbers between 4 and 7. Because of the d^0 configuration, the +5 oxidation state is diamagnetic and therefore not suitable for EPR, magnetic, and $d–d$ absorption studies. Investigations that take advantage of the joint application of CT electronic absorption, vibrational, and NMR spectroscopic techniques are currently carried out to elucidate the structure of the metal complexes formed by vanadium in this oxidation state.

4.2. ^{51}V NMR

A wide range of information can be gleaned by use of ^{51}V NMR, which probes the metal center directly. Insight is accessible not only into the metal geometry and coordination set but also into the thermodynamic and kinetic properties of simple and complex systems. A briefing on the background theory for multinuclear NMR indicates that, for nuclei heavier than fluorine, the chemical shift is affected mainly by variations in the paramagnetic (σ_p) contribution, which is a second-order effect depending on the mixing of the excited into ground state by the external magnetic field. This term may be very large when there is an asymmetric distribution of p and d electrons, according to the equation:

$$\sigma_p \propto - [\langle r^{-3}\rangle_p P_i + \langle r^{-3}\rangle_d D_i]\Delta E^{-1} \tag{3}$$

where P_i and D_i measure the degree of charge imbalance in the valence p- and d-shells, which yields the orbital angular moment, and $\langle r\rangle$ is the average radius of the valence orbitals and ΔE the average energy of low-lying excited states.

In the case of vanadium(V) equation 3 can assume the form

$$\sigma_p \propto - [\langle r^{-3}\rangle_{4p}C^2_{4p} + \langle r^{-3}\rangle_{3d}C^2_{3d}]\Delta E^{-1} \tag{4}$$

where C_{3d} and C_{4p} are the vanadium LCAO coefficient of the molecular orbitals taking part in electronic transitions and ΔE is the mean HOMO-LUMO energy gap (Rehder et al., 1988). Since the shift $\delta(^{51}V)$, usually quoted relative to $VOCl_3$, and nuclear shielding σ are defined with opposite signs, an increase in the absolute magnitude of σ_p corresponds to an increase in $\delta(^{51}V)$.

For VOX_3 (X = Br, Cl, F) complexes $\delta(^{51}V)$ increases in the order F < Cl < Br according to the increasing ΔE value between LUMO, the nonbonding $3d$ orbitals, and HOMO. The latter orbitals are $\sigma(VX)$, $\pi(VX)$, and $\pi(V{=}O)$, for X = Br or Cl, and $\pi(V{=}O)$ only, for X = F (Rehder et al., 1988). These conclusions may be extrapolated to other ligands. Indeed, the "inverse electronegativity dependence" was established for δ. In fact, as the electronegativity of the donors increases, ΔE increases, the covalence decreases, and therefore $\delta(^{51}V)$ decreases. On this basis, the $\delta(^{51}V)$ value can establish various types of donor sets, as shown by the good correlation found between $\delta(^{51}V)$ and the electronegativities of the atoms bound to vanadium (Rehder et al., 1988; Rehder, 1990).

The dependence of ^{51}V shifts on the coordination number is less straightforward, even if some trends were reported (Rehder et al., 1988). It was found that a downfield shift (increase in δ) takes place with increasing coordination number, although this effect could be counterbalanced by the overall increase of electronegativity of ligands. This effect proves useful for the assignment of the coordination number of complexes (Table 5). Strained structures (due to bulky substituents or three- and four-membered chelated rings) give rise to more negative chemical shifts than unstrained complexes with the same coordination set. The effect is particularly important for the bidentate coordination of peroxo, hydroxylamido, and carboxylato groups.

Empirical correlations between ^{51}V chemical shifts and line width were found by an analysis of a number of vanadium(V) complexes formed by nitrogen- and oxygen-containing multidentate ligands (Crans and Shin, 1994). Plotting the inverse of NMR line width measured at half-height (Δv^{-1}) as a function of the chemical shift, two distinct clusters were found. The first was defined by the 5–8 ms and −480 to −490 ppm ranges for Δv^{-1} and $\delta(^{51}V)$, respectively; the second cluster included the complexes exhibiting Δv^{-1} from 1 to 4 ms and $\delta(^{51}V)$ from −490 to −526 ppm. The first set of data appeared to be specific for five-coordinate complexes containing a cis-dioxo vanadium(V) center and polyfunctional ligands chelated through both nitrogen and oxygen donors. The second cluster contained exclusively six-coordinate cis-dioxo vanadium(V) complexes.

Table 5 ^{51}V Chemical Shifts (Relative to VOCl$_3$) of Vanadium(V) Complexes^a

Complex	Donor Set	$\delta(^{51}V)$
$[VO(O\text{-}n\text{-}C_5H_{11})_3]$	O_4	-543
$[VOCl(O\text{-}n\text{-}C_5H_{11})_2]$	O_3Cl	-466
$[VO_2(ox)_2]^{3-}$	O_6	-529
$[VO(O\text{-}t\text{-}Bu)(ac)_2]$	O_6	-584
$[\{VO(O_2)_2\}_2(\mu\text{-}O)]^{4-}$	O_6	-746
$[VO(NR_2)(pbha)_2]$	O_5N	-430
$[VOCl(pbha)_2]$	O_5Cl	-280
$[VO_2(edta)_2]^{3-}$	O_4N_2	-519
$[VO(ac)_3]$	O_7	-589
$[VO(O_2)(ox)_2]^{3-}$	O_7	-590
$[VO(O_2)_2ox]^{3-}$	O_7	-733
$[VO(O_2)_2phen]^-$	O_5N_2	-733

^{a} ac, η^2-acetato(1$-$); ox, oxalato(2$-$); pbha, N-phenylbenzo-hydroxamato(1$-$); edta, ethylenediaminetetraacetato(4$-$); phen, o-phenanthroline; (O$_2$) η^2-peroxo(2$-$) (Rehder et al., 1988).

Vanadate solutions contain complex mixtures of mono- and oligo-vanadates. Since each vanadate species gives resolved resonances in the ^{51}V NMR spectrum, the species distribution can be calculated from the integrated ^{51}V NMR signals (Crans, 1994). This makes it possible to correlate the enzymatic activity to the concentration of the different oxovanadate species present in solution. For instance, the changes in activity of the enzyme 6-phosphogluconate dehydrogenase were followed by varying total vanadate concentration (Crans et al., 1990). A linear correlation of the reciprocal enzymatic activity was found only with the vanadate tetramer (V$_4$) concentration, as measured by ^{51}V NMR. Therefore, V$_4$ was identified as the species responsible for the observed enzymatic inhibition.

The quadrupole moment of the ^{51}V nucleus, even if it is low, can hamper NMR detection through shortening of spin–spin relaxation times. Line widths are affected by the magnitude and the symmetry of the electric field gradient at the nucleus (depending on the nature and arrangement of bound ligands) and the correlation time for the molecular tumbling (depending on molecular size, medium viscosity, and temperature). Relaxation is least effective for regular metal environments. As the site symmetry decreases, owing, for example, to differences in ligand properties, the line width increases accordingly. Rapid relaxation can be a problem also when vanadium is coordinated to a large or bulky molecule. Broad resonances are observed for vanadium(V)–protein complexes. Nevertheless, the interaction can be established and the binding constants determined if the vanadate population in solution is monitored by ^{51}V NMR (Crans, 1994).

With low-molecular-weight ligands, the formation of V^V complexes can be followed by the appearance of resonances distinct from those of "free" vanadates. Rehder (1988) examined the formation of $1:1$ vanadium(V) complexes of dipeptides at physiological pH values and concluded that the terminal amine, deprotonated amide nitrogen, and terminal carboxylate are involved in coordination. In particular, the downfield shift of the ^{51}V resonance relative to vanadates and also to complexes with hydroxycarboxylates is strongly supportive of deprotonated amide binding to the metal ion.

4.3.   Combined Spectroscopic Studies on Vanadium(V) Complexes

A survey of selected applications of spectroscopic methods to complex systems involving vanadium(V) is presented in the following. However, excellent reviews are already available in the literature (Rehder, 1990).

Considerable interest is devoted to peroxo complexes of vanadium(V), because of their strong insulin-mimetic activity observed both in vivo and in vitro (Shaver et al., 1993), and to their relevance to vanadium haloperoxidases discovered in marine algae and in a lichen. The coordination of hydrogen peroxide to vanadium(V) in neutral or basic solution yields anionic peroxovanadates, with one to four coordinated peroxides, and peroxodivanadates. Under acidic conditions the red oxomonoperoxo-$[VO(O_2)]^+$ and the yellow oxodiperoxo-$[VO(O_2)_2]^-$ species are formed. The two species are distinguished by ^{51}V NMR ($\delta = -540$ and -698 ppm, for $[VO(O_2)]^+$ and $[VO(O_2)_2]^-$, respectively). A third broad resonance at ca. -670 ppm indicates the dioxotriperoxodivanadium(V) dimer, $[(VO)_2(O_2)_3]$, which seems to act as the oxidant in the bromide oxidation by hydrogen peroxide that is catalyzed by cis-dioxovanadium(V) in acidic aqueous and aqueous/ethanolic solution (Clague and Butler, 1995).

IR spectroscopy proves useful for identifying $V^V{=}O$ functions. When compared with their $V^{IV}{=}O$ analogues, monooxovanadium(V) complexes show a small but detectable increase of the stretch wave number value reflecting the increase in the metal oxidation state (Keramidas et al., 1996).

Differences were found for five- and six-coordinate complexes containing cis-dioxo V^V units (Crans and Shin, 1994). The $\nu(V{=}O)$ modes fall below 900 cm^{-1} for six-coordinate complexes and shift to higher wave number values in five-coordinate species. Vanadate at pH 8.9 exhibits a $\nu(V{=}O)$ band at 921 cm^{-1} (Crans and Shin, 1994). The V–O stretching mode falls around 350 cm^{-1}.

In peroxo complexes of V^V there is an overlap of the $\nu(O{-}O)$ and $\nu(V{=}O)$ bands. Therefore broad absorptions, for example, between 1,000 and 850 cm^{-1}, may be observed for such complexes. Isotopic shift studies are therefore suggested in order to assign the absorptions.

The potent inhibition of ribonuclease A by a $1:1$ complex of uridine has elicited considerable interest in the structure of the major uridine–vanadate complex and of related vanadate–diol complexes. Vanadate reacts with simple diols like ethylene glycol to yield a cyclic diester dimer with a 2:2 stoichiometry produced by the condensation of two vanadates with two molecules of the diol. Several inferences were made for the structure of this complex, for example, those involving one or two μ-oxo bridges and terminal oxo and or hydroxo ligands. The structure of this dimer in aqueous solution was solved in the study by Ray et al. (1995) by use of complementary spectroscopic techniques. The Raman spectrum of the dimer showed two different V–O stretching modes. An envelope of bands around 930 cm^{-1} was consistent with the symmetrical stretching mode of an O–V–O unit formed by two terminal VO bonds, therefore involving only non-ester oxygens. The assignment was substantiated by the 50 cm^{-1} ^{18}O isotopic shift and the comparison with the analogous absorption of $[(CH_3O)_2VO_2]^-$ (Deng et al., 1993). In contrast, the band at 455 cm^{-1} exhibited a wave number value and a modest isotope effect (6–8 cm^{-1}) consistent with a symmetric mode of a V_2O_2 unit in which the bridging oxygens belong to the diol ligand. The antisymmetric stretch of the terminal O–V–O modes was located at 917 cm^{-1} in the IR spectra (the ^{18}O isotope shift is ca. 50 cm^{-1}). The results were compared with those of related dimers and suggested that the ethylene glycolate(2−) dimer involves the ester oxygens as bridges to form the V_2O_2 cage. Each vanadium is further bound to two terminal oxo and to a nonbridging ester oxygen in a distorted trigonal–bipyramidal environment (**29**). The structural motif is very similar to that of dimeric 2-ethyl-2-hydroxybutyric acid–vanadate complex, in which each of the vanadium atoms is bound to two bridging alcoholate oxygens, to a monodentate carboxylate group, and to two terminal oxo ligands. Accordingly, the compound exhibits Raman and IR features very similar to those of the dimeric complex of ethylene glycol (**30**).

^{17}O NMR spectroscopy is a convenient and informative tool for studies on the oxo ligands of vanadium(V). ^{17}O NMR spectroscopy is able to distinguish

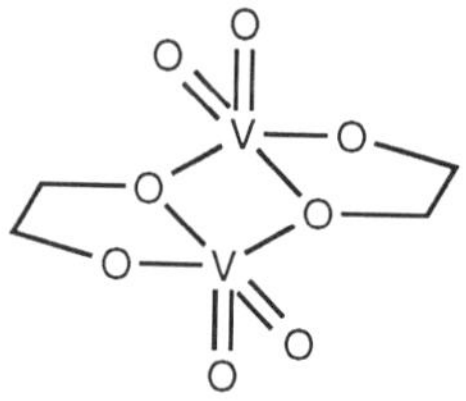

29

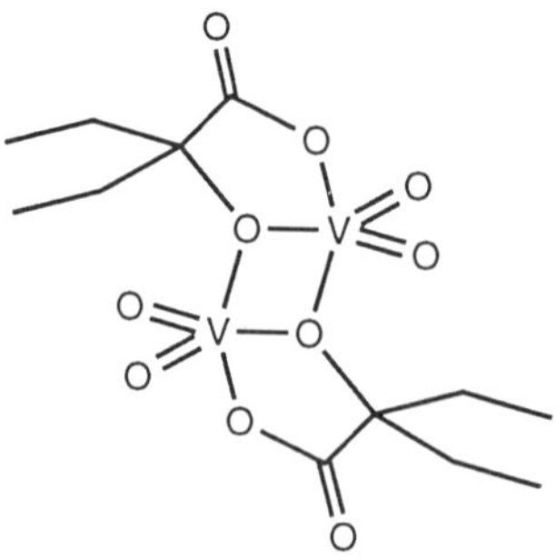

30

Figure 12.

between terminal and bridging oxygens in vanadates because of their different bond strengths. In the case of the ethylene glycolate(2−) dimer, ^{17}O NMR signals were measured after exchanging with $H_2{}^{17}O$. The chemical shift values (ca. 1,000 ppm relative to $H_2{}^{17}O$) were indicative of terminal oxygens with a VO bond strength of about 1.55 valence units, which is consistent with the proposed structure (Ray et al., 1995). The chemical shifts of bridging oxygen atoms in vanadium(V) oligomers differ from those of terminal oxygens by at least 200 ppm (Heath and Howarth, 1981; Ray et al., 1995). The line widths are typically of 500 and 1,000 Hz for terminal and bridging oxo groups (Heath and Howarth, 1981). In the case of cis-dioxo units the spectra distinguish nonequivalent oxo moieties and therefore provide information on the geometric arrangement of ligands (Crans et al., 1993; Crans and Shin, 1994).

Ligand-to-metal charge transfer bands in the UV-visible spectra are common for vanadium(V) complexes due to the high oxidation state of the metal center. It was reported that oxygen-to-vanadium(V) CT absorptions could distinguish six-coordinate from five-coordinate complexes formed by polydentate ligands (Crans and Shin, 1994). The absorption maxima shift from 367 nm (for vanadate at pH 8.9), to 380–400 mn (in five-coordinate complexes) and 410–430 nm (in six-coordinate complexes).

$[VO(SCH_2CH_2)_3N]$, which is the first oxovanadium(V) thiolate complex so far reported, is characterized by two strong charge-transfer transitions at 500 and 375 nm, which have been found to arise from S_π and S_σ ground states, respectively (Nanda et al., 1996).

Vanadium(V) o-catecholates exhibit intense absorption bands in the visible region (Cooper et al., 1982; Cass et al., 1986). The molar extinction coefficient of O(phenolate) $\rightarrow$ V^V CT bands may be remarkably high (up to 37,400 M^{-1} cm^{-1}) in low-symmetry complexes (Auerbach et al., 1990; Kabanos et al., 1992).

In monoperoxovanadates(V) peroxo $\rightarrow$ V(V) charge-transfer bands occur typically between 415 and 430 nm. For instance, it is observed at 425 nm in the monoperoxo vanadium(V) complex of nitrilotriacetate(3−) anion (NTA) $[VO(O_2)NTA]^{2-}$ (Djordjevic et al., 1995). In acidic solution the complexes decompose and the absorption distinctive of the red species $[VO(O_2)]^+$ gradually develops. The red monoperoxo species absorbs at 455 nm. On adding an excess of hydrogen peroxide the yellow diperoxo species $[VO(O_2)_2]^-$ is formed A maximum at 350 nm distinguishes the diperoxo species, whereas an isosbestic point at 404 nm indicates the equilibrium (Orhanovic and Wilkins, 1967).

Clague et al. (1993) reported that the conversion of $[LVO_2]^-$, where L is the dianionic form of N-(2-hydroxyphenyl)salicylideneamine, to $[LVO(O_2)]^-$, according to the reaction

$$[LVO_2]^- + H_2O_2 \rightleftharpoons [LVO(O_2)]^- + H_2O \qquad (5)$$

resulted in the appearance of an absorption maximum at 366 nm and a well-defined isosbestic point. ^{51}V NMR clearly indicated this process through a shift of the signal from −529 to −519 ppm.

Analogously to vanadium(IV), vanadium(V) can form complexes of rather high stability with ligands that are able to displace, partly or fully, oxygen(s) from the stable oxovanadium(V) or vanadate(V) ions. For instance, o-catechol and deferoxamine B (**31**) are effective in such a reaction. In aqueous solutions, in the very acidic pH range, the following complex formation processes were substantiated by Buglyò et al. (1995a):

$$VO_2^+ + H_4DFA^+ + H^+ \rightleftharpoons [VHDFA]^{3+} + 2H_2O \qquad (6)$$

$$VO_2^+ + H_4DFA^+ \rightleftharpoons [VOHDFA]^+ + H_2O + H^+ \qquad (7)$$

A well-defined isosbestic point supported an equilibrium between the two deeply colored species [VOHDFA]$^+$ (**32**) and [VHDFA]$^{3+}$ (**33**) in the very acidic pH range. The detachment of oxo ligand leads to a considerable shift in the ^{51}V NMR signals: -517 ppm for the "normal" cis-dioxo vana-

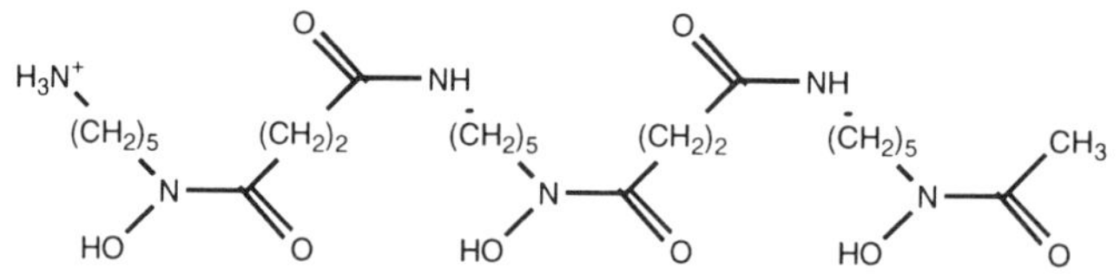

31

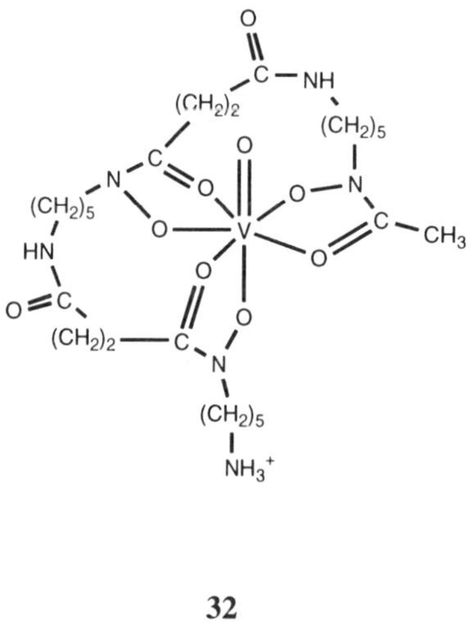

32

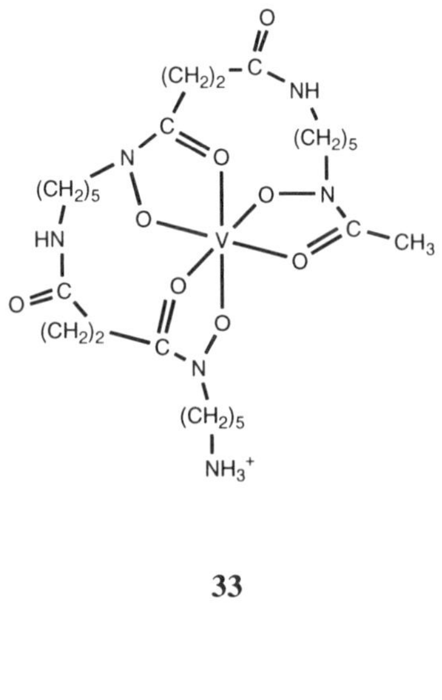

33

Figure 13.

dium(V) complex(es) existing at pH 7, -447 ppm for the mono-oxo species [VOHDFA]$^+$ and -199 ppm for the non-oxo complex.

Chemical shifts of similar magnitude (-202 and -235 ppm) were reported by Kabanos et al. (1992) for non-oxo species of vanadium(V), [V(dtbc)$_2$(phen or bipy)]$^+$, where dtbc is 3,5-di-t-butylcatecholato(2$-$). The higher deshielding in these complexes is presumably due to low-energy electronic excitations arising from CT transitions, most likely those dominating the UV-visible region of the absorption spectra.

^{1}H and ^{13}C NMR chemical shifts were used to determine the number and type of coordinate moieties in vanadium(V) complexes with some polyamino–polycarboxylic ligands (Crans and Shin, 1994). The coordination-induced shift (CIS), defined as the difference between the chemical shift in the complex versus that in the free ligand, was measured for the ^{1}H and ^{13}C nuclei of the ligands. The ^{13}C CIS values of carboxylate carbons ranged from -2.2 to 11.3 ppm. The larger positive values distinguished the coordinate groups; the smaller positive and the negative values indicated groups in noncoordinating arms. Detectable, although smaller, effects were also measured for carbon nuclei adjacent to coordinating groups; among these, the greatest CIS values were observed for carbons adjacent to two coordinating functionalities or for carbons adjacent to deprotonated hydroxyl groups. On the whole, it was concluded that the CIS values are very sensitive to the coordination of vanadium(V) and to the strength of the metal–ligand bonds.

REFERENCES

Alberico, E., and Micera, G. (1994). Phosphate complexation of oxovanadium(IV). Evidence of bidentate chelation of orthophosphate. *Inorg. Chim. Acta* **215**, 225–227.

Alberico, E., Dewaele, D., Kiss, T., and Micera, G. (1995). Oxovanadium(IV) complexation by adenosine-5′-di- and -triphosphate and nucleotide building-blocks. *J. Chem. Soc. Dalton Trans.*, pp. 425–430.

Atherton, N. M., and Winscom, C. J. (1973). Electron spin resonance and electronic structures of vanadyl bis(maleonitriledithiolene) and vanadium tris(maleonitriledithiolene). *Inorg. Chem.* **12**, 383–390.

Auerbach, U., Della Vedova, B. S. P. C., Wieghardt, K., Nuber, B., and Weiss, J. (1990). Syntheses and characterization of stable pseudo-octahedral tris-phenolato complexes of vanadium(III), $-$(IV), and $-$(V). *J. Chem. Soc. Dalton Trans.*, pp. 457–462.

Azuma, R., and Ozawa, T. (1994). Spectroscopic studies of the trans-[VOCl$_2$(H$_2$O)$_2$] chromophore in its hydrogen bonding adduct with 18-crown-6. *Inorg. Chim. Acta* **227**, 91–97.

Ballhausen, C. J., and Gray, H. B. (1962). Electronic structure of the vanadyl ion. *Inorg. Chem.* **1**, 111–122.

Batinic-Haberle, I., Birus, M., and Pribanic, M. (1991). Siderophore chemistry of vanadium. Kinetics and equilibrium of interaction between vanadium(IV) and desferrioxamine B in aqueous acidic solutions. *Inorg. Chem.* **30**, 4882–4887.

Belford, R. L., Chasteen, N. D., So H., and Tapscott, R. E. (1969). Triplet state of vanadyl tartrate binuclear complexes and EPR spectra of the vanadyl α-hydroxycarboxylates. *J. Am. Chem. Soc.* **91**, 4675–4680.

Boeri, E., and Ehrenberg, A. (1954). On the nature of vanadium in vanadocyte hemolysate from ascidians. *Arch. Biochem. Biophys.* **50**, 404–416.

Bond, M. R., Czernuszewicz, R. S., Dave, B. C., Yang, Q., Mohan, M., Verastegue, R., and Carrano, C. J. (1995). Spectroscopic and magnetostructural correlations in oxo-bridged dinuclear vanadium complexes of potential biological significance. *Inorg. Chem.* **34**, 5857–5869.

Branca, M., Micera, G., and Dessì, A. (1990a). Proton electron nuclear double resonance spectra on oxovanadium(IV) complexes formed by salicylic acid and *o*-diphenolic ligands in aqueous solution. *J. Chem. Soc. Dalton Trans.*, pp. 457–462.

Branca, M., Micera, G., Dessì, A., and Kozlowski, H. (1989). Proton electron nuclear double resonance study on oxovanadium(IV) complexes of D-galacturonic and polygalacturonic acids. *J. Chem. Soc. Dalton Trans.*, pp. 1283–1287.

Branca, M., Micera, G., Dessì, A., and Sanna, D. (1990b). Complexation of oxovanadium(IV) by humic and tannic acids. *J. Inorg. Biochem.* **39**, 109–115.

Branca, M., Micera, G., Sanna, D., Dessì, A., and Kozlowski, H. (1990c). Stabilization of the open chain structure of D-galacturonic acid in a dimeric complex with oxovanadium(IV). *J. Chem. Soc. Dalton Trans.*, pp. 1997–1999.

Branca, M., Micera, G., Sanna, D., Dessì, A., and Raymond, K. N. (1990d). Formation and structure of the tris(catecholato)vanadate(IV) complex in aqueous solution. *Inorg. Chem.* **29**, 1586–1589.

Bridgland, B. E., Fowles, G. W. A., and Walton, R. A. (1965). Some complexes of vanadium(IV) chloride. *J. Inorg. Nucl. Chem.* **27**, 383–389.

Buglyò, P., Culeddu, N., Kiss, T., Micera, G., and Sanna, D. (1995a). Vanadium(IV) and vanadium(V) complexes of deferoxamine B in aqueous solution. *J. Inorg. Biochem.* **60**, 45–59.

Buglyò, P., Dessì, A., Kiss, T., Micera, G., and Sanna, D. (1993). Formation of tris-chelated vanadium(IV) complexes by interaction of oxovanadium(IV) with catecholamines, 3-(3,4-dihydroxyphenyl)alanine and related ligands in aqueous solution. *J. Chem. Soc. Dalton Trans.*, pp. 2057–2063.

Buglyò, P., and Kiss, T. (1991). Formation of a tris complex in the vanadium(IV)–tiron system. *J. Coord. Chem.* **22**, 259–268.

Buglyò, P., Kiss, T., Alberico, E., Micera, G., and Dewaele, D. (1995b). Oxovanadium(IV) complexes of di- and triphosphate. *J. Coord. Chem.* **35**, 105–116.

Buglyò, P., Kiss, T., Micera, G., and Sanna, D. (1996). Oxovanadium(IV) complexes of ligands containing phosphonic acid moieties. *J. Chem. Soc. Dalton Trans.*, pp. 87–92.

Cass, M. E., Gordon, N. R., and Pierpont, C. G. (1986). Catecholate and semiquinone complexes of vanadium. Factors that direct charge distribution in metal–quinone complexes. *Inorg. Chem.* **25**, 3962–3967.

Castro, S. L., Cass M. E., Hollander, F. J., and Bartley, S. L. (1995). The oxovanadium(IV) dimer of 3-hydroxy-3-methylglutarate: X-ray crystal structure, solid state, magnetism, and solution spectroscopy. *Inorg. Chem.* **34**, 466–472.

Chasteen, N. D. (1981). Vanadyl(IV) EPR spin probes, inorganic and biochemical aspects. In L. J. Berliner, and J. Reuben (Eds.), *Biological Magnetic Resonance*. Plenum Press, New York, Vol. 3, pp. 53–119.

Chasteen, N. D., Belford, R. L., and Paul, I. C. (1969). The crystal structure and electronic spectrum of a vanadyl α-hydroxycarboxylate complex. Sodium tetraethylammonium bis(benzilato)oxovanadium(IV) di-2-propanolate. *Inorg. Chem.* **8**, 408–418.

Clague, M. J., and Butler, A. (1995). On the mechanism of cis-dioxovanadium(V)-catalyzed oxidation of bromide by hydrogen peroxide: Evidence for a reactive, binuclear vanadium(V) peroxo complex. *J. Am. Chem. Soc.* **117**, 3475–3484.

Clague, M. J., Keder, N. L., and Butler, A. (1993). Biomimics of vanadium bromoperoxidase: Vanadium(V)–Schiff base catalyzed oxidation of bromine by hydrogen peroxide. *Inorg. Chem.* **32**, 4754–4761.

Collison, D., Gahan, B., Garner, C. D., and Mabbs, F. E. (1980). Electronic absorption spectra of some oxovanadium(IV) compounds. *J. Chem. Soc. Dalton Trans.*, pp. 667–674.

Cooper, S. R., Koh, Y. B., and Raymond K. N. (1982). Synthetic, structural, and physical studies of bis(triethylammonium) tris(catecholato)vanadate(IV), potassium bis(catecholato)oxovanadate(IV), and potassium tris(catecholato)vanadate(III). *J. Am. Chem. Soc.* **104,** 5092–5102.

Cornman, C. R., Zovinka, E. P., Boyajian, Y. D., Geiser-Bush, K. M., Boyle, P. D., and Singh, P. (1995). Structural and EPR studies of vanadium complexes of deprotonated amide ligands: Effects on the ^{51}V hyperfine coupling constant. *Inorg. Chem.* **34,** 4213–4219.

Crans, D. C. (1994). Enzyme interactions with labile oxovanadates and other polyoxometalates. *Comments Inorg. Chem.* **16,** 35–76.

Crans, D. C., Chen, H., Anderson, O. P., and Miller, M. M. (1993). Vanadium(V)–protein model studies: Solid-state and solution structure. *J. Am. Chem. Soc.* **115,** 6769–6776.

Crans, D. C., and Shin, P. K. (1994). Characterization of vanadium(V) complexes in aqueous solutions: ethanolamine- and glycine-derived complexes. *J. Am. Chem. Soc.* **116,** 1305–1315.

Crans, D. C., Willging, E. M., and Butler, S. R. (1990). Vanadate tetramer as the inhibiting species in enzyme reactions in vitro and in vivo. *J. Am. Chem. Soc.* **112,** 427–432.

Czernuszewicz, R. S., Yan, Q., Bond, M. R., and Carrano, C. J. (1994). Origin of the unusual bending distortion in the (μ-oxo)divanadium(III) complex [V$_2$O(l-his)$_4$]: A reinvestigation. *Inorg. Chem.* **33,** 6116–6119.

Deeth, R. J. (1991). Electronic structure and *d–d* spectra of vanadium(IV) and VO^{2+} complexes: Discrete variational Xα calculations. *J. Chem. Soc. Dalton Trans.*, pp. 1467–1476.

Degani, H., Gochin, M., Karlish, S. J. D., and Shechter, Y. (1981). Electron paramagnetic resonance studies and insulin-like effects of vanadium in rat adipocytes. *Biochemistry* **20,** 5795–5799.

Deiana, S., Erre, L., Micera, G., Piu, P., and Gessa, C. (1980). Coordination of transition-metal ions to polygalacturonic acid: A spectroscopic study. *Inorg. Chim. Acta* **46,** 249–253.

Deng, H., Ray, W. J. Jr., Crans, D. C., Burgner, J. W. II, and Callender, R. (1993). Comparison of vibrational frequencies of critical bonds in ground-state complexes and in a vanadate-based transition-state analog complex of muscle phosphoglucomutase. Mechanistic implications. *Biochemistry* **32,** 12984–12992.

Desideri, A., Raynor, J. B., and Diamantis, A. A. (1978). ESR spectra of trigonal-prismatic bis[pentane-2,4-dione benzoylhydrazonato(2−)]vanadium(IV) and bis[4-phenylbutane-2,4-dione benzoylhydrazonato(2−)]vanadium(IV). *J. Chem. Soc. Dalton Trans.*, pp. 423–426.

Dessì, A., Micera, G., Sanna, D., and Strinna Erre, L. (1992). Vanadium(IV) and oxovanadium(IV) complexes of hydroxamic acids and related ligands. *J. Inorg. Biochem.* **48,** 279–287.

Djordjevic, C., Wilkins, P. L., Sinn, E., and Butcher, R. J. (1995). Peroxo aminopolycarboxylato vanadate(V) of an unusually low toxicity: Synthesis and structure of the very stable K$_2$[VO(O$_2$)(C$_6$H$_6$NO$_6$)]·2H$_2$O. *Inorg. Chim. Acta* **230,** 241–244.

Dunhill, R. H., and Smith, T. D. (1968). Electron spin resonance of vanadyl citrate and tartrate chelates. *J. Chem. Soc. A.*, pp. 2189–2192.

Dunhill, R. H., and Symons, M. C. R. (1968). Triplet state ESR spectra of vanadyl tartrate chelates in aqueous solution. *Mol. Phys.* **15,** 105–107.

Dutton, J. C., Fallon, G. D., and Murray, K. S. (1988). Synthesis, structure, ESR spectra, and redox properties of (N,N′-ethylenebis(thiosalicylideneaminato)oxovanadium(IV) and of related {S,N} chelates of vanadium(IV). *Inorg. Chem.* **27,** 34–38.

Forrest, J. G., and Prout, C. K. (1967). The crystal and molecular structure of ammonium vanadyl (+)-tartrate monohydrate. *J. Chem. Soc. A.*, pp. 1312–1317.

Frank, P., Carlson, R. M. K., and Hodgson, K. O. (1986). Vandyl ion EPR as a noninvasive probe of pH in intact vanadocytes from *Ascidia ceratodes*. *Inorg. Chem.* **25,** 470–478.

Garcia-Jaca, J., Insausti, M., Cortés, R., Rojo, T., Pizarro, J. L., and Arriortua, M. (1994). A new perspective of vanadyl-tartrate dimers. Synthesis, crystal structure, spectroscopic and magnetic properties of the chain compound: {[BaVO(C$_4$H$_2$O$_6$)(H$_2$O)$_4$]$_2$}$_n$. *Polyhedron* **13,** 357–364.

Garcia-Jaca, J., Rojo, T., Pizarro, J. L., Goni, A., and Arriortua, M. I. (1993). A new perspective on vanadyl tartrate dimers. Part II. Structure and spectroscopic properties of calcium vanadyl tartrate tetrahydrate. *J. Coord. Chem.* **30**, 327–336.

Gessa, C., Deiana, S., Erre, L., Micera, G., and Piu, P. (1981). The adsorption sites of oxovanadium(IV) ions in amorphous aluminium hydroxide: A spectroscopic characterization. *Colloids Surfaces* **2**, 293–297.

Hahn, C. W., Rasmussen, P. G., and Bayón, J. C. (1992). A dinuclear vanadyl(IV) complex of 3,5-dicarboxypyrazole: Synthesis, crystal structure and ESR. *Inorg. Chem.* **31**, 1963–1965.

Hambley, T. W., Hawkins, C. J., and Kabanos, T. A. (1987). Synthetic, structural and physical studies of tris(2,4-pentanedionato)vanadium(IV) hexachloroantimonate(V) and tris(1-phenyl-1,3-butanedionato)vanadium(IV) hexachloroantimonate(V). *Inorg. Chem.* **26**, 3740–3745.

Hasegawa, A. (1971). Electron spin resonance of a trinuclear vanadyl pyrophosphate complex. *J. Chem. Phys.* **55**, 3101–3104.

Hasegawa, A., Yamada, Y., and Miura, M. (1969). Electron spin resonance of the trinuclear vanadyl pyrophosphate complex. *Bull. Chem. Soc. Jpn.* **42**, 846.

Heath, E., and Howarth, O. W. (1981). Vanadium-51 and oxygen-17 nuclear magnetic resonance study of vanadate(V) equilibria and kinetics. *J. Chem. Soc. Dalton Trans.*, pp. 1105–1110.

Hitchman, M. A., Olson, C. D., and Belford, R. L. (1969). Behavior of the in-plane g tensor in low-symmetry d^1 and d^9 systems with application to copper and vanadyl chelates. *J. Chem. Phys.* **50**, 1195–1203.

Kabanos, T. A., Slawin, A. M. Z., Williams, D. J., and Woollins, J. D. (1992). New non-oxo vanadium-(IV) and -(V) complexes. *J. Chem. Soc. Dalton Trans.*, pp. 1423–1427.

Kanamori, K., Ookubo, Y., Ino, K., Kawai, K., and Michibata, H. (1991). Raman spectral study on the structure of a hydrolytic dimer of the aquavanadium(III) ion. *Inorg. Chem.* **30**, 3832–3836.

Kanamori, K., Teraoka, M., Maeda, H., and Okamoto, K. (1993). Preparation and structure of oxo-bridged dinuclear vanadium(III) complex $[V^{III}_2(L\text{-his})_4(\mu\text{-O})] \cdot 2H_2O$. *Chem. Lett.* **10**, 1731–1734.

Keramidas, A. D., Papaioannu, A. B., Vlahos, A., Kabanos, T. A., Bonas, G., Makriyannis, A., Rapropoulou, C. P., and Terzis, A. (1996). Model investigations for vanadium–protein interactions. Synthetic, structural, and physical studies of vanadium(III) and oxovanadium(IV/V) complexes with amidate ligands. *Inorg. Chem.* **35**, 357–367.

Kiss, T., Buglyò, P., Sanna, D., Micera, G., Decock, P., and Dewaele, D. (1995). Oxovanadium(IV) complexes of citric and tartaric acids in aqueous solution. *Inorg. Chem. Acta* **239**, 145–153.

Kiss, T., Micera, G., Sanna, D., and Kozlowski, H. (1994). Copper(II) and oxovanadium(IV) complexes of 2,3-dihydroxyterephthalic and 2,3,4-trihydroxybenzoic acids. *J. Chem. Soc. Dalton Trans.*, pp. 347–353.

Klich, P. R., Daniher, T., Challen, P. R., McConville D. B., and Youngs, W. J. (1996). Vanadium(IV) complexes with mixed O,S donor ligands. Syntheses, structures and properties of the anions tris(2-mercapto-4-methylphenolato)vanadate(IV) and bis(2-mercaptophenolato)oxovanadate(IV). *Inorg. Chem.* **35**, 347–356.

Knopp, P., and Wieghardt, K. (1991). Switching the mechanism of spin-exchange coupling in (μ-oxo)bis(μ-carboxylato)divanadium(III) complexes by protonation of the oxo bridge. *Inorg. Chem.* **30**, 4061–4066.

Kozlowski, H., Bouhsina, S., Decock, P., Micera, G., and Swiatek, J. (1991). Vanadyl(IV) complexes of lactobionic acid: Potentiometric and spectroscopic studies. *J. Coord. Chem.* **24**, 319–323.

Kwik, W.-L., and Stiefel, E. I. (1973). Single-crystal electron paramagnetic resonance study of a trigonal vanadium tris(dithiolate) complex. *Inorg. Chem.*, **12**, 2337–2342.

James, P. G., and Luckhurst, G. R. (1970). The electron resonance of exchange coupled vanadyl ions. *Mol. Phys.* **18**, 141–144.

Jezierski, A., and Raynor, J. B. (1981). Electron spin resonance spectra of dibromo- and dichloro-complexes of vanadium(IV). *J. Chem. Soc. Dalton Trans.*, pp. 1–7.

Jiang, F. S., and Makinen, M. W. (1995). NMR and ENDOR conformational studies of the vanadyl guanosine 5′-monophosphate complex in hydrogen-bonded quartet assemblies. *Inorg. Chem.* **34**, 1736–1744.

Lever, A. B. P. (1984). *Inorganic Electronic Spectroscopy*. Elsevier, Amsterdam, pp. 384–417.

Meier, R., Boddin, M., Schönherr, T., and Schmid, V. (1995). UV-Vis-investigation of V(III) complexes with biomimetic ligands. Elucidation of coordination numbers without knowing the crystal structure. *J. Inorg. Biochem.* **59**, 597.

Micera, G., and Dessì, A. (1989). Oxovanadium(IV) adsorption by plant roots. ESR identification of mobile and immobilized species. *J. Inorg. Biochem.* **35**, 71–78.

Micera, G., Sanna, D., and Dessì, A. (1996). Binding of oxovanadium(IV) to guanosine-5′-monophosphate. *Inorg. Chem.* **35**, 6349–6352.

Micera, G., Sanna, D., Dessì, A., Kiss, T., and Buglyò, P. (1993). Complex-forming properties of α-hydroxycarboxylic acids with oxovanadium(IV) ion. *Gazz. Chim. Ital.* **123**, 573–577.

Miller, G. A., and McClung, R. E. D. (1973). Electron spin resonance study of coordination of Lewis bases to vanadyl dithiophosphinate complexes. *Inorg. Chem.* **12**, 2552–2561.

Mukherjee, R. N., and Shastri, B. B. S. (1989). ESR studies of bis[(2,4,5-trimethylphenyl)dithiophosphinato]oxovanadium(IV). Calculation of bonding parameters. *J. Coord. Chem.* **20**, 135–139.

Mulks, C. F., Kirste, B., and van Willigen, H. (1982). ENDOR study of VO^{2+}-imidazole complexes in frozen aqueous solution. *J. Am. Chem. Soc.* **104**, 5906–5911.

Mustafi, D., Telser, J., and Makinen, M. (1992). Molecular geometry of vanady-ladenine nucleotide complexes determined by EPR, ENDOR, and molecular modeling. *J. Am. Chem. Soc.* **114**, 6219–6226.

Nanda, K. K., Sinn, E., and Addison, A. W. (1996). The first oxovanadium(V)–thiolate complex, $[VO(SCH_2CH_2)_3N]$. *Inorg. Chem.* **35**, 1–2.

Orhanovic, M., and Wilkins, R. G. (1967). Kinetic studies of the reactions of peroxy compounds of chromium(VI), vanadium(V), and titanium(IV) in acid media. *J. Am. Chem. Soc.* **89**, 278–282.

Pajdowski, L., and Jezowska-Trzebiatowska, B. (1996). A magnotochemical study of the hydrolysis of vanadium(III) ion. *J. Inorg. Nucl. Chem.* **28**, 443–446.

Parker, C. C., Reeder, R. R., Richards, L. B., and Rieger, P. H. (1970). A novel vanadyl pyrophosphate trimer. *J. Am. Chem. Soc.* **92**, 5230–5231.

Ray, W. J. Jr., Crans, D. C., Zheng, J., Burgner, J. W. II, Deng, H., and Mahroof-Tahir, M. (1995). Structure of the dimeric ethylene glycol–vanadate complex and other 1,2-diol–vanadate complexes in aqueous solution: Vanadate-based transition-state analog complexes of phosphotransferases. *J. Am. Chem. Soc.* **117**, 6015–6026.

Rehder, D. (1988). Interaction of vanadate ($H_2VO_4^-$) with dipeptides investigated by ^{51}V NMR spectroscopy. *Inorg. Chem.* **27**, 4312–4316.

Rehder, D. (1990). Biological applications of ^{51}V NMR spectroscopy. In N. D. Chasteen (Ed.), *Vanadium in Biological Systems*. Kluwer Academic Publishers, Dordrecht, pp. 173–197.

Rehder, D., Weidemann, C., Duch, A., and Priebsch, W. (1988). ^{51}V shielding in vanadium(V) complexes. A reference scale for vanadium binding sites in biomolecules. *Inorg. Chem.* **27**, 584–587.

Robles, J. C., Matsuzaka, Y., Inomata, S., Shimoi, M., Mori, W., and Ogino, H. (1993). Syntheses and structures of vanadium(III) complexes containing 1,3-diaminopropane-N,N,N′,N′-tetraacetate ([V(trda)]⁻) and 1,3-diamino-2-propanol-N,N,N′,N′-tetraacetate ([V₂(dpot)₂]²⁻). *Inorg. Chem.* **32**, 13–17.

Sakurai, H., Hirata, J., and Michibata, H. (1987). ESR spectra of vanadyl species in blood cells of ascidians. *Biochem. Biophys. Res. Commun.* **149**, 411–416.

Sanna, D., Micera, G., Buglyò, P., Kiss, T., Gajda, T., and Surdy, P. Oxovanadium(IV) complexes of imidazole-4-acetic, imidazole-4,5-dicarboxylic and pyrazole-3,5-dicarboxylic acids. *Inorg. Chim. Acta,* in press.

Selbin, J. (1966). Oxovanadium(IV) complexes. *Coord. Chem. Rev.* **1,** 293–314.

Shaver, A., Ng, J. B., Hall, D. A., Lum, B. S., and Posner, B. I. (1993). Insulin-mimetic peroxovanadium complexes: Preparation and structure of potassium oxodiperoxo(pyridine-2-carboxylato)vanadate(V), $K_2[VO(O_2)_2(C_5H_4 NCOO)]\cdot2H_2O$, and potassium oxodiperoxo(3-hydroxypyridine-2-carboxylato)vanadate(V), $K_2[VO(O_2)_2(OHC_5H_3NCOO)] \cdot 3H_2O$, and their reactions with cysteine. *Inorg. Chem.* **32,** 3109–3113.

Von Dreele, R. B., and Fay, R. C. (1972). Octahedral vanadium(IV) complexes. Synthesis and stereochemistry of vanadium(IV) β-diketonates. *J. Am. Chem. Soc.* **94,** 7935–7936.

Welch, J. H., Bereman, R. D., and Singh, P. (1988). Syntheses and characterization of two vanadium tris complexes of the 1,2-dithiolene 5,6-dihydro-1,4-dithiin-2,3-dithiolate. Crystal structure of $[(C_4H_9)_4N][V(DDDT)_3]$. *Inorg. Chem.* **27,** 2862–2868.

Zhou, D.-H., Wan, H.-L., Hu, S.-Z., and Tsai, K.-R. (1995). Syntheses and structures of the potassium-ammonium dioxocitratovanadate(V) and sodium oxocitratovanadate(IV) dimers. *Inorg. Chim. Acta* **237,** 193–197.

8

BIOACCUMULATION AND TRANSFER OF VANADIUM IN MARINE ORGANISMS

Pierre Miramand

Laboratoire de Biologie et Biochimie Marines (EA 1220), Université de La Rochelle, Pôle Sciences et Technologie, Avenue Marillac, 17042 La Rochelle, Cedex 1, France

Scott W. Fowler

International Atomic Energy Agency, Marine Environment Laboratory, P.O. Box 800, MC 98012, Monaco

Vanadium in the Environment. Part 1: Chemistry and Biochemistry, Edited by Jerome O. Nriagu.
ISBN 0-471-17778-4. © 1998 John Wiley & Sons, Inc.

---

1. SOURCES AND CONCENTRATIONS OF VANADIUM IN SEA WATER AND SEDIMENTS

The trace element vanadium enters the ocean mainly through natural weathering processes and atmospheric fallout (Bertine and Goldberg, 1971; Duce and Hoffman, 1976). The natural source of vanadium in sea water is both volcanic activity and erosion of the earth's crust; Zoller et al. (1973) have estimated this contribution to be 37×10^9 g per year. The main anthropogenic source of this element to the oceans is combustion of fossil fuels that contain a relatively high proportion of vanadium. As an example, Venezuela crude oils contain 200–1,000 ppm of vanadium (Zoller et al., 1973), and there is about 200 ppm vanadium in the Canadian Athabasca oil sands (Funk and Gomez, 1977). In addition, vanadium concentrations in Saudi and Kuwaiti crudes vary between 29 and 60 mg/kg oil (Sadiq and Zaidi, 1984).

Calculations show that fossil fuel combustion introduces from 12,000 to 24,000 tons of vanadium each year into the sea, of which roughly 10–15% is deposited in the ocean as atmospheric fallout (Bertine and Goldberg, 1971; Duce and Hoffman, 1976; Walsh and Duce, 1976). Once in the sea, vanadium may be rapidly solubilized from the particles enriched with this element (Walsh and Duce, 1976). Recently, Hope (1994) estimated the contribution of anthropogenic vanadium in particulate emissions to be about 53% of total atmospheric vanadium loading. Hope reported that oceanic deposition of vanadium from this source was ≈5% of total ocean vanadium loading and could increase the sea water vanadium concentration to 44 pg L^{-1} per year. This addition does not constitute a significant environmental threat on a global scale.

Industrial and domestic effluents may be an additional source of local input; however, relatively little quantitative information on this pathway is available. Vanadium may occur in high concentration in the acid effluents of factories producing titanium dioxide (Grange, 1974). For example, in the Bay of Seine (France), it has been reported that about 317 tons per year of vanadium are discharged into the coastal environment in this manner (Anonymous, 1990). Other important sources are products of leachates and effluents from the mining and milling of uranium (Linsted and Kruger, 1969; Van Zinderen Bakker and Jaworski, 1980).

Although highly dependent on its various sources, the chemical form of vanadium in sea water may vary. At the pH of sea water, vanadium is potentially soluble in two oxidation states: +4 and +5. Ladd (1974) has shown that in sea water this metal is in steady state between the vanadate anion (V), which is the dominant species, and the cationic form (IV), which represents only about 1%. Thus, vanadium in oxic sea water should be present as vanadium(V) which can be hydrolyzed to $VO_2(OH)^{2-}$ at neutral pH. Under moderately reducing conditions, vanadium(IV) is stable as the oxovanadium cation VO_2^+ which should also hydrolyze at sea water pH to $VO(OH)_3^-$ (reviewed by Emerson and Huested, 1991).

Concentrations of dissolved vanadium in the ocean range from approximately 1 to 3 μg/L (20–60 nM) (Burton, 1966; Goldberg et al., 1971; Weiss et al., 1977). In coastal areas, measurements have shown large geographical variations in vanadium concentrations, ranging from 0.61 μg/L (12 nM) in the Black Sea to 7.1 μg/L (140 nM) near the coasts of Great Britain (Burton, 1966; Chan and Riley, 1966). Recently, Yeats (1992) reported for the Gulf of St. Lawrence an average vanadium concentration of 1.22 μg/L (24 nM) in deep waters. This value is in good agreement with provisional measurements in the open ocean which show a more homogenous content. Thus, Morris (1975) reported for ocean waters west of the United Kingdom a vanadium concentration of 1.19 μg/L (23.4 nM), whereas Huizenga and Kester (1982) and Weisel et al. (1984), respectively, gave values of around 1.78 μg/kg (35 nM) and 1.22 μg/kg (24 nM) in the Sargasso Sea. Collier (1984) reports vanadium concentrations in the Pacific Ocean of 1.53–2.03 μg/kg (30–40 nM), and Jeandel et al. (1987), in four vertical profiles in the Atlantic and Pacific Oceans and the Mediterranean Sea, have determined mean vanadium concentrations of 1.78 μg/kg (35 nM), 1.63 μg/kg (32 nM), and 2.49 μg/kg (49 nM), respectively. Furthermore Sherrel and Boyle (1988) have shown that a maximum total dissolved concentration of vanadium, 1.78 μg/kg (35 nM), occurs in the core of the Atlantic inflow jet in the Alboran Sea.

Compared with dissolved concentrations, particulate vanadium in sea water is low. Jeandel et al. (1987) estimate the contribution of particulate vanadium to be 0.001–0.01 μg/kg in open sea water, values close to the results of Buat-Ménard and Chesselet (1979) and Collier (1984), who did not find more than 8 pmol/kg vanadium of sea water in particulate suspended matter. In coastal sea water, the concentrations of particulate vanadium appear to be 1–3 orders of magnitude higher. For example, Kalogeropoulos et al. (1989) observed in the vicinity of the Pireus coast values from 3 to 260 ng/L, with maximum concentrations observed near the main anthropogenic source of vanadium in the area (Athens sewage outfall). In the Bay of Seine, an area receiving a relatively large input of vanadium, Miramand et al. (1993) have found comparable levels of vanadium (50–100 ng/L) associated with particulate matter. In conclusion, if dissolved vanadium is the main source of vanadium for marine organisms, particulate vanadium may be a vector of contamination for suspension feeders in coastal areas receiving anthropogenic inputs.

Vanadium fixed on particles can sink with the particulates, eventually reaching the deep-sea sediments (Bertine and Goldberg, 1977; Buat-Ménard and Chesselet, 1979). For example, Loring (1976) found vanadium mainly in the detrital fractions of sediments in the Sagenay Fjord, which accounted for 71–91% of the total elemental concentration. Furthermore, Bertine and Goldberg (1977) reported that anthropogenic fluxes of vanadium to sediments were one order of magnitude greater in inshore areas near the source of pollution than were found further offshore. Yeats (1992) has confirmed this trend and stated, "Vanadium shows extensive depletion in coastal waters indicative of an important removal process that occurs in these regions." As a result of this process, vanadium concentrations in sediments are generally high, varying with granulometry and location from 20 to 200 $\mu g/g$ dry weight (Loring, 1976; Kalogeropoulos et al., 1979; Boust, 1981; Anderlini et al., 1982). Sediment-bound vanadium could be an another source of contamination for benthic organisms, particularly detrivorous species.

2.   LEVELS OF VANADIUM IN MARINE ORGANISMS

In Tables 1–5, we have listed data on levels of vanadium in both planktonic and benthic marine organisms, which were obtained mainly after 1950. The older values are mainly of only historical interest. The data are scarce (≈150) compared with those for other heavy metals (for example, (Cd, Cu, Pb, Zn), for which during the same period some 1,000 values are reported in the literature (see reviews by Bryan, 1984, and Eisler, 1981). One reason for the general lack of data is the difficulty of analyzing vanadium in organisms.

2.1.   Benthic Species

Seaweeds have similar vanadium concentrations ranging from 1 to 8 $\mu g/g$ dry wt for chlorophytes and 0.2 to 6 $\mu g/g$ dry wt for pheophytes (Table 1).

Among benthic organisms, the ascidians have been the most widely studied species. The ability of certain ascidians to concentrate vanadium was first reported by Henze (1911). Since that time, a great number of articles, including analyses of various species of ascidians and experimental studies on the possible physiological function of vanadium, have been devoted to this group of invertebrates. In Table 2 recent data are reported for vanadium in different species of ascidians. Vanadium is always high in Phlebobranchia, varying from 1,000 to 9,000 $\mu g/g$ dry weight. In Aplousobranchia, the ability to concentrate this element varies between families and species. A special chapter of this

Table 1   Vanadium Concentrations (μg/g Dry Weight) in Seaweeds

Species	Origin	[V]	Authors
Bryophyta			
Fontinalis sp.	Baltic sea	3.7–5.7	Söderlung et al., 1988
Chlorophyta			
Codium fragile	Japan	1.72	Yamamoto et al., 1970
Enteromorpha spp.	Channel	6.3 ± 2.1	Present work
Ulva lactuca	Channel	1.9 ± 0.7	Present work
U. fasciata	Japan	0.99–8.85	Yamamoto et al., 1970
Pheophyta			
Ascophyllum nodosum	Atlantic	0.7–1.9	Black and Mitchell, 1952
Desmaretia firma	South Africa	1.5–2.5	Cockerill et al., 1978
Fucus serratus	Atlantic	3.3	Black and Mitchell, 1952
F. spiralis	Atlantic	1.9	Black and Mitchell, 1952
F. vesiculosus	Atlantic	1.9	Black and Mitchell, 1952
F. vesiculosus	Atlantic	0.26–2.15[a]	Forsberg et al., 1988
F. vesiculosus	Baltic sea	0.17–2.15[a]	Söderlung et al., 1988
Laminaria cloustoni	Atlantic	1.1	Black and Mitchell, 1952
L. digitata	Atlantic	0.4–1.9	Black and Mitchell, 1952
L. digitata	Channel	0.6 ± 0.3	Present work
Pelvetia canaliculata	Atlantic	0.7–2.6	Black and Mitchell, 1952
Sagarssum spp.	Japan	0.43–6.15	Yamamoto et al., 1970

[a] Values from contaminated areas.

volume (Michibita and Kanamori, Chapter 10) is consecrated to the important topic of selective accumulation of vanadium by ascidians.

Besides ascidians, certain species of tubicole annelids are known for their enhanced capacity to concentrate vanadium in their tissues, levels of which approach those reported for tunicates. The first data were obtained for a sabellid annelid by Popham and D'Auria (1982) (Table 3). Recently, Ishii et al. (1993, 1994) found comparable levels in another species of tubicole annelid, *Pseudopotamilla occelata,* from the Sea of Japan (Table 3), the highest concentrations (5,500 ± 1,800 μg/g dry wt) being located in the worm's branchial crown, which is composed of many bipinnate radioles. The detailed characterization of vanadium in this worm species is discussed by Ishii (this volume, Chapter 9).

Except for the Phlebobranchia, certain species of Aplousobranchia and these particular tubicole worms, vanadium concentrations in benthic invertebrates are generally low and very similar for all the groups analyzed, with levels varying from 1 to 4 μg/g dry wt (Table 3). However, there are limited data that indicate that bivalves collected near polluted areas can reach concentrations of a few tens of micrograms per gram dry wt. Unfortunately, there is no comparable information concerning important phyla or invertebrate

Table 2   Vanadium Concentrations (μg/g Dry Weight) in Ascidians

Species	Origin	[V]	Authors
Aplousobranchia			
Aplidium sp.	Pacific	9,075	Hawkins et al., 1983
Ciona intestinalis	Pacific	100	Goldberg et al., 1951
C. intestinalis	Atlantic	29	Kustin et al., 1975
Clavelina huntsmani	Pacific	30	Swinehart et al., 1974
C. spp.	Pacific	5,500	Hawkins et al., 1983
Cystodytes lobatus	Pacific	50	Swinehart et al., 1974
Distaphia occidentalis	Pacific	600–1,200	Swinehart et al., 1974
Eudistoma amphus	Pacific	880	Hawkins et al., 1983
E. diaphanes	Pacific	25	Swinehart et al., 1974
Leptoclinides dubius	Pacific	6,800–8,700	Hawkins et al., 1983
L. lissus	Pacific	10,000	Hawkins et al., 1983
L. reticulatus	Pacific	6,300–8,000	Hawkins et al., 1983
Podoclavella maluccensis	Pacific	150	Hawkins et al., 1983
Polyclinum spp.	Pacific	250	Hawkins et al., 1983
Rhopalaea crassa	Pacific	7,000	Hawkins et al., 1983
R. abdominalis	Pacific	1,800	Swinehart et al., 1974
Phlebobranchia			
Ascidia ceratodes	Pacific	1,300	Swinehart et al., 1974
A. sydneyensis	Pacific	2,000	Hawkins et al., 1983
Ecteinascidia diaphanes	Pacific	2,700	Hawkins et al., 1983
E. hataii	Pacific	1,200	Hawkins et al., 1983
E. nexa	Pacific	4,500	Hawkins et al., 1983
Perophora annectens	Pacific	700–9,000	Swinehart et al., 1974
Phallusia juliinea	Pacific	3,000	Hawkins et al., 1983
Stolidobranchia			
Halocynthia roretzi	Japan	0.12 (FW)	Ikebe and Tanaka, 1979

FW: Values in μg/g fresh weight.

groups such as sponges, cnidaria, lophophora, gastropod mollusks, polyplacophora mollusks, or peracaridean crustaceans. Vanadium data for fish are even more sparse; however, concentrations appear to be of the same order of magnitude as those measured in invertebrates (Table 4).

2.2.   Plankton

Vanadium concentrations have been measured in phytoplankton, both in chlorophytes grown in artificial media and in diatoms collected in situ. In

Table 3   Vanadium Concentrations (μg/g Dry Weight) in Benthic Invertebrates Other Than Ascidians

Species	Origin	[V]	Authors
ANNELIDAE			
Arenicola marina	Channel	1.7 ± 0.2	Present work
Eudistylia	British		Popham and
vancouveri	Columbia	94–786.1[a]	D'Auria, 1982
Nereis diversicolor	Atlantic	1.8 ± 0.6	Present work
Perinereis cultrifera	Atlantic	0.7	Bertrand, 1950
Pseudopotamilla			
occelata	Japan	510 ± 330	Ishii et al., 1994
CRUSTACEANS			
Cirripeds			
Lepas anatifera	Atlantic	1.2	Bertrand, 1943
Decapods			
			Blotcky et al.,
Callinectes sapidus	Texas coast	1.09–1.84	1979
Carcinus maenas	Atlantic	0.4	Bertrand, 1943
C. maenas	Mediterranean	3 ± 1	Miramand et al., 1981
Crangon crangon	Channel	1.4 ± 0.5	Present work
Lysmata	Mediterranean	2.9	Miramand et al.,
seticaudata			1981
Pandalus sp.	Japan	0.07	Fukai and Meinke, 1962
Parapenaeus	Mediterranean	0.004–0.039 (FW)	Papadopoulou et
longirostris			al., 1980
Penaeus setiferous	Texas coast	0.4–3.05[a]	Blotcky et al., 1979
Tetraclita	Japan	0.208–0.501 (FW)	Ikebe and
squamosa			Tanaka, 1979
japonica			
ECHINODERMS			
Holothuria forskali	Mediterranean	4.0	Miramand et al., 1982
Marthasterias	Mediterranean	1.5	Miramand et al.,
glacialis			1982
Ophiothrix fragilis	Channel	0.5 ± 0.2	Present work
Paracentrotus	Mediterranean	3.7	Miramand et al.,
lividus			1982
MOLLUSKS			
Bivalvia			
Crassostrea	Texas coast	0.53–1.42[a]	Blotcky et al.,
virginica			1979
Cyrtodaria	Pacific	4.3–20.3[a]	Bourgoin and
hurriana	(Arctic)		Risk, 1987

Table 3   (*Continued*)

Species	Origin	[V]	Authors
Malleus regula	Persian Gulf	1.4	Fowler et al., 1993
Meretrix meretrix	Persian Gulf	0.13–0.35 (FW)	Sadiq et al., 1992
M. meretrix	Persian Gulf	1.7–2.1	Fowler et al., 1993
Mytilus edulis	New Zealand	2–8	Brooks and Rumsby, 1965
M. edulis	Atlantic coast (U.S.)	0.82–1.06	Latouche et al., 1981
M. edulis	Japan	0.184 (FW)	Ikebe and Tanaka, 1979
M. edulis	English Channel	0.9 ± 0.3	Present work
M. edulis	English Channel	1.5–5.5[a]	Coulon et al., 1987
M. edulis	Mediterranean	0.61	Ünsal, 1978a
M. galloprovincialis	Mediterranean	0.4 ± 0.2	Miramand et al., 1980
Ostrea sinuata	New Zealand	2–8	Brooks and Rumsby, 1965
Pecten maximus	English Channel	1.9 ± 0.1	Present work
P. novae-zelandiae	New Zealand	5–14	Brooks and Rumsby, 1965
Perna perna	Persian Gulf	2.0–4.5	Fowler et al., 1993
Pinctada margaritifera	Persian Gulf	1–34[a]	Fowler et al., 1993
Placeopecten magellanicus	Atlantic coast	11.4–45.7[a]	Pesh et al., 1977
Saccostrea cucullata	Persian Gulf	0.06–2.2	Fowler et al., 1993
Spondylus sp.	Persian Gulf	3.7–10.7[a]	Fowler et al., 1993
Tapes sulcarius	Persian Gulf	1.1	Fowler et al., 1993
Trachycardium lacunosum	Persian Gulf	2.2	Fowler et al., 1993
CEPHALOPODIA			
Eledone cirrhosa	English Channel	1 ± 0.2	Miramand and Bentley, 1992
Octopus vulgaris	Mediterranean	0.7 ± 0.1	Miramand and Guary, 1980
Ommastrephes illicebrosa	Atlantic	4	Nicholls et al., 1959
Sepiola rondeleti	Mediterranean	1.1–5.3	Fowler et al., 1985

[a] Values from contaminated area.

Table 4   Vanadium Concentrations (μg/g Dry Weight) in Fish

Species	Origin	[V]	Authors
Anguilla anguilla (*Leptocephale*)	Mediterranean	0.15–2.3	Fowler et al., 1985
≪Barracuda≫	Japan	0.112 (FW)	Ikebe and Tanaka, 1979
≪Black rockfish≫	Japan	0.075 (FW)	Ikebe and Tanaka, 1979
Gobius minutus	Mediterranean	2.6 ± 0.4	Miramand et al., 1992
Merlangus vulgaris	Atlantic	0.14	Bertrand, 1942
Mictophum glaciale	Mediterranean	0.08	Fowler, 1986
Mullus barbatus	Mediterranean	0.002–0.02 (FW)	Papadopoulou et al., 1980
M. barbatus	Mediterranean	2 ± 1	Miramand et al., 1980
Scomber japonicus	Mediterranean	0.09	Fowler et al., 1985
Vinciguerra sp.	Mediterranean	0.15	Fowler et al., 1985

FW: Values in μg/g fresh weight.

either case, average vanadium levels were similar at approximately 3 μg/g dry wt (Table 5).

The majority of vanadium concentration data in zooplankton are for crustacean species. For other planktonic groups, data are very sparse (Table 5). If we disregard the concentrations reported by Nicholls et al. (1959), which are 1–2 orders of magnitude greater than those found by other authors, vanadium concentrations in plankton are in general ≤1 μg/g dry wt. Only pteropods (mainly their shells) display higher concentrations (16–290 μg/g dry wt). Nevertheless, as with benthic species, there is an overall lack of data for vanadium concentrations in plankton, in particular gelatinous forms and planktonic larvae.

2.3. Sea Mammals

Levels of vanadium in liver tissues from ten species of cetaceans and pinnipeds have been measured (Table 6). Bioaccumulation of vanadium in the samples was observed for all species studied. The concentrations of vanadium in hepatic tissue from marine mammals collected in Alaskan sea waters were higher than those measured in species living along the Atlantic coast (Mackey et al., 1996). According to these authors this observation deserves additional studies including measurement of the levels of vanadium in bone and kidney tissues, which are known to accumulate vanadium in terrestrial mammals (Waters, 1977, Parker and Sharma, 1978; Chasteen, 1983).

2.4. Vanadium in Food Chains

Environmental concentration factors (CF) can be estimated by dividing vanadium concentrations in organisms (μg/g wet wt) by an average dissolved concentration of 2 ppb vanadium in sea water. The data obtained are reported

Table 5   Vanadium Concentrations (μg/g Dry Weight) in Planktonic Species

Species	Origin	[V]	Authors
PHYTOPLANKTON			
Chlorophyta			
Chlorella salina	Culture	3.7	Riley and Roth, 1971
Dunaliella marina	Culture	4.4	Ünsal, 1978a
D. primolecta	Culture	2.4	Riley and Roth, 1971
D. tertiolecta	Culture	2.9	Riley and Roth, 1971
Monochrysis lutheri	Culture	3.1	Riley and Roth, 1971
Stichococcus bacillaris	Culture	2.4	Riley and Roth, 1971
Diatoms			
Biddulphia spp.	English Channel	1.5–4.7[a]	Miramand et al., 1993
Coscinodiscus spp.	English Channel	1.9–3.0[a]	Miramand et al., 1993
ZOOPLANKTON			
Chetognatha			
Sagitta elegans	Atlantic	13[b]	Nicholls et al., 1959
S. hexaptera	Mediterranean	1.0	Fowler et al., 1985
Crustaceans			
Amphipods			
Phronima sedentaria	Mediterranean	0.45	Fowler, 1986
Copepods			
Acartia clausi	English Channel	0.7 ± 0.1	Miramand et al., 1993
Calanus finmarchus	Atlantic	21[b]	Nicholls et al., 1959
Centropages typicus + *C. hamatus*	Atlantic	16[b]	Nicholls et al., 1959
Microplankton (mainly copepods)	Mediterranean	1.45	Fowler, 1986
Decapods			
Acanthephyra purpurea	Atlantic	0.37 ± 0.42	Ridout et al., 1989
Gennadas valens	Atlantic	0.3–1.00	Ridout et al., 1989
Pasiphae sivado	Mediterranean	0.07	Fowler, 1986
Sergia robustus	Atlantic	1.1 ± 0.4	Ridout et al., 1989
Systellaspis debilis	Atlantic	1.49–1.84	Ridout et al., 1989
Euphausiids			
Euphausia krohnii	Atlantic	45[b]	Nicholls et al., 1959
E. spp.	Mediterranean	0.48 ± 0.45	Fowler, 1986
Meganyctiphanes norvegica	Atlantic	0.17 ± 0.11	Ridout et al., 1989
M. norvegica	Mediterranean	0.23	Fowler, 1986
Nematobrachion boopis	Atlantic	0.19 ± 0.06	Ridout et al., 1989
Nematoscelis megalops	Atlantic	0.18 ± 0.09	Ridout et al., 1989
Thysanopoda microphtalma	Atlantic	0.36 ± 0.12	Ridout et al., 1989

Table 5 (*Continued*)

Species	Origin	[V][a]	Authors
Mysids			
Eucopia unguiculata	Atlantic (NE)	0.32 ± 0.23	Ridout et al., 1989
E. sculpticauda	Atlantic (NE)	0.24 ± 0.27	Ridout et al., 1989
Mysidopsis slabberi	Channel	3.6–5.1[a]	Miramand et al., 1993
Cnidaria			
Abylopsis tetragona	Mediterranean	0.03–0.27	Fowler et al., 1985
Aurelia aurita	Mediterranean	0.44	Fowler et al., 1985
Cyanea capillata	Atlantic	5[b]	Nicholls et al., 1959
Ctenaria			
Beroe cucumis	Atlantic	8[b]	Nicholls et al., 1959
Pleurobrachia pileus	Channel	0.4 ± 0.2	Miramand et al., 1993
Pteropod Mollusks			
Test of different species	Gulf of Mexico	50–290	Pyle and Tieh, 1970
Cliona limacina	Atlantic	16[b]	Nicholls et al., 1959
Euclio pyramidata	Mediterranean	0.8	Fowler et al., 1985
Limacina retroversa	Atlantic	85[b]	Nicholls et al., 1959
Polychaeta			
Alciopa cantrainii	Mediterranean	0.08–0.3	Fowler et al., 1985
Tunicates			
Pyrosoma atlanticum	Mediterranean	0.3–0.52	Fowler et al., 1985
Salpa maxima	Mediterranean	0.12–0.85	Fowler et al., 1985
S. fusiformis	Atlantic	7[b]	Nicholls et al., 1959

[a] Values from contaminated areas.
[b] Earlier spectrographic method, giving very high values that are questionable.

Table 6   Vanadium Concentrations (μg/g Wet Weight) in Marine Mammal Liver Tissue

Species	Origin	[V]	Authors
CETACEANS			
Balaena mysticetus	Alaskan	0.084–1.2	Mackey et al., 1996
Delphinapterus leucas	Alaskan	0.03–0.19	Mackey et al., 1996
Globicephala melas	U.S. Atlantic coast	<0.01–0.02	Mackey et al., 1995
Lagenorhynchus acutus	U.S. Atlantic coast	<0.01–0.06	Mackey et al., 1995
Phocoena phocoena	U.S. Atlantic coast	<0.01–0.02	Mackey et al., 1995
PINNIPEDS			
Callorhinus ursinus	Bering Sea	0.11–0.84	Zeisler et al., 1993
Erignathus barbatus	Alaskan	0.15–1.04	Mackey et al., 1996
Halichoerus grypus	Atlantic	0.024–0.172	Frank et al., 1992
Phoca hispada	Atlantic	0.023–0.264	Frank et al., 1992
P. hispada	Alaskan	0.019–0.47	Mackey et al., 1996
P. vitulina	Atlantic	0.003–0.282	Frank et al., 1992

in Figure 1. It is clear that, in passing along the food chain from micro or macroalgae to tertiary carnivores, concentration factors display no noticeable trend in benthic food chains, whereas clearly they decrease in pelagic food chains. Thus, there is no indication of a biomagnification effect for vanadium in marine food chains.

3.   VANADIUM DISTRIBUTION IN MARINE ORGANISMS OTHER THAN ASCIDIANS

3.1   External Contamination in Invertebrates

Reported in Table 7 is the percentage distribution of vanadium found in some tissues of benthic invertebrates obtained from both in situ measurements and

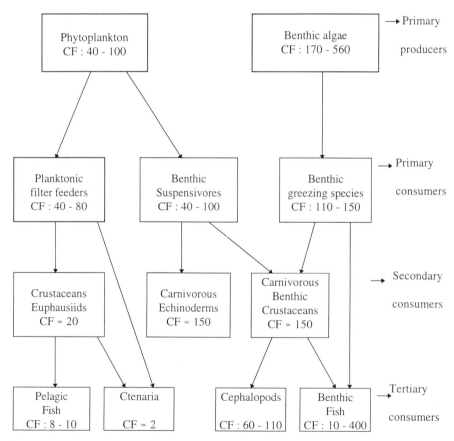

Figure 1. Environmental vanadium concentration factors (CF) in organisms in typical marine food chains. CF were estimated by dividing stable vanadium concentrations in organisms (μg/g wet wt) by an average of 2 ng/g of vanadium in sea water (concentration data taken from Tables 1–5).

Table 7 Percentage Distribution of Vanadium in Some Tissues of Benthic Invertebrates

Species	Body or Shell or Test		Digestive Gland		Muscle		Digestive Tract	
	Exp	In situ	Exp	In situ	Exp	In situ	Exp	In situ
CRUSTACEANS								
Carcinus maenas[1]	90	90	1.3	6.5	0.8	1.7	0.2	0.7
Lysmata seticaudata[1]	58	43.5	8.3	44.2	3.8	2.4	5.4	7.9
ECHINODERMATA								
Holothuria forskali[1]	77.1	83.7	—	—	—	—	11.3	3.6
Marthasterias glacialis[1]	70	95.9	25.1	3.4	—	—	4.9	0.7
Paracentrotus lividus[1]	63.3	96.6	—	—	—	—	13.9	0.9
BIVALVIA MOLLUSKS								
Mytilus galloprovincialis[1]	96	97.2	1.1	1.7	0.3	<0.05	—	—
CEPHALOPOD MOLLUSKS								
Eledone cirrhosa[2]	—	—	—	62	—	—	—	>0.5
Octopus vulgaris[3]	—	—	—	32.2	—	—	—	1.3
Sepia officinalis[2]	—	<2	—	71	—	—	—	1.4

Data from: (1) Miramand et al., 1980, 1981, 1982; (2) Miramand and Bentley, 1992; (3) Miramand and Guary, 1980.
Exp: Values obtained experimentally after uptake from sea water labeled with ^{48}V.
In situ: Based on stable vanadium in individuals collected in situ.

uptake after experimental exposure of individuals to ^{48}V. In most of the invertebrates, except for cephalopods, vanadium content (from 44 to 97%) resides in external tissues, (body wall or test), which are in contact with sea water. Although the mechanism for vanadium uptake from water by marine organisms is not well understood, it is clear that passive adsorption processes may predominate in hard-shelled, invertebrate forms. Both experimental studies and in situ analyses suggest that, in these organisms, exchanges between vanadium in water and external tissues are very slow (Miramand et al., 1980, 1981, 1982). This slow rate of uptake into external tissues is most probably due to the fact that vanadium is deposited in these tissues throughout lifetime exposure to this element.

Following short-term exposure, the highest amount of vanadium was found in mussel (*Mytilus edulis, M. galloprovincialis*) byssus and shell (Miramand et al., 1980; Edel and Sabbioni, 1993). Gills and exoskeleton were the most contaminated tissues noted in crustaceans (crab and shrimp) (Miramand et al., 1981). The strong adsorption of vanadium in these tissues is probably due to the affinity of the metal for polysaccharides, which are a constituent of conchioline, byssal threads, and chitin. For example, after uptake of vanadium from sea water, mussel shell periostracum contains a sizeable pool ($\approx 40\%$) of firmly bound vanadium that is not readily leached, even after much of the calcareous layer has been dissolved by acid (Miramand et al., 1980).

Considering the large percentage of vanadium associated with external tissues of invertebrates, it seems clear that deposition of these tissues into sediments may mobilize a substantial fraction of vanadium in the oceans. In this context, considering the large fraction of vanadium associated with crustacean exuviae (from 40% to 80% body burden) and the relatively high frequency at which many species molt, it seems clear that cast molts will play an equally important role in the marine biogeochemical cycle of vanadium. As a consequence of molting, the large amount of vanadium associated with periodically shed hard tissue indicates that vanadium levels measured in whole crustaceans will strongly depend upon the intermolt period of the organism at the time of sampling. Thus, crustaceans would not be ideal bioindicator organisms for monitoring vanadium in marine waters. On the other hand, mussels are often employed as bioindicator organisms (Goldberg et al., 1978). Experimental studies have shown that environmental factors (e.g., salinity, temperature) can influence uptake and loss of vanadium in mussels. For example, measurement of vanadium levels in mussels during a 9-month period indicated that gonadal tissue burdens display seasonal variations that correspond to the gametogenesis period, whereas levels in somatic tissues fluctuate in an irregular manner (Latouche and Mix, 1981). Thus, we conclude that caution must be exercised in attempting to use mussels as bioindicators of ambient vanadium levels. Byssus, on the other hand, may prove to be a useful tissue in this context because of both its very rapid uptake response to increased vanadium levels in water and its relatively high vanadium concentration (Miramand et al., 1980; Edel and Sabbioni, 1993). In fact, in mussels collected in the vicinity of a titanium dioxide industrial effluent containing 0.14 g/L of vanadium, vanadium in the byssus threads was significantly enhanced (Coulon et al., 1987).

3.2. Internal Deposition in Invertebrates

Vanadium in the digestive system, particularly in the digestive gland, is enriched relative to other internal tissues of marine invertebrates; however, levels are low overall (Table 8). Experimental studies showed that direct uptake from water into internal tissues is also a very slow process (Miramand et al., 1980, 1981, 1982). Vanadium uptake into soft parts is generally 1–2 orders of magnitude lower than has been reported for other heavy metals. This is illustrated for mussels in Table 9.

Digestive glands of invertebrates, especially crustaceans and mollusks, are believed to play a role in the storage and detoxification of many heavy metals (Coombs and George, 1978; Bryan, 1984), and likewise this organ may be involved in the metabolism of vanadium. To date, the subcellular distribution of vanadium in this organ has been studied only in mussels. In the exposure of *Mytilus edulis* to different ^{48}V-vanadate concentrations in the water, the digestive gland showed distribution of vanadium in pellets and cytosol to be similar. In cytosol vanadium was found to be associated with high-molecular-weight components (Edel and Sabbioni, 1993), which represent for those authors the form that would remain in the tissues for long periods and would not be easily excreted. This finding is in good agreement with the observation reported by Miramand et al. (1980) that mussels contaminated with ^{48}V retain a large fraction of the element for a relatively long period of time,—that is, 20% of the original dose was retained after a 4-month depuration period with biological half-time of 103 days.

Beside digestive tissues, the kidney of mussels also concentrates vanadium at least under experimental contamination (Coulon et al., 1987; Edel and Sabbioni, 1993). Mussels experimentally contaminated with industrial effluent that contained a high concentration of vanadium showed an increase in vanadium excretion by the kidney (Martoja et al., 1986). Also noteworthy is the enhanced accumulation of vanadium in the branchial hearts of cephalopods. Concentrations are much higher than those in the digestive gland; for example, they range from approximately 25 μg/g dry weight in *Octopus vulgaris* to 6–7 μg/g dry wt in *Eledone cirrhosa* and *Sepia officinalis* (Miramand and Guary, 1980; Miramand and Bentley, 1992). This organ, which performs circulatory and excretory functions (Cuénot et al., 1908; Martin and Harrison, 1966; Schipp and von Boletzky, 1975) constitutes only 0.2% of the whole animal weight but contains 1–6% of the total vanadium body burden. Vanadium, like Co (Ueda et al., 1979; Nakahara et al., 1979), Cu and Fe (Schipp and Hevert, 1978; Miramand and Guary, 1980), Ni (Miramand and Bentley, 1992), and the transuranic elements Am and Pu (Guary et al., 1981; Guary and Fowler,

Table 8 Vanadium Concentrations (μg/g Dry Weight) in Internal Tissues of Some Marine Invertebrates

Species[a]	Muscle	Digestive Gland	Digestive Tract	Gonad	Kidney
CRUSTACEANS					
Carcinus maenas[1]	1 ± 0.5	4 ± 2	3 ± 1	<0.5	—
Lysmata seticaudata[1]	0.3	14	17	—	—
ECHINODERMATA					
Holothuria forskali[2]	—	1.6	2.4	1.2	—
Marthasterias glacialis[2]	—	0.4	0.7	—	—
Paracentrotus lividus[2]	—	—	2.5	1.1	—
MOLLUSKS (BIVALVIA)					
Mytilus galloprovincialis[3]	<0.1	3 ± 1[b]	—	—	—
MOLLUSKS (CEPHALOPODS)					
Eledone cirrhosa[4]	<0.5	2.9–3.6	<0.5	<0.5	<0.5
Loligo vulgaris[5]	<0.5	1.2–2.7	<0.5	<0.5	<0.5
Octopus vulgaris[6]	0.3 ± 0.2	4.5 ± 1	0.4 ± 0.1	0.2–0.3	0.4 ± 0.2
Sepia officinalis[4]	<0.5	4.1–5.9	0.7 ± 0.3	<0.5	0.5 ± 0.1

[a] Data from: (1) Miramand et al., 1981; (2) Miramand et al., 1982; (3) Miramand et al., 1980; (4) Miramand and Bentley, 1992; (5) Miramand and Guary, 1980; (6) Miramand, unpublished results.
[b] Digestive gland + stomach.

Table 9 Comparison of Experimentally Derived Concentration Factors in Tissues of *Mytilus galloprovincialis* and *M. edulis*

	T, °C	Number of Days	Shell	Visceral Mass	Mantle	Gills	Muscle	Authors
[74]As	13	6	0.4	24	3	4	2	Ünlu and Fowler, 1979[a]
[109]Cd	13	20	10	700	150	150	200	Fowler and Benayoun, 1974[a]
[58]Co	10	21	100	1,000	20	50	8	Pentreath, 1973a[b]
[59]Fe	10	21	80	3,000	90	300	50	Pentreath, 1973a[b]
[54]Mn	10	21	60	110	40	110	30	Pentreath, 1973a[b]
[75]Se	13	20	50	300	40	100	70	Fowler and Benayoun, 1976[a]
[65]Zn	10	21	30	500	80	200	80	Pentreath, 1973a[b]
[48]V	13	21	70	20	3	3	2	Miramand et al., 1980[a]

[a] Data from *M. galloprovincialis*.
[b] Data from *M. edulis*.

1982), is concentrated in this tissue, which probably plays a role in storage and detoxification of these elements.

In general, the low degree of vanadium uptake in the soft parts of marine invertebrates will minimize any risk that might arise in the event that contaminated invertebrates are used for human consumption. In this context its is interesting to note that muscle, which is the tissue essentially used for human consumption, is among those tissues with the lowest vanadium concentration.

3.3. Fish

Short-term experiments with ^{48}V have shown that vanadium in sea water is accumulated to only a very limited degree in the internal organs of marine fish. After 2–3 weeks of uptake from sea water, little vanadium can be detected in liver, kidney, spleen, or muscle of exposed fish (Table 10). In this respect, vanadium behaves very differently from other heavy metals (e.g., cadmium, cobalt, iron, mercury, silver, and zinc), which tend to accumulate in fish tissues with concentration factors ranging from 10° to 5×10^2 (Table 10). As with invertebrates, the highest concentration factors (<10) were found in tissues in direct contact with sea water or in the digestive tract. Marine fish show a similar response to vanadium as do freshwater fish. For example, the goldfish *Carassius auratus* exposed to 50 μg/L of ^{48}V vanadate for 4 days concentrates the highest amounts of vanadium in the intestine (Edel and Sabbioni, 1993). Nevertheless in long-term experiments (8 and 13 weeks) following exposures to relatively high concentrations (10^{-5} M orthovanadate), Bell et al. (1980, 1981) found that vanadium can penetrate into internal organs of the common eel, *Anguilla anguilla,* with the following order of concentration: liver > kidney > bone > blood > carcass (in their studies, the digestive tract was not taken into account). These observations confirm that, as with invertebrates, accumulation in fish of vanadium directly from water is a gradual process that results in a relatively low degree of contamination. Vanadium analyses of organisms from the natural environment have confirmed this trend (Table 11). For example, vanadium concentrations in fish liver are low in comparison with those in muscle. As in the experimental studies, the skin and digestive tract contain the highest concentrations for stable vanadium.

3.4. Relationship Between Distribution of Vanadium in Tissue and Toxicity

Vanadium LC_{50} values for some marine organisms such as mussels and crustaceans vary from 2.5 to 65 mg/L (Table 12). Furthermore, there is a definite relationship between vanadium uptake from sea water and vanadium toxicity. In these species, when the salinity decreases from 30‰ to 28‰ and 19‰, both the uptake rate (Miramand et al., 1980, 1981) and the vanadium toxicity increase. Ünsal (1978b) reported an LC_{50} (9 days) of 65 mg/L for *Mytilus*

Table 10   Comparison of Experimentally Derived Concentration Factors of ^{48}V in Tissues of Fish, by Organ, with Those Determined for Selected Heavy Metals[a]

	Exposure Time, days	Species	Whole Fish	Muscle	Skin	Gill	Digestive Tract	Liver	Kidney	Spleen	Authors[b]
^{110}Ag	63	Pleuronectes platessa	2	1	10	14	12	16	8	11	A
Cd	60	Anguilla anguilla	4	1.5	3	15	52	37	123	11	B
^{115}Cd	59	P. platessa	3.5	0.5	1	125	46–115	16	1	1	A
^{58}Co	90	P. platessa	0.8	0.6	10.4	3.8	5.5	5.9	10.6	7.3	C
^{59}Fe	90	P. platessa	1	0.4	2	13	62	12	12	28	C
Hg	32	A. anguilla	153	134	188	669	145	485	1,157	1,100	B
^{65}Zn	20	Gobius minutus	2.5	—	—	—	—	—	—	—	D
^{48}V	21	G. minutus	1.0	0.1	1.1	0.7	9.3	0	0	0	E
^{48}V	15	Scorpaena porcus	0.5	0	0.2	0.3	7.5	0	0	0	F

[a] Exposure to stable metals and radionuclides occurred in contaminated sea water.

[b] Authors: (A) Pentreath (1977a, 1977b); (B) Noël-Lambot and Bouquegneau (1977); (C) Pentreath (1973b); (D) Renfro et al. (1975); (E) Miramand et al. (1992); (F) Miramand and Fowler (unpublished results).

Table 11 Vanadium Concentrations (μg/g Dry Wt) in Some Benthic Fish Tissues

Species	Origin	Muscle	Liver	Digestive Tract	Skin	Authors
Auxis rochei	Mediterranean	0.01–0.24	0.03–0.5	—	—	Andreotis and Papadopoulou, 1980
Gobius minutus	Mediterranean	0.5 ± 0.2	0.5 ± 0.2	15.1 ± 2.0	1.9 ± 0.1	Miramand et al., 1992
Morone saxatilis	Atlantic	0.03 ± 0.03[a]	0.04 ± 0.01[a]	—	—	Heit, 1979
Mullus barbatus	Mediterranean	0.04–0.45	0.3–1.1	0.4–1.6[b]	1.2–5.2	Miramand et al., 1991
Scorpaena porcus	Mediterranean	0.4	0.5	0.9[b]	0.5 ± 0.1	Miramand and Fowler, unpublished results

[a] Values in μg/g fresh wt.
[b] Pyloric caeca only.

Table 12   Lethal Concentration (LC$_{50}$) Observed for Some Marine Benthic Organisms

Species	Exposure, days	LC$_{50}$ Values, mg/L	Authors
Invertebrates			
Carcinus maenas	9	35	Miramand and Ünsal, 1978
Mytilus edulis	9	65	Miramand and Ünsal, 1978
Nereis diversicolor	9	10	Miramand and Ünsal, 1978
Fish			
Limanda limanda	4	27.8	Taylor et al., 1985
Scorpaena porcus	9	2.5–5	Ünsal, 1978b

edulis and of 35 mg/L for *Carcinus maenas* at salinity of 38‰. At a salinity of 20‰, the LC$_{50}$ (9 days) are 53 mg/l and 22 mg/l, respectively. In fact, in both marine invertebrates and fish, vanadium is only weakly accumulated from sea water into internal tissues (see above). From these observations, it appears that vanadium is relatively unreactive with biological substrates, probably because of its anionic form in sea water. The very slow penetration of vanadium into the internal organs of marine organisms could also explain why, over the short term, high vanadium concentrations in sea water are necessary to produce a lethal effect in organisms. A delay in the expression of toxicity also occurs with fish (Holdway et al., 1983).

For invertebrates, vanadium toxicity is of the same order of magnitude as that for lead, nickel, and zinc, but it is considerably less than the toxicity of cadmium, copper, chromium, mercury, or silver (see Miramand and Ünsal, 1978). In the case of fish, the toxicity observed at such very high levels may be due to cellular interaction with vanadium adsorbed to the gill surface, similar to that proposed for freshwater fish by Anderson et al. (1979), who recorded histopathological effects of high concentrations of vanadium on the branchial lamellae of young rainbow trout (*Salmo gairdneri*, LC$_{50}$ (96 h), 10 mg/L).

Vanadium toxicity is very low following short-term exposure; nevertheless, there is a general lack of data concerning chronic toxicity and sublethal effects of this element in marine organisms. For example, following longer exposures (23 days), the LC$_{50}$ for freshwater organisms is low, about 2 mg/L for *Daphnia magna* (Beusen and Neven, 1987). According to Holdway and Sprague (1979), the sublethal threshold for vanadium toxicity in freshwater flag fish (*Jordanella floridae*) is 0.08 mg/L, a value similar to 0.16 mg/L for the zebrafish, *Brachydanio rerio* (Beusen and Neven, 1987). Moreover, it is well known that embryos and larvae are more sensitive to the toxic action of heavy metals than are adults of the same species. However, comparable data for vanadium are not available. Miramand (unpublished data) has found an LC_{50} (9 days) value of 0.2–0.3 mg/L for *Artemia salina* larvae. Likewise for the pluteus of the sea urchin, *Arbaccia lixula,* Miramand (unpublished data) determined a 100% mortality after 72 h of exposure to 0.5 mg/L vanadium levels. These values are 1–2 orders of magnitude lesser than those found for adult organisms. From the limited data available, it is clear that we need further observations on vanadium toxicity in marine larvae before conclusions can be drawn on the overall toxicity of this metal in coastal waters.

3.5. Transfer Pathways

To better understand the cycling of vanadium through marine biota, it is important to know the relative importance of the different pathways of contamination that occur under natural conditions,—that is, accumulation from water and from food, and for benthic organisms direct or indirect contamination from sediments. To our knowledge, vanadium transfer from contaminated sediments has not been studied. Nevertheless, in a preliminary experiment, Miramand (1979) has estimated a transfer factor at about 0.02 for *Nereis diversicolor* after 50 days of contact with a sediment heavily labeled with ^{48}V. This value may appear to be negligible; however, in view of the relatively high concentrations of vanadium that occur in coastal sediments (see above), this pathway for vanadium bioaccumulation merits further investigation.

More information is available for other bioaccumulation pathways. For carnivorous/omnivorous invertebrates, assimilation coefficients for vanadium, estimated experimentally following a single ingestion of labeled food, are high—ranging from about 30% in crustaceans to 88% in asteroids (Table 13). In all animals tested, a large fraction (>70%) of assimilated vanadium is incorporated into the digestive gland. This assimilated fraction is not rapidly lost and appears to be firmly bound to this organ. In the invertebrates listed in Table 13, vanadium transfer from food results in storage in internal organs. The digestive gland, which is the organ involved in digestion, absorption, and storage of nutrients, seems under these conditions to be the main organ of vanadium accumulation via the food chain. This observation merits further discussion. For asteroids, both the high degree of assimilation and the strong retention of assimilated vanadium in their pyloric caeca imply that transfer from food is a major pathway for vanadium uptake by sea stars. Despite this, the pyloric caeca of *Marthasterias glacialis* from the Mediterranean Sea display low vanadium concentrations (0.4 μg/g dry weight) (Miramand et al, 1982). This apparent paradox is probably due to the relatively low levels of vanadium in bivalves, particularly mussels (Table 3), normally ingested by these asteroids. Thus, in an uncontaminated situation, direct adsorption of vanadium from water appears to be the principal route by which these sea stars accumulate vanadium in their tissues, mainly the body wall, which is in direct contact with sea water. Alternatively, under pollution conditions in which bivalves could become highly contaminated, the food chain could play an important role

Table 13   Assimilation of Vanadium by Some Marine Organisms Following a Single Ingestion of Labeled Food, Biological Half-life for Vanadium Turnover of the Assimilated Fraction, and Percentage of Assimilated Vanadium Retained in the Digestive Gland

Species	Assimilation Coefficient, %	Biological Half-life, days	Digestive Gland % Body Burden	Authors
CARNIVOROUS				
Asteroids				
Marthasterias glacialis	88	57	98.8	Miramand et al., 1982
Cephalopods				
Sepia officinalis	40	7	96.3	Present study
Crustaceans				
Carcinus maenas	38	10	73–82	Miramand et al., 1981
Lysmata seticaudata	25	12	63–87	Miramand et al., 1981
Fish				
Gobius minutus	2–3	3	—	Miramand et al., 1992
SUSPENSIVOROUS				
Bivalvia				
Mytilus galloprovincialis	7	7	—	Miramand et al., 1980

in vanadium accumulation in internal tissues such as the pyloric caeca of carnivorous sea stars.

The situation is quite different with crustaceans. Miramand et al. (1981) observed a 90–150-fold difference in concentration factors in the digestive gland of crab and shrimp in direct uptake experiments from those calculated for the same animals collected along the Mediterranean coast. The observed differences indicate that steady-state vanadium levels in these species would be difficult to achieve from the water pathway alone. This hypothesis was confirmed by other experiments, with the crab *Carcinus maenas* (Ünsal, 1983), in which animals exposed for 30 days both to contaminated food (*Nereis diversicolor*) and to sea water exhibited a higher vanadium accumulation than those exposed to sea water alone. Thus, food is probably the principal source of vanadium accumulation in the digestive gland of crustaceans, and the water pathway is responsible for the adsorption and accumulation of vanadium in their external tissues such as exoskeleton and gills.

With cephalopods the situation is clearer. Relatively high concentrations of vanadium in the digestive gland of animals collected in situ (3–6 μg/g dry weight, Table 7), combined with a strong retention of vanadium in the digestive gland following ingestion of contaminated food (Table 13), suggest that transfer of vanadium from food is the principal pathway for vanadium accumulation in these carnivorous mollusks. This observation clearly merits further study, for example, experiments conducted with animals exposed to sea water alone.

There is a lack of similar data for marine fish. The single study reported in the literature concerns the Mediterranean benthic fish *Gobius minutus* (Miramand et al., 1992). This fish displayed a low degree of gut assimilation after ingestion of brine shrimp *Artemia salina* labeled with [48]V, as well as a rapid turnover of the fraction assimilated (Table 12). In fact, little vanadium is absorbed across the digestive tract, and that accumulated is rapidly eliminated through soluble excretion. This situation is similar to that for the freshwater fish *Carassius auratus* labeled only by water (Edel and Sabbioni, 1993) and terrestrial vertebrates (Waters, 1977). Therefore it appears that ingestion of food in the natural environment is unlikely to represent a major pathway for vanadium uptake in *Gobius minutus*. Nevertheless, it is clear that, in view of a single experiment conducted with one species of fish, no firm conclusions can be drawn about the relative importance of food and water routes as they affect vanadium accumulation by marine fish.

Ünsal (1978a,b, 1982) has shown an important accumulation of vanadium in soft parts of mussels fed the phytoplankton *Dunaliella marina* previously contaminated with sodium metavanadate. After 7 days exposure, mussels (*Mytilus edulis*) contaminated only through food displayed vanadium concentrations about 8–10 times higher than uncontaminated control mussels. The difference was approximately 30-fold for mussels exposed to both contaminated sea water and food. Thus, food appears to be a primary source of vanadium for mussels. This observation was confirmed by Miramand et al. (1980), who suggested that vanadium uptake from water alone might not be sufficient to account for the highest stable vanadium concentration factors found in mussels collected in situ, and they concluded that vanadium input through the food chain is probably the most important pathway in achieving equilibrium concentration factors under natural conditions. Even though the assimilation coefficient of vanadium (~7%) measured in mussels following ingestion of labeled phytoplankton is relatively low (Table 13), it is probably compensated by the relatively high concentration of vanadium that occurs in phytoplankton (Table 5) or by suspended particles ingested by mussels (see above) combined with their elevated filtration rate.

4. CONCLUSIONS

Despite the lack of data that concern many aspects of vanadium transfer in marine organisms, we can at present propose a simplified biological cycle of vanadium in the coastal environment (Fig. 2). In such a ecosystem, in which the biomass of marine organisms is high, vanadium that enters the sea is eventually transferred into marine sediments after a transit through a variety of marine species. In this context, marine biota contribute to an acceleration of the process of vanadium sedimentation through biological transport, mainly via shells, fecal pellets and molts; and costal sediments therefore appear to be a sink for vanadium.

Nevertheless, our knowledge is incomplete and many points of this cycle merit future investigations, for example:

• Transfer through planktonic compartments.
• Chronic inputs via food chain.
• Direct or indirect transfer from sediments.

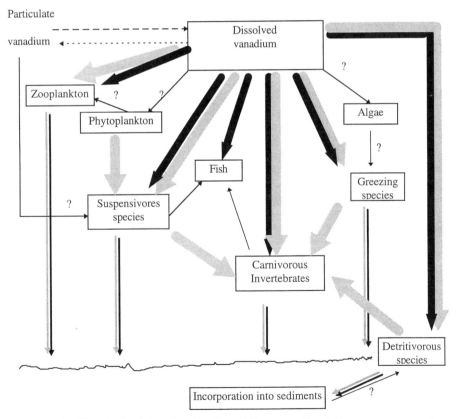

Figure 2. Simplified Cycle of Vanadium in a Coastal Ecosystem. Key: Thin arrow, poor transfer; ?, mechanism of transfer not studied; gray arrow, internal penetration; black arrow, adsorption; black arrow with gray shadow, sedimentation by biogenic transport.

- Transfer from macroalgae to herbivorous species.
- Transfer from suspended materials to suspensivores.
- Bioaccumulation in top marine predators such as mammals and birds.

More important, in general only a limited number of species have been investigated and some processes are not, or are only poorly, understood, for example:

- Mechanism of accumulation in phytoplankton.
- Mechanism of accumulation in benthic macroalgae.
- Subcellular localization in tissues.
- Toxicity to larvae or embryos of marine organisms.
- Sublethal effects in ecosystems.

Therefore, much additional research will be necessary before the biological fate of vanadium in the marine environment is as well known as that of other trace elements.

ACKNOWLEDGMENTS

The International Laboratory of Marine Radioactivity operates under an agreement between the International Atomic Energy Agency and the government of the Principality of Monaco. The authors would like to thank C. Huan for the editing work.

REFERENCES

Anderlini, V. C., Mohammad, O. S., Zarba, M. A., Fowler, S. W., and Miramand, P. (1982). Trace metals in marine sediments of Kuwait. *Bull. Environ. Contam. Toxicol.* **28,** 75–80.

Anderson, P. D., Spear, P., d'Appollinia, S., de Luca, J., and Dick, J. (1979). The multiple toxicity of vanadium, nickel and phenol to fish. *Alberta Oils Sands Environmental Research Program.* Report No. 79. Alberta Environment/Environment Canada, Alberta, pp. 1–109.

Andreotis, J. L., and Papadopoulou, C. (1980). A comparison of trace element content in muscle (dark and white) and liver of *Auxis rochei. V*[25] *Journées Etud. Pollutions, Cagliari,* CIESM, pp. 313–315.

Anonymous. (1990). *Commission chargée de contrôler l'évolution de la pollution en estuaire et en baie de Seine.* Agence de Bassin Seine Normandie, Rouen, 36 pp.

Bell, M. V., Kelly, K. F., and Sargent, J. R. (1980). The uptake of orthovanadate into various organs of the common eel, *Anguilla anguilla,* maintained in fresh water. *Sci. Total Environ.* **16,** 99–108.

Bell, M. V., Kelly, K. F., and Sargent, J. R. (1981). The uptake from fresh water and subsequent clearance of a vanadium burden by the common eel (*Anguilla anguilla*). *Sci. Total Environ.* **19,** 215–222.

Bertine, K. K., and Goldberg, E. D. (1971). Fossil fuel combustion and the major sedimentary cycle. *Science* **173,** 233–235.

Bertine, K. K., and Goldberg, E. D. (1977). History of heavy metal pollution in southern California coastal zone—Reprise. *Environ. Sci. Technol.* **11**, 297–299.

Bertrand, D. (1942). Le vanadium chez les vertébrés. *C.R. Acad. Sci.* **215**, p. 150.

Bertrand, D. (1943). Sur la diffusion du vanadium chez les invertébrés et chez les vertébrés. *Bull. Soc. Chim. Biol.* **25**, 36–39.

Bertrand, D. (1950). The biogeochemistry of vanadium. *Bull. Am. Mus. Nat. Hist.* **94**, 407–455.

Beusen, J. M., and Neven, B. (1987). Toxicity of vanadium to different freshwater organism. *Bull. Environ. Contam. Toxicol.* **39**, 194–201.

Black, W. A. P., and Mitchell, R. L. (1952). Trace elements in the common brown algae and in sea water. *J. Mar. Biol. Assoc. U.K.* **30**, 575–584.

Blotcky, A. J., Falcone, C., Medina, V. A., and Rack, E. P. (1979). Determination of trace-level vanadium concentrations in marine biological samples by chemical neutron activation analysis. *Anal. Chem.* **51**, 178–182.

Bourgoin, B. P., and Risk, M. J. (1987). Vanadium contamination monitored by an arctic bivalve *Cyrtodaria kurriana. Bull. Environ. Contam. Toxicol.* **39**, 1063–1068.

Boust, D. (1981). Les métaux traces dans l'estuaire de la Seine et ses abords. Rapport CEA-R-5104, 207 pp.

Brooks, R. R., and Rumsby, M. G. (1965). The biogeochemistry of trace element uptake by some New Zealand bivalves. *Limnol. Oceanog.* **10**(4), 521–527.

Bryan, G. W. (1984). Pollution due to heavy metals and their compounds. In O. Kinne (Ed.), *Marine Ecology.* Wiley-Interscience, Chichester, Vol. 5, Part 3, pp. 1289–1431.

Buat-Ménard, P., and Chesselet, R. (1979). Variable influence of the atmospheric flux on the trace metal chemistry of oceanic suspended matter. *Earth Planet. Sci. Lett.* **42**, 399–411.

Burton, J. D. (1966). Some problems concerning the marine geochemistry of vanadium. *Nature* (London) **212**, 976–978.

Chan, K. M., and Riley, J. P. (1966). Determination of vanadium in sea and natural waters, biological materials and silicate sediments and rocks. *Anal. Chim. Acta* **34**, 337–345.

Chasteen, D. (1983). The biochemistry of vanadium. In M. J. Clarke (Ed.), *Copper, Molybdenum and Vanadium in Biological Systems, Structure and Binding,* Vol. 53. Springer-Verlag, Berlin, pp. 105–139.

Cockerill, B. M., finch, P., and Percival, E. (1978). Vanadium in the brown seaweed, *Desmarestia firma. Phytochemistry* **17**, 2129.

Collier, R. W. (1984). Particulate and dissolved vanadium in the North Pacific Ocean. *Nature* **309**, 441–44.

Coombs, T. L., and George, S. J. (1978). Mechanisms of immobilization and detoxification of metals in marine organisms. In D. S. McLusky and A. J. Berry (Eds.), *Physiology and Behaviour of Marine Organisms.* Pergamon Press, Oxford and New York, pp. 179–187.

Coulon, J., Truchet, M. and Martoja, R. (1987). Chemical features of mussels (*Mytilus edulis*) *in situ* exposed to an effluent of the titanium dioxide industry. *Ann. Inst. Océnogr. Paris N. S.* **63**(2), 89–100.

Cuénot, K., Gonet, V. and Bruntz, L. (1908). Recherches chimiques sur les coeurs branchiaux des céphalopodes. Démonstration du rôle excréteur des cellules qui éliminent le carmin ammoniacal des injections physiologiques. *Arch. Zool. Exp. Gén.* (Notes Rev.) **9**(3), 49–53.

Duce, R. A., and Hoffman, G. L. (1976). Atmospheric vanadium transport to the ocean. *Atmos. Environ.* **10**, 989–996.

Edel, J., and Sabbioni, E. (1993). Accumulation, distribution and form of vanadate in the tissues and organelles of the mussel *Mytilus edulis* and the goldfish *Carassius auratus. Sci. Total Environ.* **133**, 139–151.

Eisler, R. (1981). *Trace Metal Concentrations in Marine Organisms.* Pergamon Press, New York, 687 pp.

Emerson, S. R., and Huested, S. S., (1991). Ocean anoxia and the concentrations of molybdenum and vanadium in sea water. *Mar. Chem.* **34**, 177–196.

Forsberg, A., Söderlund, S., Frank A., Petersson, L. R., and Pedersén, M., (1988). Studies on metal content in the brown seaweed, *Fucus vesiculosus,* from the Archipelago of Stockholm. *Environ. Pollut.* **49**, 245–263.

Fowler, S. W. (1986). Trace metal monitoring of pelagic organisms from the open Mediterranean Sea. *Environ. Monit. Assessment.* **7,** 59–78.

Fowler, S. W., and Benayoun, G. (1974). Experimental studies on cadmium flux through marine biota. In: *Comparative Studies of Food and Environmental Contamination.* IAEA, Vienna, pp. 159–178.

Fowler, S. W., and Benayoun, G. (1976). Accumulation and distribution of selenium in mussel and shrimp tissues. *Bull. Environ. Contam. Toxicol.* **16,** 339–346.

Fowler, S. W., Papadopoulou, C., and Zafiropoulos, D. (1985). Trace elements in selected species of zooplankton and nekton from the open Mediterranean Sea. In T. D. Lekkas, (Ed.), *Heavy Metals in the Environment,* Vol. 1. CEP Consultants, Ltd, Edinburgh, pp. 670–672.

Fowler, S. W., Readman, J. W., Oregioni, B., Villeneuve, J. P., and Mc Kay, K. (1993). Petroleum hydrocarbons and trace metals in nearshore gulf sediments and biota before and after the 1991 war: An assessment of temporal and spatial trends. *Mar. Pollut. Bull.* **27,** 171–182.

Frank, A., Galgan, V., Roos, A., Olsson, M., Peterson, L. R., and Bignert, A. (1992). Metal concentrations in seals from Swedish waters. *Ambio* **21,** 529–538.

Fukai, R., and Meinke, W. W. (1962). Activation analyses of vanadium, arsenic, molybdenum, tungsten, rhenium and gold in marine organisms. *Limnol. Oceanogr.* **7,** 186–200.

Funk, E. W., and Gomez, E. (1977). Determination of vanadium in Athabasca bitumen and other heavy hydrocarbons by visible spectrometry. *Anal. Chem.* **49,** 972–973.

Goldberg, E. D., Bowen, V. T., Farrington, J. W., Harvey, G., Martin, J. H., Parker, P. L., Risebrough, R. W., Robertson, W., Schneider, E., and Gamble, E. (1978). The mussel watch. *Environ. Conserv.* **5,** 101–125.

Goldberg, E. D., Broecker, W. S., Gross, M. G., and Turekian K. K., (1971). Marine chemistry. In National Academy of Science (Eds.), *Radioactivity in the Marine Environment.* Washington, D.C., pp. 137–146.

Goldberg, E. D., Mc Blair, W., and Taylor, K. M. (1951). The uptake of vanadium by tunicates. *Biol. Bull. Mar. Biol. Lab. Woods Hole* **110,** 84–94.

Grange, L. (1974). Le livre blanc consacré aux rejets en Méditerranée des résidus industriels de la Société Montedisson. *J. Fr. Hydrol.* **13,** 9–29.

Guary, J. C., and Fowler, S. W. (1982). Experimental studies on the biokinetics of plutonium and americium in the cephalopod *Octopus vulgaris. Mar. Ecol. Prog. Ser.* **7,** 327–335.

Guary, J. C., Higgo, J. J. W., Cherry, R. D., and Heyraud, M. (1981). High concentrations of transuranics and natural radioactive elements in the branchial hearts of the cephalopod *Octopus vulgaris. Mar. Ecol. Prog. Ser.* **4,** 123–126.

Hawkins, C. J., Kott, P., Parry, D. L., and Swinehart, J. H. (1983). Vanadium content and oxidation state related to ascidian phylogeny. *Comp. Biochem. Physiol.* **76B**(3), 555–558.

Heit, M. (1979). Variability of the concentrations of seventeen trace elements in the muscle and liver of a single striped bass, *Morone saxatilis. Bull. Environ. Contam. Toxicol.* **23,** 1–5.

Henze, M. (1911). Die vanadium Verbindung der Blutkorperchen. *Hoppe-Seylers Z. Phys. Chem.* **72,** 494–501.

Holdway, D. A., and Sprague, J. W. (1979). Chronic toxicity of vanadium to flagfish. *Water Res.* **13,** 905–910.

Holdway, D. A., Sprague, J. B., and Dick, J. G. (1983). Bioconcentration of vanadium in American flagfish over one reproductive cycle. *Water Res.* **17,** 937–941.

Hope, B. K. (1994). A global biogeochemical budget for vanadium. *Sci. Total Environ.* **141,** 1–10.

Huizenga, D. L., and Kester, D. R. (1982). The distribution of vanadium in the North Western Atlantic. *Ocean. Trans. Am. Geophys. Union Abstr.* **63,** 990.

Ikebe, K., and Tanaka, R. (1979). Determination of vanadium and nickel in marine samples by flameless and flame atomic absorption spectrophotometry. *Bull. Environ. Contam. Toxicol.* **21,** 526–532.

Ishii, T., Nakai, I., Numako, C., Okoshi, K., and Otake, T. (1993). Discovery of a new vanadium accumulator, the fan worm *Pseudopotamilla occelata. Naturwissenschaften* **80,** 268–270.

Ishii, T., Otake, T., Okoshi, K., Nakahara, M., and Nakamura, R. (1994). Intracellular localization of vanadium in the fan worm *Psuedopotamilla occelata. Mar Biol.* **121,** 143–151.

Jeandel, C., Caisso, M., and Minster, J. F. (1987). Vanadium behaviour in the global ocean and in the Mediterranean Sea. Mar. Chem. **21**, 51–74.

Kalogeropoulos, N., Scoullos, M., Vassilaki-Grimani, M., and Grimanis, A. P. (1989). Vanadium in particles and sediments of the northern Saronikos Gulf, Greece. Sci. Total Environ. **79**, 241–252.

Kustin, K., Ladd, K. V., and Mc Leod, G. C. (1975). Site and rate of vanadium assimilation in the tunicate Ciona intestinalis. J. Gen. Physiol. **65**(3), 315–328.

Ladd, K. V. (1974). The distribution and assimilation of vanadium with respect to the tunicate Ciona intestinalis. Ph.D. thesis, Brandeis University, 108 pp.

Latouche, Y. D., Bennett, C. W., and Mix, M. C. (1981). Determination of vanadium in a marine mollusc using a chelating ion exchange resin and neutron activation. Bull. Environ. Contam. Toxicol. **26**, 224–227.

Latouche, Y. D, and Mix, M. C. (1981). Seasonal variation in soft tissue weights and trace metal burdens in the bay mussel, Mytilus edulis. Bull. Environ. Contam. Toxicol. **27**, 821–828.

Linstedt, K. D., and Kruger, P. (1969). Vanadium concentrations in Colorado River basin waters. J. Am. Water Works Assoc. **61**, 85–88.

Loring, D. H. (1976). Distribution and partition of cobalt, nickel, chromium and vanadium in the sediments of the Saguenay Fjord. Can. J. Earth Sci. **13**(12), 1706–1718.

Mackey, E. A., Becker, P. R., Demiralp, R., Greenberg, R. R., Koster, B. J., and Wise, S. A. (1996). Bioaccumulation of vanadium and other trace metals in livers of Alaskan cetaceans and pinnipeds. Arch. Environ. Contam. Toxicol. **30**, 503–512.

Mackey, E. A., Demiralp, R., Becker, P. R., Greenberg, R. R., Koster, B. J., and Wise, S. A. (1995). Trace element concentrations in cetacean liver tissues archived in the National Marine Mammal Tissue Bank. Sci. Total Environ. **175**, 21–41.

Martin, A. W., and Harrison, F. M. (1966). Excretion. In K. M. Wilbur and C. M. Yonge (Eds.), Physiology of Mollusca, Vol 2., Academic Press, New York, pp. 353–386.

Martoja, R., Martin, J. L., Ballan-Dufrançais, C., Jeantet, A. Y., and Truchet, M. (1986). Effets d'un effluent de fabrication du bioxyde de titane sur un mollusque (Mytilus edulis). Comparaison d'animaux traités expérimentalement et prélevés à proximité d'un rejet d'usine. Marine Environ. Res. **18**, 1–27.

Miramand, P. (1979). Contribution à l'étude de la toxicité et des transferts du vanadium chez quelques organismes marins. Thèse doctorat 3ème cycle, Université des Sciences et Techniques du Languedoc, 115 pp.

Miramand, P., and Bentley, D., (1992). Concentration and distribution of heavy metals in tissues of two cephalopods, Eledone cirrhosa and Sepia officinalis, from the French coast of the English Channel. Mar. Biol. **114**, 407–414.

Miramand, P., Bentley, D., Guary, J. C., and Brylinski, J. M. (1993). Rôle du plancton dans le cycle biogéochimique du cadmium et du vanadium en baie de Seine orientale: Premiers résultats. Oceanol. Acta. **16**(5-6), 625–632.

Miramand, P., Fowler, S. W., and Guary, J. C. (1982). Comparative study of vanadium biokinetics in three species of echinoderms. Mar. Biol. **67**, 127–134.

Miramand, P., Fowler, S. W., and Guary, J. C. (1992). Experimental study on vanadium transfer in the benthic fish Gobius minutus. Mar. Biol. **114**, 349–353.

Miramand, P., and Guary, J. C. (1980). High concentrations of some heavy metals in tissues of the Mediterranean octopus. Bull. Environ. Contam. Toxicol. **24**, 783–788.

Miramand, P., Guary, J. C., and Fowler, S. W. (1980). Vanadium transfer in the mussel Mytilus galloprovincialis. Mar. Biol. **56**, 281–293.

Miramand, P., Guary, J. C., and Fowler, S. W. (1981). Uptake, assimilation, and excretion of vanadium in the shrimp Lysmata seticaudata (Risso), and the crab Carcinus maenas (L.). J. Exp. Mar. Biol. Ecol. **49**, 267–287.

Miramand, P., Lafaurie, M., Fowler, S. W., Lemaire, P., Guary, J. C., and Bentley, D. (1991). Reproductive cycle and heavy metals in the organs of red mullet Mullus barbatus (L.), from the northwestern Mediterranean. Sci. Total Environ. **103**, 47–56.

Miramand, P., and Ünsal, M. (1978). Toxicité aigue du vanadium vis-à-vis de quelques espèces benthiques et phytoplanctoniques marines. *Chemosphere* **7**(10), 827–832.

Morris, A. W. (1975). Dissolved molybdenum and vanadium in the northeast Atlantic Ocean. *Deep Sea Res.* **22**, 49–54.

Nakahara, M., Koyanagi, T., Ueda, T., and Shimizu, C. (1979). Peculiar accumulation of cobalt-60 by the branchial hearts of *Octopus. Bull. Jpn. Soc. Scient. Fish.* **45**, 539.

Nicholls, G. D., Curl, H. Jr., and Bowen, V. T. (1959). Spectrographic analyses of marine plankton *Limnol. Oceanog.* **4**, 472–476.

Noël-Lambot, F., and Bouquegneau, J. M. (1977). Comparative study of toxicity, uptake and distribution of cadmium and mercury in the sea water adapted eel *Anguilla anguilla. Bull. Environ. Contam. Toxicol.* **18**, 418–424.

Papadopoulou, C., Zafiropoulos, D., and Grimanis, A. P. (1980). Arsenic, copper and vanadium in *Mullus barbatus* and *Parapenaeus longirostris* from Saronikos Gulf, Greece. V^{es} *Journées Etud. Pollutions, Cagliari,* CIESM. pp. 419–422.

Parker, R. D. R., and Sharma, R. P. (1978). Accumulation and depletion of vanadium in selected tissues of rats treated with vanadyl sulfate and sodium orthovanadate. *J. Environ. Pathol. Toxicol.* **2**, 235–245.

Pentreath, R. J. (1973a). The accumulation from water of ^{65}Zn, ^{54}Mn, ^{58}Co and ^{59}Fe by the mussel *Mytilus edulis. J. Mar. Biol. Assoc. U. K.* **53**, 127–143.

Pentreath, R. J. (1973b). The accumulation and retention of ^{59}Fe and ^{58}Co by the plaice *Pleuronectes platessa. J. Exp. Mar. Biol. Ecol.* **12**, 315–326.

Pentreath, R. J. (1977a). The accumulation of ^{110m}Ag by the plaice *Pleuronectes platessa* L. and the thornback ray *Raja clavata* L. *J. Exp. Mar. Biol. Ecol.* **29**, 315–325.

Pentreath, R. J. (1977b). The accumulation of cadmium by the plaice *Pleuronectes platessa* L. and the thornback ray *Raja clavata* L. *J. Exp. Mar. Biol. Ecol.* **30**, 223–232.

Pesh, G., Reynolds, P., and Rogerson, P. (1977). Trace metals in scallops from within and around two ocean disposal sites. *Mar. Pollut. Bull.* **8**(10), 224–228.

Popham, J. D., and D'Auria, J. M. (1982). A new sentinel organism for vanadium and titanium. *Mar. Pollut. Bull.* **13**(1), 25–27.

Pyle, T. E., and Tieh, T. T. (1970). Strontium, vanadium and zinc in the shells of pteropods. *Limnol. Oceanogr.* **15**(1), 153–154.

Renfro, W. C., Fowler, S. W., Heyraud, M., and La Rosa, J. (1975). Relative importance of food and water in long-term ^{65}Zn accumulation by marine biota. *J. Fish. Bd. Can.* **32**, 1339–1345.

Ridout, P. S., Rainbow, P. S., Roe, H. S. J., and Jones, H. R. (1989). Concentration of V, Cr, Mn, Fe, Ni, Co, Cu, Zn, As and Cd in mesopelagic crustaceans from the North East Atlantic Ocean. *Mar. Biol.* **100**, 465–471.

Riley, J. P., and Roth, I. (1971). The distribution of trace elements in some species of phytoplankton grown in culture. *J. Mar. Biol. Assoc. U.K.* **51**(1), 63–72.

Sadiq, M., Alam, I. A., and Al-Mohanna, H. (1992). Bioaccumulation of nickel and vanadium by clams (*Meretrix meretrix*) living in different salinities along the Saudi coast of the Arabian Gulf. *Environ. Pollut.* **7**, 225–231.

Sadiq, M., and Zaidi, T. H. (1984). Vanadium and nickel content of Nowruz oil spill tar flakes on Saudi Arabian coastline and its probable environmental impact. *Bull. Environ. Contam. Toxicol.* **32**, 635–639.

Schipp, R., and von Boletzky, S. (1975). Morphology and function of the excretory organs in dibranchiate cephalopods. *Fortschr. Zool.* **23**, 89–11.

Schipp, R., and Hevert, F. (1978). Distribution of copper and iron in some central organs of *Sepia officinalis* (Cephalopoda). A comparative study by flameless atomic absorption and electron microscopy. *Mar. Biol.* **47**, 391–399.

Sherrel, R. M., and Boyle, E. A. (1988). Zinc, chromium, vanadium and iron in the Mediterranean Sea. *Deep Sea Res.* **35**(8), 1319–1334.

Söderlund, S., Forsberg, A., and Pedersén, M. (1988). Concentrations of cadmium and other metals in *Fucus vesiculosus* L. and *Fontinalis dalecarlica* Br. Eur. from the northern Baltic Sea and the southern Bothnian Sea. *Environ. Pollut.* **51**, 197–212.

Swinehart, J. H., Biggs, W. R., Malko, D. J., and Schroeder, N. C. (1974). The vanadium and selected metal contents of some ascidians. *Biol. Bull. Mar. Biol. Lab. Woods Hole* **146**, 302–312.

Taylor, D., Maddock, B. G., and Mance, G. (1985). The acute toxicity of nine "grey list" metals (arsenic, boron, chromium, copper, lead, nickel, tin, vanadium and zinc) to two marine fish species: dab (*Limanda limanda*) and grey mullet (*Chelon labrosus*). *Aquat. Toxicol.* **7**, 135–144.

Ueda, T., Nakahara, M., Ishii, T., Suzuki, T., and Suzuki, M. (1979). Amounts of trace elements in marine cephalopods. *J. Radiat. Res.* **20**, 338–342.

Ünlu, M. Y., and Fowler, S. W. (1979). Factors affecting the flux of arsenic through the mussel *Mytilus galloprovincialis. Mar. Biol.* **51**, 209–219.

Ünsal, M. (1978a). Etudes des voies de transfert et des phénomènes d'accumulation du vanadium chez les mollusques: *Mytilus edulis (L.). Rev. Int. Océanogr. Méd.* **51–52**, 71–81.

Ünsal, M. (1978b). Contribution à l'étude de la toxicité du vanadium vis-à-vis des organismes marins. Thése doctorat 3ème cycle, Université Paris VI, 85 pp.

Ünsal, M. (1982). The accumulation and transfer of vanadium within the food chain. *Mar. Pollut. Bull.* **13**(4), 139–141.

Ünsal, M. (1983). Transfer pathways and accumulation of vanadium in the crab *Carcinus maenas. Mar. Biol.* **72**, 279–282.

Van Zinderen Bakker, E. M., and Jaworski, J. F. (1980). Effects of vanadium in the Canadian environment. Report No. 18132, National Research Council of Canada, 94 pp.

Walsh, P. R., and Duce, R. A. (1976). The solubilization of anthropogenic atmospheric vanadium in sea water. *Geophys. Res. Lett.* **3**, 375–378.

Waters, M. D. (1977). Toxicology of vanadium. *Adv. Mod. Toxicol.* **2**, 147–189.

Weisel, C. P., Duce, R. A., and Fashing, J. L. (1984). Determination of aluminium, lead and vanadium in north Atlantic seawater after coprecipitation with ferric hydroxide. *Anal. Chem.* **56**, 1050–1052.

Weiss, H. V., Guttman, M. A., Korkish J., and Steffan, I. (1977). Comparison of methods for the determination of vanadium in sea water. *Talanta* **24**, 509.

Yamamoto, T., Fujita, T., and Ishibashi, M. (1970). Chemical studies on the seaweeds (25) vanadium and titanium contents in seaweeds. *Rec. Oceanogr. Works Jpn.* **10**(2), 125–135.

Yeats, P. A. (1992). The distribution of dissolved vanadium in eastern Canadian coastal waters. *Estuarine Coastal Shelf Sci.* **34**, 85–93.

Zeisler, R., Demiralp, R., Koster, B. J., Becker, P. R., Burow, P. R., Ostapczuk, P., and Wise, S. A. (1993). Determination of inorganic constituents in marine mammal tissues. *Sci Total Environ.* **139/140**, 365–386.

Zoller, W. H., Gordon, E. S., and Jones, A. G. (1973). The sources and distribution of vanadium in the atmosphere. Trace elements in the environment. In American Chemical Society (Eds.), *Advances in Chemistry*, Series No. 123 American Chemical Society, Washington, D.C., pp. 31–47.

9

CHARACTERIZATION OF VANADIUM IN THE FAN WORM, *Pseudopotamilla occelata*

Toshiaki Ishii

Division of Marine Radioecology, National Institute of Radiological Sciences, Hitachinaka 311-12, Japan

Vanadium in the Environment. Part 1: Chemistry and Biochemistry, Edited by Jerome O. Nriagu.
ISBN 0-471-17778-4.   © 1998 John Wiley & Sons, Inc.

1. INTRODUCTION

1.1. Accumulators of Specific Elements

Some accumulators, which have a special ability to concentrate a specific element at a high level, have been discovered in the animal kingdom. Accumulators are regarded as useful indicator organisms for monitoring marine pollution due to heavy metals and radionuclides. Strontium in the skeletons of the celestite radiolaria of Protozoa (Odum, 1951), zinc in the jaws of polychaete worms of Annelida (Bryan and Gibbs, 1980), manganese in the kidneys of clams of bivalve Mollusca (Carmichael et al., 1979; George et al., 1980; Ishii et al., 1986; Ishii et al., 1992), cobalt (Ueda et al., 1979) and uranium (Ishii et al., 1991) in the branchial hearts of the octopus of Cephalopoda Mollusca, iron in the radula teeth of limpets and chitons of Gastropoda Mollusca (Kim et al., 1986; Mann et al., 1986), iodine in the operculum of conches of Gastropoda Mollusca (Ishii, unpublished) are good examples. It is also well known that a remarkably high level of vanadium is contained in the blood cells of ascidians classified as Protochordata (Henze, 1911; Macara et al., 1979; Michibata et al., 1986).

1.2. Discovery of a New Vanadium Accumulator

In order to determine the concentrations of vanadium in marine organisms, approximately 500 species of marine animals such as echinoderms, mollusks, crustaceans, fishes, and so on, which were collected from the shallow water off the coast of Japan, were analyzed by flameless atomic absorption spectroscopy (AAS), inductively coupled plasma atomic emission spectrometry (ICP-AES), and neutron activation analysis (NAA). The vanadium concentrations (0.1–25 μg/g dry wt) determined for the whole bodies of marine animals other than ascidians did not reach the levels found in ascidians.

Ishii et al. (1993) discovered that the fan worm *Pseudopotamilla occelata* (Fig. 1) had a very high level of vanadium when the screening studies were focused on Polychaeta worms, which are classified as Annelida. The vanadium concentrations in the whole soft body of *P. occelata* ranged from 320 to 1,350 μg/g on a dry weight basis. Since *P. myriops* (350 $\pm$ 110 μ/g dry wt) and *P. ehlersi* (420 $\pm$ 160 μg/g dry wt) also showed a high concentration of vanadium, and the fan worms *Sabellastarte japonica* (12 $\pm$ 5 μg/g dry wt) and *Hydroides ezoensis* (17 $\pm$ 8 μg/g dry wt), which belong to other genera, showed very low vanadium contents, the specific accumulation of vanadium by the genus *Pseudopotamilla* is thought to be a widespread phenomenon. It is very interesting that new vanadium accumulators were found in the Teloblast series of the phylogenetic tree although Prochordata is classified in the Enterocoel series. This suggests that some vanadium accumulators are possibly present in the animal kingdom, regardless of their taxonomic ranking (Chapter 10). Extensive screening studies are necessary to find vanadium accumulators in addition to ascidians and the fan worms of the genus *Pseudopotamilla*.

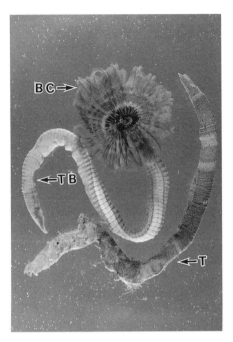

Figure 1. Lateral view of *P. occelata*. This fan worm has a trunk body (TB) and a branchial crown (BC) that is composed of many bipinnate radioles. The fan worm constructs a tube (T) of sand grains embedded in mucus.

In this chapter, the concentration, distribution, and chemical states of vanadium in *P. occelata* are reported and the possible physiological roles are described.

2. VANADIUM CONCENTRATION IN *P. occelata*

2.1. Qualitative and Quantitative Analyses

Qualitative analyses for finding vanadium accumulators among many species of marine organisms were carried out by inductively coupled plasma mass spectrometry (ICP-MS) and X-ray fluorescence analysis by synchrotron radiation. A large peak indicating the presence of a significant amount of vanadium appeared at an atomic mass unit of 51 when the solution of the branchial crown (see Fig. 1) of *P. occelata* digested by HNO_3 and $HClO_4$ was analyzed by ICP-MS. It was also found from the X-ray fluorescence analysis by synchrotron radiation that a high level of vanadium was contained in the branchial crown.

The concentrations of vanadium and other elements in *P. occelata* were measured by means of flameless AAS, NAA, and ICP-AES, because the quantitative analyses of mid-mass elements by ICP-MS were compromised by matrix elements like Ca, Cl, K, Mg, Na, and P, or Ar gas. For example, the measurement of ^{51}V by ICP-MS is affected by a molecular ion of $^{35}Cl^{16}O^+$ that occurred in the ICP plasma. Furthermore, quantitative analysis of elements by X-ray fluorescence analysis using synchrotron radiation is not systematically possible.

2.2.   Elemental Concentrations in *P. occelata*

The concentrations of 12 elements in the branchial crown composed of many bipinnate radioles, the trunk body, and the whole soft body of *P. occelata* are listed in Table 1. Wide differences in concentrations of most elements except iron, vanadium, and zinc were not observed between the branchial crown and the trunk body. In contrast, the concentrations of iron, vanadium, and zinc in the branchial crown were higher than those in the trunk body. In particular, the vanadium concentration (mean $\pm$ SD = 5,500 $\pm$ 1,800 μg/g dry wt, n = 30) in the branchial crown was about 100 times higher than that (60 $\pm$ 25 μg/g dry wt, n = 30) in the trunk body. Approximately 90% of the total body burden was concentrated in the branchial crown, which occupies only 7.6 $\pm$ 1.9% (n = 30) of the whole soft body weight.

The average vanadium concentration (510 $\pm$ 130 μg/g dry wt, n = 55) in the whole body of *P. occelata* was comparable to that of three species of ascidians, *Ascidia ahodori* (1550 μg/g dry wt), *Ascidia sydneiensis samae* (260 μg/g dry wt), and *Ciona saviqnyi* (220 μg/g dry wt).

Table 1   Concentrations (μg/g Dry Wt) of Vanadium in the Branchial Crown, Trunk Body, and Whole Soft Body of *Pseudopotamilla occelata*

Element	Branchial Crown, n = 30	Trunk Body, n = 30	Whole Soft Body, n = 55
Al	180 $\pm$ 42	160 $\pm$ 31	200 $\pm$ 40
Ca	1,370 $\pm$ 170	1,400 $\pm$ 180	1,480 $\pm$ 150
Cu	7.0 $\pm$ 3.9	4.3 $\pm$ 1.7	5.5 $\pm$ 2.2
Fe	300 $\pm$ 39	160 $\pm$ 26	190 $\pm$ 30
K	11,800 $\pm$ 2,540	11,200 $\pm$ 1,410	12,600 $\pm$ 2,640
Mg	2,150 $\pm$ 270	1,980 $\pm$ 120	2,130 $\pm$ 180
Mn	14.5 $\pm$ 3.4	9.7 $\pm$ 2.7	11.8 $\pm$ 5.1
Na	12,200 $\pm$ 4,020	6,900 $\pm$ 1,520	9,880 $\pm$ 3,120
P	6,990 $\pm$ 1,410	5,520 $\pm$ 1,860	6,980 $\pm$ 1,730
Sr	34 $\pm$ 16	52 $\pm$ 21	51 $\pm$ 25
V	5,500 $\pm$ 1,800	60 $\pm$ 25	510 $\pm$ 130
Zn	200 $\pm$ 29	71 $\pm$ 18	99 $\pm$ 22

3. DISTRIBUTION OF VANADIUM IN THE BRANCHIAL CROWN

3.1. Pretreatment for EPMA and AEM Measurements

In order to examine the distribution of vanadium in the bipinnate radioles, specimens for a local analysis by electron probe X-ray microanalysis (EPMA) and analytical electron microscopy (AEM) were prepared in accordance with conventional procedures (i.e., fixing with glutaraldehyde and osmic solutions and dehydrating in a series of ethanol solutions). However, little vanadium was detected by both EPMA and AEM from any part of the specimens. It was presumed that vanadium was eluted from its localized sites and moved into the medium during the pretreatment. Therefore, freeze-cracking and cryosectioning methods were adopted as preferable sample preparation techniques.

Specimens for EPMA were made by the following procedure. The whole soft bodies were quick-frozen in liquid nitrogen. Longitudinal sections of the bipinnate radioles 10 μm thick were cut with a Leitz 1720 digital cryostat and placed on a high-purity quartz plate. Sections were then dried in a chamber (-30 °C) for 2 days. In addition, the transverse fracture surface of the frozen radioles was exposed by a cracking method and was freeze-dried in a vacuum. After carbon-coating of the specimens area, line, and point analyses were carried out with a JEOL JXA-8600MX EPMA (accelerating voltage of 15 kV and probe current of 2×10^{-8} A). The characteristic X-ray (wavelength = 0.2505 nm) of vanadium was measured with a PET analyzing crystal.

In the case of AEM, the bipinnate radioles were rapidly frozen in liquid propane. Cryosections 100 nm thick were cut with a Reichert Ultracut S/FCS cryomicrotome and put on a copper grid. They were moved to the AEM with a JEOL EM-CTM cryotransfer holder and then dried at -100 °C for 24 h. In order to examine the subcellular distribution of vanadium, cryosections without any staining were analyzed with the JEOL JEM-2000FXII AEM equipped with an EDAXPV-9800 X-ray analyzer operating at an acceleration voltage of 200 kV, with a probe size of 10 nm and live time of 100 s.

3.2. Vanadium Distribution in the Bipinnate Radiole

Ishii et al. (1993) have reported that most of the vanadium in the branchial crown of *P. occelata* was concentrated only in the epidermis covering the bipinnate radioles, whereas little vanadium was contained in the muscular, connecting tissues or in the blood plasma. Line analysis by EPMA for a longitudinal section of the bipinnate radiole indicates that vanadium is not uniformly distributed in the epidermis and that a clear peak indicating the highest vanadium content is observed in the outer position of the epidermis (Fig. 2).

It was also found from area analysis for a transverse fracture surface of a bipinnate radiole that vanadium is not uniformly distributed in the epidermis

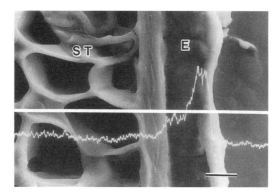

Figure 2. Line analysis by EPMA apparatus for a longitudinal section of the bipinnate radiole. A clear peak indicating the highest vanadium content is observed in the outer portion of the epidermis (E). ST, supporting tissue. Scale bar = 10 μm.

(Fig. 3b). Furthermore, significant differences in the vanadium content were observed among the vanadium-concentrated places in the epidermis. Point analyses (probe size = 5 μm) for nine high-intensity positions of three individual specimens indicated that the vanadium concentration ranged from 46.5 to 100.5 mg/g dry wt (mean $\pm$ SD = 64.6 $\pm$ 21.0 mg/g dry wt) in the epidermis. Usually the dry weight percentage of soft tissues of marine animals ranges from 10% to 30%. However, the dry weight percentage of the vacuoles in which vanadium was contained (AEM disclosed that vanadium was localized in the vacuoles, as described below) was assumed to be less than 10%, if the dry weight percentage of the vacuoles was approximately similar to that of the supernatant of the homogenate of the bipinnate radiole. Therefore, the vanadium concentration in the vacuoles was estimated to be more than 6.46 $\pm$ 2.10 mg/g on a wet weight basis.

On the other hand, among the several types of blood cells of ascidians, the vanadium-containing cell of *Ascidia ahodori* (Michibata et al., 1987), *A. sydneiensis samae* (Michibata and Uyama, 1990), and *A. gemmata* (Hirata and Michibata, 1991) has been identified as the signet ring cell. It is of great interest that the vanadocytes (epidermal cell and blood cell) are quite different between the two vanadium accumulators. The similarity and the difference in vanadium distribution between *P. occelata* and ascidians are expected to provide important information for the investigation of mechanisms of bioconcentration and the physiological roles of vanadium.

3.3.   Subcellular Distribution in the Epidermal Cells

An electron micrograph of the cryosection of biological materials without osmic fixation and staining usually gives low contrast. However, as shown in Figure 4b of a cryosection (100 nm), many electron-dense deposits suggesting

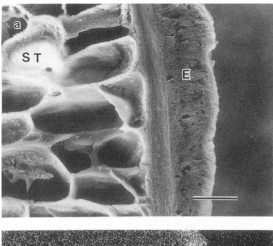

Figure 3. a. Scanning electron micrograph of a transverse fracture surface of the bipinnate radiole. E, epidermis. ST, supporting tissue. Scale bar = 15 μm. b. Area analysis by EPMA for the same position of the transverse fracture surface as in Figure 3a. Vanadium concentrations are denoted by white spots. Some places (arrow heads) have an extremely high level of vanadium.

the localization of certain elements are present in the apical portion of the cytoplasm. The transmission electron micrograph of Figure 4a reveals that electron-dense deposits are situated at the same places where many vacuoles with low electron densities are located.

As illustrated in Figure 4c, clear peaks that are characteristic of X-ray diffraction of sulfur (2.31 keV) and vanadium (4.95 keV) appear from the electron-dense deposits. Ishii et al. (1994) concluded that most of the vanadium in the branchial crown of *P. occelata* was localized in the vacuoles of the epidermal cells. A small quantity of vanadium was also found to be contained in the cuticle, but no vanadium was detected in the other cellular components such as the nucleus, cytoplasm, and so forth.

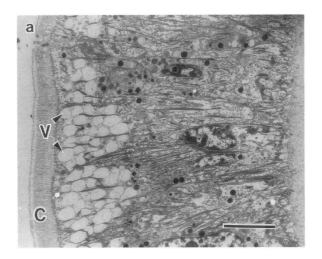

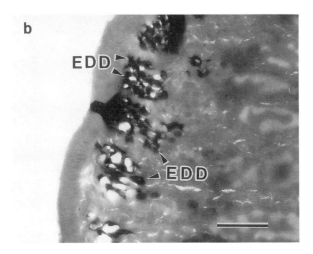

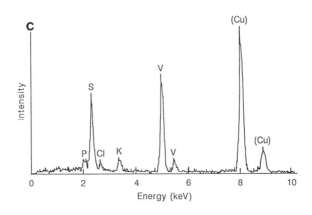

Scippa et al. (1988) applied X-ray microanalysis in scanning transmission electron microscopy to vacuolated and granular amoebocytes, signet ring cells, and a new type of compartment cell of the ascidian *Phallusia mammillata*. They found vanadium deposits on the vacuolar membranes and in the intravacuolar and cytoplasmic electron-dense inclusions.

Most of the vanadium in ascidians is thought to be concentrated in the vacuole of the signet ring cell, from results based on the separation of the vanadocyte (Michibata et al., 1987) and the confirmation of the locality of vanadium deposits (Scippa et al., 1988). It is very interesting that the subcellular distribution of vanadium is the same, although the vanadocytes are quite different between the two vanadium accumulators.

4. CHEMICAL SPECIATION OF VANADIUM IN LIVING ORGANISMS

4.1. Application of XAFS to in Vivo Measurement

Knowledge about the oxidation state and coordination form of vanadium in *P. occelata* is important to speculation about the physiological roles of vanadium in this animal. X-ray absorption fine structure (XAFS) spectroscopy using synchrotron radiation is a powerful technique for examining the electronic structure and the local environment of specific atoms in biological samples (Powers, 1982). This technique was successfully applied to the study of vanadium in living ascidian blood cells (Tullius et al., 1980). It was revealed that the vanadium in the living cells is present as vanadium(III) ions surrounded by six oxygen atoms. It was postulated that XAFS is particularly useful for the study of *P. occelata* because it literally allows nondestructive analysis of living organisms; the oxidation state of vanadium is very sensitive to aerooxidation and in vivo analysis is often a crucial requirement. Therefore, the XAFS technique was utilized to clarify the oxidation state and local structure of vanadium in *P. occelata* (Ishii et al., 1993). This is possibly the first successful application of the XAFS technique to the whole soft body of a living marine animal. Since the principle and theory of the XAFS technique have already been explained in the literature (Powers, 1982; Teo, 1986), they are not repeated here.

Figure 4. a. Transmission electron micrograph of an ultrathin section of the bipinnate radiole that was fixed and dehydrated by a conventional method. Many vacuoles (V) with low electron density are observed in the apical portion of the epidermal cells. C, cuticle. Scale bar = 3 μm.
b. Transmission electron micrograph of the cryosection of the bipinnate radiole. Many electron-dense deposits (EDD) are present in the apical portion of the epidermis. Scale bar = 3 μm.
c. X-ray spectrum of the electron-dense deposit obtained by analytical electron microscopy. The peaks of copper are derived from a copper grid.

XAFS spectral data on the living *P. occelata* were measured at the Photon Factory of the National Laboratory for High-Energy Physics, Tsukuba, Japan, with a ring energy of 2.5 GeV. Data were collected in the fluorescence mode using a Si(111) two-crystal monochromator and a Lytle-type fluorescence detector (Lytle et al., 1984) at a bending magnet beam line.

The specimens of *P. occelata* were collected on the Sanriku coast of the Pacific Ocean in the northeastern region of Japan. They were kept in fresh seawater, transported to the Photon Factory within 24 h, and subjected to the XAFS measurement as a live specimen. The whole soft bodies of the animals were pulled out from their tubes and three specimens were placed in a polyethylene bag. The bipinnate radioles were exposed to X-ray irradiation. The X-ray absorption spectra were measured form 4.96 keV to 6.43 keV with a step size of approximately 1.2 eV in the edge region and 4 s per point (total 545 steps). It was remarkable that the animals were still alive even after the X-ray irradiation. For comparison, the XAFS spectra of six vanadium compounds of known structures were also measured: $V_2(SO_4)_3$, V(acetylacetonate = acac)$_3$, and VO (acac)$_2$ were prepared by the methods of Claunch and Jones (1963), Grdenic and Korpar-Colig (1964), and Rowe and Jones (1957), respectively; V_2O_5 and NH_4VO_3 were purchased from Kanto Chem. Co. and $VO(C_2O_4)_2 \cdot nH_2O$ was from Wako Chem. Co. in Japan.

4.2. Oxidation State of Vanadium in *P. occelata*

Figure 5 shows the vanadium K-edge spectra of the specimens and the reference compounds obtained by X-ray absorption near-edge structure (XANES) analysis. It is found that the K-edge spectra of certain compounds exhibit a strong pre-edge peak at around 5.47 keV. This peak is ascribed to the 1s-to-3d transition, which is formally forbidden by dipole selection rules if the coordination sphere of vanadium has octahedral (O_h) symmetry with a center of inversion (Wong et al., 1984). When the symmetry of the ligand is lowered from O_h, the inversion center is broken, and the pre-edge absorption becomes dipole, which is allowed because of a combination of stronger 3d-4p mixing and overlap of the metal 3d orbitals with the 2p orbitals of the ligand. Therefore, if a vanadium compound contains a terminal oxo ligand (V=O) as in the vanadium(IV) compounds, the octahedral symmetry is broken and the compound exhibits intense pre-edge absorption. It was reported that the intensity of the pre-edge absorption is roughly proportional to the number of oxo groups present in the molecule (Tullius et al., 1980).

As shown in Figure 5, the intensity variation of the pre-edge peaks across the series of the reference compounds from vanadium(III) to vanadium(V) compounds is remarkable. On the other hand, the pre-edge absorption in the spectrum of *P. occelata* is almost negligible and resembles that of the symmetrically coordinated V^{III}(acac)$_3$. This indicates that most of the vanadium exists in the vanadium(III) state in the living condition with lack of any significant quantity of a VO^{2+} component.

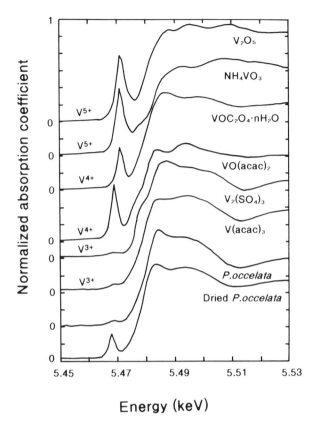

Figure 5. Vanadium K-edge XANES spectra of the bipinnate radioles of living *P. occelata*, dried *P. occelata*, and reference vanadium compounds. Intense pre-edge peaks at around 5.47 keV were observed for dried *P. occelata* and compounds containing vanadium(IV) and vanadium-(V) ions.

It is also noteworthy that a significant increase in the pre-edge intensity was observed for the dried sample (Fig. 5). A similar effect was obtained when the bipinnate radioles were removed from the body and were ground in air with an agate mortar. This suggests that vanadium was partly oxidized by air to produce the VO^{2+} component. Therefore, it is postulated that in vivo measurement of the XAFS spectra is essential for the determination of the chemical state of vanadium in *P. occelata*.

4.3. Coordination Structure of Vanadium in *P. occelata*

The analysis of the data was performed following the procedure reported in the literature (Ozutumi and Kawashima, 1991). The extended X-ray absorption fine structure (EXAFS) oscillation, $\chi(k)$, was extracted from the spectra by pre-edge and background subtraction, and energy conversion into the photoelectron wave vector $k(k = 2\pi[2m(E - E_0)/h^2]^{1/2})$, where m is the mass of the electron, h is Planck's constant, E is the incident photon energy and E_0 is the threshold energy (i.e., the energy at $k = 0$) of that particular absorption edge (Teo, 1986). Figure 6 shows the k^2-weighted $\chi(k)$. It can be seen that the difference in amplitude of the $\chi(k)$ function between living *P. occelata* and dried specimen is significant. The Fourier transforms of the $k^2\chi(k)$ EXAFS oscillation yielded the radial structure functions given in Figure 7, which are not corrected for any phase shift. This gives a single main peak with the absence of any ordered second shell of scattering atoms at longer distances. Considering the result of the XANES analysis, the lack of the second-coordination atoms would suggest that vanadium is present in solution in the form of an aqua complex with symmetrical octahedral coordination.

The structural parameters of vanadium were obtained by the Fourier filtering technique. This was accomplished by an inverse Fourier transformation of the main peak in the radial structure function. This peak defines the contribution of the first coordination shell belonging to the vanadium(III) atom to the EXAFS oscillation. The resulting "observed" $\chi(k)$ function was then

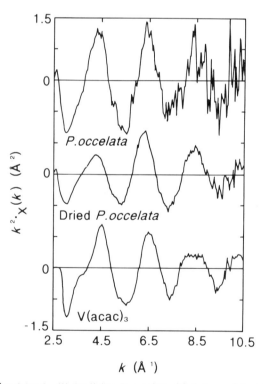

Figure 6. k^2-weighted $\chi(k)$ for living *P. occelata,* dried *P. occelata,* and V(acac)$_3$.

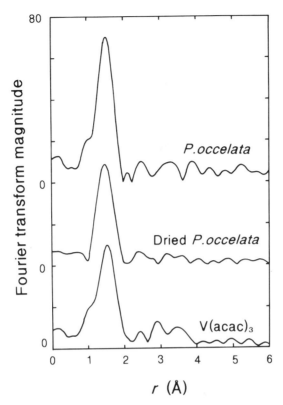

Figure 7. Radial structure functions for the bipinnate radioles of living *P. occelata*, dried *P. occelata*, and V(acac)₃ calculated without correcting for the phase shifts.

analyzed by iterative nonlinear least square curve fitting with the $\chi(k)$ function calculated from a coordination model and with backscattering amplitude and phase parameters given by Teo and Lee (1979). As a model, the single oxygen wave fit one reported for the vanadocyte of ascidians and was used in the fitting analysis and the coordination number was fixed at 6. The agreement between the experimental and calculated data was fair ($R = 0.072$), considering that the data were obtained from the living animal. The V–O distance thus obtained was 2.00 Å with a Debye-Waller factor σ of 0.021.

There have been many studies concerning the chemical form of vanadium in ascidians. Electron paramagnetic resonance (EPR) analysis disclosed that the whole blood of *Ascidia nigra* contained V(III) as the major component, along with a minor amount of vanadyl ion (Dingley et al., 1981; Kustin et al., 1976). A magnetic study of the vanadocyte with a superconducting quantum interference device (SQUID) susceptometer also supported the predominance of vanadium(III) in ascidians; that is, the intact blood cells of the tunicate *A.*

nigra contained 90% vanadium(III) and 10% vanadium(IV) (Lee et al., 1988). XAFS examination of living *Ascidia ceratodes* vanadocytes revealed that the first coordination sphere of vanadium(III) consisted of six oxygen atoms at an average distance of 1.99 Å (Tullius et al., 1980). No evidence for a ligand second shell was found, and the possibility that the vanadium was confined either by a rigid low-molecular-weight chelate or a highly ordered protein binding site was excluded. In addition, XANES analysis of sulfur within the plasma cells of *A. ceratodes* disclosed the presence of a large endogenous sulfate concentration, as well as smaller significant amounts of thiol or thio-ether sulfur (Frank et al., 1987). EPR linewidth/pH correlation showed that the internal vanadophoric pH in *A. ceratodes* vanadocytes was 1.8 ± 0.1 (Frank et al., 1986). This pH is sufficiently low to prevent hydrolysis of endogenous $V(H_2O)_6^{3+}$ or its autoxidation. Furthermore, nuclear magnetic resonance (NMR) analysis indicated the presence of four to five ligating waters (Carlson, 1975). From these observations, it was concluded that the most reasonable form to be deduced for vanadophoric vanadium(III) in the presence of high (SO_4^{2-}) is the $V(SO_4)(H_2O)_{4-5}^+$ complex ion (Lee et al., 1988). On the other hand, our preliminary study of the sulfur K-edge XANES spectrum of the bipinnate radiole of *P. occelata* has indicated that the predominant chemical form of sulfur is sulfate ion. Moreover, our local analysis by AEM disclosed the coexistence of vanadium and sulfur (Fig. 4c). It was reported that the signal strengths of the two elements in the cells from *Ascidia mentula* and *Ascidiella aspersa* determined by their X-ray microanalysis were usually similar (Bell et al., 1982). As can be seen in Figure 4c, this tendency was also observed in the case of *P. occelata*. These apparent similarities between the two animals suggest the possibility that vanadium in *P. occelata* exists as a chemical form similar to that in ascidians, that is, an aqua complex of vanadium(III) in sulfuric acid media. As for the existence of a vanadium(IV) component in the ascidians, EPR analysis indicated the presence of vanadyl ion in the whole-cell preparation of *A. ceratodes*, although there is the possibility that the observed EPR signal of vanadyl ion arose from the signet ring cells or compartment cells rather than from vanadocytes, which are the major constituent of the plasma cell (Bell et al., 1982). This existence of vanadyl ion is also supported by the magnetic study, as previously mentioned (Lee et al., 1988). The presence of vanadyl ion as a minor component of *P. occelata* could not be ruled out because the vanadium K-edge XANES spectrum of *P. occelata* exhibited a faint pre-edge peak assignable to a VO^{2+} component. In order to solve this problem, EPR analysis of this animal will be essential.

5. PHYSIOLOGICAL ROLES OF VANADIUM

Many biologists and chemists have been interested in the biochemical signifi-cance of the high vanadium concentration in ascidians since the first report by Henze (1911). In particular, physiologists have studied the biological functions of vanadium in ascidians and presented hypotheses such as oxygen binding, tunicin synthesis, antimicrobial defenses, and antipredation mechanisms. But as yet there is no widespread agreement that supports any one of these hypotheses.

Reasonable data have not yet been obtained, although I am intensively studying the physiological roles of vanadium in *P. occelata.* However, results on the localization and chemical states of vanadium in the vacuoles might provide important suggestions towards solving the mystery of the high vanadium concentration in *P. occelata.*

I propose several ideas as to the physiological roles of vanadium in *P. occelata.*

1. Generally speaking, the localization of vanadium in the apical vacuoles of the epidermal cells and in the cuticle implies that vanadium itself or vanadium-containing substances are absorbed, secreted, or excreted at the surface of the bipinnate radioles.

2. Vanadium may be related to the regulation of oxidation-reduction reactions on the surface of the bipinnate radioles.

3. Since the branchial crown serves as a site of gas exchange (Barnes, 1980), it is probable that vanadium is related to the absorption of O_2 through many cytoplasmic processes in the cuticle.

4. There is a possibility that vanadium in *P. occelata* does not play any important role in its physiology after vanadium reaches the vacuoles. This is because the vacuoles are often thought to be used as an internal disposal site to isolate or confine toxic substances, as indicated by the examples of metal-containing granules that intracellularly appear in the vacuoles of the kidney cells of clams or the branchial heart of Cephalopoda Mollusca.

5. Vanadium absorbed from the surrounding sea water is thought to be temporarily associated with organic materials such as proteins. When vanadium-containing substances pass through the vacuole membrane and change into an aqua complex ion, a certain amount of energy may be produced.

6. *P. occelata,* which is used as a fishing bait, has several defense mechanisms against predators like fishes. The most important defense is swift withdrawal into its tube when predators approach (Barnes, 1980). According to my observations, *P. occelata* in a colony could live with three species of starved carnivorous fishes in an aquarium for over 2 weeks. I thought that a specific chemical defense substance, including vanadium, may be secreted from the bipinnate radioles.

7. Vanadium is known to be an integral component of some enzymes such as bromoperoxidase of brown algae, nirogenase of nitrogen fixation bacteria, and so forth. Since free peroxide would oxidatively destroy many cellular components, peroxidase plays an important role in the

detoxification of hydrogen peroxide. Although the presence of vanadium enzymes in the vacuoles of the epidermal cells of *P. occelata* has not been examined, vanadium may have some biological functions during the process of catalysis of certain enzymes in relation to detoxification.

More than 80 years have passed since Henze (1911) first reported that a high level of vanadium was contained in the blood of ascidians. However, its biological function still remains unknown in spite of many efforts by biologists.

I hope that detailed studies comparing *P. occelata* and ascidians will be helpful in clarifying the mechanism of the specific accumulation or physiological roles of vanadium in the two vanadium accumulators.

REFERENCES

Barnes, R. D. (1980). *Invertebrate Zoology.* Saunders College, Philadelphia, pp. 467–566.

Bell, M. V., Pirie, B. J. S., McPhail, D. B., Goodman, B. A., Falk-Petersoen, I. B., and Sargent, J. R. (1982). Contents of vanadium and sulfur in the blood cells of *Ascidia mentula* and *Ascidiella aspersa. J. Mar. Biol. Assoc. U.K.* **62,** 709–716.

Bryan, G. W., and Gibbs, P. E. (1980). Metals in nereid polychaetes: The concentration of metals in the jaws to the total burden. *J. Mar. Biol. Assoc. UK.* **60,** 641–654.

Carlson, R. M. K. (1975). Nuclear magnetic resonance spectrum of living tunicate blood cells and the structure of the native vanadium chromogen. *Proc. Natl. Acad. Sci. USA* **72,** 2217–2221.

Carmichael, N. G., Squibb, K. S., and Fowler, B. A. (1979). Metals in the molluscan kidney: A comparison of two closely related bivalve species (*Argopecten*), using X-ray microanalysis and atomic absorption spectroscopy. *J. Fish. Res. Bd. Can.* **36,** 1149–1155.

Claunch, R. T., and Jones, M. M. (1963). *Inorganic Syntheses,* Vol. 7. McGraw-Hill Book Company, New York, p. 92.

Dingley, A. L., Kustin, K., Macara, I. G., and McLeod, G. C. (1981). Accumulation of vanadium by tunicate blood cells occurs via a specific anion transport system. *Biohim. Biophys. Acta* **649,** 493–502.

Frank, P., Carlson, R. M. K., and Hodgson, K. O. (1986). Vanadyl ion EPR as a non-invasive probe of pH in intact vanadocytes from *Ascidia ceratodes. Inog. Chem.* **25,** 470–478.

Frank, P., Hedman, B., Carlson, R. M. K., Tyson, T. A., Roe, A. L., and Hodgson, K. O. (1987). A large reservoir of sulfate and sulfonate resides within plasma cells from *Ascidia ceratodes* revealed by X-ray absorption near-edge structure spectroscopy. *Biochemistry* **26,** 4975–4979.

George, S. G., Pirie, B. J. S., and Cooms, T. L. (1980). Isolation and elemental analysis of metal-rich granules from the kidney of scallop, *Pecten maximus. J. Exp. Mar. Biol. Ecol.* **42,** 143–156.

Grdenic, D., and Korpar-Colig, B. (1964). Another preparation method of vanadium(III) and uranium(IV) 1,3-diketone complexes. *Inorg. Chem.* **3,** 1328–1329.

Henze, M. (1911). Untersuchungen über das Blut der Ascidien. I. Mitteilung. Die Vanadiumverbindung der Blutkorperchen. *Hoppe-Selyer's Z. Physiol. Chem.* **72,** 494–501.

Hirata, J., and Michibata, H. (1991). Valency of vanadium in the vanadocytes of *Ascidia gemmata* separated by density-gradient centrifugation. *J. Exp. Zool.* **257,** 160–165.

Ishii, T. (Submitted for publication.) The specific accumulation of iodine by strawberry conch *Strombus luhuanus. Fisheries Science.*

Ishii, T., Nakahara, M., Matsuba, M., and Ishikawa, M. (1991). Determination of U-238 in marine organisms by inductively coupled plasma mass spectrometry. *Nippon Suisan Gakkaishi* **57,** 779–787.

Ishii, T., Nakai, I., Numako, C., Okoshi, K., and Otake, T. (1993). Discovery of a new vanadium accumulator, the fan worm *Pseudopotamilla occelata. Naturwissenschaften* **80,** 268–270.

Ishii, T., Okoshi, K., Otake, T., and Nakahara, M. (1992). Concentrations of elements in tissues of four species of Tridacnidae. *Nippon Suisan Gakkaishi* **58,** 1285–1290.

Ishii, T., Otake, T., Hara, M., Ishikawa, M., and Koyanagi, T. (1986). High accumulation of elements in the kidney of the marine bivalve *Cyclosunetta menstrualis. Bull. Japan Soc. Sci. Fish.* **52,** 147–154.

Ishii, T., Otake, T., Okoshi, K., Nakahara, M., and Nakamura, R. (1994). Intracellular localization of vanadium in the fan worm *Pseudopotamilla occelata. Mar. Biol.* **121,** 143–151.

Kim, K. S., Webb, J., Macey, D. J., and Cohen, D. D. (1986). Compositional changes during biomineralization of the radula of the chiton *Clavarizona hirtosa. J. Inorg. Biochem.* **28,** 337–345.

Kustin, K., Levine, D. S., McLeod, G. C., and Curby, W. A. (1976). The blood of *Ascidia nigra:* Blood cell frequency distribution, morphology, and the distribution and valence of vanadium in living blood cells. *Biol. Bull.* **150,** 426–441.

Lee, S., Kustin, K., Robinson, W. E., Frankel, R. B., and Spartalian, K. (1988). Magnetic properties of tunicate blood cells. I. *Ascidia nigra. J. Inorg. Biochem.* **33,** 183–192.

Lytle, F. W., Greegor, R. B., Sandstom, D. R., Marques, E. C., Wong, J., Spiro, C. L., Huffman, G. R., and Huggins, F. E. (1984). Measurement of soft X-ray absorption spectra with a fluorescention chamber detector. *Nucl. Instrum. Methods.* **226,** 542–548.

Macara, I. G., McLeod, C., and Kustin, K. (1979). Tunichromes and metal ion accumulation in tunicate blood cells. *Comp. Biochem. Physiol.* **63B,** 299–302.

Mann, S., Perry, C. C., Webb, J., Luke, B., and Williams, R. J. P. (1986). Structure, morphology, composition and organization of biogenic in limpet teeth. *Proc. R. Soc. Lond.* **B227,** 179–190.

Michibata, H., Hirata, J., Uesaka, M., Numakunai, T., and Sakurai, H. (1987). Separation of vanadocytes: Determination and characterization of vanadium ion in the separated blood cells of the ascidian *Ascidia ahodori. J. Exp. Zool.* **244,** 33–38.

Michibata, H., Terada, T., Anada, N., Yamakawa, K., and Numakunai, T. (1986). The accumulation and distribution of vanadium, iron and manganese in some solitary ascidians. *Biol. Bull.* **171,** 672–681.

Michibata, H., and Uyama, T. (1990). Extraction of vanadium-binding substances (vanadobin) from a subpopulation of signet ring cells newly identified as vanadocytes in ascidians. *J. Exp. Zool.* **254,** 132–137.

Odum, H. T. (1951). Notes on the strontium content of sea water, celestite radiolaria, and strontianite snail shells. *Science* **114,** 211–213.

Ozutumi, K., and Kawashima. T. (1991). Structure of copper(II) -bpy and -phen complexes: EXAFS and spectrophotometric studies on the structure of copper(II) complexes with 2,2'-bipyridine and 1,10-phenanthroline solution. *Inorg. Chim. Acta.* **180,** 231–238.

Powers, L. (1982). X-ray absorption spectroscopy application to biological molecules. *Biochim. Biophys. Acta* **683,** 1–38.

Rowe, R. A., and Jones, M. M. (1957). *Inorganic Syntheses,* Vol. 5. McGraw-Hill Book Company, New York, p. 114.

Scippa, S., Zierold, K., and Vincentiis, M. (1988). X-ray microanalytical studies on cryofixed blood cells of the ascidians. *J. Submicrosc. Cytol. Pathol.* **20,** 719–730.

Teo, B. K. (1986). *EXAFS: Basic Principles and Data Analysis.* Springer-Verlag, Berlin.

Teo, B. K., and Lee, P. A. (1979). Ab initio calculations of amplitude and phase functions for extended X-ray absorption fine structure spectroscopy. *J. Am. Chem. Soc.* **101**, 2815–2832.

Tullius, T. D., Gillum, W. O., Carlson, R. M. K., and Hodgson, K. O. (1980). Structual study of the vanadium complex in living ascidian blood cells by X-ray absorption spectrometry. *J. Am. Chem. Soc.* **102**, 5670–5676.

Ueda, T., Nakahara, M., Ishii, T., Suzuki, Y., and Suzuki, H. (1979). Amounts of trace elements in marine Cephalopoda. *J. Radiat. Res.* **20**, 338–342.

Wong, J., Lytle, F. W., Messmer, R. P., and Maylotte, D. H. (1984). K-edge absorption spectra of selected vanadium compounds. *Phys. Rev.* **30**, 5596–5610.

10

SELECTIVE ACCUMULATION OF VANADIUM BY ASCIDIANS FROM SEA WATER

Hitoshi Michibata

*Mukaishima Marine Biological Laboratory,
Faculty of Science and Laboratory of Marine Molecular
Biology, Graduate School of Science, Hiroshima University,
Mukaishima-cho 2445, Hiroshima 722, Japan*

Kan Kanamori

*Department of Chemistry, Faculty of Science, Toyama
University, Gofuku 3190, Toyama 930, Japan*

Vanadium in the Environment. Part 1: Chemistry and Biochemistry, Edited by Jerome O. Nriagu.
ISBN 0-471-17778-4. © 1998 John Wiley & Sons, Inc.

1. INTRODUCTION

Organisms born in the primitive sea have evolved and differentiated into a huge number of species through adaptive radiation. At present, more than 10^8 species of organisms are estimated to live on the earth. Many extant organisms classified into phyla still live in the sea. However, less is known about marine organisms than terrestrial organisms because of various difficulties in studying the former. New research that falls between biology and chemistry has focused on the unknown and unusual physiological functions of certain marine organisms.

Ascidians, known as tunicates or sea squirts, are one of the organisms being studied. Ascidians are phylogenically classified into the phylum Chordata, which consists of subphyla Urochordata, Cephalochordata, and Vertebrata. All animals belonging to the phylum Chordata have a notochord, a dorsal nerve cord, and pharyngeal gill slits. Ascidians are an evolutionary link between Invertebrata and Vertebrata.

In 1911 a German chemist, Martin Henze, discovered high levels of vanadium in the blood cells (coelomic cells) of an ascidian collected from the Bay of Naples (Henze, 1911). His finding attracted not only analytical chemists, but also physiologists, biochemists, and chemists of natural products, in part because of the initial interest in the extraordinarily high level of vanadium never before reported in other organisms but also because of the considerable interest in the possible participation of vanadium in oxygen transport as a third candidate of the prosthetic group in respiratory pigment in addition to iron and copper. Since then, many scientists have endeavored to analyze the contents of vanadium, to isolate the organs participating in the accumulation of vanadium, to reveal the respiratory pigment, to determine the route of accumulation of vanadium, and to elucidate the physiological function of vanadium. Progress, however, has been very slow. One of the reasons for this is insufficient cooperation between biologists and chemists in resolving such interdisciplinary problems.

Although many review articles on the accumulation of vanadium by ascidians and the characterization of vanadium have been reported (Biggs et al., 1976; Boas et al., 1987; Boyd and Kustin, 1985; Chasteen, 1981, 1983; Goodbody, 1974; Kustin et al., 1983; Kustin and Robinson, 1995; Michibata, 1989, 1993, 1996; Michibata and Sakurai, 1990), we describe mainly our own results obtained from interdisciplinary studies in this review article.

2. OCCURRENCE OF VANADIUM

The abundance of vanadium in the earth's crust is 135 ppm, which is higher than that of copper (55 ppm) (Mason, 1966). In this sense, vanadium is not rare. However, vanadium is considered a rare metal because the vanadium ores are lean and localized. The important vanadium ores are vanadinite, $Pb_5(VO_4)_3Cl$ and carnotite, $K_2(UO_2)_2(V_2O_8) \cdot 3H_2O$; the latter is also an important uranium ore (Chapter 1).

Vanadium is also known to be present in crude oil and is the source of environmental pollution, catalytic poisoning in the petroleum cracking process, and corrosion of refinery equipment (Chapter 4). Saudi Arabian crude oil contains about 10 mg/g of vanadium (Al-Swaidan, 1996). Vanadium can be recovered as V_2O_5 from the smoke dust after combustion of oil. Interestingly, about 50% of vanadium in oil is in the form of vanadyl porphyrin complex (Barbooti et al., 1989). Vanadium in coal, tar sand, and crude oil was briefly reviewed by Wever and Kustin (1990).

The concentration of vanadium in sea water is 3.5×10^{-8} mol/dm^3, which is similar to that of zinc, iron, and nickel. The oxidation state of vanadium in sea water is believed to be +5 (McLeod et al., 1975) but this remains to be confirmed (Sugimura et al., 1978). The behavior of vanadium(V) in aqueous solution is very complex because of the oligomerization and protonation equilibria (Chapter 4). Figure 1 shows the predominant vanadium(V) species in

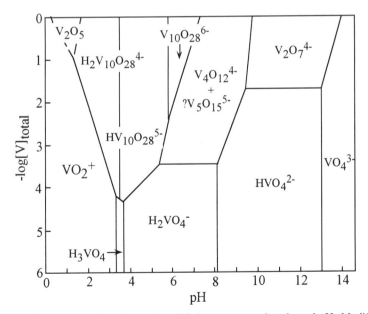

Figure 1. Predominant species of vanadium(V) in water as a function of pH. Modified and redrawn from Pope (1983, Fig. 3.1).

water as a function of pH and concentration (Pope, 1983). Considering the very low concentration of vanadium in sea water, it is reasonable to assume that the species predominantly present are HVO_4^{2-} and $H_2VO_4^-$ and that oligomeric species are negligibly present. However, if the preconcentration of vanadium(V) is due to the accumulation of vanadium from sea water by ascidians, the participation of the oligomers can not be ruled out.

3. CHEMICAL NATURE OF VANADIUM

Vanadium forms a diversity of chemical compounds over a wide range of oxidation states from -3 to $+5$. Of these oxidation states, $+3$, $+4$, and $+5$ are biologically the most relevant. Vanadium(II) is proposed for the reduced form of vanadium nitrogenases from *Azotobacter* (Smith et al., 1988). Vanadium(V) and (IV) are stable oxidation states under ordinary conditions. Vanadium(III) compounds are generally oxidized by air. Vanadium(II) compounds are strongly reducing and oxidized even by water.

Vanadium ion adopts a structure characteristic of its oxidation state (Chapter 7). Vanadium(V) exists as tetrahedral vanadate anion, VO_4^{3-}, in strongly basic solution. On addition of acid to vanadate solution, vanadate undergoes successive condensation from dimeric pyrovanadate, $V_2O_7^{4-}$, in weakly basic solution to orange decavanadate, $V_{10}O_{28}^{6-}$, in weakly acidic solution. Several polyvanadate species such as tetramer and pentamer are formed as intermediates. The cation VO_2^+ exists at very low pH. Many coordination compounds with VO_2^+ moiety have been crystallographically characterized, including $[VO_2(C_2O_4)_2]^{3-}$ (Scheidt et al., 1971a) and $[VO_2(edta)]^{3-}$ (Scheidt et al., 1971b). In these d^0 complexes, VO_2^+ cations take a bent structure, which is distinct from the linear structure found in several d^n dioxo compounds such as RuO_2^{2+}.

The coordination chemistry of vanadium(IV) is dominated by oxovanadium(IV), VO^{2+}, commonly called "vanadyl" ion. The vanadyl group retains its identity in many compounds. Vanadyl complexes usually take a square pyramidal or distorted octahedral geometry. A less common trigonal bipyramidal structure has been also found. $[VOX_4]$ ($X = Cl$ or Br) undergoes facile temperature-dependent interconversion between square pyramidal and trigonal bipyramidal structures (Nicholls and Seddon, 1973; Cave et al., 1978).

The short VO distance (ca. 1.6 Å) of the vanadyl group indicates a significant π-bonding character. In hexacoordinate complexes the coordination bond trans to the oxo group is significantly elongated owing to the strong trans influence of the oxo atom, resulting in a tetragonally distorted octahedral structure. The substitution at the trans position is very rapid compared with that at the cis positions.

[VO(salen)] (salen: N,N'-disalicylideneethylenediamine) loses the vanadyl oxygen by reaction with oxophylic reagents such as $SOCl_2$, PCl_5 (Pasquali et al., 1979), and Ph_3PBr_2 (Callahan and Durand, 1980) to produce $[VX_2(salen)]$

(X = Cl or Br). The deoxygenation occurs also in strongly acidic nonaqueous media (Tsuchida et al., 1994). Non-oxo vanadium(IV) compounds are called "bare" vanadium(IV). The tris(maleonitrile dithiolato) complex, $[V(mnt)_3]^{2-}$, was the bare tris(chelate)vanadium(IV) complex to be characterized (Davison et al., 1964; Stiefel et al., 1968). Several bare vanadium(IV) complexes containing other dithiolene ligands have since been prepared and well characterized (Welch et al., 1988; Matsubayashi et al., 1988; Kondo et al., 1996). Cathecol (cat), 2,4-pentanedione (acac), and their derivatives also produce bare tris(chelate)vanadium(IV) complexes (Cooper et al., 1982; Hambley et al., 1987; Diamantis et al., 1976). A unique example of a bare vanadium(IV) compound is the complex with the cage-type hexamine ligand (Comba et al., 1985). The above tris(chelate)vanadium(IV) complexes adopt a trigonal prismatic, octahedral, or intermediate geometry.

The chemistry of vanadium(III) is less developed than that of vanadium(IV) and vanadium(V) because vanadium(III) compounds are usually unstable toward air oxidation. The acid dissociation constant of $[V(H_2O)_6]^{3+}$ (pKa = 2.85) is much smaller than that of $[VO(H_2O)_5]^{2+}$ (pKa = 6.0) and comparable to that of HF (pKa = 3.17). Thus, the aqueous solution of simple vanadium(III) salts is fairly acidic. The aqua vanadium(III) species develop a dark brown coloring upon hydrolysis at around pH 2–3 depending on the concentration. The visible absorption spectrum of the hydrolytic species is quite similar to that of so-called "Henze's solution" (lysate of blood cells of ascidians). Resonance Raman study demonstrated that the oxo-bridged dinuclear vanadium(III) species, $[V–O–V]^{4+}$, is responsible for the dark brown coloring (Kanamori et al., 1991). The well-characterized oxo-bridged dinuclear vanadium(III) complexes include $[V_2OCl_4(thf)_6]$ (Chandrasekhar and Bind, 1984), $[V_2OCl_2(bpy)_4]Cl_2$ (Brand et al., 1990), $[V_2O(SCH_2CH_2NMe_2)_4]$ (Money et al., 1987), $[V_2O(L\text{-}his)_4]$ (Kanamori et al., 1993; Czernuszewicz et al., 1994), and $[V_2OBr_2(tpa)_2]Br_2$ (Kanamori et al., 1995).

The most common coordination geometry of vanadium(III) complexes is a hexacoordinate octahedron exemplified by $[V(C_2O_4)_3]^{3-}$ (Fenn et al., 1967). Heptacoordinate vanadium(III) compounds are also not rare and the number of examples characterized has increased. Three appropriate geometries of heptacoordination—namely, pentagonal bipyramid, capped trigonal prism, and capped octahedron (Fig. 2)—are all realized in the aminopolycarboxylato vanadium(III) complexes (Robles et al., 1993; Shimoi et al., 1989; Okamoto et al., 1992). The flexibility of the coordination geometry may relate to the biological function of vanadium(III).

4. LEVELS OF VANADIUM IN ASCIDIANS

After the first finding of vanadium in ascidian blood cells, many species of ascidians were analyzed for the presence not only of vanadium but also of other heavy metals. Transition metals including vanadium are, difficult to

pentagonal bipyramid   capped octahedron   capped trigonal prism

Figure 2. Polyhedrons adopted by heptacoordinate vanadium(III) complexes with aminopolycarboxylates.

analyze quantitatively however. A variety of analytical methods, such as colorimetry, emission spectrometry, and atomic absorption spectrometry have been applied. Consequently, niobium, chromium, tantalum, tungsten, and titanium have also been reported to be present in ascidians, but these results have not been reproducible in most cases (Cantacuzène and Tchekirian, 1932; Vinogradov, 1934; Kobayashi, 1935; Webb, 1939; Noddack and Noddack, 1939; Bertrand, 1950; Boeri, 1952; Lybing, 1953; Boeri and Ehrenberg, 1954; Webb, 1956; Levine, 1961; Bielig et al., 1954, 1961a, b, c, 1966; Kalk, 1963a, b; Ciereszko et al., 1963; Rummel et al., 1966; Carlisle, 1968; Swinehart et al., 1974; Danskin, 1978; Botte et al., 1979a, b; Hawkins et al., 1980a). Techniques vary widely in sensitivity and in precision as do data on dry weight, wet weight, ash weight, inorganic dry weight, or amount of protein (Chapter 9). Thus, early data are not useful for resolving how ascidians accumulate vanadium physiologically because values cannot be compared. Redetermination of the contents of vanadium was therefore required.

We collected numerous species of ascidians belonging to two of three suborders, Phlebobranchia and Stolidobranchia, mainly from the waters around Japan and the Mediterranean. Samples of blood cells (coelomic cells), plasma, tunic, mantle (muscle), branchial basket, stomach, hepatopancreas, and gonad were submitted to neutron-activation analysis in a nuclear reactor, which is an extremely sensitive method for the quantification of vanadium (Michibata, 1984; Michibata et al., 1986).

The data obtained are summarized in Table 1. Although vanadium was detectable in samples from almost all species examined, the ascidian species belonging to the suborder Phlebobranchia apparently contained higher levels of vanadium than those belonging to Stolidobranchia. We also confirmed that blood cells contained the highest amounts of vanadium among the tissues examined. Furthermore, the highest concentration of vanadium, 350 mM (mmol/dm^3), was found in the blood cells of *Ascidia gemmata* (Fig. 3), which belongs to the suborder Phlebobranchia (Michibata et al., 1991a); 350 mM corresponds to 10^7 times the vanadium concentration in sea water (Cole et al., 1983; Collier, 1984). Levels of iron and manganese, determined simultaneously, did not vary significantly among the members of the two suborders.

Table 1   Concentrations of Vanadium in Tissues of Several Ascidians (mM)[a]

	Tunic	Mantle	Branchial Basket	Serum	Blood Cells
Phlebobranchia					
Ascidia gemmata	nd	nd	nd	nd	347.2
A. ahodori	2.4	11.2	12.9	1.0	59.9
A. sydneiensis	0.06	0.7	1.4	0.05	12.8
Phallusia mammillata	0.03	0.9	2.9	nd	19.3
Ciona intestinalis	0.003	0.7	0.7	0.008	0.6
Stolidobranchia					
Styela plicata	0.005	0.001	0.001	0.003	0.007
Halocynthia roretzi	0.01	0.001	0.004	0.001	0.007
H. aurantium	0.002	0.002	0.002	nd	0.004

[a] nd: not determined

5. ISOLATION OF VANADIUM-CONTAINING BLOOD CELLS

Blood cells were found to contain the highest amounts of vanadium among the tissues examined in ascidians. Ascidian blood cells are classified into nine to eleven different types that can be grouped into six categories on the basis of their morphology: hemoblasts, lymphocytes, leukocytes, vacuolated cells,

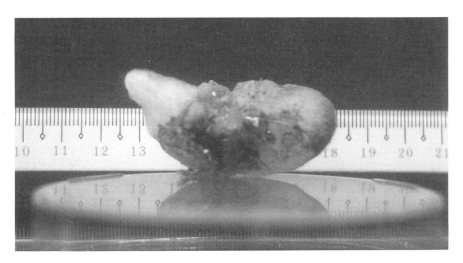

Figure 3. *Ascidia gemmata,* which belongs to the suborder Phlebobranchia, has the highest concentration of vanadium, 350 mM, in the blood cells (coelomic cells). All ascidians are marine animals. More than 2,500 species are distributed from polar waters to the tropics. They are classified into a phylum of the Chordata, between the invertebrates and the vertebrates. Several species belonging to the Phlebobranchia are known to accumulate high levels of vanadium in their blood cells.

pigment cells, and nephrocytes (Wright, 1981). The vacuolated cells can be further divided into at least four different types: morula cells, signet ring cells, compartment cells, and small compartment cells. For many years the morula cells were thought to be the so-called vanadocytes (Webb, 1939; Endean, 1960; Kalk, 1963a, b; Kustin et al., 1976), because their pale-green color resembles that of a vanadium complex dissolved in aqueous solution. The dense granules in morula cells, which can be observed under an electron microscope after fixation with osmium tetroxide, were assumed to be deposits of vanadium.

At the end of the 1970s, with the increasing availability of scanning transmission electron microscopes equipped with an energy-disperse X-ray detector, it became possible to address with greater confidence the question of whether morula cells are the true vanadocytes. An Italian group was the first to demonstrate that the characteristic X-ray due to vanadium was not detected from morula cells but from granular ameobocytes, signet ring cells, and compartment cells and, moreover, that vanadium was selectively concentrated in the vacuolar membranes of these cells, where vanadium granules were present inside the vacuoles (Botte et al., 1979b; Scippa et al., 1982, 1985; Rowley, 1982). Identification of the true vanadocytes became a matter of the highest priority to those concerned with the mechanism of accumulation of vanadium by ascidians.

To end the controversy and identify the true vanadocytes, we used not X-ray microanalysis, but a combination of density gradient centrifugation, for the isolation of specific types of blood cells, and neutron activation analysis, for the quantification of the vanadium contents of the isolated subpopulations of blood cells (Michibata et al., 1987). *Ascidia ahodori,* a vanadium-rich ascidian, was used. Blood cells separated from the serum by centrifugation were loaded onto a discontinuous gradient consisting of four different concentrations of Ficoll in artificial sea water, and then the gradient was centrifuged at $100 \times g$. The blood cells were partitioned into four discrete layers and the subpopulation recovered from each layer was submitted to neutron activation analysis. The subpopulation of cells in layer 4, where signet ring cells were dominant, contained the highest level of vanadium. The distribution pattern of vanadium coincided with that of signet ring cells but not with that of morula cells or compartment cells, as shown in Figure 4. These results proved that the signet ring cells are the true vanadocytes (Fig. 5). The same conclusion was obtained for the other ascidian species and it was finally confirmed that signet ring cells are the true vanadocytes (Michibata et al., 1990, 1991a; Hirata and Michibata, 1991).

6.   PREPARATION OF MONOCLONAL ANTIBODIES AGAINST ASCIDIAN BLOOD CELLS

The identification of vanadocytes had been a subject of controversy, as stated above, in part because of difficulties associated with morphological discrimina-

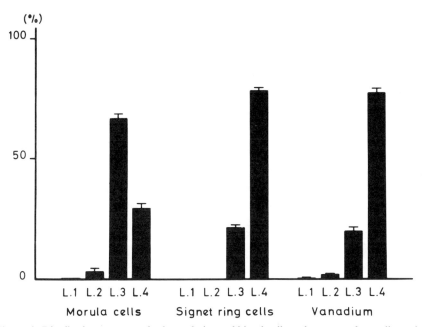

Figure 4. Distribution patterns of subpopulations of blood cells and content of vanadium after density gradient centrifugation. Histograms revealed that the pattern of distribution of vanadium coincided with that of signet ring cells but not with that of morula cells or that of compartment cells. Namely, signet ring cells were identified as vanadium-containing blood cells (vanadocytes) in ascidians (Michibata et al., 1987). L.1, layer 1; L.2, layer 2; L.3, layer 3; and L.4, layer 4 from the top to bottom of discontinuous density gradient.

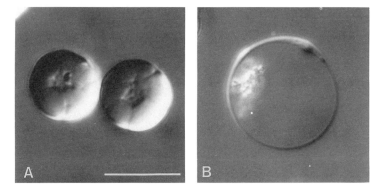

Figure 5. A. Morula cells, misidentified initially as vanadocytes. B. Signet ring cells, identified newly as vanadocytes, in the ascidian *Ascidia ahodori*. Scale bar indicates 10 μm.

tion between several types of blood cells and in part because of an inadequate knowledge of cell lineages from the so-called stem cells to the peripheral cells. For example, Scippa et al. (1988) reported that the vacuolated and granular amoebocytes and a variety of compartment cells in *Phallusia mammillata,* in addition to the signet ring cells, could be considered to be vanadocytes.

The establishment of reliable cell markers for the recognition of different types of blood cells is necessary to clarify not only the function but also the lineage of each type of cell. We prepared a monoclonal antibody, which we hoped might serve as a powerful tool for solving these problems, using a homogenate of the subpopulation of signet ring cells from *Ascidia sydneiensis samea* as the antigen (Uyama et al., 1991). The monoclonal antibody obtained, S4D5, was shown to react specifically with the vanadocytes not only from *A. sydneiensis samea* (Fig. 6) but also from two additional species, *A. gemmata* and *A. ahodori.* Immunoblotting analysis showed that this antibody recognizes a single polypeptide of about 45 kDa from all three species. Another monoclonal antibody, C2A4, recently prepared, also reacts specifically with the vacuolar amoebocytes and recognizes a single band of a protein of about 200 kDa (Kaneko et al., 1995).

There are several reports on the localization of blood cells in ascidians (Kalk, 1963a; Smith, 1970a; Ermak, 1975, 1976). According to these earlier reports, hematogenic activity is observed in three main areas: (1) in the connective tissues around the alimentary canal, (2) in the pharyngeal wall and transverse vessels of the branchial basket, and (3) in discrete nodules located in

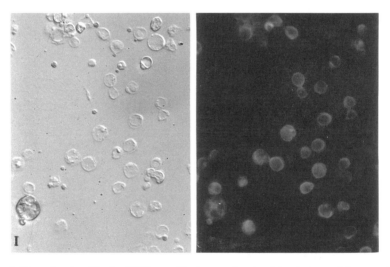

Figure 6. Monoclonal antibody, S4D5, specific to vanadocytes (signet ring cells) was raised. The blood cells from four different ascidian species were stained immunologically with S4D5. Antigenicity was observed specifically in vanadocytes among about 10 types of blood cells. Scale bar indicates 10 μm.

the body wall. Employing these monoclonal antibodies and the autonomous fluorescence emitted by each type of cell as cell markers (Wuchiyama and Michibata, 1995), we tried to identify the hematopoietic sites of vanadocytes. We found vanadocytes in the connective tissues around the alimentary canal, separate from morula cells, compartment cells, and amoebocytes. It seems that the precursor cells of vanadocytes are formed in the connective tissues, whereas other types of blood cells are formed at other sites (Kaneko et al., 1995).

7. CHEMICAL SPECIES IN ASCIDIAN BLOOD CELLS

Henze (1911) was the first to suggest the existence of vanadium in the +5 oxidation state. Later, Lybing (1953), Bielig et al. (1954), Boeri and Ehrenberg (1954), and Webb (1956) reported the +3 oxidation state of vanadium. More recently, noninvasive physical methods, including electron spin resonance spectrometry (ESR), X-ray absorption spectrometry (XAS), nuclear magnetic resonance spectrometry (NMR), and superconducting quantum interference device (SQUID), have been used to determine the intracellular oxidation state of vanadium. Such studies indicated that the vanadium ions in ascidian blood cells were predominantly in the +3 oxidation state, a small amount being in the +4 oxidation state (Carlson, 1975; Tullius et al., 1980; Dingley et al., 1981; Frank et al., 1986; Lee et al., 1988a; Brand et al., 1989).

These results, however, were derived not from the vanadocytes but from the entire population of blood cells. Thus, some questions remained to be answered. In particular, does vanadium exist in two oxidation states in one type of blood cell, or is each state formed in a different cell type? After separation of the various types of blood cells of *A. gemmata,* we made noninvasive ESR measurements of the oxidation state of vanadium in the fractionated blood cells under a reducing atmosphere (Hirata and Michibata, 1991). Consequently it was revealed that vanadium in vanadocytes is predominantly in the +3 oxidation state, a small amount being in the +4 oxidation state. The ratio of vanadium(III) to vanadium(IV) was 97.6 : 2.4.

A considerable amount of sulfate has always been found in association with vanadium in ascidian blood cells (Henze, 1932; Califano and Boeri, 1950; Bielig et al., 1954; Levine, 1961; Botte et al., 1979a, b; Scippa et al., 1982; 1985, 1988; Bell et al., 1982; Pirie and Bell, 1984; Lane and Wilkes, 1988; Frank et al., 1986, 1987, 1994, 1995; Anderson and Swinehart, 1991), suggesting that sulfate might be involved in the biological function and/or the accumulation and reduction of vanadium.

Hodgson's group has been studying the status and the chemical forms of sulfur-containing substances in ascidian blood cells using new analytical techniques such as (XAS) in near-edge regions (XANES) and in fine structure regions (EXAFS). They reported that of the total sulfur in the cell lysate analyzed by means of inductively coupled plasma (ICP) emission spectroscopy,

63% exists as sulfate and the remaining 37% is present as nonsulfate sulfur that is not precipitated by Ba^{2+} (Frank et al., 1986). They observed sulfur K-edge X-ray absorption spectra of a variety of sulfur compounds in order to obtain the spectral characteristics depending on the nature of sulfur-containing substances, and indicated that a solution containing cysteine (50 mM), cysteic acid (0.5 M), and sodium sulfate (0.5 M) reproduced nicely the spectral features of the blood cells. It is thus likely that most of the nonsulfate sulfur compounds are aliphatic sulfonic acid, which is analogous to cysteic acid, whereas the remaining are thiol or thioether (Frank et al., 1987).

Raman spectroscopy can also be used to detect sulfate ion selectively in ascidian blood cells, because sulfate ion gives a very intense Raman band at the diagnostic position, 983 cm^{-1}. We observed fairly good Raman spectra of the blood cell lysate from *Ascidia gemmata,* which has the highest concentration of vanadium(III) among ascidians (Fig. 7). Vanadium(III) ions in the blood cells were converted to vanadyl(IV) ions by air-oxidation prior to Raman measurements so as to facilitate detection based on V=O stretching vibration. From analysis of the band intensities due to $V=O^{2+}$ and SO_4^{2-}

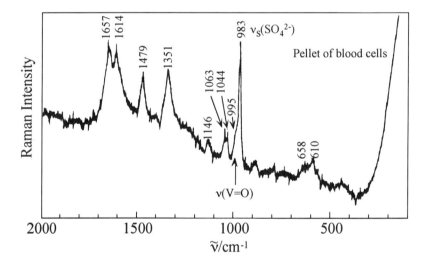

Raman Spectrum of Pellet of Blood Cells of *Ascidia gemmata*

$$C(SO_4^{2-}) = 510 \text{ mmol/dm}^3$$

Figure 7. Raman spectra of the blood cells of *Ascidia gemmata*. High levels of sulfate and/or sulfur compounds were found in association with vanadium in ascidian blood cells. Raman spectrometry can determine noninvasively the amounts of sulfate and vanadium in the vanado-cytes. A band at 983 cm^{-1} and a shoulder at 995 cm^{-1} are derived from SO_4^{2-} symmetric stretching vibration and V=O stretching vibration, respectively. Based on the intensities of the peaks, the concentration ratio of SO_4^{2-} and V^{3+} was calculated to be 1.47 (Kanamori and Michibata, 1994).

ions, we estimated the content ratio of sulfate to vanadium to be approximately 1.5, as would be predicted if sulfate ions were present as the counterions of vanadium(III). Carlson (1975) reported a similar value of the content ratio for *Ascidia ceratodes,* but lower values were obtained by Bell et al. (1982) and Frank et al. (1986). Further studies are planned to ascertain that the content ratio obtained by us was not fortuitous. We have also observed the Raman band assignable to alkyl sulfonate, the presence of which was proposed by Frank et al. (1987).

The status of vanadium(III) ions in ascidian blood cells has also been examined by means of several spectroscopic techniques. Carlson found a paramagnetically shifted signal at 21 ppm in the ^{1}H NMR (100 MHz) spectrum of living ascidian blood cells; the signal was assigned to water protons in rapid exchange between the bulk and the coordination sphere of vanadium(III) because paramagnetic species other than vanadium(III) are negligibly present in ascidian blood cells (Carlson, 1975). He concluded that at least five of the six available coordination sites of vanadium(III) are occupied by water molecules. This indicates that most of the vanadium(III) in ascidian blood cells is not tightly bound to polydentate organic ligand.

Evidence of coordination of sulfate to vanadium(III) in ascidian blood cells was obtained by vanadium K-edge X-ray absorption spectroscopy (Tullius et al., 1980; Frank et al., 1995). According to the delicate analysis of the XAS envelope of whole blood cell samples from *Ascidia ceratodes,* at least 90% of vanadium exists as a mixture of $[V(H_2O)_6]^{3+}$ and $[V(SO_4)(H_2O)_{4-5}]^+$ ions. Because of the restricted sensitivity of the vanadium K-edge XAS spectral feature, the presence of up to 10% of vanadium(III) compound chelated by organic substances cannot be excluded yet. It is possible that this remaining smaller portion of vanadium(III) in ascidian blood cells plays a physiologically important role; however, the physiological function of simple aqua species of vanadium(III) has yet to be elucidated.

The presence of sulfatovanadium(III) complex in ascidian blood cells is consistent with the formation constant of $V(SO_4)^+$ in aqueous solution (Kimura et al., 1972). We also provided direct evidence for the formation of $V(SO_4)^+$ complex in acidic solutions by means of Raman spectroscopy (Kanamori et al., 1991). Interaction between vanadium(III) and sulfate in aqueous solution have been examined in some detail (Meier et al., 1995).

The role of sulfate ions in ascidian blood cells is still puzzling. With regard to this problem, Frank et al. (1994) postulated that the choice of sulfate as the endogenous counterion has the effect of reducing the osmotic gradient against the cytosol or surrounding plasma because the complexation between V^{3+} and SO_4^{2-} markedly lowers the ionic strength within the blood cells. We noticed another aspect of sulfate coordination. Since sulfate, but not halide, is a potential bidentate ligand, the coordination of sulfate would influence the coordination geometry and consequently affect the properties of vanadium(III) complexes. We have found that sulfate yields heptacoordinate vanadium(III) complexes with bispicen (bis(2-pyridylmethyl)ethylenediamine) and tpen (tetrakis(2-pyridylmethyl)ethylenediamine) (Kanamori et al., 1996). More interestingly, the V^{3+}–tpen–halogen ion system tends to yield an oxo-bridged dinuclear vanadium(III) complex, while, the V^{3+}–tpen–SO_4^{2-} system does not. This suggests that sulfate has a pronounced effect of suppressing the formation of the oxo-bridged dimer formation of vanadium(III).

An approach based on a combination of solution and structural chemistry, and the physiology of sulfur-containing compounds, should shed light on the function of sulfate in ascidian blood cells.

8. REDUCING SUBSTANCES OF VANADIUM

It is evident that some reducing agent must play an important role in the accumulation of vanadium. Potential reducing agents that have attracted the most attention are tunichromes. Tunichromes are an organic chromogen first isolated in a crude form from *Ascidia nigra* by Macara et al. (1979a, b), who suggested linkage to the vanadium uptake system (Robinson et al., 1984). Later, Nakanishi and co-workers successfully purified these agents, identifying them as a class of hydroxy-Dopa-containing tripeptides (Bruening et al., 1985; Oltz et al., 1988), as shown in Figure 8. The presence of pyrogallol (1,2,3-trihydroxybenzene) moiety in tunichromes suggested that tunichromes would act as a reducing agent as well as a complexing agent. Early in vitro studies showed, however, that tunichromes are only able to reduce vanadium(V) to

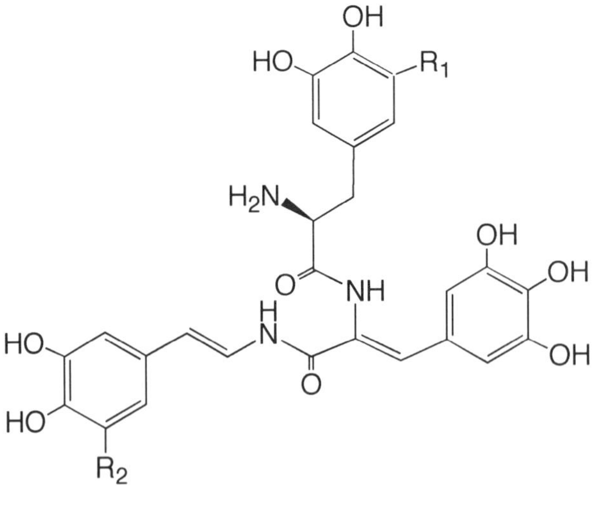

An-1, $R_1 = R_2 = $ OH; An-2, $R_1 = $ H, $R_2 = $ OH; An-3, $R_1 = R_2 = $ H

Figure 8. Structure of tunichromes isolated from *Ascidia nigra*. Reprinted from Oltz et al. (1988), with permission.

vanadium(IV) and not further to vanadium(III) (Macara et al., 1979a). Similar behavior was observed for a model compound of tunichromes (Kime-Hunt et al., 1988; Bulls et al., 1990). More recently, the reduction of vanadium(V) by tunichromes was re-examined by EPR spectroscopy and indirect evidence of the reduction to vanadium(III) was presented (Ryan et al., 1992). However, an attempt to obtain direct evidence by X-ray absorption spectroscopy was not successful (Smith et al., 1995; Ryan et al., 1996a, b).

Bonadies et al. (1986) studied the behavior of the vanadyl complex of salen, which is a hydroxybenzene-containing ligand similar to tunichrome, in acidic solution and found that when $HClO_4$ was added to an acetonitrile solution of [VO(salen)] in anaerobic conditions, the rotating platinum electrode voltammogram exhibits two waves; +0.47 and +0.73V (vs. SCE). They concluded that two species were formed by the acid-promoted reaction. They also noticed that the original EPR signal attributed to [VO(salen)] disappeared on addition of 2 equiv of $HClO_4$ (Bonadies et al., 1987). The above observations were interpreted to be a result of acid-promoted disproportionation of [VO(salen)] to give EPR-silent vanadium(V) and vanadium(III) species.

$$2[VO(salen)] + 4H^+ \rightarrow \text{"}V^{III}(H_2salen)\text{"} + V^VO(salen)^+ + H_2O$$

This reaction suggested that vanadium(IV) produced by tunichrome reduction could produce vanadium(III) through the disproportionation. The postulated vanadium(III) species, however, has not been well characterized to date. Recently, Tsuchida et al. (1994) claimed that the acid-promoted reaction of [VO(salen)] is the proton-induced deoxygenation to give V(salen)$^{2+}$ rather than disproportionation. The two electrochemical waves were then assigned to the redox couples of V(salen)$^{3+}$/V(salen)$^{2+}$ and V(salen)$^{2+}$/V(salen)$^+$. They also mentioned that the presence of water would be responsible for the disappearance of the EPR response observed by Bonadies et al. (1987). An alternative pathway for tunichrome-induced reduction of vanadium(V) was suggested by Bayer et al. (1992). The reaction includes a proposed precursor of tunichrome that contains $a-C_6H_2(OH)_2(OSO_3H)$ group.

(R = L-Topa-L-Topa)

Although this mechanism is of interest in the context that it can explain not only the reduction to vanadium(III) but the occurrence of sulfate ions in the ascidian blood cells, further experiments are clearly needed to demonstrate that such a reaction is practical.

The finding that vanadium and tunichrome are located in separate blood cells raised further doubt as to the participation of tunichromes in the vanadium reduction process (Michibata et al., 1988; Michibata et al., 1990). Furthermore, a very acidic environment in ascidian blood cells (see below) is not good for coordination of phenolic ligands. It is also predicted that catechol (cat: 1,2-dihydroxybenzene) does not stabilize the vanadium(III) oxidation state. Conversely, hard catechol-type ligands stabilize the highest oxidation state of the metal and in fact $[V(III)(cat)_3]^{3-}$ is very sensitive to oxidation (Cooper et al., 1982). It is therefore very unlikely that tunichromes simultaneously play the two roles as a reducing agent and a complexing agent, though the possibility as a reducing agent cannot be precluded.

As outlined above, the problem as to the reduction of vanadium(V) to vanadium(III) in ascidians is yet to be solved despite the isolation of tunichromes as a potential reducing agent. Under these circumstances, it may be of importance to survey the redox chemistry of vanadium with or without regard to the direct relation to the vanadium-reducing system employed by ascidians.

Two-electron reduction of vanadium(V) or successive one-electron reductions via vanadium(IV) would be the most probable pathway to yield vanadium(III). Zhang and Holm (1990) reported that the strong oxophilic regent Ph_3P abstracted the oxo ligand from $[VOCl_4]^-$ in acetonitrile to afford the vanadium(III) product $[VCl_4(CH_3CN)_2]^-$ and Ph_3PO. Another example of a direct reduction of vanadium(V) to vanadium(III) was reported by Neumann and Assael (1989). They obtained the vanadium(III) complex $[V(teg)Br_2]Br$ (teg: tetraethylene glycol) by the reaction between a vanadium(V) species and hydrogen bromide using tetraglyme, $CH_3O(CH_2CH_2O)_4CH_3$, as a complexing agent in 1,2-dichloroethane. Although such a reaction is not likely to occur in ascidians, it is important to be aware that if an appropriate reaction media and complexing agent are chosen, vanadium(V) can be reduced to vanadium(III) rather easily even with a weak reducing agent.

In the case that vanadium(V) is first reduced to vanadium(IV) and then further to vanadium(III), the key step is a reduction of vanadium(IV) to vanadium(III), because vanadium(V) is easily reduced to vanadium(IV), but it is hard to reduce vanadium(IV) to vanadium(III). The reduction potential of vanadium(IV) compounds is strongly dependent on the structure and properties of the vanadium(IV) species. Table 2 summarizes the redox potential of V^{4+}/V^{3+} couples with a variety of ligands. It can be recognized from Table 2 that non-oxo or bare vanadium(IV) compounds are far more easily reducible than normal vanadyl complexes. In other words, abstraction of the oxo ligand of vanadyl compounds should give easy access to vanadium(III). Protonation of the oxo ligand would bring about a similar effect. The nature of the ligand is also an important factor that affects the redox potential. Lee et al. (1988b) showed that an excess of pyrogallol can reduce $[VO(acac)_2]$ to vanadium(III) compound in tetrahydrofuran. In this reaction the oxo ligand might be protonated by a hydroxyl proton of pyrogallol. Recently we have found that cysteine derivative can reduce vanadium(IV) to vanadium(III) in the presence of

Table 2   Redox Chemistry of Vanadium(IV) and V(III) Compounds Studied by Cyclic Voltammetry

Reaction[a]	Potential, V vs. SCE	Electrolyte	Solvent	Reference
$[V(acac)_3]^+ + e^- = [V(acac)_3]$	+0.76	$[Et_4N][ClO_4]$	DMSO	Nawi and Riechel (1981)
$[V(bzac)_3]^+ + e^- = [V(bzac)_3]$	+0.66[c]	$[Et_4N][ClO_4]$	CH_2Cl_2	Hambley et al. (1987)
$[V(salen)]^{2+} + e^- = [V(salen)]^+$	+0.49[d]	$[Bu_4N][BF_4]$	CH_3CN	Tsuchida et al. (1994)
$[V(oxin)_3]^+ + e^- = [V(oxin)_3]$	+0.45	$[Et_4N][ClO_4]$	CH_3CN	Riechel and Sawyer (1975)
$[V(sal-NSO)_2]^+ + e^- = [V(sal-NSO)_2]^-$	−0.24	$[Bu_4N][ClO_4]$	DMF	Dutton et al. (1988)
$[V(SBu^t)_4] + e^- = [V(SBu^t)_4]^-$	−0.68	$[Bu_4N][ClO_4]$	CH_2Cl_2	Heinrich et al. (1991)
$[V(cat)_3]^{2-} + e^- = [V(cat)_3]^{3-}$	−0.86	$[Et_4N][ClO_4]$	CH_3CN	Cooper et al. (1982)
$[V(dtbc)_2(phen)] + e^- = [V(dtbc)_2(phen)]^-$	−0.89	$[Bu_4N][BF_4]$	CH_2Cl_2	Kabanos et al. (1992)
$[VO(tsalen)] + e^- = [VO(tsalen)]^{-b}$	−1.29	$[Bu_4N][ClO_4]$	DMF	Dutton et al. (1988)
$[VO(dtc)_2] + e^- \rightarrow V(III)$	−1.35	$[Et_4N][ClO_4]$	CH_3CN	Riechel et al. (1976)
$[VO(salen)] + e^- \rightarrow V(III)$	−1.58	$[Bu_4N][ClO_4]$	DMF	Dutton et al. (1988)
$[VO(pycac)] + e^- = [VO(pycac)]^-$	−1.78[c]	$[Et_4N][ClO_4]$	CH_3CN	Hanson et al. (1992)
$[VO(acac)_2] + e^- \rightarrow V(III)$	−1.90	$[Et_4N][ClO_4]$	DMSO	Nawi and Riechel (1981)

[a] = and → denote reversible and irreversible reaction, respectively.
[b] Quasi-reversible reaction

Ligands

233

aminopolycarboxylate in water (Kanamori et al., 1997). Further experiments are in progress.

9. EXTREMELY LOW pH IN VANADOCYTES

Henze (1911, 1912, 1913, 1932) was the first to report that the homogenate of ascidian blood cells is extremely acidic. Almost all subsequent investigations have supported his observation; however, Kustin's group recently disputed the earlier reports. They claimed that the previously used methods would give spurious results in an application to ascidian blood cells, because the interior of the blood cells that contained high levels of vanadium and iron was probably a strongly reducing environment (Dingley et al., 1982; Agudelo et al., 1983). In their view, the possibility remained that the data obtained in the earlier works did not reflect the pH but the intracellular redox potential. They reported that the intracellular pH was neutral on the basis of measurements made by a new technique, which involved the transmembrane equilibrium of ^{14}C-labeled methylamine. Hawkins et al. (1983) and Brand et al. (1987) also reported nearly neutral values for the pH of the interior of ascidian blood cells, which they determined noninvasively from the chemical shift of ^{31}P-NMR. However, Frank et al. (1986), basing their results on the new finding that the ESR line width accurately reflects the intracellular pH, demonstrated that the interior of the blood cells of *Ascidia ceratodes* has a pH of 1.8. Thus, the reported pH within ascidian blood cells (summarized in Table 3) has excited considerable controversy.

We consider that the main reason for the extreme variations is that the measurements of pH were made with entire populations of blood cells and not with the subpopulation of vanadocytes specifically. Thus, one or two specific types of blood cells might have a highly acidic solution within their

Table 3   Reported pH Inside Ascidian Blood Cells

pH	Ascidian Species	Analytical Methods
1 N sufuric acid	*Phallusia mammillata*	Titration (Henze, 1911)
1.8 N acid	*P. mammillata*	Titration (Webb, 1939)
0.4 N acid	*Pyura stolonifera*	Titration (Endean, 1955a)
pH 1.5	*Eudistoma ritteri*	Electrode (Levine, 1961)
pH 7.2	*Ascidia nigra*	^{14}C-Methylamine (Dingley et al., 1982)
pH 7.19	*Boltenia ovifera*	^{14}C-Methylamine (Agudelo et al., 1983)
pH 6.4	*Pyura stolonifera*	^{31}P-NMR (Hawkins et al., 1983)
pH 1.8	*Ascidia ceratodes*	ESR (Frank et al., 1986)
pH 6.5	*Phallusia julinea*	Electrode under N_2 (Brand et al., 1987)
pH 6.98	*Ascidia ceratodes*	^{14}C-dimethyloxazolidine (Lee et al., 1990)

vacuoles, in which vanadium would be present in a reduced state. With this possibility in mind, we designed an experiment in which we combined the separation of each type of blood cell, measurement of pH with a microelectrode under anaerobic conditions to avoid air-oxidation, and ESR spectrometry as a noninvasive method for the measurement of pH to confirm the results obtained with the microelectrode (Michibata et al., 1991a). Three species of vanadium-rich ascidians, *Ascidia gemmata*, *A. ahodori*, and *A. sydneiensis samea*, were used. Blood cells drawn from each species were fractionated by density-gradient centrifugation, as described above. The distribution of each type of blood cell, the concentrations of protons ($[H^+]$), and the levels of vanadium in each layer of cells are compared in Figure 9. It is clear from Figure 9 that the distribution patterns of protons and vanadium were similar. Thus, the signet ring cells contain high levels of both vanadium and proton in all three species. In *A. sydneiensis samea* a different type of blood cell in addition to signet ring cells was found to be acidic. These large and spherical cells with a diameter of 30–80 μm have a single fluid-filled vacuole. This type of cell is probably analogous to the so-called nephrocyte, as judged by the criteria proposed by Wright (1981). However, we propose to call these cells

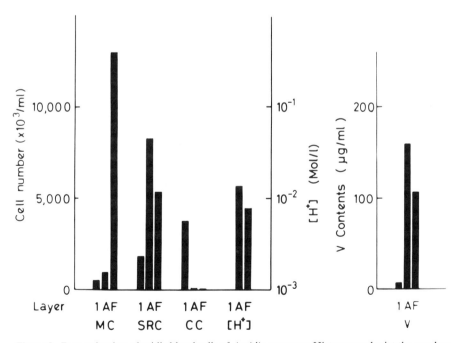

Figure 9. Determination of acidic blood cells of *Ascidia gemmata*. Histograms depict the number of each type of blood cell and the concentrations of H^+ and vanadium in three different layers (layers 1, A, and F; cf. Michibata et al., 1991a) of blood cells that were fractionated by density gradient centrifugation in Percoll. MC, morula cells; SRC, signet ring cells; CC, compartment cells; $[H^+]$, concentration of protons; V, vanadium.

"giant cells" until their function is revealed. Giant cells have the lowest specific gravity among all the blood cells, and although they contain no vanadium, they have the very low pH of 1.48 (Michibata et al., 1991a). The contents and function of these cells remain to be clarified.

ESR spectrometry was also used for noninvasive measurements of the intracellular acidity of blood cells (Michibata et al., 1991a). The method is based on the ESR line width due to oxovanadium [VO(IV)] ions, which increases (cf. Frank et al., 1986) almost linearly from pH 1.4 to pH 2.3. That the low pH values obtained with a microelectrode were not artifacts was confirmed by the fact that the ESR line width also indicated a low pH for the contents of signet ring cells from *A. gemmata.* To avoid any misunderstanding, we have to note the assumption that the acidic solution was contained in the vacuole of each signet ring cell. However, the greater part of each signet ring cell is, in fact, occupied by the vacuole itself, so the contents of the vacuole are almost equivalent to the contents of the cell.

10. POSSIBLE ENERGETICS OF ACCUMULATION OF VANADIUM

A comparison of pH values with the levels of vanadium in the signet ring cells of three different species, as shown in Table 4, suggested that there might be a close correlation between a higher level of vanadium and lower pH, namely, a higher concentration of protons. It is well known that H^+-ATPases can generate a proton-motive force by hydrolyzing ATP. This enzyme plays a role in pH homeostasis in various intracellular organelles, including clathrin-coated vesicles, endosomes, lysosomes, Golgi-derived vesicles, multivesicular bodies, and chromaffin granules that belong to the central vacuolar system (Forgac, 1989, 1992; Nelson, 1992).

Therefore, we examined the presence of H^+-ATPase in the signet ring cells of the ascidian *Ascidia sydneiensis samea* (Uyama et al., 1994). The vacuolar-type H^+-ATPase is composed of several subunits, and subunits of 72 kDa and 57 kDa have been reported to be common to all eukaryotes examined. Antibodies prepared against the 72 kDa and 57 kDa subunits of a vacuolar-type H^+-ATPase from bovine chromaffin granules did indeed react with the vacuolar membranes of signet ring cells. Immunoblotting analysis confirmed

Table 4   Correlation between Concentrations of Vanadium and pH Values in Ascidian Blood Cells

Species	Concentration of Vanadium	pH
Ascidia gemmata	350 mM	1.86
A. ahodori	60 mM	2.67
A. sydneiensis samea	13 mM	4.20

that the antibodies reacted with specific antigens in ascidian blood cells. Furthermore, addition of bafilomycin A_1, a specific inhibitor of vacuolar-type H^+-ATPases (Bowman et al., 1988), inhibited the pumping function of the vacuoles of signet ring cells, resulting in neutralization of the contents of the vacuoles, as shown in Figure 10.

The acidification of vacuolar compartments in eukaryotic cells has been shown to have a number of important functions in neuronal and endocrine systems (Moriyama et al., 1992) and to be important for the degradation of proteins (Yoshimori et al., 1991). It is likely that the activity of the enzyme is linked to the accumulation of vanadium in signet ring cells. We are trying to obtain direct evidence for such an association.

11. VANADIUM-ASSOCIATED PROTEINS

With respect to the pathway for the accumulation of vanadium from sea water, the involvement of certain proteins seems likely, even though the results

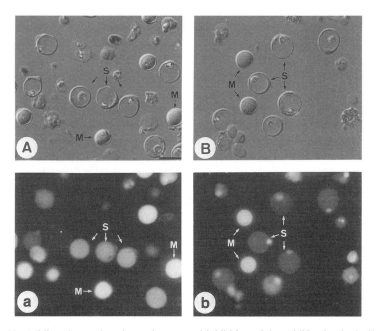

Figure 10. Acidity of vacuoles of vanadocytes and inhibition of the acidification by bafilomycin A_1. Signet ring cells (vanadocytes) were revealed to emit a brilliant vermilion, indicating acidic pH, after incubation of blood cells of *Ascidia sydneiensis samea* with 2 μM acridine orange for 1 h, but the other types of blood cells did not indicate acidic pH. However, addition of 1 μM bafilomycin A_1, a specific inhibitor of vacuolar H^+-ATPase, neutralized vanadocytes (showing green fluorescence) with resultant inhibiting pump function of the H^+-ATPase. No changes in color of autonomous fluorescence emitted from morula cells were found with or without bafilomycin A_1 (Uyama et al., 1994). S, Signet ring cells (vanadocytes). M, morula cells. Scale bar indicates 10 μm.

reported to date indicate that in ascidians most of vanadium ions are likely to be present in a free, noncomplexed form or associated with low-molecular-weight components. The route for the accumulation of vanadium ions from sea water in the blood system has not yet been revealed. The uptake of vanadium ions was studied with radioactive vanadium ions (^{48}V). However, previous studies were, with a few exceptions (Hawkins et al., 1980b; Roman et al., 1988), commonly designed with an assumption of the direct uptake of vanadium ions from the surrounding sea water and were, therefore, limited in their determination of how much vanadium was incorporated directly into some tissues (Goldberg et al., 1951; Bielig et al., 1963; Dingley et al., 1981; Michibata et al., 1991b). The majority of the vanadium incorporated by ascidians was thought to be dissolved as ionic species or associated with low-molecular-weight substances rather than proteins (Kustin and Robinson, 1995). However, generally, heavy metal ions incorporated into the tissues of living organisms are known to bind to macromolecules such as proteins. Using a combination of SDS-PAGE and flameless atomic absorption spectrometry under a working hypothesis that there should be at least three types of proteins (namely, vanadium-transfer, vanadium-receptor and vanadium-channel proteins), we first succeeded in isolating a vanadium-associated protein composed of 12.5-kDa and 15-kDa peptides (Kanda et al., 1997).

12. ACCUMULATION OF VANADIUM DURING EMBRYOGENESIS

Monoclonal antibodies are also useful tools with which to determine the time at which the accumulation of vanadium starts during embryogenesis. Since the amount of vanadium stored in embryos is beneath the limits of detection of conventional analytical methods, such as atomic absorption spectrometry, there have been no reports of the direct determination of vanadium accumulated during ascidian embryogenesis. Using neutron-activation analysis and an immunofluorescence method, we found that the amount of vanadium per individual increased markedly 2 weeks after fertilization. Within 2 months, the amount accumulated in larvae was about 600,000 times greater than that in the unfertilized eggs of *A. sydneiensis samea* (Michibata et al., 1992), as shown in Figure 11. A vanadocyte-specific antigen, recognized by a monoclonal antibody specific to the signet ring cells, first became apparent in the body wall at the same time as the first significant accumulation of vanadium (Uyama et al., 1993).

13. PHYSIOLOGICAL ROLES OF VANADIUM IN ASCIDIANS

Although the unusual phenomenon whereby some ascidians accumulate vanadium to levels more than 10 million times higher than that in sea water has

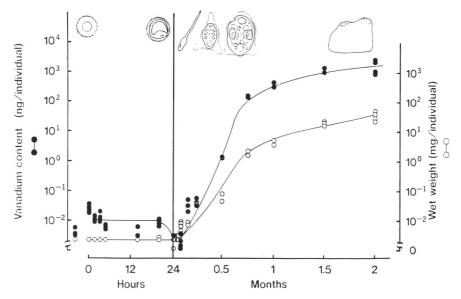

Figure 11. Accumulation of vanadium during embryogenesis of the ascidian *Ascidia sydneiensis samea.* To determine the time at which the accumulation of vanadium commences during embryogenesis, eggs and embryos were submitted for neutron activation analysis. The levels of vanadium began to increase and the amount in larvae reached 2.3 μg per individual, which was about 600,000 times higher than the amount in unfertilized eggs (Michibata et al., 1992). Furthermore, a vanadocyte-specific antigen first became apparent in the body wall at the same time as the first significant accumulation of vanadium (Uyama et al., 1993).

attracted researchers in various fields, the physiological roles of vanadium remain to be explained. Endean (1955a, b, c, 1960) and Smith (1970a, b) proposed that the cellulose of the tunic might be produced by vanadocytes. Carlisle (1968) suggested that vanadium-containing vanadocytes might reversibly trap oxygen under conditions of low oxygen tension. A hypothesis has also been proposed that vanadium in ascidians acts to protect them against fouling or as an antimicrobial agent (Stoecker, 1978; Rowley, 1983). However, most of the proposals presented so far do not seem to be supported by sufficient evidence.

Recently we observed an unexpected phenomenon of great interest. We found that the number of vanadocytes increased when ascidians were immersed in a solution that contained 10 mM or 20 mM NH_4Cl (Hayashi et al., 1996). The increase in size of the population of signet ring cells might be interpreted as a self-defense response. Of course, the reason why the number of signet ring cells increased in response to NH_4Cl is still a matter of conjecture.

On the other hand, the polychaeta *Pseudopotamilla occelata* was reported to be a new accumulator of high levels of vanadium (Ishii et al., 1993). We have disclosed that *P. occelata* has the same antigens as those in the ascidian

Ascidia sydneiensis samea, which were recognized by two types of antibodies, a polyclonal antibody against vanadium-associated proteins extracted from blood cells and a monoclonal antibody against vanadocytes in the vanadium-rich ascidian *A. sydneiensis samea.* There is, therefore, a possibility that a similar mechanism works on the accumulation of vanadium among the Polychaeta and the Ascidiidae (Uyama et al., 1997).

Characterization of these phenomena can be expected to help elucidate the reason for the unusual accumulation of vanadium by one class of marine organisms.

ACKNOWLEDGMENTS

H. M. would like to express a hearty thanks to his associates, Dr. Y. Moriyama, Dr. T. Uyama, Dr. J. Hirata, Ms. J. Wuchiyama, Mr. Y. Nose, Mr. T. Kanda, and Ms. M. Hayashi, for their unstinting assistance. Particular thanks are due to Dr. T. Numakunai and the staff of the Marine Biological Station of Tohoku University; Mr. K. Morita of Ootsuchi Marine Research Center, Ocean Research Institute, University of Tokyo; Prof. M. Yamamoto and his staff at the Marine Biological Station of Okayama University; and Mr. N. Abo of our laboratory for their kind cooperation in collecting the materials for so many years. K. K. also would like to express a sincere thanks to his colleagues, Ms. Y. Ookubo, Ms. H. Maeda, Mr. K. Ino, Ms. M. Fukagawa, and Mr. E. Kameda. Thanks are also due to Prof. K. Okamoto, University of Tsukuba, for the X-ray crystallographic analysis. Neutron activation analysis was done in cooperation with the Institute for Atomic Energy of Rikkyo University. Early work on preparing the monoclonal antibodies were performed in collaboration with Prof. N. Satoh of Kyoto University. This work was supported in part by grants-in-aid from the Ministry of Education, Science and Culture of Japan (to H. M. and K. K.), by the Asahi Glass Foundation, and by the Salt Science Research Foundation (to H. M.).

REFERENCES

Agudelo, M. I., Kustin, K., and McLeod, G. C. (1983). The intracellular pH of the blood cells of the tunicate *Boltenia ovifera. Comp. Biochem Physiol.* **75A,** 211–214.

Al-Swaidan, H. M. (1996). The determination of lead, nickel and vanadium in Saudi Arabian crude oil by sequential injection analysis/inductively-coupled plasma mass spectrometry. *Talanta* **43,** 1313–1319.

Anderson, D. H., and Swinehart, J. H. (1991). The distribution of vanadium and sulfur in the blood cells, and the nature of vanadium in the blood cells and plasma of the ascidian, *Ascidia ceratodes. Comp. Biochem. Physiol.* **99A,** 585–592.

Barbooti, M. M., Said, E. Z., Hassan, E. B., and Abdul-Ridha, S. M. (1989). Separation and spectrophotometric investigations of the distribution of nickel and vanadium in heavy crude oils. *Fuel,* **68,** 84–87.

Bayer, E., Schiefer, G., Waidelich, D., Scippa, S., and de Vincentiis, M. (1992). Structure of tunichrome of tunicates and its role in concentrating vanadium. *Angew. Chem. Int. Ed. Engl.* **31,** 52–54.

Bell, M. V., Pirie, B. J. S., McPhail, D. B., Goodman, B. A., Falk-Petersen, I.-B., and Sargent, J. R. (1982). Contents of vanadium and sulphur in the blood cells of *Ascidia mentula* and *Ascidiella aspersa. J. Mar. Biol. Assoc. UK* **62,** 709–716.

Bertrand, D. (1950). Survey of contemporary knowledge of biogeochemistry. 2. The biogeochemistry of vanadium. *Bull. Am. Mus. Natl. Hist.* **94**, 403–455.

Bielig, H.-J., Bayer, E., Califano, L., and Wirth, L. (1954). The vanadium containing blood pigment. II. Hemovanadin, a sulfate complex of trivalent vanadium. *Publ. Staz. Zool. Napoli* **25**, 26–66.

Bielig, H.-J., Bayer, E., Dell, H. D., Robins, G., Möllinger, H., and Rüdiger, W. (1966). Chemistry of Haemovanadium. *Protides Biol. Fluids* **14**, 197–204.

Bielig, H.-J., Joste, E., Pfleger, K., Rummel, W., and Seifen, E. (1961a). Aufnahme und Verteilung von Vanadin bei der Tunicate *Phallusia mammillata* Cuvier. *Hoppe-Seyler's Z. Physiol. Chem.* **325**, 122–131.

Bielig, H.-J., Joste, E., Pfleger, K., and Rummel, W. (1961b). Sulfataufnahme bei *Phallusia mammillata* Cuvier. Verteilung und Schicksal von Sulfat-und Aminosäure-Schwefel im Blut (Untersuchungen über Hämovanadin, VI). *Hoppe-Seyler's Z. Physiol. Chem.* **325**, 132–145.

Bielig, H.-J., Pfleger, K., Rummel, W., and Seifen, E. (1961c). Sulfataufnahme bei *Ciona intestinalis* L. und deren Beeinflussung durch Vanadin. *Hoppe-Seyler's Z. Physiol. Chem.* **327**, 35–40.

Bielig, H.-J., Pfleger, K., Rummel, W., and de Vincentiis, M. (1963). Beginning of the accumulation of vanadium during the early development of the ascidian *Phallusia mammillata* Cuvier. *Nature* **197**, 1223–1224.

Biggs, W. R., and Swinehart, J. H. (1976). Vanadium in selected biological systems. In H. Sigel (Ed.), *Metal Ions in Biological Systems*. Marcel Dekker, New York, Vol. 6, pp. 141–196.

Boas, L. V. and Pessoa, J. C. (1987). Vanadium. In *Comprehensive Coordination Chemistry*. Pergamon Press, Oxford, Vol. 2, pp. 453–583.

Boeri, E. (1952). The determination of hemovanadin and its oxidation potential. *Arch. Biochem. Biophys.* **37**, 449–456.

Boeri, E., and Ehrenberg, A. (1954). On the nature of vanadium in vanadocytes hemolysate from ascidians. *Arch. Biochem. Biophys.* **50**, 404–416.

Bonadies, J. A., Pecoraro, V. L., and Carrano, C. J. (1986). The acid promoted disproportionation of a vanadium(IV) phenolate: Implication for vanadium uptake in tunicates. *J. Chem. Soc. Chem. Commun.*, pp. 1218–1219.

Bonadies, J. A., Butler, W. M., Pecoraro, V. L., and Carrano, C. J. (1987). Novel reactivity patterns of (N,N'-ethylenebis(salicylideneaminato))oxovanadium(IV) in strongly acidic media. *Inorg. Chem.* **26**, 1218–1222.

Botte, L., Scippa, S., and de Vincentiis, M. (1979a). Content and ultrastructural localization of transitional metals in ascidian ovary. *Dev. Growth Differ.* **21**, 483–491.

Botte, L. S., Scippa, S., and de Vincentiis, M. (1979b). Ultrastructural localization of vanadium in the blood cells of Ascidiacea. *Experientia* **35**, 1228–1230.

Bowman, E. J., Siebers, A., and Altendorf, K. (1988). Bafilomycins: A class of inhibitors of membrane ATPases from microorganisms, animal cells, and plant cells. *Proc. Natl. Acad. Sci. USA*, **85**, 7972–7976.

Boyd, D. W., and Kustin, K. (1985). Vanadium: A versatile biochemical effector with an elusive biological function. *Adv. Inorg. Biochem.* **6**, 311–365.

Brand, S. G., Edelstein, N., Hawkins, C. J., Shalimoff, G., Snow, M. K., and Tiekink, E. R. T. (1990). An oxo-bridged binuclear vanadium(III) 2,2′-bipyridine complex and its vanadium(IV) and vanadium(V) oxidation products. *Inorg. Chem.* **29**, 434–438.

Brand, S. G., Hawkins, C. J., Marshall, A. T., Nette, G. W., and Parry, D. L. (1989). Vanadium chemistry of ascidians. *Comp. Biochem. Physiol.* **93B**, 425–436.

Brand, S. G., Hawkins, C. J., and Parry, D. L. (1987). Acidity and vanadium coordination in vanadocytes. *Inorg. Chem.* **26**, 627–629.

Bruening, R. C., Oltz, E. M., Furukawa, J., Nakanishi, K., and Kustin, K. (1985). Isolation and structure of tunichrome B-1, a reducing blood pigment from the tunicate *Ascidia nigra* L. *J. Am. Chem. Soc.* **107**, 5298–5300.

Bulls, A. R., Pippin, C. G., Hahn, F. E., and Raymond, K. N. (1990). Synthesis and characterization of a series of vanadium-tunichrome B1 analogues. Crystal structure of a tris(catecholamide) complex of vanadium. *J. Am. Chem. Soc.* **112**, 2627–2632.

Califano, L., and Boeri, E. (1950). Studies on haemovanadin. III. Some physiological properties of haemovanadin, the vanadium compound of the blood of *Phallusia mamillata* Cuv. *J. Exp. Zool.* **27**, 253–256.

Callahan, K. P., and Durand, P. J. (1980). Reactions of the vanadyl group: Synthesis of $V=S^{2+}$ and VBr_2^{2+} from $V=O^{2+}$. *Inorg. Chem.* **19**, 3211–3217.

Cantacuzène, J., and Tchekirian, A. (1932). Sur la présence de vanadium chez certains tuniciers. *Acad. Sci. Paris* **195**, 846–849.

Carlisle, D. B. (1968). Vanadium and other metals in ascidians. *Proc. R. Soc. B* **171**, 31–42.

Carlson, R. M. K. (1975). Nuclear magnetic resonance spectrum of living tunicate blood cells and the structure of the native vanadium chromogen. *Proc. Natl. Acad. Sci. USA* **72**, 2217–2221.

Cave, J., Dixon, P. R., and Seddon, K. R. (1978). Preparation and properties of tetrachlorooxovanadate(IV) and vanadium(IV) oxide dichloride adducts: The existence of two isomers of pyridinium tetrachlorooxovanadate(IV). *Inorg. Chim. Acta* **30**, L349–L352.

Chandrasekhar, P., and Bird, P. H. (1984). Synthesis and characterization of an oxygen-bridged vanadium(III) dimer. *Inorg. Chem.* **23**, 3677–3679.

Chasteen, N. D. (1981). Vanadyl(IV) EPR spin probes. Inorganic and biochemical aspects. *Biol. Magnetic Resonance* **3**, 53–119.

Chasteen, N. D. (1983). The biochemistry of vanadium. *Structure and Bonding* **53**, 105–138.

Ciereszko, L. S., Ciereszko, E. M., Harris, E. R., and Lane, C. A. (1963). Vanadium content of some tunicates. *Comp. Biochem. Physiol.* **8**, 137–140.

Cole, P. C., Eckert, J. M., and Williams, K. L. (1983). The determination of dissolved and particular vanadium in sea water by X-ray fluorescence spectrometry. *Anal. Chim. Acta* **153**, 61–67.

Collier, R. W. (1984). Particulate and dissolved vanadium in the North Pacific Ocean. *Nature* **309**, 441–444.

Comba, P., Engelhardt, L. M., Harrowfield, J. M. Lawrence, G. A., Martin, L. L., Sargeson, A. M., and White, A. H. (1985). Synthesis and characterization of a stable hexa-amine vanadium(IV) cage complex. *J. Chem. Soc. Chem. Commun.*, pp. 174–176.

Cooper, S. R., Koh, Y. B., and Raymond, K. N. (1982). Synthetic, structural, and physical studies of bis(triethylammonium) tris(catecholato)vanadate(IV), potassium (catecholato)oxovanadate(IV), and potassium tris(catecholat)vanadate(III). *J. Am. Chem. Soc.* **104**, 5092–5102.

Czernuszewicz, R. S., Yan, Q., Bond, M. R., and Carrano, C. J. (1994). Origin of the unusual bending distortion in the (μ-oxo)divanadium(III) complex [$V_2O(\text{L-his})_4$]: A reinvestigation. *Inorg. Chem.* **33**, 6116–6119.

Danskin, G. P. (1978). Accumulation of heavy metals by some solitary tunicates. *Can. J. Zool.* **56**, 547–551.

Davison, S., Edelstein, N., Holm, R. H., and Maki, A. H. (1964). Synthetic and electron spin resonance studies of six-coordinate complexes related by electron-transfer reactions. *J. Am. Chem. Soc.* **86**, 2799–2805.

Diamantis, A. A., Snow, M. R., and Vanzo, J. A. (1976). A trigonal prismatic vanadium(IV) complex: Bis(acetylacetonebenzoylhydrozonato)vanadium(IV); X-ray crystal structure. *J. Chem. Soc. Chem. Commun.* pp. 264–265.

Dingley, A. L., Kustin, K., Macara, I. G., and McLeod, G. C. (1981). Accumulation of vanadium by tunicate blood cells occurs via a specific anion transport system. *Biochim. Biophys. Acta* **649**, 493–502.

Dingley, A. L., Kustin, K., Macara, I. G., McLeod, G. C., and Roberts, M. F. (1982). Vanadium-containing tunicate blood cells are not highly acidic. *Biochim. Biophys. Acta* **720**, 384–389.

Dutton, J. C., Fallon, G. D., and Murray, K. S. (1988). Synthesis, structure, ESR spectra, and redox properties of (N,N'-ethylenebis(thiosalicylideneaminato) oxovanadium(IV) and of related {S,N} chelates of vanadium(IV). *Inorg. Chem.* **27**, 34–38.

Endean, R. (1955a). Studies of the blood and tests of some Australian ascidians. I. The blood of *Pyura stolonifera* (Heller). *Aust. J. Mar. Freshwater Res.* **6**, 35–59.

Endean, R. (1955b). Studies of the blood and tests of some Australian ascidians. II. The test of *Pyura stolonifera* (Heller). *Aust. J. Mar. Freshwater Res.* **6**, 139–156.

Endean, R. (1955c). Studies of the blood and tests of some Australian ascidians. III. The formation of the test of *Pyura stolonifera* (Heller). *Aust. J. Mar. Freshwater Res.* **6**, 157–164.

Endean, R. (1960). The blood cells of the ascidian, *Phallusia mammillata. Q. J. Microsc. Sci.* **101**, 177–197.

Ermak, T. H. (1975). Autoradiographic demonstration of blood cell renewal in *Styela clava* (Urochordata: Ascidiacea). *Experientia* **31**, 837–839.

Ermak, T. H. (1976). The hematogenic tissues of tunicate. In R. K. Wright and E. L. Cooper (Eds.), *Phylogeny of Thymus and Bone Marrow—Bursa Cells.* Elsevier/North-Holland Biomedical Press, Amsterdam, pp. 45–56.

Fenn, R. H., Gragam, A. J., and Gillard, R. D. (1967). Structure of tris-oxalato-complexes of trivalent metals. *Nature* **213**, 1012–1013.

Forgac, M. (1989). Structure and function of the vacuolar class of ATP-driven proton pumps. *Physiol. Rev.* **69**, 765–796.

Forgac, M. (1992). Structure, function and regulation of the coated vesicle V-ATPase. *J. Exp. Biol.* **172**, 155–169.

Frank, P., Carlson, R. M. K., and Hodgson, K. O. (1986). Vanadyl ion EPR as a non-invasive probe of pH in intact vanadocytes from *Ascidia ceratodes. Inorg. Chem.* **25**, 470–478.

Frank, P., Hedman, B., Carlson, R. M. K., and Hodgson, K. O. (1994). Interaction of vanadium and sulfate in blood cells from the tunicate *Ascidia ceratodes:* Observations using X-ray absorption edge structure and EPR spectroscopies. *Inorg. Chem.* **33**, 3794–3803.

Frank, P., Hedman, B., Carlson, R. K., Tyson, T. A., Row, A. L., and Hodgson, K. O. (1987). A large reservoir of sulfate and sulfonate resides within plasma cells from *Ascidia ceratodes,* revealed by X-ray absorption near-edge structure spectroscopy. *Biochemistry* **26**, 4975–4979.

Frank, P., Kustin, K., Robinson, W. E., Linebaugh, L., and Hodgson, K. O. (1995). Nature and ligation of vanadium within whole blood cells and Henze solution from the tunicate *Ascidia ceratodes,* as investigated by using X-ray absorption spectrometry. *Inorg. Chem.* **34**, 5942–5949.

Goldberg, E. D., McBlair, W., and Taylor, K. M. (1951). The uptake of vanadium by tunicates. *Biol. Bull.* **101**, 84–94.

Goodbody, I. (1974). The physiology of ascidians. *Adv. Mar. Biol.* **12**, 1–149.

Hambley, T. W., Hawkins, C. J., and Kabanos, T. A. (1987). Synthetic, structural, and physical studies of tris(2,4-pentanedionato)vanadium(IV) hexachloroantimonate(V) and tris(1-phenyl-1,3-butanedionato)vanadium(IV) hexachloroantimonate(V). *Inorg. Chem.* **26**, 3740–3745.

Hanson, G. R., Kabanos, T. A., Keramidas, A. D., Mentzafos, D., and Terzis, A. (1992). Oxovanadium(IV)-amide binding. Synthetic, structural, and physical studies of {N-[2-(4-oxopent-2-en-2-ylamino)phenyl]pyridine-2-carboxamido}oxovanadium(IV) and {N-[2-(4-phenyl-4-oxobut-2-en-2-ylamino)phenyl]pyridine-2-carbo xamido}oxovanadium(IV). *Inorg. Chem.* **31**, 2587–2594.

Hawkins, C. J., Kott, P., Pary, D. L., and Swinehart, J. H. (1983). Vanadium content and oxidation state related to ascidian phylogeny. *Comp. Biochem. Physiol.* **76B**, 555–558.

Hawkins, C. J., Parry, D. L., and Pierce, C. (1980a). Chemistry of the blood of the ascidian *Podoclavella moluccensis. Biol. Bull.,* **159**, 669–680.

Hawkins, C. J., Merefield, P. M., Parry, D. L., Biggs, W. R., and Swinehart, J. H. (1980b). Comparative study of the blood plasma of the ascidians *Pyura stolonifera* and *Ascidia ceratodes. Biol. Bull.* **159**, 656–668.

Hayashi, M., Nose, Y., Uyama, T., Moriyama, Y., and Michibata, H. (1996). Rapid increases in number of vanadocytes in the vanadium-rich ascidian, *Ascidia sydneiensis samea*, upon treatment of live animals with NH_4Cl. *J. Exp. Zool.* **275**, 1–7.

Heinrich, D. D., Folting, K., Huffman, J. C., Reynolds, J. G., and Christou, G. (1991). Preparation and properties of mononuclear vanadium thiolates: Structural characterization of the [V(S-Bu^t)$_4$]$^{0,-}$ pair and C–S bond cleavage in V(SBu^t)$_4$ in the gas phase. *Inorg. Chem.* **30**, 300–305.

Henze, M. (1911). Untersuchungen über das Blut der Ascidien. I. Mitteilung. Die Vanadiumverbindung der Blutkörperchen. *Hoppe-Seyler's Z. Physiol. Chem.* **72**, 494–501.

Henze, M. (1912). Untersuchungen über das Blut der Ascidien. II. Mitteilung. *Hoppe-Seyler's Z. Physiol. Chem.* **79**, 215–228.

Henze, M. (1913). Über das Vorkommen freier Schwefelsäure im Mantel von *Ascidia mentula*. *Hoppe-Seyler's Z. Physiol. Chem.* **88**, 345–346.

Henze, M. (1932). Über das Vanadiumchromogen des Ascidienblutes. *Hoppe-Seyler's Z. Physiol. Chem.* **213**, 125–135.

Hirata, J., and Michibata, H. (1991). Valency of vanadium in the vanadocytes of *Ascidia gemmata* separated by density-gradient centrifugation. *J. Exp. Zool.* **257**, 160–165.

Ishii, T., Nakai, I., Numako, C., Okoshi, K., and Otake, T. (1993). Discovery of a new vanadium accumulator, the fan worm *Pseudopotamilla occelata*. *Naturwissenschaften* **80**, 268–270.

Kabanos, T. A., White, A. J. P., Williams, D. J., and Woolins, J. D. (1992). Synthesis and X-ray structures of bis(3,5-di-*tert*-butylcatecholato) (phenanthroline)vanadium(IV) and its vanadium(V) analogue [V(dtbc)$_2$(phen)][SbF$_6$]. *J. Chem. Soc. Chem. Commun.* pp. 17–18.

Kalk, M. (1963a). Intracellular sites of activity in the histogenesis of tunicate vanadocytes. *Q. J. Microsc. Sci.* **104**, 483–493.

Kalk, M. (1963b). Absorption of vanadium by tunicates. *Nature* **198**, 1010–1011.

Kanamori, K., Kameda, E., Kabetani, T., Suemoto, T., Okamoto, K., and Kaizaki, S. (1995). Syntheses, characterization, and properties of vanadium(III) complexes containing tripodal tetradentate ligands with single pyridyl groups. *Bull. Chem. Soc. Jpn.* **68**, 2581–2589.

Kanamori, K., Kameda, E., and Okamoto, K. (1996). Heptacoordinate vanadium(III) complexes containing a didentate sulfate ligand. X-ray structures of [V$_2$(SO$_4$)$_3$\{*N,N'*-bis(2-pyridylmethyl)-1,2-ethanediamine\}$_2$] and [V(SO$_4$)\{*N,N,N',N'*-tetrakis(2-pyridylme thyl)-1,2-ethanediamine\}]$^+$ and their solution properties. *Bull. Chem. Soc. Jpn.* **69**, 2901–2909.

Kanamori, K, and Michibata, H. (1994). Raman spectroscopic study of the vanadium and sulphate in blood cell homogenates of the ascidian, *Ascidia gemmata*. *J. Mar. Biol. Assoc UK* **74**, 279–286.

Kanamori, K., Ookubo, Y., Ino, K., Kawai, K., and Michibata, H. (1991). Raman spectral study on the structure of a hydrolytic dimer of the aquavanadium(III) ion. *Inorg. Chem.* **30**, 3832–3836.

Kanamori, K., Teraoka, M., Maeda, H., and Okamoto, K. (1993). Preparation and structure of oxo-bridged dinuclear vanadium(III) complex [V^{III}$_2$(L-his)$_4$(μ-O)]·2H$_2$O. *Chem. Lett.*, pp. 1731–1734.

Kanamori, K., Kinebuchi, Y., and Michibata, H. (1997). Reduction of vanadium(IV) to vanadium(III) by cysteine methyl ester in water in the presence of amino polycarboylates. *Chem. Lett.*, pp. 423–424.

Kanda, K., Nose, Y., Wuchiyama, J., Uyama, T., Moriyama, Y., and Michibata, H. (1997). Identification of a vanadium-associated protein from the vanadium-rich ascidian, *Ascidia sydneiensis samea*. *Zool. Sci.*, **14**, 37–42.

Kaneko, A., Uyama, T., Moriyama, Y., and Michibata, H. (1995). Localization, with monoclonal antibodies and by detection of autonomous fluorescence, of blood cells in the tissues of the vanadium-rich ascidian, *Ascidia sydneiensis samea*. *Zool. Sci.* **12**, 733–739.

Kime-Hunt, E., Spartalian, K., and Carrano, C. J. (1988). Models for vanadium–tunichrome interactions. *J. Chem. Soc. Chem. Commun.*, pp. 1217–1218.

Kimura, T., Morinaga, K., and Nakano, K. (1972). The determination of association constant of vanadium(III) with sulfate ion. *Nippon Kagaku Kaishi,* pp. 664–667.

Kobayashi, S. (1935). On the presence of vanadium in certain Pacific ascidians. *Sci. Rep. Tohoku Univ. Fourth Ser.* **18,** 185–193.

Kondo, M., Minakoshi, S., Iwata, K., Shimizu, T., Matsuzaka, H., Kamigata, N., and Kitagawa, S. (1996). Crystal structure of a tris(dithiolene) vanadium(IV) complex having unprecedented D_{3h} symmetry. *Chem. Lett.,* pp. 489–490.

Kustin, K., Levine, D. S., McLeod, G. C., and Curby, W. A. (1976). The blood of *Ascidia nigra:* Blood cell frequency distribution, morphology, and the distribution and valence of vanadium in living blood cells. *Biol. Bull.* **150,** 426–441.

Kustin, K., McLeod, G. C., Gilbert, T. R., and Briggs 4th L. B. R. (1983). Vanadium and other metal ions in the physiological ecology of marine organisms. *Struct. Bonding* **53,** 139–160.

Kustin, K., and Robinson, W. E. (1995) Vanadium transport in animal systems. In H. Sigel and A. Sigel (Eds.), *Metal Ions in Biological Systems.* Marcel Dekker, New York, Vol. 31, pp. 511–542.

Lane, D. J. W., and Wilkes, S. L. (1988). Localization of vanadium, sulphur and bromine within the vanadocytes of *Ascidia mentula* Müller: A quantitative electron probe X-ray microanalytical study. *Acta Zool. (Stockholm)* **69,** 135–145.

Lee, S., Kustin, K., Robinson, W. E., Frankel, R. B., and Spartalian, K. (1988a). Magnetic properties of tunicate blood cells. I. *Ascidia nigra. Inorg. Biochem.* **33,** 183–192.

Lee, S., Nakanishi, K., Chiang, M. Y., Frankel, R., and Spartalian, K. (1988b). Preparation, crystal structure, and physical properties of a pyrogallol-bridged vanadium(II) complex. *J. Chem. Soc. Chem. Commun.* pp. 785–786.

Lee, S., Nakanishi, K., and Kustin, K. (1990). The intracellular pH of tunicate blood cells: *Ascidia ceratodes* whole blood, morula cells, vacuoles and cytoplasm. *Biochim. Biophys. Acta* **1033,** 311–317.

Levine, E. P. (1961). Occurrence of titanium, vanadium, chromium, and sufuric acid in the ascidian *Eudistoma ritteri. Science* **133,** 1352–1353.

Lybing, S. (1953). The valence of vanadium in hemolysates of blood from *Ascidia obliqua* Alder. *Arkiv. Kemi.* **6,** 261–269.

Macara, I. G., McLeod, G. C., and Kustin, K. (1979a). Tunichromes and metal ion accumulation in tunicate blood cells. *Comp. Biochem. Physiol.* **63B,** 299–302.

Macara, I. G., McLeod, G. C., and Kustin, K. (1979b). Isolation, properties and structural studies on a compound from tunicate blood cells that may be involved in vanadium accumulation. *Biochem. J.* **181,** 457–465.

Mason, B. (1966). *Principles of Geochemistry.* 3rd ed. John Wiley & Sons, New York.

Matsubayashi, G., Akiba, K., and Tanaka, T. (1988). Spectroscopic and electrical properties of VO(dmit)$_2$ and V(dmit)$_3$ anion complexes and X-ray crystal structure of [NMP]$_2$[V(dmit)$_3$] (dmit = 2-thioxo-1,3-dithiole-4,5-dithiolate, NMP = N-methylphenazinium. *Inorg. Chem.* **27,** 4744–4749.

McLeod, G. C., Ladd, K. V., Kustin, K., and Toppen, D. L. (1975). Extraction of vanadium(V) from seawater by tunicate: A revision of concepts. *Limnol. Oceanogr.* **20,** 491–493.

Meier, R., Boddin, M., Mitzenheim, S., and Kanamori, K. (1995). Solution properties of vanadium-(III) with regard to biological systems. In H. Sigel and A. Sigel (Eds.), *Metal Ions in Biological Systems. Vanadium and Its Role in Life.* Marcel Dekker, New York, Vol. 31, pp. 45–88.

Michibata, H. (1984). Comparative study on amounts of trace elements in the solitary ascidians, *Ciona intestinalis* and *Ciona robusta. Comp. Biochem. Physiol.* **78A,** 285–288.

Michibata, H. (1989). New aspects of accumulation and reduction of vanadium ions in ascidians, based on concerted investigation for both a chemical and biological viewpoint. *Zool. Sci.* **6,** 639–681.

Michibata, H. (1993). The mechanism of accumulation of high levels of vanadium by ascidians from seawater: Biophysical approaches to a remarkable phenomenon. *Adv. Biophys.* **29**, 103–131.

Michibata, H. (1996). The mechanics of accumulation of vanadium by ascidians: Some progress towards an understanding of this unusual phenomenon. *Zool. Sci.* **13**, 489–502.

Michibata, H., Hirata, J., Terada, T., and Sakurai, H. (1988). Autonomous fluorescence of ascidian blood cells with special reference to identification of vanadocytes. *Experientia* **44**, 906–907.

Michibata, H., Hirata, J., Uesaka, M., Numakunai T., and Sakurai, H. (1987). Separation of vanadocytes: Determination and characterization of vanadium ion in the separated blood cells of the ascidian, *Ascidia ahodori. J. Exp. Zool.* **224**, 33–38.

Michibata, H., Iwata, Y., and Hirata, J. (1991a). Isolation of highly acidic and vanadium-containing blood cells from among several types of blood cell from Ascidiidae species by density gradient centrifugation. *J. Exp. Zool.* **257**, 306–313.

Michibata, H., and Sakurai, H. (1990). "Vanadium in ascidians." in N. D. Chasteen (Ed.), *Vanadium in Biological Systems.* Kluwer Academic Publishers, Dortrecht, pp. 153–171.

Michibata, H., Seki, Y., Hirata, J., Kawamura, M., Iwai, K., Iwata, R., and Ido, T. (1991b). Uptake of ^{48}V-labeled vanadium by subpopulations of blood cells in the ascidian, *Ascidia gemmata. Zool. Sci.* **8**, 447–452.

Michibata, H., Terada, T., Anada, N., Yamakawa, K., and Numakunai, T. (1986). The accumulation and distribution of vanadium, iron, and manganese in some solitary ascidians. *Biol. Bull.* **171**, 672–681.

Michibata, H., Uchiyama, J., Seki, Y., Numakunai, T., and Uyama, T. (1992). Accumulation of vanadium during embryogenesis in the vanadium-rich ascidian *Ascidia gemmata. Biol. Trace Element Res.* **34**, 219–223.

Michibata, H., Uyama, T., Hirata, J. (1990). Vanadium containing cells (vanadocytes) show no fluorescence due to the tunichrome in the ascidian *Ascidia sydneiensis samea. Zool. Sci.* **7**, 55–61.

Money, J. K., Folting, K., Huffman, J. C., and Christou, G. (1987). A binuclear vanadium(III) complex containing the linear $[VOV]^{4+}$ unit: Preparation, structure, and properties of $[V_2O(SCH_2CH_2NMe_2)_4]$. *Inorg. Chem.* **26**, 944–948.

Moriyama, Y., Maeda, M., and Futai, M. (1992). The role of V-ATPase in neuronal and endocrine system. *J. Exp. Biol.* **172**, 171–178.

Nawi, M. A., and Riechel, T. L. (1981). Electrochemical studies of vanadium(III) and vanadium(IV) acetylacetonate complexes in dimethyl sulfoxide. *Inorg. Chem.* **20**, 1974–1978.

Nelson, N. (1992). The vacuolar H^+-ATPase—One of the most fundamental ion pumps in nature. *J. Exp. Biol.* **172**, 19–27.

Neumann, R., and Assael, I. (1989). Vanadium(V)/vanadium(III) redox couple in acidic organic media. Structure of a vanadium(III)-tetraethylene glycol pentagonal-bipyramidal complex $([V(teg)(Br)_2]^+Br^-)$. *J. Am. Chem. Soc.* **111**, 8410–8413.

Nicholls, D., and Seddon, K. R. (1973). Reactions of vanadium(V) oxide tribromide: Preparation and properties of complex oxobromovanadates(IV). *J. Chem. Soc. Dalton Trans.*, pp. 2747–2750.

Noddack, I., and Noddack, W. (1939). Die Häufigkeiten der Schwermetalle in Meerestieren. *Arkiv. Zool.* **32A**, 1–35.

Okamoto, K., Hidaka, J., Fukagawa, M., and Kanamori, K. (1992). Structure of triaqua(nitrilotriacetato)vanadium(III) tetrahydrate. *Acta Crystallogr.* **C48**, 1025–1027.

Oltz, E. M., Bruening, R. C., Smith, M. J., Kustin, K., and Nakanishi, K. (1988). The tunichromes. A class of reducing blood pigments from sea squirts: Isolation, structure, and vanadium chemistry. *J. Am. Chem. Soc.* **110**, 6162–6172.

Pasquali, M., Marchetti, F., and Floriani, C. (1979). Deoxygenation of oxovanadium(IV) complexes: A novel synthetic route to dichlorovanadium (IV) chelate complexes. *Inorg. Chem.* **18**, 2401–2404.

Pirie, B. J. S., and Bell, M. V. (1984). The localization of inorganic elements, particularly vanadium and sulphur, in haemolymph from the ascidians *Ascidia mentula* (Müller) and *Ascidiella aspersa* (Müller). *J. Exp. Mar. Biol. Ecol.* **74,** 187–194.

Pope, M. T. (1983). *Heteropoly and Isopoly Oxometallates.* Springer-Verlag, New York.

Riechel, T. L., De Hayes, L. J., and Sawyer, D. T. (1976). Electrochemical studies of vanadium(III), -(IV), and -(V) complexes of diethyldithiocarbamate in acetonitrile. *Inorg. Chem.* **15,** 1900–1904.

Riechel, T. L., and Sawyer, D. T. (1975), Electrochemical studies of vanadium(III), -(IV), and -(V) complexes of 8-quinolinol in acetonitrile. Formation of a binuclear mixed-valence(IV,V) complex. *Inorg. Chem.* **14,** 1869–1875.

Robinson, W. E., Agudelo, M. I., and Kustin, K. (1984). Tunichrome content in the blood cells of the tunicate, *Ascidia callosa* Stimpson, as an indicator of vanadium distribution. *Comp. Biochem. Physiol.* **78A,** 667–673.

Robles, J. C., Matsuzaka, Y., Inomata, S., Shimoi, M., Mori, W., and Ogino, H. (1993). Syntheses and structures of vanadium(III) complexes containing 1,3-diaminopropane-$N,N,N'N'$-tetraacetate ([V(trdta)]$^-$) and 1,3-diamino-2-propanol-N,N,N',N'-tetraacetate ([V$_2$(dpot)$_2$]$^{2-}$). *Inorg. Chem.* **32,** 13–17.

Roman, D. A., Molina, J., and Rivera, L. (1988). Inorganic aspects of the blood chemistry of ascidians. Ionic composition, and Ti, V, and Fe in the blood plasma of *Pyura chilensis* and *Ascidia dispar.* *Biol. Bull.* **175,** 154–166.

Rowley, A. F. (1982). The blood cells of *Ciona intestinalis:* An electron probe X-ray microanalytical study. *J. Mar. Biol. Assoc. UK,* **62,** 607–620.

Rowley, A. F. (1983). Preliminary investigations on the possible antimicrobial properties of tunicate blood cell vanadium. *J. Exp. Zool.* **227,** 319–322.

Rummel, W., Bielig, H.-J., Forth, W., Pfleger, K., Rudiger, W., and Seifen, E. (1966). Absorption and accumulation of vanadium by tunicates. *Protides Biol. Fluids* **14,** 205–210.

Ryan, D. E., Ghatlia, N. D., McDermott, A. E., Turro, N. J., and Nakanishi, K. (1992). Reactivity of tunichromes: Reduction of vanadium(V) and vanadium(IV) to vanadium(III) at neutral pH. *J. Am. Chem. Soc.* **114,** 9659–9660.

Ryan, D. E., Grant, K. B., and Nakanishi, K. (1996a). Reactions between tunichrome Mm-1, a tunicate blood pigment, and vanadium ions in acidic and neutral media. *Biochemistry,* **35,** 8640–8650.

Ryan, D. E., Grant, K. B., Nakanishi, K., Frank, P., and Hodgson, K. O. (1996b). Reactions between vanadium ions and biogenic reductants of tunicates: Spectroscopic probing for complexation and redox products in vitro. *Biochemistry* **35,** 8651–8661.

Scheidt, W. R., Tsai, C., and Hoard, J. L. (1971a). Stereochemistry of dioxovanadium(V) complexes. I. The crystal and molecular structure of triammonium bis(oxalato)dioxovanadate(V) dihydrate. *J. Am. Chem. Soc.* **93,** 3867–3872.

Scheidt, W. R., Countryman, R., and Hoard, J. L. (1971b). Stereochemistry of dioxovanadium(V) complexes. III. The crsytal and molecular structures of trisodium(ethylenediaminetetra-acetato)dioxovanadate(V) tetrahydrate. *J. Am. Chem. Soc.* **93,** 3878–3882.

Scippa, S., Botte, L., and de Vincentiis, M. (1982). Ultrastructure and X-ray microanalysis of blood cells of *Ascidia malaca.* *Acta Zool. (Stockholm)* **63,** 121–131.

Scippa, S., Botte, L., Zierold, K., and de Vincentiis, M. (1985). X-ray microanalytical studies on cryofixed blood cells of the ascidian *Phallusia mammillata.* I. Elemental composition of morula cells. *Cell Tissue Res.* **239,** 459–461.

Scippa, S., Zierold, K., and de Vincentiis, M. (1988). X-ray microanalytical studies on cryofixed blood cells of the ascidian *Phallusia mammillata.* II. Elemental composition of the various blood cell types. *J. Submicrosc. Cytol. Pathol.* **20,** 719–730.

Shimoi, M., Saito, Y., and Ogino, H. (1989). Syntheses of M[V(edta)(H$_2$O)]·nH$_2$O (M = Na, K, NH$_4$) and X-ray crystal structure of Na[V(edta)(H$_2$O)]·3H$_2$O. *Chem. Lett.,* pp. 1675–1678.

Smith, B. E., Eady, R. R., Lowe, D. L., and Gormal, C. (1988). The vanadium–iron protein of vanadium nitrogenase from *Azotobacter chroococcum* contains an iron–vanadium cofactor. *Biochem. J.* **250**, 299–302.

Smith, M. J. (1970a). The blood cells and tunic of ascidian *Halocynthia aurantium* (Pallas). I. Hematology, tunic morphology, and partition of cells between blood and tunic. *Biol. Bull.* **138**, 345–378.

Smith, M. J. (1970b). The blood cells and tunic of ascidian *Halocynthia aurantium* (Pallas). II. The histochemistry of the blood cells and tunic. *Biol. Bull.* **138**, 379–388.

Smith, M. J., Ryan, D. E., Nakanishi, K., Frank, P., and Hodgson, K. O. (1995). Vanadium in ascidians and the chemistry of tunichromes. In H. Sigel and A. Sigel (Eds.), *Metal ions in Biological Systems.* Marcel Dekker, New York, Vol. 31, pp. 423–490.

Stiefel, E. I., Dori, Z., and Gray, H. B. (1968). Octahedral *vs.* trigonal-prismatic coordination. The structure of $(Me_4N)_2[V(mnt)_3]$. *J. Am. Chem. Soc.* **89**, 3353–3354.

Stoecker, D. (1978). Resistance of a tunicate to fouling. *Biol. Bull.* **155**, 615–626.

Sugimura, Y., Suzuki, Y., and Miyake, Y. (1978). Chemical forms of minor metallic elements in the ocean. *J. Oceanogr. Soc. Jpn.* **34**, 93–96.

Swinehart, J. H., Biggs, W. R., Halko, D. J., Schroeder, N. C. (1974). The vanadium and selected metal contents of some ascidians. *Biol. Bull.* **146**, 302–312.

Tsuchida, E., Yamamoto, K., Oyaizu, K., Iwasaki, N., and Anson, F. C. (1994). Electrochemical investigations of the complexes resulting from the acid-promoted deoxygenation and dimerization of (*N,N'*-ethylenebis(salicylideneaminato)oxovanadium(IV). *Inorg. Chem.* **33**, 1056–1063.

Tullius, T. D., Gillum, W. O., Carlson, R. M. K., and Hodgson, K. O. (1980). Structural study of the vanadium complex in living ascidian blood cells by X-ray absorption spectrometry. *J. Am. Chem. Soc.* **102**, 5670–5676.

Uyama, T., Moriyama, Y., Futai, M., and Michibata, H. (1994). Immunological detection of a vacuolar-type H^+-ATPase in the vanadocytes of the ascidian *Ascidia sydneiensis samea. J. Exp. Zool.* **270**, 148–154.

Uyama, T., Nishikata, T., Satoh, N., and Michibata, H. (1991). Monoclonal antibody specific to signet ring cells, the vanadocytes of the tunicate, *Ascidia sydneiensis samea. J. Exp. Zool.* **259**, 196–201.

Uyama, T., Nose, Y., Wuchiyama, J., Moriyama, Y., and Michibata, H. (1997). Finding of the same antigens in the Polychaeta *Pseudopotamilla occelata* as those in the vanadium-rich ascidian *Ascidia sydneiensis samea. Zool. Sci.,* **14**, 43–47.

Uyama, T., Uchiyama, J., Nishikata, T., Satoh, N., and Michibata, H. (1993). The accumulation of vanadium and manifestation of an antigen recognized by a monoclonal antibody specific to vanadocytes during embryogenesis in the vanadium-rich ascidian, *Ascidia sydneiensis samea. J. Exp. Zool.* **265**, 29–34.

Vinogradov, A. P. (1934). Distribution of vanadium in organisms. *C. R. (Doke.) Acad. Sci. URSS* **3**, 454–459.

Webb, D. A. (1939). Observations on the blood of certain ascidians, with special reference to the biochemistry of vanadium. *J. Exp. Biol.* **16**, 499–523.

Webb, D. A. (1956). The blood of tunicates and the biochemistry of vanadium. *Publ. Staz. Zool. Napoli* **28**, 273–288.

Welch, J. H., Bereman, R. D., and Singh, P. (1988). Syntheses and characterization of two vanadium tris complexes of the 1,2-dithiolene 5,6-dihydro-1,4-dithiin-2,3-dithiolate. Crystal structure of $[(C_4H_9)_4N][V(DDDT)_3]$. *Inorg. Chem.* **27**, 2862–2868.

Wever, R., and Kustin, K. (1990) Vanadium: A biologically relevant element. In A. G. Sykes (Ed.), *Advances in Inorganic Chemistry.* Academic Press, New York, Vol. 35, pp. 81–115.

Wright, R. K. (1981). Urochordata. In N. A. Ratcliffe and A. F. Rowley (Eds.), *Invertebrate Blood Cells*. Academic Press, London, Vol. 2, pp. 565–626.

Wuchiyama, J., and Michibata, H. (1995). Classification, based on autonomous fluorescence, of the blood cells of several ascidians that contain high levels of vanadium. *Acta Zool. (Stockholm)* **76,** 51–55.

Yoshimori, T., Yamamoto, A., Moriyama, Y., Futai, M., and Tashiro, Y. (1991). Bafilomycin A_1, a specific inhibitor of vacuolar-type H^+-ATPase, inhibits acidification and protein degradation in lysosomes of cultured cells. *J. Biol. Chem.* **266,** 17707–17712.

Zhang, Y., and Holm, R. H. (1990). Vanadium-mediated oxygen atom transfer reactions. *Inorg. Chem.* **29,** 911–917.

11

STRUCTURE, FUNCTION, AND MODELS OF BIOGENIC VANADIUM COMPOUNDS

Dieter Rehder and Sven Jantzen

Institut für Anorganische und Angewandte Chemie der Universität Hamburg, D-20146 Hamburg, Germany

Vanadium in the Environment. Part 1: Chemistry and Biochemistry, Edited by Jerome O. Nriagu.
ISBN 0-471-17778-4.   © 1998 John Wiley & Sons, Inc.

1. INTRODUCTION

Under physiological conditions (pH ca. 7, vanadium concentration 0.1–1 μM, ionic strength ca. 0.15 M), the main soluble "free" (i.e., uncomplexed) vanadium form available is $H_2VO_4^-$ in equilibrium with some HVO_4^{2-} (the pK_a is 8.17; Elvingson et al., 1996). $H_2VO_4^-$ is also believed to be the main species present in sea water. The vanadate concentration in sea water amounts to 20–35 nM; vanadium is hence the most abundant transition metal present in this medium, where life is thought to have originated. At concentrations approaching 0.1 mM, nucleation of monovanadate to di-, tetra-, and pentavanadate, and (below pH 6.5) decavanadate occurs. Nucleation is also favored at higher ionic strengths. Vanadate(V), in many respects a close analog to phosphate, interacts with various biogenic molecules such as proteins, peptides, nucleosides, and sugars, and hence may be attributed a general role in physiological processes mainly where phosphate transfer is concerned. In contrast to phosphate, however, vanadate is redox-labile, and direct action of "free" vanadate is restricted to aerobic conditions. The oxidation state +5 of vanadium is of relevance in the vanadate-dependent haloperoxidases, which contain vanadate coordinated to $N\varepsilon$ of a histidine residue, and it may be maintained in V^V complexes of biogenic ligands formed under physiological conditions.

Vanadate(V) is coupled to oxovanadium(IV) by the redox couple $VO^{2+} + 3H_2O \rightleftharpoons H_2VO_4^- + 4H^+ + e^-$, the electrochemical potential E of which—at pH 7 and a (physiologically about reasonable) ratio $H_2VO_4^-/VO^{2+} = 10^3$—amounts to -0.17 V. Comparison with the redox pairs $2H_2O \rightleftharpoons O_2 + 4H^+ + 4e^-$ ($E = +0.82$ V) and $NADH + H^+ \rightleftharpoons NAD^+ + 2H^+ + 2e^-$ ($E = -0.32$ V) shows that, in the intracellular medium, vanadate(V) will be reduced to oxovanadium(IV). The reduction may well proceed to vanadium(III), which is present in several marine organisms. Apart from NADH, cysteine residues of peptides and proteins are active in vanadium redox chemistry, as are ascorbate and several sugars, just to mention those compounds, which have been investigated in this respect. Quite generally, there is a pronounced tendency for vanadium(V) to pass over to vanadium(IV) in the presence of organics, and for vanadium(IV) to lose its single electron in the presence of oxygen. Whether a redox-active system prevails is a matter, inter alia, of the special condition associated with the formation and the nature of the vanadium coordination compound.

Both forms of reduced vanadium, VO^{2+} (vanadyl) and V^{3+}, can exist at physiological pH as sufficiently stable complexes only. The same is true for the non-oxo form of vanadium(IV), V^{4+}, which is a constituent in a molecular vanadium complex ("amavadin") that is present in mushrooms of the genus *Amanitae*. The "free" forms of VO^{2+} and V^{3+}, $[VO(H_2O)_{6-n}(OH)_n]^{(2-n)+}$ and $[V(H_2O)_{6-n}(OH)_n]^{(3-n)+}$, may be present in the presumably acidic media of specialized cells or vacuoles of several genera of Ascidiae (sea squirts) and Polychaetae fan worms (Chapters 9 and 10). Starting at about pH 4, insoluble vanadium dioxide separates from aqueous vanadyl solutions in the absence of ligands. In the presence, however, of suitable ligands, such as those provided by many biogenic molecules, stable complexes can form that are soluble at physiological pH, and quite often reversibly oxidized to the corresponding vanadium(V) or reduced to the vanadium(III) complexes. Many of the physiological roles of vanadium may in fact be traced back to the formation, in competition with other predominantly di- and trivalent metal ions, of complexes with the formally V^{5+}, VO^{3+}, VO_2^+, VO^{2+}, V^{4+}, and V^{3+} cations. The existence of stable vanadium(V) forms other than vanadate, as well as the ability of V^V and V^{IV} to pick up or lose O^{2-} (oxo vs. non-oxo, monooxo vs. dioxo complexes; see, e.g., Zhang and Holm, 1990; Kelm and Krüger, 1996), points to another distinct difference between the elements phosphorus and vanadium.

Stable vanadium coordination compounds may exist in a variety of coordination geometries, depending on the coordination number, the formal oxidation state, and the electronic and steric nature of the ligand system (Rehder 1991, 1992, 1995). Many of these complexes will be introduced in the context of "structural models" in the following sections of this chapter. A few biogenic examples are picked out here for the sake of an overview (for references see the respective sections): Coordination numbers may vary from eight (in the non-oxo complex amavadin) to six (in vanadium nitrogenase, vanadyl-transferrin), five (vanadate-dependent peroxidases, the vanadate–uridine–ribonuclease A complex, or vanadyl-porhyrins), and four (in the vanadate–ribonuclease T_1 complex). For a particular coordination number, the geometry may vary (e.g., octahedral vs. trigonal prismatic, or trigonal-bipyramidal vs. tetragonal-pyramidal). As shown by model compounds, many intermediate geometries are possible, and there is a remarkable flexibility both with respect to the interconversion of geometries and the number of ligand functions linked to vanadium by a strong covalent bond, a weak covalent bond, or a hydrogen or contact ion-pair interaction. This flexibility of the arrangement of the coordination sphere, and the concomitant ease of changes of oxidation states of the vanadium center with and without alterations in the coordination sphere, which is well known for a number of vanadium-based catalytic processes conducted in vitro, may be the clue for an understanding of the physiological and environmental roles of vanadium in the framework of a structure–function synergism.

2. BIOGENIC VANADIUM COMPOUNDS

2.1. Vanadium Nitrogenase

2.1.1. Structure and Function of the FeV Protein

Vanadium nitrogenase (Eady and Leigh, 1994; Eady, 1995) is expressed together with molybdenum nitrogenase and an iron-only nitrogenase in microorganisms such as *Azotobacter vinelandii, A. chroococcum,* and *Anabaena variabilis.* According to X-ray absorption and electron paramagnetic resonance (EPR) spectroscopies (Chen et al., 1993; Arber et al., 1989a), vanadium nitrogenase, including the cofactor, is very similar to its molybdenum analog, for which a recent X-ray diffraction analysis has revealed the structural details (Chan et al., 1993). The nitrogenases consist of an iron protein containing a 4Fe–4S ferredoxin, and an iron-heterometal protein. In molybdenum nitrogenase, the FeMo protein is an $\alpha_2\beta_2$ tetramer with the two FeMo-cofactors (*M* clusters) embedded in the α subunits und the *P* clusters (4Fe–4S double cubanes bridged by sulfide and cysteinate) at the α/β interface. One of the forms of vanadium nitrogenase isolated from *Azotobacter* has an $\alpha_2\beta_2\delta_2$ subunit structure (the δ subunits being minor ones) and the same arrangement of *M* and *P* clusters as in molybdenum nitrogenase. According to Blanchard and Hales (1996), a second form of vanadium nitrogenase exists that lacks one of the β subunits and thus contains only half of the *P* cluster, that is, a 4Fe–4S cubane. The probable structure of the FeV cofactor is depicted in Figure 1: Vanadium is a constituent of a complex iron–sulfur cluster, occupying one of the corners of a cuboidal iron–sulfur moiety, and is thus linked to three trebly bridging sulfides, and additionally coordinated to an imidazole nitrogen of histidine and the vicinal carboxylate and alkoxide functions of homocitrate. The oxidation state of vanadium is between +2 and +4 (Eady, 1989).

The main reaction catalyzed by vanadium nitrogenase is the reductive protonation of dinitrogen. However, about 50% of the reduction equivalents are consumed in the concomitant production of hydrogen. Eq 1 summarizes the overall reaction. In contrast to molybdenum nitrogenase, a small amount of hydrazine is also formed—in accord with a step-wise reduction of N_2 via

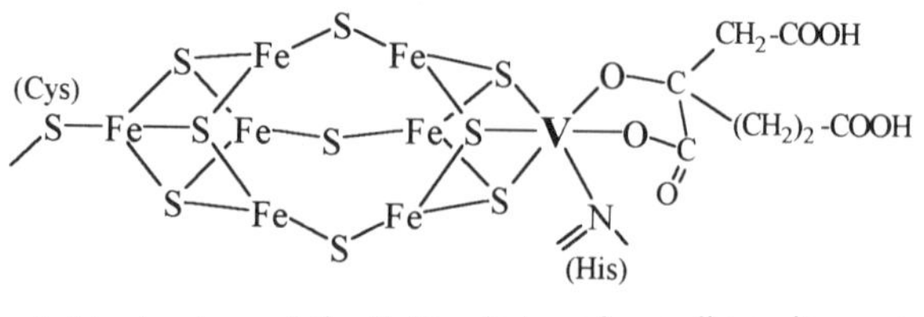

Figure 1. The probable structure of the FeV cofactor of vanadium nitrogenase, based on the assumtion that it is closely related to the FeMo cofactor of molybdenum nitrogenase, the structure of which has been revealed by a 2.2-Å resolution X-ray analysis (Chan et al., 1993).

hydrazine or, as proposed on the basis of model studies by Gailus et al. (1994a), a hydrazide$(1-)$ or hydrazinyl intermediate (eq 2). Substrates other than N_2 and H^+ are also reduced. An example is acetylene, which is converted to ethylene and further to ethane (Eady, 1995). In either case, the electrons for the reduction are delivered by the iron protein of the nitrogenase and probably picked up by the P cluster (two ATPs are hydrolyzed per electron transferred) to be pumped further into the reaction center, the M cluster. Whether the heterometal (Mo or V) or an Fe_2 site is the actual site for the reductive protonation is still under debate. The availability of histidine and homocitrate at the heterometal may facilitate the proton transfer accompanying the reduction, and hence support the direct involvement of the heterometal, as do synthetic cluster compounds which has been demonstrated in model studies (see below).

$$N_2 + 14H^+ + 12e^- \rightarrow 2NH_4^+ \text{ (including ca. 0.5\% } N_2H_5^+) + 3H_2 \quad (1)$$

$$\begin{aligned} V^q &\leftarrow N_2 + 3H^+ \rightarrow V^{(q+4)} \leftarrow {}^-NHNH_2 \\ &\leftrightarrow V^{(q+3)+} - NHNH_2 \rightarrow V^{(q+3)+} + N_2 + NH_3 \end{aligned} \quad (2)$$

2.1.2. Structural Models: Sulfide and Thiolate Coordination; Homocitrate and Analogs

The nature of the coordination environment of vanadium in the FeV cofactor (Fig. 1) has initiated various studies aiming towards mimics of the structural and functional features of nitrogenase. In this section we will emphasize structural models for the binding of sulfur ligands (inorganic and organic sulfide, thiolate) on the one hand, and vicinal carboxylate–alkoxide binding on the other hand. Enamine binding as exhibited by the coordination of histidine will be dealt with in some detail in the context of vanadium-dependent haloperoxidases (Section 2.2.2). Thio-ligation to vanadium does also have implications for the redox interaction between vanadium and glutathion or proteins with cysteine in their active site (Banabe et al., 1987). It is further related to the in vitro enantioselective catalysis of thioethers to sulfoxides by vanadium complexes (Schmidt et al., 1996), and to the thiophenogenic vanadium compounds formed during the processing (hydrodesulfurization) of crude oil (Beaton and Bertolacini, 1991).

Selected heterometal–sulfide clusters containing vanadium are shown in Figure 2. The cluster **1** (Liu et al., 1995) is a recent example for vanadium(V) complexes derived from the tetrahedral tetrathiovanadate $[VS_4]^{3-}$ by coordination of metal fragments—copper(I)thiophenolate in this case—to the sulfido ligands. The resulting cluster contains the five metal ions in an almost ideal coplanar arrangement. Clusters **2** (Cen et al., 1994) and **3** (Malinak et al., 1995) represent closer approaches to the FeV cofactor (compare Fig. 1). Cluster **2** in part matches the $Fe_3(FeCys)S_6$ structural unit of the cofactor, while **3** mimics the hexa-coordination of the vanadium site in the cuboidal Fe_3VS_4 moiety. Cluster **3** catalyzes the reduction of hydrazine (eq 3a), which,

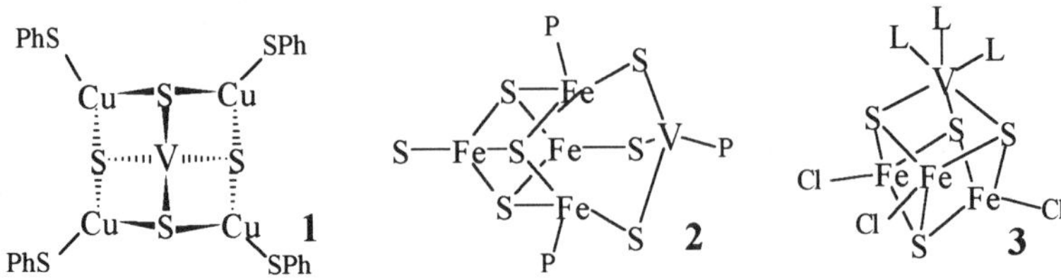

Figure 2. Selected mixed metal sulfide clusters containing vanadium. Sources: **1** (Liu et al., 1995), **2** (Cen et al. 1994; *P* denotes PEt$_3$, omitted on the back-side Fe for clarity), **3** (Malinak et al., 1995). Note the similarities between the cofactor depicted in Figure 1 and the Fe$_3$(FeSR)S$_6$ core in **2**, on the one hand, and the VS$_3$L$_3$ moiety of **3**, on the other hand. L in the structurally characterized version of **3** is {tris(pyrazolyl)borate}(1-).

as pointed out above, is an intermediate in N_2 fixation, in the presence of external proton and electron sources. If these are absent, the disproportionation of hydrazine is catalyzed (Cen et al., 1994) (eq 3b). Comparison of the reactivity patterns of **3** and its molybdenum counterpart, and comparison of the activity as the ligands L are varied, supports the view that the heterometal attains a central role in substrate reduction.

$$N_2H_4 + 2e^- + 4H^+ \rightarrow 2NH_4^+ \tag{3a}$$

$$3N_2H_4 \rightarrow 4NH_3 + N_2 \tag{3b}$$

The examples collated in Figure 3 demonstrate the versatility of the coordination of sulfur-functional ligands with respect to the nature of the function, the oxidation number of vanadium, and the coordination geometries. For the coordination number 6, octahedral (**4** and **5**) as well as trigonal-prismatic (**6**) ligand arrangements are realized. The latter is not restricted to the dithiolene complex **6** but has also been noted for tris(thiocatecholate) complexes of vanadium (see, e.g., Eisenberg and Gray, 1967). While thiophenolate forms stable complexes throughout the range of oxidation states (**4** with V^{II}, **5** with V^{IV}, and **7** with V^V), aliphatic thiolates (including cysteine) are usually readily oxidized by V^V, and stable complexes are restricted to those with vanadium in the oxidation states +4 (see [V(Cys)$_2$], **8**) and +3. The thiolate complex **9** is of particular interest in as much as oxovanadium($+V$) and thiolate coexist in the same compound, probably as a consequence of the stabilizing effect imparted by the tripodal, tetradentate ligand. A biogenic counterpart, vanadate(V) complexed to a cysteinate of a phosphotyrosyl phosphatase, has recently been structurally characterized (Zhang et al., 1997). Even V^{IV} may act as an oxidizing agent towards thiolates: The formation of **5** is accompanied by a two-electron oxidation of the ligand to a heterocycloeicosane containing two disulfide units (Tsagkalidis et al., 1995b), and hence is a reaction reminiscent of the oxidation of cyteine to cystine, possibly a key reaction in the inhibition by vanadium of those enzymes, which carry cysteine at their active

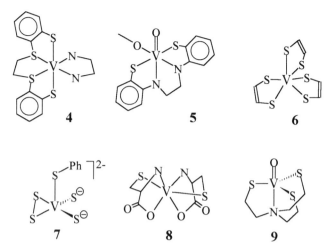

Figure 3. Vanadium complexes with various modes of thio ligation: cysteine (**8**, Maeda et al., 1993), aliphatic thiolate only (**9**, Nanda et al., 1996), dithiolene only (**6**, Kondo et al., 1996), thiophenolate and organic sulfide (**4**, Tsagkalidis et al., 1995a), thiophenolate and amine (**5**, one of the vanadium centers shown only; the compound is an oxo-bridged, cyclic tetramer; Tsagkalidis et al., 1995b), and thiophenolate + inorganic sulfide and disulfide (**7**, Sendlinger et al., 1993). The formal oxidation states of vanadium vary from +5 (**9, 7**) to +4 (**5, 6, 8**) through +2 (**4**). For the oxidation state +4 see also **56** in Figure 14; for +3 see **2** in Figure 2. The oxidation state +4 for **6** implies that, on an average scale, one of the three dithiolenato ligands is subject to a two-electron oxidation.

site. In the case of glyceraldehyde 3-phosphate dehydrogenase, such a role of vanadium, resulting in the inhibition of glycolysis, has been verified (Banabe et al., 1987). On the other hand, V^{II} can act as a reducing agent: Along with the formation of the V^{II} complex **4** with a bis(thioether)–bis(thiophenolate) ligand system, the tetradentate ligand is reductively split into ethylene and dithiocatecholate (Tsagkalidis et al., 1995a). Cluster **4** is reversibly oxidized to the V^{III} complex ($E^0 = -0.38$ V vs. SCE).

Vicinal carboxylate/alkoxide coordination has been shown to take place in the aqueous vanadate–lactate system (Tracey et al., 1987). Relatively weak trigonal-bipyramidal and octahedral 1:1 complexes have been proposed on the basis of a ^{51}V NMR evaluation. Citrate forms dimeric oxoperoxo (Djordjevic et al., 1989) and dioxo complexes (**10** in Fig. 4; Zhou et al., 1995) with V^V; homocitrate forms dimeric dioxo complex (**11**; Wright et al., 1996). In all three cases, the alkoxide bridges the two vanadium centers. A closer approach to the vanadium site in nitrogenase is provided by the mononuclear complexes **12a** (V^V; Hambley et al., 1992) and **12b** (V^{IV}; Barr-David et al., 1992) with penta-coordinated vanadium. An even "better" structural model is the benzilate complex **13**, which contains V^{IV} in a slightly distorted octahedral array, lacks (as in nitrogenase) the otherwise common oxo group, and mimics histidine binding by means of the enamine nitrogens provided by the supporting

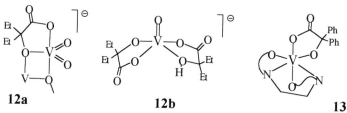

Figure 4. Coordination of *vicinal* carboxylate/alkoxide groups as mimics of homocitrate coordination at the vanadium site in nitrogenase. **10** (Zhou et al., 1995), **11** (Wright et al., 1996) and **12a** (Hambley et al., 1992), are dinuclear complexes; for clarity, the complete environments for just one of the vanadium centers is shown in the case of **11** and **12a**. The indicated charge of these anionic complexes corresponds to one half of the dimer. The ligands in **10** and **11** are citrate (3-) and homocitrate(2-), respectively. The chelating ligands are 2-oxo-2-ethylbutanoate(2-) in **12a** and **12b** (Barr-David et al., 1992) and benzilate(2-) in **13** (Vergopoulos et al., 1995). The *ONNO*(2-) ligand in the non-oxo complex **13** derives from the double Schiff base formed between ethylenediamine and two equivalents of salicylaldehyde.

Schiff-base ligand. Cluster **13** is capable of both, a reversible one-electron oxidation at +0.92 V and a reversible one-electron reduction at −0.34 V (vs. SCE) (Vergopoulos et al., 1995).

2.1.3. Functional Models: Dinitrogen and Hydrazine Complexes; Hydrido and Alkyne Complexes

We pointed out in the previous section that compounds with structural similarity to the vanadium site in nitrogenase, such as cluster **3** in Figure 2 (L = dimethylformamide, cluster charge −1) are able to catalytically convert hydrazine to ammonia and acetylene to ethylene (plus a trace of ethane) in the presence of external H^+ and e^-, underlining the view that vanadium is the primary catalytic site also in nitrogenase (Malinak et al., 1995). However, the fixation of dinitrogen has been achieved with functional models only. Among these are bulk reactions containing V^{II} in an undefined coordination mode (for reviews see Hidai and Mizobe, 1995; Rehder, 1995), as well as defined dinitrogen complexes, some of which are shown in Figure 5. According to bond distances and magnetic properties, **15–18** should be described as diazenido(2−) rather than as dinitrogen complexes. The only ingenious dinitrogen complex, **14,** contains vanadium in the (unphysiological) oxidation

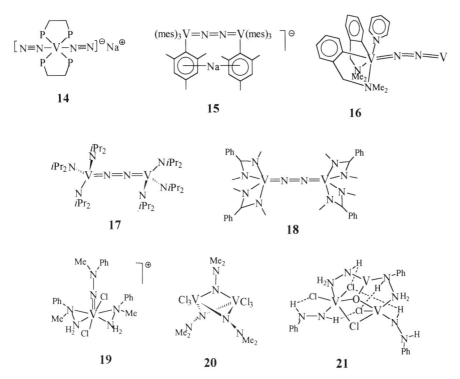

Figure 5. Dinitrogen and hydrazine complexes as models for the functional site of vanadium nitrogenase. Sources: **14** (Gailus et al., 1994a; $PP = [(Me/Ph)_2PCH_2]_2$), **15** (Ferguson et al., 1993), **16** (Leigh et al., 1991), **17** (Song et al., 1994), **18** (Hao et al., 1995), **19** (Bultitute et al., 1986), **20** (Le Floc'h et al., 1993), **21** (Murray et al., 1993). For the sake of clarity, parts of the molecules have been omitted in the case of **16** (a dimeric complex) and **21** $\{[V_3\mu_3\text{-}O(\mu\text{-}Cl)(\mu\text{-PhHNNH}_2)_2(\text{PhHNNH})_2\text{thf}]\cdot5\text{THF}\}$. Dashed lines in **21** indicate hydrogen bonds (see Section 4).

state -1. Na$^+$ (or alternatively Li$^+$) is in close-contact ion-pair interaction with the coordinated N$_2$ (Gailus, 1994a). Both the V–N *and* the N–N bonds in **14** are relatively weak (Deeth and Langford, 1995) and hence activated. Complexes **14–16** have been shown to evolve ammonia with strongly protic media such as HCl, using up reduction equivalents directly provided by the low-valent vanadium center. These complexes may therefore be considered true functional models. In contrast, a coordination sphere exclusively built up by nitrogen functionalities, as in the case of the dinuclear, N$_2$-bridged amide complex **17** and the amidinate complex **18**, does not provide any N$_2$-fixing capacity. Hydrazine, which is formed in low yields along with ammonia by reductive protolysis of N$_2$ in **14** as well as in native vanadium nitrogenase itself, very probably is an intermediate in N$_2$ fixation (cf. also eq 2). Hydrazine may be coordinated to vanadium as a neutral ligand (**19, 21**) or as an anionic ligand [hydrazido(2−)] (**19, 20**), either end-on (**19**), end-on bridging (**20, 21**) or side-on (**19**).

The nitrogenases also exhibit a hydrogenase activity (eq 1) and an alkyne reductase activity. If the reduction of acetylene is carried out in D_2O, the product is Z-ethylene (eq 4; Dilworth et al., 1988), indicating that this reduction is preceded by coordination of the alkyne to the metal (presumably the hetero metal) center, activating the alkyne and predetermining the stereoselectivity of the reaction. The 2-butyne complex of V^I, **22,** shown in Figure 6, models this kind of side-on coordination (Gailus et al., 1994b). The alkyne acts as a four-electron donor, underlining the extent of activation by coordination. The coordination of hydrogen to vanadium is exemplified in the hydrido complex **23** (Greiser et al., 1979), which reacts with alkynes to form σ-alkenyl complexes of the kind represented by compound **24** (Süssmilch et al., 1994). Cluster **24** may be taken as an intermediate in reductive protonation of alkynes, the alkenylo ligand being split off as alkene by protolysis.

$$HC\equiv CH + 2D^+ + 2e^- \longrightarrow \underset{D}{\overset{H}{>}}C=C\underset{D}{\overset{H}{<}} \quad (\longrightarrow \text{ethane}) \quad (4)$$

2.2.   Vanadate-Dependent Haloperoxidases

2.2.1.   Structure and Function of Chloro- and Bromoiodoperoxidases

Vanadate(V)-dependent haloperoxidases are abundant in marine (brown and red) algae but have also been isolated from a lichen and a fungus. The enzyme from the latter, *Curvularia inaequalis,* is a chloroperoxidase of molecular weight 67 kDa, an azido and peroxo derivative of which has recently been structurally characterized by X-ray diffraction spectroscopy to a resolution of 2.1 Å (Messerschmidt and Wever, 1996; Messerschmidt et al., 1997). Vanadium, in the oxidation state +5, is in a trigonal-bipyramidal coordination environment (Fig. 7), with three nonprotein oxygens at a distance of about 1.65 Å in the trigonal plane, and Nε of histidine in the axis at a distance of 2.25 Å. The second axial position is occupied by azide (which was a constituent

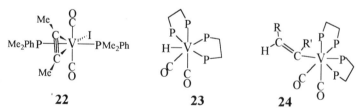

Figure 6. Models for the hydrogenase and alkyne reductase activity (eq 4) of vanadium nitrogenase: The σ-alkenyl complex **24** (Süssmilch et al., 1994) is a possible intermediate in reductive protonation of activated alkyne (**22;** Gailus et al., 1994b) by vanadium-mediated transfer of H⁺ + 2e⁻ (**23;** Greiser et al., 1979). The *PP* ligands in **23** and **24** stand for $Ph_2PCH_2CH_2PPh_2$.

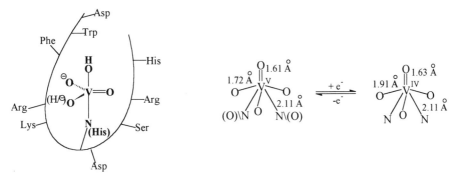

Figure 7. Structures of the active center of vanadate-dependent chloroperoxidase from *Curvularia inaequalis*, as revealed by X-ray diffraction (left; Messerschmidt and Wever, 1996), and bromoperoxidase from *Ascophyllum nodosum*, as proposed on the basis of X-ray absorption spectroscopy (right; Arber et al., 1989b). Recent results (Weyand 1996) on the *A. nodosum* bromoperoxidase suggest a coordination of V^V as in the *C. inaequalis* chloroperoxidase. Amino acids in hydrogen or counterion (salt) interaction with the vanadate oxygens are indicated.

of the buffer solution) and probably is occupied by water or OH^- in the native product. In addition to the one covalent bond ($N\varepsilon$), the vanadate is hydrogen-bonded to, inter alia, arginine-N, lysine-N, serine-O, and the main-chain amide-N of a glycine.

The first authentic haloperoxidase containing V^V in the active center was a bromo/iodoperoxidase, a homodimer of molecular weight 120 kDa, from the sea weed knobbed wrack (*Ascophyllum nodosum*), described more than a decade ago by Vilter (1984). EXAFS (Arber et al., 1989b) and XANES investigations (Hormes et al., 1988; Weidemann et al., 1989) on the native (V^V) and reduced (V^{IV}) enzyme, as well as ^{51}V NMR on the native form (Rehder et al., 1987) and EPR/ESEEM studies on the inactive, reduced form (deBoer et al., 1988a, 1988b), have led to the general assumption that the coordination sphere of vanadium is dominated by oxygen functionalities, including one vanadyl oxygen, with at least one nitrogen function also present in the first coordination sphere. The structures proposed on the basis of X-ray absorption spectroscopy are included in Figure 7. Latest results by Weyand (1996) and Weyand et al. (1996) based on X-ray diffraction (2.8 Å resolution) of the *A. nodosum* bromoperoxidase revealed that the binding of vanadium in the active center of bromoperoxidase is very much the same as in *C. inaequalis* chloroperoxidase: Vanadate is embedded at the bottom of a substrate channel of ca. 15 Å depth, coordinated to a His directly and by hydrogen or salt bridges (at a distance < 3.5 Å), to 2 Arg, 1 Lys, 2 His, 1 Gly, and 1 Ser.

The enzymes catalyze the oxidation of halides (X^-) in the presence of hydrogen peroxide. The intermediately formed hypohalous acids (or other X^+-containing species; Everett et al., 1990) halogenate organic substrates or, in the absence of such a substrate, react with hydrogen peroxide to form oxygen (singlet oxygen in the presence of bromide) and water (eq 5).

$$H_2O_2 + X^- + H^+ \xrightarrow{\text{(enzymatic)}} XOH + H_2O$$

$$\xrightarrow{+ RH} RX + H_2O$$

$$\xrightarrow{+ H_2O_2} {}^{(1)}O_2 + X^- + H_2O + H^+ \qquad (5)$$

The halogenated substrates possess a variety of metabolic and physiological functions, among which are antifungal, antibacterial, and antitumor properties (Butler and Walker, 1993). In the case of chloroperoxidase, it has been proposed that the enzyme and its products are used to oxidize plant cell walls to facilitate penetration of the fungus into the host (Simons et al., 1995). As to the mechanism of halogenation, a peroxovanadium intermediate has been proposed (Colpas et al., 1996). Vanadium does not cycle between the oxidation states +5 and +4 during turnover. Rather, the vanadium center acts as a Lewis acid that activates the peroxo or hydroperoxo ligand for the attack of the halide. Kinetic studies indicate that both the halogenation and the disproportionation reactions (cf. eq 5) proceed through a common "X^+" intermediate (Butler, 1993). We will come back to mechanistic aspects in Section 2.2.3.

2.2.2. Structural Models: Coordination of Enamines and Aromatic Amines

Even prior to the proof of its involvement by structure analysis, the presence of histidyl residues at the vanadium site has prompted efforts towards the preparation of model compounds with aromatic amines as ligands: imidazole, pyrazole, and pyridine in particular. Histidine itself has been shown to bind in a tridentate and/or bidentate fashion to V^{IV} and V^{III} (cf. the complexes 50 and 51 in Fig. 13, Section 3.1.2). Complexes 25 to 27 in Figure 8 are examples of structurally characterized imidazole complexes. Neglecting the fact that the vanadium center in peroxidase is in a trigonal-bipyramidal environment, the octahedral vanadyl complexes 25 (Cornman et al., 1992a) and 26 (Calviou et al., 1992) may be considered to approach structural models for the catalytically inactive, reduced form of vanadium haloperoxidase. In 25, the imidazole plane is perpendicular to the $V=O$ axis, indicating that π-bonding between vanadium and imidazole is not an important feature for the electronic structure in this system. Comparison of the bond lengths of the V^{IV} complex 25 and the V^V complex 27 (Cornman et al., 1992b) does not reveal any significant dependence between bond lengths and the vanadium oxidation state. Especially noteworthy in this context is the common V–N(imidazole) bond length of 2.1 Å, which corresponds to the reported EXAFS-based bond lengths for both native and reduced *A. nodosum* bromoperoxidase. In all three cases (25–27), $N\delta$ of imidazole binds to vanadium, contrasting with the binding of $N\varepsilon$ of the imidazole moiety of His in the peroxidases. Dikanov et al. (1993) have shown, based on an ESEEM study of VO(imidazole)$_4$, that there is also

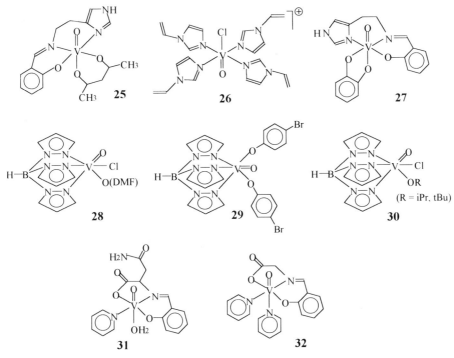

Figure 8. Models for the histidine binding in vanadium-dependent haloperoxidases: imidazole, pyrazole, and pyridine complexes. Sources: **25** (Cornman et al., 1992a), **26** (Calviou et al., 1992), **27** (Cornman et al., 1992b), **28** (Kime-Hunt et al., 1989), **29** (Holmes and Carrano, 1991), **30** (Carrano et al., 1994), **31** (Cavaco et al., 1996), **32** (Cavaco et al., 1994).

an interaction (of the single electron) of the VO^{2+} group with the remote imidazole nitrogen.

The complexes **28–30** in Figure 8, again with an overall octahedral geometry, represent examples for the (*facial*) coordination of tris(pyrazolyl)borate to VO^{2+} (**28**; Kime-Hunt et al., 1989) and VO^{3+} (**29**: Holmes and Carrano, 1991; **30**: Carrano et al., 1994). The V–N bonds to the nitrogen trans to the oxo ligand is, as expected for the effective trans effect of this group, significantly longer than that to the cisoid nitrogens. The respective data for **28** are 2.33 and 2.12. Complex **30** contains the shortest V–OR bond yet reported (1.719 Å), a feature that may be of relevance to the short V–O single bond of 1.72 Å found by EXAFS for bromoperoxidase (Fig. 7). The five-coordinated pyridine complexes **31** (Cavaco et al., 1996) and **32** (Cavaco et al., 1994) are essentially square pyramidal, with the tetragonal plane formed by the ONO Schiff-base ligand and a pyridine, and with a weakly bound water or pyridine in the apex.

Enamine functions as provided by Schiff bases may also be considered to model vanadium–histidine interaction. Among the large variety of Schiff-base complexes of vanadium reported in the literature we have picked out some of those that contain V^V—in accordance with the oxidation state +5 of vanadium in the peroxidases. Nine examples are depicted in Figure 9. Complex **33** (Cornman et al., 1992c) contains a tridentate ONN Schiff-base ligand in addition to a bidentate "non-innocent" salicylhydroxamate(2−). Non-innocent ligands affect low-energy charge-transfer transitions and hence unusual electronic conditions particularly in V^V (d^0) systems. One of the most surprising features of this category of complexes is the strong deshielding of the ^{51}V nucleus, namely $\delta(^{51}V) = -132$ ppm for **33**, as compared with δ values around -530 ppm in "normal" hexacoordinated V^V complexes with ON ligand sets (Rehder et al., 1988). Complex **34a** (Asgedom et al., 1996) is one of the very few examples of a V^V complex of trigonal-bipyramidal geometry. The complex contains two oxo groups in the equatorial plane. Five-coordinated V^V complexes tend to dimerize. In the case of **34a,** this is prevented by intermolecular hydrogen-bonding interaction. In **34b** (Mokry and Carrano, 1993),

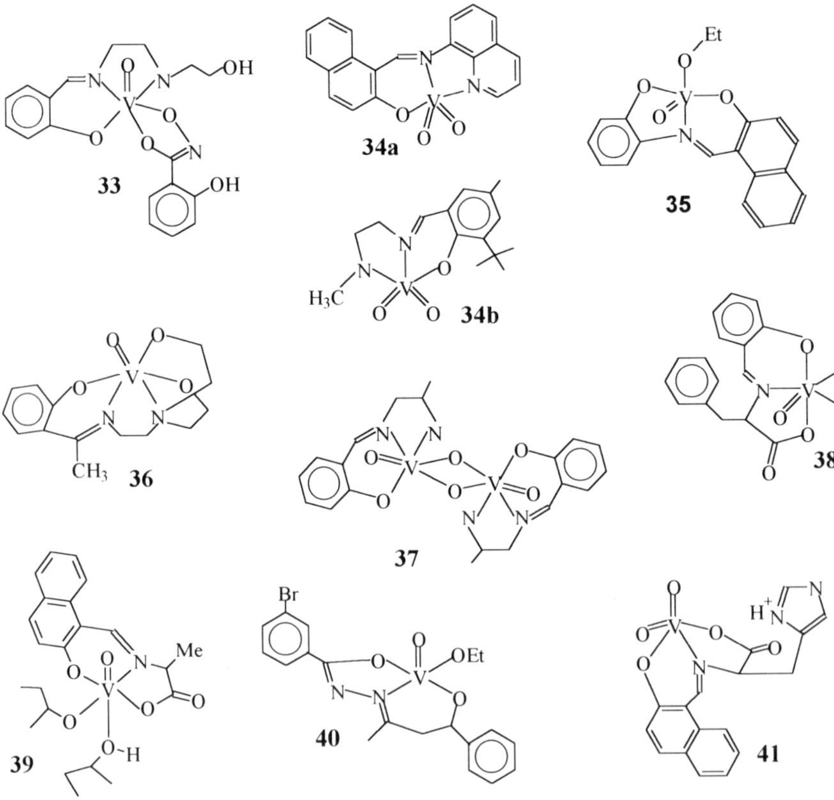

Figure 9. Models for histidine binding in vanadium-dependent peroxidases: Enamine complexes. Sources: **33** (Cornman et al., 1992c), **34a** (Asgedom et al.,1996), **34b** (Mokry and Carrano, 1993), **35** (Bashirpoor et al., 1997), **36** (Plass, 1994), **37** (Root et al., 1993), **38** (Mondal et al., 1996), **39** (Fulwood et al., 1995), **40** (Wang et al., 1996), **41** (Vergopoulos et al., 1993).

another example of trigonal-bipyramidal coordination to a V^V center, the bulky tert-butyl substituents hinder dimerization. In **36** (Plass, 1994), a trivalent pentadentate Schiff base provides, together with an oxo ligand, a distorted octahedral environment. In the dimeric, severely distorted octahedral complex **37** (Root et al., 1993), vanadium is surrounded by six different ligand functions: terminal O^{2-}, bridging O^{2-} at a short (1.665 Å) and a long distance (2.427 Å), phenolate-*O*, imine-*N*, and amine-*N*. In DMSO, the compound dissociates into monomers, stressing the lability of the long $V-(\mu\text{-}O)$ bond. In **38** (Mondal et al., 1996) and **39** (Fulwood et al., 1995), which are, again, strongly distorted octahedral complexes, the vanadium center is displaced from the equatorial plane spanned by the Schiff base and one of the alcoholic oxygen functions towards the oxo group. The axial *sec*-butanol in **39** (a complex with four centers of chirality, including vanadium itself) is at a rather long distance (2.328 Å). Complex **39** may thus be considered a transient case towards **40** (Wang et al., 1996) and **41** (Vergopoulos et al., 1993), which are examples of five-coordinated complexes with square-pyramidal coordination geometries. In the case of **41**, extended intermolecular hydrogen bonds lead to a pseudo-polymeric character in the solid state (see Section 4). It is noteworthy that five-coordinate dioxovanadium complexes can attain both the trigonal-bipyramidal (**34a**) and the square-pyramidal (**41**) arrangements. This flexibility is underlined by complex **35,** the geometry of which is in between these two limiting cases (Bashirpoor et al., 1997).

2.2.3. *Functional Models and Peroxo Complexes*

The octahedral Schiff-base complex **42** (Clague et al., 1993) in Figure 10, a close relative to **39** in Figure 9, appears to be a catalyst precursor in the oxidation of bromide by hydrogen peroxide in dimethyl formamide. Using UV-VIS spectrophotometry, [51]V NMR spectroscopy, and kinetic data, a mechanism has been proposed by which a catalytically active LVO(OH) (L is the Schiff base), upon binding peroxide and release of H_3O^+, oxidizes bromide and binds another equivalent of peroxide, generating $[LVO(O_2)]^-$. The oxidized bromine species, finally trapped by an organic substrate, has been shown to

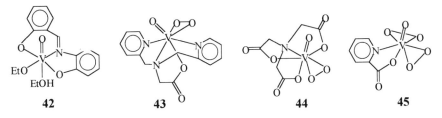

42   **43**   **44**   **45**

Figure 10. The complexes **42** (Clague et al., 1993) and **43** (Colpas et al., 1996) are functional models for vanadate-dependent haloperoxidases. This might also apply to **44** (Wei et al., 1994) and **45** (Shaver et al., 1993). Complex **45,** as many other oxo- and peroxovanadium complexes (and vanadate itself), exhibits insuline-mimetic properties (see, e.g., the chapter by Elberg et al.).

derive from Br⁻ by a two-electron oxidation ("Br⁺"; e.g., HOBr, Br_2, or Br_3^-), excluding a process involving radicals. The exchange of oxide by peroxide is a well-known phenomenon in aqueous vanadium chemistry, and it is challenging to assume that during action of vanadate-dependent peroxidases the axial OH⁻ (Fig. 7) is replaced by HO_2^- or O_2^{2-}. Direct binding of peroxide to the vanadium center in the enzyme has recently been evidenced (Messerschmidt et al., 1997).

Another liable model is the peroxovanadium complex **43** (Colpas et al., 1996) in Figure 10, with vanadium in a trigonal-bipyramidal environment (taking the peroxide as a unidentate, side-on coordinating ligand). In the presence of acid, this complex catalyzes the two-electron oxidation of bromide and iodide by peroxide. The kinetic analysis of the halide oxidation supports a mechanism that is first-order in halide and in protonated **43.** The rate constant for the bromide oxidation amounts to 21 $M^{-1}s^{-1}$. A proton is consumed in each oxidation step and is required for the activation of the peroxo complex. Furthermore, the conditions under which halogenation with **43**/H⁺ occurs have been interpreted in terms of the presence of a hydrophobic active site with a nearby acid/base catalytic site in the vanadium enzyme. Figure 11 illustrates the proposed mechanism for the in vitro halide oxidation and a catalytic cycle for the in vivo action of the enzyme.

Two additional examples, **44** (Wei et al., 1994) and **45** (Shaver et al., 1993) of structurally characterized peroxovanadium complexes are shown in Figure 10. Complex **45,** which contains two peroxo ligands along with picolinate, has been proven to exhibit an insuline-mimetic effect in tests with rats. It has been proposed that this effect is due to the ability of the complex to oxidize cysteine to cystine: The α-subunit of the membrane-bound insuline receptor contains cycteine-rich regions.

2.3.   Amavadin: Structure; Hydroxylamine Complexes; Non-Oxo Complexes

Amavadin (**46** in Fig. 12; Armstrong et al., 1993) is a molecular compound present in the fly agaric toadstool (*Amanita muscaria*) and other mushrooms belonging to the genus *Amanita* (Meisch et al., 1978). It contains vanadium in the oxidation state +4, lacking the otherwise common doubly bonded oxo group. Depending on point of view, the coordination number is six or eight. There are two chelating ligands present, derived from organic hydroxylamide $R_2NO(1-)$, where R is (*S,S*)-α-propionate(1-). Each ligand coordinates through the two carboxylates and the hydroxylamide moiety, the latter in the side-on fashion, a well-known coordination mode for the isoelectronic peroxo (O_2^{2-}) group (see above). The function of amavadin is not known. There are good reasons, however, to assume that it acts as a mediator in electron transfer. Criteria supporting this assumption are the electrochemically reversible oxidation ($V^{IV} \rightarrow V^V$) at +0.49 V (vs. SCE), and the electrocatalytic oxidation of

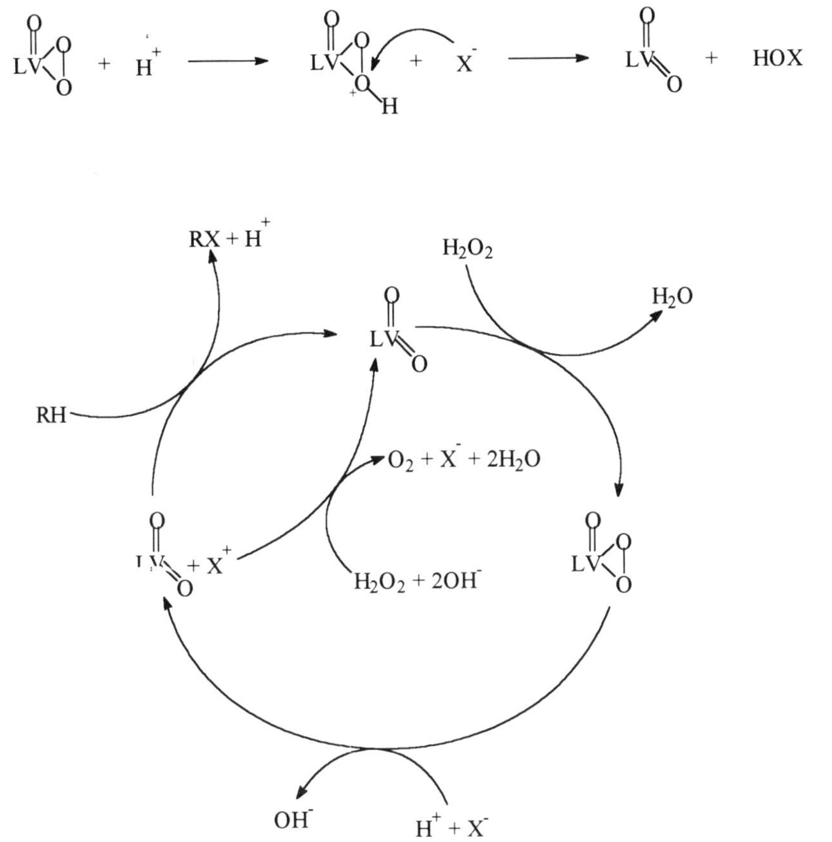

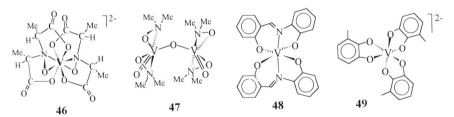

The reaction scheme at the top of the page:

$$LV\overset{O}{\underset{O}{\langle}}O + H^+ \longrightarrow LV\overset{O}{\underset{O}{\langle}}O\cdot O\cdot H + X^- \longrightarrow LV\overset{O}{\underset{O}{}} + HOX$$

Catalytic cycle diagram:

$RX + H^+$

H_2O_2

H_2O

RH

$O_2 + X^- + 2H_2O$

$H_2O_2 + 2OH^-$

OH^- $H^+ + X^-$

$X^+ = (Enz\text{-}X^+, VO\text{-}X^+, HOX, or\ X_3^-)$

Figure 11. The mechanism of halide peroxidation by compound **43** (cf. Fig. 10), top, and in the catalytic cycle involving vanadate-dependent peroxidases, as proposed by Colpas et al. (1996).

46   **47**   **48**   **49**

Figure 12. The non-oxo, hydroxylamido complex amavadin (**46**; Armstrong et al., 1993) has been isolated from *Amanita muscaria*. Complex **47** (Paul et al., 1997), containing coordinated organic hydroxylamide, is a model compound for amavadin. The Schiff-base complex **48** (Hefele et al., 1995) and the catecholate (enterobactin) complex **49** (Karpishin et al., 1993) are examples of the still small group of non-oxo vanadium complexes.

thiols such as cystein and mercaptoacetic acid in the presence of synthetic amavadin and closely related model compounds (da Silva et al., 1996).

A recent study of the system vanadate/dimethylhydroxylamine by Angus-Dunne et al. (1997) has revealed the presence, in solution, of two main mono-nuclear V^V species, containing one or two hydroxylamides, respectively, and possibly derived from $H_2VO_4^-$ by replacement of one or two OH^-/O^{2-} by Me_2NO^-—again in analogy with the formation of peroxovanadates. Various other vanadium complexes containing organic and inorganic hydroxylamides are known (Rehder and Wieghardt, 1981). A homoleptic, dinuclear V^V com-plex with Me_2NO^-, **47** in Figure 12, has recently been isolated by Paul et al. (1996). The lacking oxo group in amavadin has initiated some research into the chemistry of "non-oxo" complexes of vanadium with N,O ligand sets. The VI^V-benzilate complex **13** in Figure 4 is an example. Two additional examples, containing V^{IV}, namely, **48** (Hefele et al., 1995) and **49** (Karpishin et al., 1993), are displayed in Figure 12. The ligand system in **49** is the hexadentate tris(catecholate) enterobactin(6-), a biogenic ionophore that usually acts as a scavenger for Fe^{III}. See also **71** and **72** in Figure 16.

3.   INTERACTION OF VANADIUM WITH BIOGENIC MOLECULES

3.1   Binding to Proteins

3.1.1.   Selected Examples

Apart from the presence of vanadium in the active site of vanadium nitroge-nase (Section 2.1) and vanadate-dependent haloperoxidases (Section 2.2), interaction of vanadates and vanadyl with various proteins has been docu-mented. The vanadium species may be hydrogen-bonded to a protein, as in $H_2VO_4^-$/ribonuclease-T_1 (Kostrewa et al., 1989) and vanadate-uridine/ribonuclease-A (Borah et al., 1985), or bound covalently to functional groups of amino acid side chains of the protein matrix. In any case, the binding of vanadium is of substantial interest in the context of enzyme inhibition or, to a lesser extent, enzyme stimulation (Crans, 1995). The inhibition essentially of phosphate-metabolizing enzymes may involve redox chemistry at the vana-dium site, as in the case of uteroferrin (a phosphatase; Crans et al., 1992) and glyceraldehyde 3-phosphate dehydrogenase (Banabe et al., 1987). In the latter case, a cystein appears to be the redox partner for vanadium, although the presence of cystein at the active site does not necessarily go along with a redox-based inactivation, as has been shown by Crans and Simone (1991) for glycerol 3-phosphate dehydrogenase and by Zhang et al. (1997) for a phosphotyrosyl phosphatase. Thioligation, which should preceed the oxidation of cystein, has been dealt with in Section 2.1.2. In the following treatment, we will emphasize ligation to other amino acid functions. Well-established examples, in which this coordination has been verified, are the binding of *carboxylate* (Asp, Glu) in transferrins (Saponja and Vogel, 1996) and ferritin

(Hanna et al., 1991; Gerfen et al., 1991), *phenolate* (Tyr) in transferrins, *alcoholate* (Ser) in phosphoglucomutase (Percival et al., 1990), *amine* (Lys) in S-adenosylmethionine synthetase (Zhang et al., 1993) and Cu,Zn-superoxide dismutase (Wittenkeller et al., 1991), and *enamine* (His) in transferrins, ferritin, and xylose isomerase (Bogumil et al., 1991). As mentioned in the preceding sections, histidine coordination has also been established for structurally characterized haloperoxidases, and found to be plausible for vanadium nitrogenase. In addition to covalent binding to amino acid side chains, coordination to the functionalities of peptide bonds of the protein backbone may come in. This is suggested by model studies (see below).

3.1.2. Models for Protein Binding: Amino Acid, Amide, and Peptide Complexes

Complexes formed between amino acids and vanadium are restricted to a few examples. A homoleptic V^{IV} complex with cystein, in which the ligand acts in a tridentate mode, is presented in Figure 3 (complex **8**). Histidine coordinates to V^{IV} as a tridentate ligand (**50** in Figure 13; Xiaping and Kangjing,

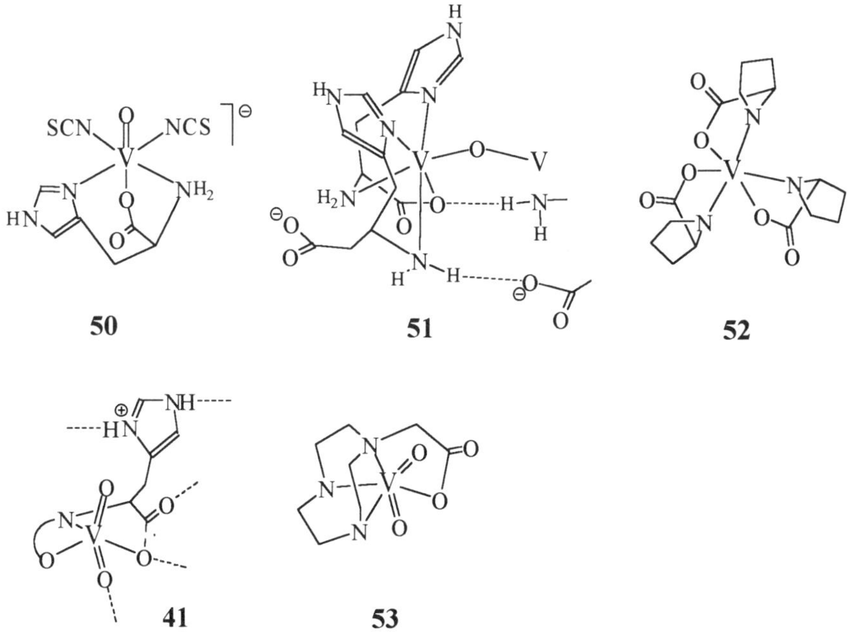

Figure 13. Complexes of vanadium with the amino acids histidine (**50**: Xiaping and Kangjing, 1986; **51**: Czernuszewicz et al., 1994) and proline (**52**: Magill et al., 1993), and two model compounds (**41** and **53**). Complex **41** (Vergopoulos et al., 1993; see also Fig. 9) contains protonated histidine as a constituent of a Schiff base (symbolized by *ON*). Complex **53** represents a vast class of polyfunctional aminocarboxylato complexes of vanadium; the ligand in **53** is 1,4,7-triazacyclononane-*N*-acetate (Schulz et al., 1995). The dashed lines in **51** (only half of the molecule is shown) and **41** indicate hydrogen bonding (see Section 4 for a more detailed discussion).

1986), and as a tri- and bidentate ligand to V^{III} (**51;** Czernuszewicz et al., 1994). Histidine coordination to vanadate in aqueous solution has been established on the basis of ^{51}V NMR spectroscopy and potentiometric studies (Fritzsche et al., 1997; Elvingson et al., 1994). Models for histidine binding, such as the coordination of enamine or of aromatic nitrogen provided by imidazole, pyrazole, and pyridine have been introduced comprehensively in the context of model complexes for vanadate-dependent halo-peroxidases (Section 2.2.2). Complex **52** in Figure 13 (Magill et al., 1993) contains three bidentate prolins coordinated to V^{III}. A manifold of vanadium complexes containing amino acids as an *integral* part of a polydentate ligand system or, alternatively, polyfunctional aminocarboxylates, in which vanadium coordinates through the amino plus the carboxylate function(s), have been reported. Two examples representing these two classes of complexes are included in Figure 13: Complex **41** is a V^V dioxo complex containing a Schiff base with histidine as the amine component and coordination of phenolate-*O,* enamine-*N,* and carboxylate-*O* of the His moiety (Vergopoulos et al., 1993). In **53,** again a dioxo complex, the ligand is a triazanonane-*N*-carboxylate, which coordinates through amine-*N* and carboxylate-*O* (Schulz et al., 1995).

On the basis of EPR investigations carried out by Cornman et al. (1995), histidine and cystein coordination to VO^{2+} has been verified for the interaction between vanadyl and the hexapeptide Gly–Ala–Ser–Cys–His–Val, an active-site peptide from a tyrosine phosphatase involved in dephosphorylation of tyrosine residues of the insulin transmembrane receptor. While, in this hexapeptide, VO^{2+} appears to coordinate to the enamine-*N* of His and the thiolate of Cys, respectively, the coordination mode of vanadate ($H_2VO_4^-$) to the dipeptides Ala-His (Elvingson et al., 1994) and Gly–Tyr (Crans et al., 1995) in aqueous solution, as revealed by ^{51}V NMR, differs in that bicyclic structures are formed via the terminal NH_2 and carboxylate plus the deprotonated amide-*N* of the peptide linkage. The side-chain functional groups are not involved. The necessity of deprotonating the amide group apparently is responsible for the very slow complex formation. The rate constant for the proton-catalyzed pathway of the formation of the vanadate–Gly–Tyr complex amounts to 0.017 $M^{-2}s^{-1}$. Complex formation constants K are relatively small: K = $1.2 \cdot 10^2$ M^{-1} in the case of Gly–Tyr; K = $2.8 \cdot 10^3$ M^{-1} and $3.6 \cdot 10^{10}$ M^{-2} for the two complexes formed with Ala–His. On the other hand, Gly–Ser, in addition to forming a comparable complex, also gives rise to a weak complex where the side-chain alkoxide is involved (Rehder, 1988; Jaswal and Tracey, 1991). Einstein et al. (1996) recently verified the bis(chelate) coordination mode of dipeptides by an X-ray characterization of a Gly–Gly complex of the oxoperoxovanadium unit (**54** in Fig. 14), while Angus-Dunne et al. (1996) established an alternative, unprecedented coordination mode for Gly–Gly in a mono(chelate) fashion, in which the dipeptide uses the terminal amine and the carbonyl-*O* of the peptide group (**55** in Fig. 14). Coordination of deprotonated amide is manifested in compound **56** (Tasiopoulos et al., 1996), which contains the dipeptide-related glycyl derivative of α-mercaptopropionic

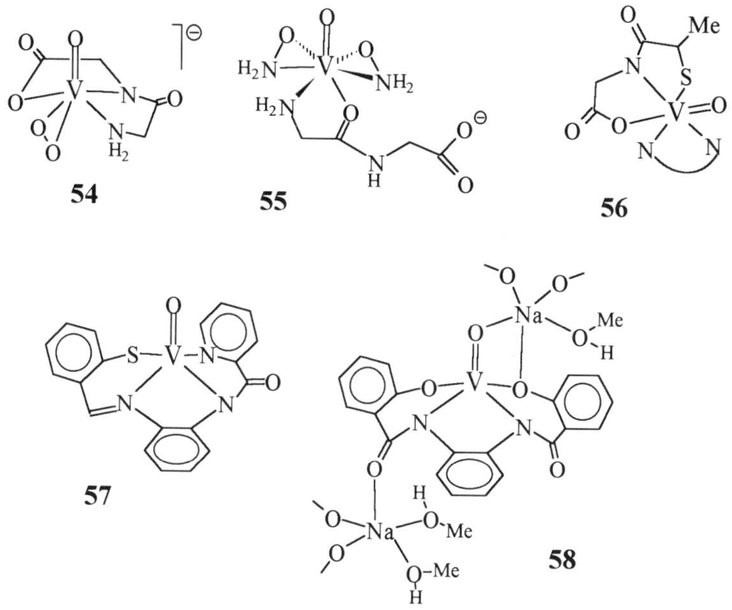

Figure 14. The complexes **54** (Einstein et al., 1996) and **55** (Angus-Dunne et al., 1996) represent two alternative coordination modes for the dipeptide glycylglycine. As in **54** (with V^V) and the V^{IV} complex **56** (Tasiopoulos et al., 1996), containing a dipeptide analog, the deprotonated amide participates in coordination. Complexes **57** and **58** (Keramidas et al., 1996) are amidate complexes that model coordination of deprotonated peptide-N. For the Na^+ coordination in **58** see Section 4.

acid amide. Amide coordination to V^{IV} and V^V has also been modeled with amidate and related ligands. Two examples, **57** and **58** (Keramidas et al., 1996), are displayed in Figure 14.

3.2. Binding to Phosphate, Sugars, Nucleosides, and Nucleotides

Vanadium, in the oxidation states 5, 4, and 3, has a high affinity to phosphates, on the one hand, and to sugar hydroxyls, on the other hand. Under physiological conditions, it may thus bind to phosphates and sugars, to the phosphate and/or sugar moieties of nucleic acids and its building blocks, or, in a more general way, to phosphate residues of substrates and intermediates in enzymatic phosphorylation reactions. Inhibition of phosphatases, ATPases, and ribonucleases as a consequence of vanadate binding is well documented, as is the stimulation of phosphomutases and phosphoisomerases by vanadate and vanadyl: A vanadate–uridine complex, in which the vanadate is coordinated to the 2' and 3' positions of the ribose moiety, has been noted (Borah et al., 1985), as well as the involvement of vanadate–sugar phosphate complexes in the stimulation of, for example, phosphoribose isomerase (Mendz, 1991). Vanadylporphyrine complexes have been shown to form adducts with the adenosine-thymidine domains in DNA (Lin et al., 1993). In many cases, the transition-state analogy between vanadate and phosphate esters is the crucial point when tracing back the activity of vanadate (Crans et al., 1996). NADV, a vanadate analog of NADP, has been recognized as a "better" cofactor for dehydrogenases (Crans et al., 1993). The importance of the vanadium–phosphate system in the in vitro catalytical oxidation of, for example, butane to maleic acid anhydride has been reported (Wroblewski, 1988; Thorn et al., 1996), and once more demonstrates the sometimes close relations between in vitro and in vivo processes catalyzed by similar catalyst systems.

The interaction between vanadate or vanadyl and phosphates, sugars, or nucleosides/nucleotides in solution, investigated during the last decade by spectroscopic and electrochemical methods, has been traced back to the formation of mixed vanadate–phosphate anhydrides in the case of phosphates, or predominantly cyclic vanadate esters in the case of sugars and nucleosides. For details the reader should consult the chapter in this book by D. C. Crans. Binding of the vanadyl group to the phosphates of guanosine-5′-monophosphate (GMP) in a complex of composition $[VO(H_2O)(GMP)_2]$ (**65** in Fig. 15) has also been noted (Jiang and Makinen, 1995).

A recent study of the system vanadyl + di/triphosphate, combining several methods of investigation, revealed the formation of a trinuclear compound of composition $\{VO(diphosphate)\}_3$ between pH 3 and 6, and mononuclear compounds of composition $VO(di/triphosphate)_2$ at physiological pH (Buglyó et al., 1995). The basic structural element in binary $V^{IV/V}$–phosphate systems is represented by **59** in Figure 15. Complex **59** shows a section from the polymer $\{VO(H_2O)_2(HPO_4) \cdot 2H_2\}_\infty$ (Wroblewski, 1988). The highly charged, dinuclear V^V peroxo complex **60** (Schwendt et al., 1995) in Figure 15 is a rare example of a structurally characterized molecular phosphate complex. As in **59,** the complex **60** lacks auxiliary ligands, the phosphate is in the bridging mode, and a vanadyl fragment is present, while two of the oxo functions are replaced by peroxide. The structures **61** (Otieno et al., 1996), **62** (Thorn et al., 1996), and **63** (Dean et al., 1996) in Figure 15 all contain a supporting ligand system in addition to a bridging phosphate monoester (**61**) or phosphate diester (**62** and **63**). The supporting ligands are dipicolinate(2-) in the V^V complex **62,** and {tris(pyrazolylborate)}(1-) in the V^{III} complexes **61** and **63.** The methylphosphonato–V^V complex **64** (Harrison et al., 1996) is an example of a larger group of phosphonate complexes of vanadium, which are related to the phosphate complexes and may have some relevance for the (so far not yet established) interaction between vanadate and biogenic *organo*phosphorus compounds.

The cyclic 2,3-coordination of sugars (2′,3′-coordination of nucleosides) to vanadate has recently been verified by the solid-state structures of the complexes formed between $H_2VO_4^-$ and a derivative of α-D-mannopyranose (see **66** in Fig. 15; Zhang et al., 1996), on the one hand, and adenosine, on the other hand (**67;** Angus-Dunne at al., 1995). In either case, a dinuclear biligate complex is formed (a composition that had already been deduced from [51]V

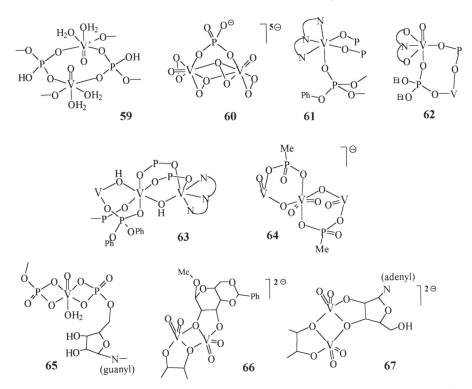

Figure 15. Complexes modeling the interaction of vanadium with phosphate (**59**: Wroblewski, 1988; **60**: Schwendt et al., 1995), phosphate esters (**61**: Otieno et al., 1996; **62**: Thorn et al., 1996; **63**: Dean et al., 1996), phosphonate (**64**: Harrison et al., 1996) and sugar derivatives (**65**: Jiang and Makinen, 1995; **66**: Zhang et al., 1996; **67**: Angus-Dunne et al., 1995). Complex **59** is a section from a polymeric structure. In the case of **61–67**, only those parts of the complex molecules are presented, that are essential for a comprehensive view of the connectivities around the vanadium center(s). See text for further discussion.

NMR analyses in solution by Tracey et al., 1990), and the sugar coordinates in its bis(2',3'-deprotonated) form, giving rise to five-membered chelate rings. One of the alkoxo groups of each of the two sugar ligands bridges the two dioxovanadium(V) centers, hence generating a $V_2(\mu\text{-}O)_2$ core.

3.3. Vanadium in Sea Squirts and Fan Worms; Catecholate Coordination

Vanadium is found in several sea squirts (ascidians, a suborder of the tunicates) belonging to the branches *Aplousobranchia* and *Phlebobranchia* (Brand et al, 1989; Smith, 1989; see also Michibata and Kanamori, this volume; Ch 10), and in the *Polychaeta* fan worm *Pseudopotamilla occelata* (Ishii et al., 1993; Ishii et al., 1994; see also Ishii, this volume, Ch 9) Ascidians may take up vanadate from sea water and accumulate it, after reduction or concomitantly with reduction, in specialized blood cells (vanadocytes) up to a factor of 10^7, or 0.1 M. *Aplousobranchia* contain about 5% of the vanadium contents in the form of EPR-active V^{IV} (VO^{2+}), the remaining part being weakly ferromagnetically coupled ($J = 3.5$ cm^{-1}) (V^{III})$_2\mu$–O (Kustin et al., 1996). The dominant form by far in *Phlebobranchia* and *Pseudopotamilla* is EPR-silent V^{III}, possibly present as the hexaqua- or pentaquahydroxovanadium cation (Tullius et al., 1980; Ishii et al, 1994), implying a rather acidic medium or matrix stabilization. In fan worms, vanadium has been localized in apical vacuoles of epidermal cells and in the cuticle, and its function has been related to the absorption or storage of oxygen. In the case of sea squirts, it has been suggested that vanadium, interacts with tunichromes. These are oligopeptides based on dopa and hydroxydopa (**68** in Fig. 16), and hence on molecules containing catecholate moieties with a high reducing capability. Tunichromes, which are building blocks of the tunic of tunicates, have also been localized in the vanadocytes of sea squirts (Parry et al., 1992). Since vanadium in the oxidation states 4, 5, and 3 tightly binds to catecholates, it has been suggested that this metal ion acts as a template in the synthesis of condensed tunichromes. The complexes **69–72** in Figure 16 illustrate the coordination of vanadium to catecholates. The complexes contain one (**69:** Cornman et al., 1992c), two (**70:** Dewey et al., 1993; **72:** Kabanos et al., 1922), or three (**71:** Cooper et al., 1982) catecholate ligands, with the metal center in the oxidation states 5 (**69**), 4 (**70**), and 3 (**71**). The non-oxo catecholate complex **72** exists in the V^V and the V^{IV} form.

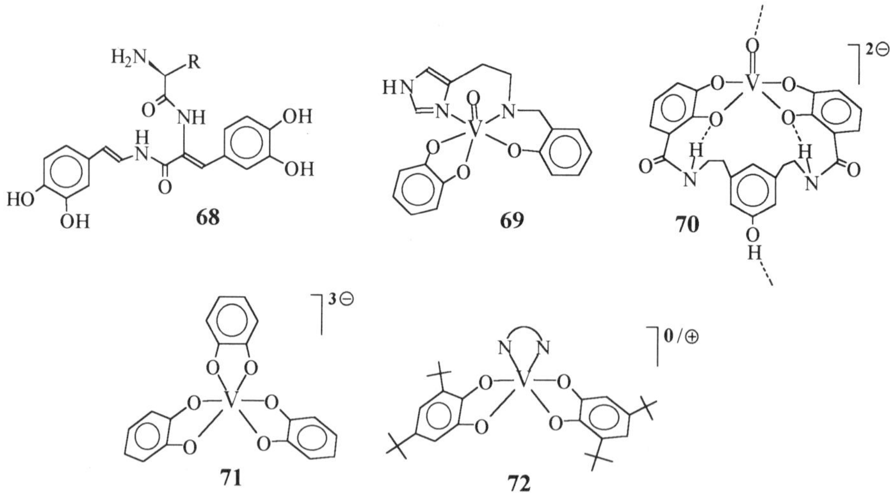

Figure 16. Molecule **68** (Parry et al., 1992; Taylor et al., 1997) is the basic structural unit of the tunichromes found in the morula-type blood cells of ascidians. Complexes **69–72** model catecholate binding to vanadium(V, IV, and III). The dashed lines in **70** (Dewey et al., 1993) indicate intra- and intermolecular hydrogen bonds (see Section 4, below).

4. VANADIUM–ORGANIC NETWORKS BY HYDROGEN BONDING AND CONTACT ION PAIRING

Apart from covalent bonding of vanadium to O, N, and S functions of biogenic molecules, vanadium species may also be associated to a matrix through hydrogen bonds and/or salt bridges, including alkali metal (Na^+, K^+) contact ion-pair interaction—either exclusively or together with covalent linkages. Examples of extensive hydrogen bonding are the intimate binding of $H_2VO_4^-$ to ribonuclease T_1 from the mold *Aspergillus oryzae* (Kostrewa et al., 1989), and the covalent (to histidine) plus hydrogen bonding of vanadate in the active center of vanadate-dependent haloperoxidase from the fungus *Curvularia inaequalis* (Fig. 7 in Section 2.2.1) and the sea weed *Ascophyllum nodosum*. In addition to external H bonds, *intra*molecular hydrogen bonding within a complex vanadium compound often plays a role in pre-determining structural patterns. The complexes **21** (Murray et al., 1993) in Figure 5, **51** (Czernuszewicz et al., 1994) in Figure 13, and **70** (Dewey et al., 1993) in Figure 16 demonstrate this kind of interaction in the ligand periphery for N–H. . . .Cl (**21**) and N–H. . . .O bonds, where the oxygen is a constituent of a carboxylate (the histidine complex **51**) or phenoxide (the catecholate complex **70**). Complex **70** additionally exerts strong *inter*molecular hydrogen bond interaction between phenolic OH and the oxo group of vanadyl, giving rise to infinite chains with relatively short interatomic distances d(O. . . .O) = 2.61 Å. An intermolecular hydrogen bonding network involving vanadyl-*O*, carboxylate, and the two nitrogens of protonated histidine, although weaker [d(N. . . .O) = 2.70–2.9 Å] than in the case of **70,** has been established for the Schiff-base complex **41** (Figs. 13 and 17; Vergopoulos et al., 1993). An example, finally, for hydrogen bonding involving *thio*phenolate is the complex **73** (Tsagkalidis and Rehder, 1996) in Figure 17. Apart from the intramolecular N–H. . . .S bond [d(N. . . .S) = 3.42 Å], there is weak external hydrogen bond between the second coordinated amino group and a THF of solvation [d(N. . . .O) = 3.12 Å].

Intermolecular Na^+ (or Li^+) ion-pair interaction has been found to be essential for the stability of the $[V(N_2)_2(dmpe)_2]^-$ anion (**14** in Figs. 5 and 17), a complex that has been introduced as a functional model for vanadium nitrogenase (Gailus et al. 1994a; see Section 2.1.3). The bond lengths d(N. . . .Na) are 2.445 Å and hence about normal for covalent bonds. In the dinuclear, mixed-valent (V^{IV}/V^V) anionic Schiff-base complex **74** (Pessoa et al., 1992) shown in Figure 17, sodium ions surrounded by seven oxygens (including two water molecules) in a distorted pentagonal-bipyramidal array are linked to each other and to either half of the dinuclear anion via the carboxylates of the serine moieties of the ligand system in such a way that the dimers by themselves are interlinked. The distances d(Na. . . .O) amount to 2.36–2.61 Å. In addition, there are intermolecular hydrogen bonds between two H_2O molecules coordinated to two different Na^+. An interesting anionic V^V network interlinked by cationic Na^+ chains has also been detected in

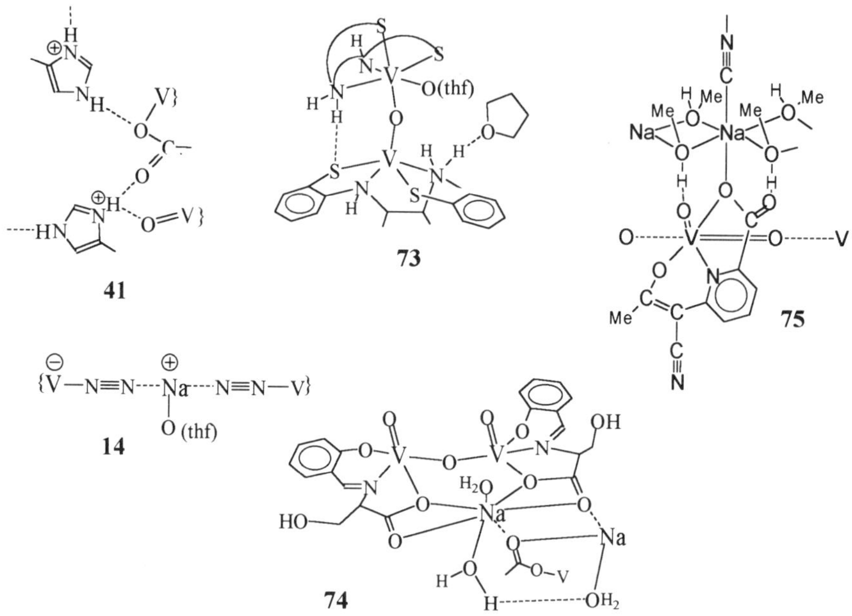

Figure 17. **41** (Vergopoulos et al., 1993) and **73** (Tsagkalidis and Rehder, 1996) are examples of intermolecular hydrogen bonding interaction, and **14** (Gailus et al., 1994a), **74** (Pessoa et al., 1992), and **75** (Julien-Cailhol et al., 1996) are examples of intermolecular-contact ion-pair interaction. For further examples, see compounds referred to in the text.

compound **75** (Julien-Cailhol et al., 1996): Here, the sodium ions are in an octahedral array, bridged by two methanol molecules and further coordinated to carboxylate-O and nitrile-N of the multifunctional ligand. Bond distances [d(N. . . .O) = 2.42–2.56 Å, d(Na. . . .N) = 2.440 Å] indicate normal covalent bonds. Again, the network is further stabilized by hydrogen bonds.

REFERENCES

Angus-Dunne, S. J., Batchelor, R. J., Tracey, A. S., and Einstein, F. W. B. (1995). The crystal and solution structures of the major products of the reaction of vanadate with adenosine. *J. Am. Chem. Soc.* **117**, 5292–5296.

Angus-Dunne, S. J., Paul, P. C., Batchelor, R. J., Einstein, F. W. B., and Tracey, A. S. (1997). Reactions of hydroxylaminovanadate with peptides: Aqueous equilibria and crystal structure of oxobis(hydroxylamino)glycylglycinatovanadium(V). *Can. J. Chem.* **75**, 183–191.

Angus-Dunne, S. J., Paul, P. C., and Tracey, A. S. (1997). A ^{51}V NMR investigation of the interaction of aqueous vanadate with hydroxylamine. *Can. J. Chem.* **75**, 1002–1010.

Arber, J. M., Dobson, B. R., Eady, R. R., Hasnain, S. S., Garner, C. D., Matsushita, T., Nomuras, M., and Smith, B. E. (1989a). Vanadium K-edge X-ray-absorption spectroscopy of the functioning and thionine-oxidized forms of the VFe-protein of the vanadium nitrogenase from *Azotobacter chroococcum*. *Biochem. J.* **258**, 733–737.

Arber, J. M., de Boer, E., Garner, C. D., Hasnain, S. S., and Wever, R. (1989b). Vanadium K-edge X-ray absorption spectroscopy of bromoperoxidase from *Ascophyllum nodosum*. *Biochemistry* **28**, 7968–7973.

Armstrong, E. M., Beddoes, R. L., Calviou, L. J., Charnock, J. M., Collison, D., Ertok, N., Naismith, J. H., and Garner, C. D. (1993). The chemical nature of amavadin. *J. Am. Chem. Soc.* **115**, 807–808.

Asgedom, G., Sveedhara, A., Kivikoski, J., Kolehmainen, E., and Rao, C. P. (1996). Structure, characterization and photoreactivity of monomeric dioxovanadium(V) Schiff-base complexes of trigonal-bipyramidal geometry. *J. Chem. Soc. Dalton Trans.*, pp 93–97.

Banabe, J. E., Echegoyen, L. A., Pastrona, B., and Martinez-Maldonado, M. (1987). Mechanism of inhibition of glycolysis by vanadate. *J. Biol. Chem.* **262**, 9555–9560.

Barr-David, G., Hambley, T. W., Irwin, J. A., Judd, R. J., Lay, P. A., Martin, B. D., Bramley, R., Dixon, N. E., Hendry, P., Ji, J.-Y., Baker, R. S. U., and Bonin, A. M. (1992). Suppression by vanadium(IV) of chromium(V)-mediated DNA cleavage and chromium(VI/V)-induced mutagenesis, synthesis and crystal structure of the vanadium(IV) complex (NH$_4$)[VO-(HOCEt$_2$COO)(OCEt$_2$COO)]. *Inorg. Chem.* **31**, 4906–4908.

Bashirpoor, M., Schmidt, H., Schulzke, C., and Rehder, D. (1997). Models for vanadate-dependent haloperoxidases: Vanadium complexes with O$_4$N-donor sets. *Chem. Ber./Receuil* **130**, 651–657.

Beaton, W. I., and Bertolacini, R. J. (1991). Resid hydroprocessing at Amoco. *Catal. Rev. Sci Eng.* **33**, 281–317.

Blanchard, C. Z., and Hales, B. J. (1996). Isolation of two forms of the nitrogenase VFe protein from *Azotobacter vinelandii*. *Biochemistry* **35**, 472–478.

Bogumil, R., Hüttermann, J., Kappl, R., Stabler, R., Sudfeldt, C., and Witzel, H. (1991). Visible, EPR and electron nuclear double-resonance spectroscopic studies on the two metal-binding sites of oxovanadium(IV)-substituted D-xylose isomerase. *Eur. J. Biochem.* **196**, 305–312.

Borah, B., Chen, C.-W., Egan, W., Miller, M., Wlodawer, A., and Cohen, J. S. (1985). Nuclear magnetic resonance and neutron diffraction studies of the complex of ribonuclease A with uridine vanadate, a transition state analogue. *Biochemistry* **24**, 2058–2067.

Brand, S. G., Hawkins, C. J., Marshall, A. T., Nette, G. W., and Parry, D. L. (1989). Vanadium chemistry of ascidians. *Comp. Biochem. Physiol. B.* **93**, 425–436.

Buglyó, P., Kiss, T., Alberico, E., Micera, G., and Dewaele, D. (1995). Oxovanadium(IV) complexes of di- and triphosphate. *J. Coord. Chem.* **36**, 105–116.

Bultitute, J., Larkeorthy, L. F., Povey, D. C., Smith, G. W., Dilworth, J. R., and Leigh, G. J. (1986). The crystal and molecular structure of bis(1-methyl-1-phenylhydrazine)dichloro-{1-methyl-1-phenylhydraz ido(2-)}vanadium(V) chloride, a complex containing two side-on-coordinated hydrazine molecules. *J. Chem. Soc. Chem. Commun.*, pp. 1748–1750.

Butler, A. (1993). Vanadium bromoperoxidase. In J. Reedijk (Ed.), *Bioinorganic Catalysis*. Marcel Dekker, New York, pp. 425–445.

Butler, A., and Walker, J. V. (1993). Marine haloperoxidases. *Chem. Rev.* **93**, 1937–1944.

Calviou, L. J., Arber, J. M., Collison, D., Garner, C. D., and Clegg, W. (1992). A structural model for vanadyl–histidine interactions: Structure determination of [VO(1-vinylimidazole)$_4$Cl]Cl by a combination of X-ray crystallography and X-ray absorption spectroscopy. *J. Chem. Soc. Chem. Commun.*, pp. 654–656.

Carrano, C. J., Mohan, M., Holmes, S. M., de la Rosa, R., Butler, A., Charnock, J. M., and Garner, C. D. (1994). Oxovanadium(V) alkoxo-chloro complexes of the hydridotrispyrazolylborates as models for the binding site in bromoperoxidase. *Inorg. Chem.* **33**, 646–655.

Cavaco, J., Pessoa, J. C., Costa, D., Duarte, M. T., Gillard, R. D., and Matias, P. (1994). N-Salicylideneamino amidate complexes of oxovanadium(IV). Part 1. Crystal and molecular structures and spectroscopic properties. *J. Chem. Soc. Dalton Trans.*, pp. 149–157.

Cavaco, J., Pessoa, J. C., Duarte, M. T., Gillard, R. D., and Matias, P. (1996). Molecular structure of [VO(sal-D,L-Asn)(py)(H$_2$O)] and reaction to produce coumarin-3-carboxamide. *Chem. Commun.*, pp. 1365–1366.

Cen, W., MacDonnell, F. M., Scott, M. J., and Holm, R. H. (1994). Heterometal clusters containing the cuboidal Fe_4S_3 fragment: Synthesis, electron distribution, and reactions. *Inorg. Chem.* **33**, 5809–5818.

Chan, M. K., Kim, J., and Rees, D. C. (1993). The nitrogenase FeMo-cofactor and P-cluster pair: 2.2 Å resolution structures. *Science* **260**, 792–794.

Chen, J., Christiansen, J., Tittsworth, R. C., Hales, B. J., George, S. J., Coucouvanis, D., and Cramer, S. P. (1993). Iron EXAFS of *Azotobacter vinelandii* nitrogenase Mo-Fe and V-Fe proteins. *J. Am. Chem. Soc.* **115**, 5509–5515.

Clague, M. J., Keder, N. L., and Butler, A. (1993) Biomimics of vanadium bromoperoxidase: vanadium(V)- Schiff base catalyzed oxidation of bromide by hydrogen peroxide. *Inorg. Chem.* **32**, 4754–4761.

Colpas, G. J., Hamstra, B. J., Kampf, J. W., and Pecoraro, V. L. (1996). Functional models for vanadium haloperoxidase: Reactivity and mechanism of halide oxidation. *J. Am. Chem. Soc.* **118**, 3469–3478.

Cooper, S. R., Koh, Y. B., and Raymond, K. N. (1982). Synthetic, structural, and physical studies of bis(triethylammonium) tris(catecholato)vanadate(IV), potassium bis(catecholato)oxo-vanadate(IV), and potassium tris(catecholato)vanadate(III). *J. Am. Chem. Soc.* **104**, 5092–5102.

Cornman, C. R., Kampf, J., Lah, M. S., and Pecoraro, V. L. (1992a) Modelling vanadium bromoperoxidase: Synthesis, structure, and spectral properties of vanadium(IV) complexes with coordinated imidazole. *Inorg. Chem.* **31**, 2035–2043.

Cornman, C. R., Kampf, J., and Pecoraro, V. L. (1992b). Structural and spectroscopic characterization of V^VO-imidazole complexes. *Inorg. Chem.* **31**, 1981–1983.

Cornman, C. R., Colpas, G. J., Hoeschele, J. D., Kampf, J., and Pecoraro, V. L. (1992c). Implications for the spectroscopic assignment of vanadium biomolecules: Structural and spectroscopic characterization of monooxovanadium(V) complexes containing catecholate and hydroximate based noninnocent ligands. *J. Am. Chem. Soc.* **114**, 9925–9933.

Cornman, C. R., Zovinka, E. P., and Meixner, M. H. (1995). Vanadium(IV) complexes of an active-site peptide of a protein tyrosine phosphatase. *Inorg. Chem.* **34**, 5009–5100.

Crans, D. C. (1995). Vanadate–protein interaction. In G. Berthon (Ed.), *Handbook of Metal–Ligand Interactions in Biological Fluids (Bioinorganic Chemistry)*. Marcel Dekker, New York, pp. 267–273.

Crans, D. C., Holst, H., Keramidas, A., and Rehder, D. (1995). A slow exchanging vanadium(V) peptide complex: Vanadium(V)–glycine–tyrosine. *Inorg. Chem.* **34**, 2524–2534.

Crans, D. C., Keramidas, A. D., and Drouza, C. (1996). Organic vanadium compounds—Transition state analogy with organic phosphorus compounds. *Phos. Sulf. Silic.* **109–110**, 245–248.

Crans, D. C., Marshman, A. W., Nielsen, R., and Felty, I. (1993). NADV: A new cofactor for alcohol dehydrogenase from *Thermoanaerobium brockii*. *J. Org. Chem.* **58**, 2244–2252.

Crans, D. C., and Simone, C. M. (1991). Nonreductive interaction of vanadate with an enzyme containing a thiol group in the active site: Glycerol-3-phosphate dehydrogenase. *Biochemistry* **30**, 6734–6741.

Crans, D. C., Simone, C. M., Holz, R. C., and Que, L. Jr. (1992). Interaction of porcine uterine fluid purple acid phosphatase with vanadate and vanadyl cation. *Biochemistry* **31**, 11731–11739.

Czernuszewicz, R. S., Yan, Q., Bond, M. R., and Carrano, C. J. (1994). Origin of the unusual bending distortion in the (μ-oxo)divanadium(III) complex [$V_2O(l$-his)$_4$]: A reinvestigation. *Inorg. Chem.* **33**, 6116–6119.

da Silva, M. F. C. G., da Silva, J. A. L., da Silva, J. J. R. F., Pombeiro, A. J. L., Amatore, C., and Verpeaux, J.-N. (1996). Evidence for a Michaelis–Menten type mechanism in the electrocatalytic oxidation of mercaptopropionic acid by an amavadin model. *J. Am. Chem. Soc.* **118**, 7568–7573.

de Boer, E., Keijzers, C. P., Klaasen, A. A. K., Reijerse, E. J., Collison, D., Garner, C. D., and Wever, R. (1988a). ^{14}N-Coordination to VO^{2+} in reduced vanadium bromoperoxidase, an electron spin echo study. *FEBS Lett.* **235**, 93–97.

de Boer, E., Boon, K., and Wever, R. (1988b). Electron paramagnetic resonance studies on conformational states and metal ion exchange properties of vanadium bromoperoxidase. *Biochemistry* **27**, 1629–1635.

Dean, N. S., Mokry, L. M., Bond, M. R., O'Connor, C. J., and Carrano, C. J. (1996). Vanadium hydridotris(pyrazolyl)borate complexes of diphenylphosphate. Homometallic trimeric complexes of the $[LV\{(PhO)_2PO_2\}_3]^-$ fragment. *Inorg. Chem.* **35**, 3541–3547.

Deeth, R. J., and Langford, S. A. (1995). Metal–dinitrogen σ- and π-bonding roles in [Fe-(dmpe)$_2$H(N$_2$)$_2$]$^-$ and *trans*-[V(dmpe)$_2$(N$_2$)$_2$]$^-$. *J. Chem. Soc. Dalton Trans.*, pp. 1–4.

Dewey, T. M., du Bois, J., and Raymond, K. N. (1993). Ligands for oxovanadium(IV): bis(catecholamide) coordination and intermolecular hydrogen bonding to the oxo atom. *Inorg. Chem.* **32**, 1729–1738.

Dikanov, S. A., Burgard, C., and Hüttermann, J. (1993). Determination of the hyperfine coupling with the remote nitrogen in the VO^{2+}-(imidazole)$_4$ complex by ESEEM spectroscopy. *Chem. Phys. Lett.* **212**, 493–498.

Dilworth, M. J., Eady, R. R., and Eldridge, M. E. (1988). The vanadium nitrogenase of *Azotobacter chroococcum*. Reduction of acetylene and ethylene to ethane. *Biochem. J.* **249**, 745–751.

Djordjevic, C., Lee, M., and Sinn, E. (1989). Oxoperoxo(citrato)- and dioxo(citrato)vanadates(V): Synthesis, spectra, and structure of a hydroxyl oxygen bridged dimer, K$_2$[VO(O$_2$)(C$_6$H$_6$O$_7$)]$_2$2H$_2$O. *Inorg. Chem.* **28**, 719–723.

Eady, R. R. (1989). The vanadium nitrogenase of *Azotobacter*. *Polyhedron* **8**, 1695–1700.

Eady, R. R. (1995). Vanadium nitrogenases of *Azotobacter*. In H. Sigel and A. Sigel (Eds.), *Vanadium and Its Role in Life. Metal Ions in Biological Systems*, Vol. 31. Marcel Dekker, New York, pp. 363–405.

Eady, R. R., and Leigh, G. J. (1994). Metals in the nitrogenases. *J. Chem. Soc. Dalton Trans.*, pp. 2739–2747.

Einstein, F. W. B., Batchelor, R. J., Angus-Dunne, S. J., and Tracey, A. S. (1996). A product formed from glycylglycine in the presence of vanadate and hydrogen peroxide: The (glycyl-*N*-hydroglycinato-$\kappa^3 N^2, N^N, O^1$)oxoperoxovanadate(V) anion. *Inorg. Chem.* **35**, 1680–1684.

Eisenberg, R., and Gray, H. B. (1967). Trigonal-prismatic coordination: The crystal and molecular structure of tris(*cis*-1,2-diphenylethene-1,2-dithiolato)vanadium. *Inorg. Chem.* **6**, 1844–1849.

Elvingson, K., Baró, A. G., and Pettersson, L. (1996). Speciation in vanadium bioinorganic systems. 2. An NMR, ESR and potentiometric study of the aqueous H$^+$-vanadate-maltol system. *Inorg. Chem.* **35**, 3388–3393.

Elvingson, K., Fritzsche, M., Rehder, D., and Pettersson, L (1994). A potentiometric and ^{51}V NMR study of aqueous equilibria in the H$^+$-vanadate(V)-L-α-alanyl-L-histidine system. *Acta Chem. Scand.* **48**, 878–885.

Everett, R. R., Soedjak, H. S., and Butler, A. (1990). Mechanism of dioxygen formation catalyzed by vanadium bromoperoxidase. *J. Biol. Chem.* **265**, 15671–15679.

Ferguson, V. R., Solari, E., Floriani, C.; Chiesi-Villa, A., and Rizzoli, C. (1993). Fixierung und Reduktion von N$_2$ durch V^{II} und V^{III}: Synthese und Struktur von Mesityl(distickstoff) vanadium-Komplexen. *Angew. Chem.* **105**, 453–455.

Fritzsche, M., Elvingson, K., Rehder, D., and Pettersson, L. (1997). A potentiometric and ^{51}V, ^{13}C and ^{1}H NMR study of the aqueous H$^+$-vanadate(V)-L-prolyl-L-alanine/L-alanyl-glycine systems. *Acta Chem. Scand.* **51**, 483–491.

Fulwood, R., Schmidt, H., and Rehder, D. (1995). [VO{N-(2-oxido-1-naphtylmethylene)-L-ala}O-Bu^s(HOBu^s)]: Characterization of a complex containing four centres of chirality. *J. Chem. Soc. Chem. Commun.*, pp. 1443–1444.

Gailus, H., Woitha, C., and Rehder, D. (1994a). Dinitrogenvanadates(-I): Synthesis, reactions and conditions for their stability. *J. Chem. Soc. Dalton Trans.*, pp. 3471–3477.

Gailus, H., Maelger, H., and Rehder, D. (1994b). A Route to η^2-alkyne complexes of vanadium(I): Crystal structure of *trans*-[IV(CO)$_2$(PMe$_2$Ph)$_2$(2-butyne)]. *J. Organomet. Chem.* **465**, 181–185.

Gerfen, G. J., Hanna, P. M., Chasteen, N. D., and Singel, D. J. (1991). Characterization of the ligand environment of vanadyl complexes of apoferritin by multifrequency electron spin-echo envelope modulation. *J. Am. Chem. Soc.* **113**, 9513–9519.

Greiser, T., Puttfarcken, U., and Rehder, D. (1979). The molecular structure of tetracarbonylhydrido(1,1,4,4-tetraphenyl-1,4-diphosphabutane)vanadium(+I). Trans. Met. Chem. **4**, 168–171.

Hambley, T. W., Judd, R. J., and Lay, P. A. (1992). Synthesis and structure of a vanadium(V) complex with a 2-hydroxy acid ligand, (NH$_4$)$_2$[V{OC(CH$_2$CH$_3$)$_2$COO}(O)$_2$]$_2$: A structural model of both vanadium(V) transferrin and ribonuclease complexes with inhibitors. *Inorg. Chem.* **31**, 343–345.

Hanna, P. M., Chasteen, N. D., Rottmann, G. A., and Aisen, P. (1991). Iron binding to horse spleen apoferritin: A vanadyl ENDOR spin probe study. *Biochemistry* **30**, 9210–9216.

Hao, S., Berno, P., Minhas, R. K., and Gambarotta, S. (1995). The role of ligand steric hindrance in determining the stability of very short V–V contacts. Preparation and characterization of a series of V(II) and V(III) amidinates. *Inorg. Chim. Acta* **244**, 37–49.

Harrison, W. T. A., Dussack, L. L., and Jacobson, A. J. (1996). Sysnthesis and properties of new layered alkali–metal/ammonium vanadium(V) methylphosphonates. *Inorg. Chem.* **35**, 1461–1467.

Hefele, H., Ludwig, E., Uhlemann, E., and Weller, F. (1995). Struktur von bis[salicylaldehyd-2-hydroxyanilato(2-)]vanadium(IV). *Z. Anorg. Allgem. Chem.* **621**, 1973–1976.

Hidai, M., and Mizobe, Y. (1995). Recent advances in the chemistry of dinitrogen complexes. *Chem. Rev.* **95**, 1115–1133.

Holmes, S., and Carrano, C. J. (1991). Models for the binding site in bromoperoxidase: Mononuclear vanadium(V) phenolate complexes of the hydridotris(3,5-dimethylpyrazolyl)borate ligand. *Inorg. Chem.* **30**, 1231–1235.

Hormers, J., Kuetgens, U., Chauvistre, R., Schreiber, W., Anders, N., Vilter, H., Rehder, D., and Weidemann, C. (1988). Vanadium K-edge absorption spectrum of bromoperoxidase from *Ascophyllum nodosum*. *Biochim. Biophys. Acta* **956**, 293–299.

Ishii, T., Nakai, I., Numako, C., Okoshi, K., and Otake, T. (1993). Discovery of a new vanadium accumulator, the fan worm *Pseudopotamilla occelata*. *Naturwissenschaften* **80**, 268–270.

Ishii, T., Otake, T., Okoshi, K., Nakahara, M., and Nakamura, R. (1994) Intracellular localization of vanadium in the fan worm *Pseudopotamilla occelata*. *Mar. Biol.* **121**, 143–151.

Jaswal, J. S., and Tracey, A. S. (1991). Stereochemical requirements for the formation of vanadate complexes with peptides. *Can. J. Chem.* **69**, 1600–1607.

Jiang, F. S., and Makinen, M. W. (1995). NMR and ENDOR conformational studies of the vanadyl guanosine 5'-monophosphate complex in hydrogen-bonded quartet assemblies. *Inorg. Chem.* **34**, 1736–1744.

Julien-Cailhol, N., Rose, E., Vaisserman, J., and Rehder, D. (1996). An unusual anionic oxo-(μ-oxo)-vanadium(V) network interlinked by cationic sodium chains. *J. Chem. Soc. Dalton Trans.*, pp. 2111–2115.

Kabanos, T. A., White, A. J. P., Williams , D. J., and Woolins, J. D. (1992). Synthesis and X-ray structure of bis(3,5-di-*tert*-butylcatecholato)(phenanthroline)vanadium(IV) and its vanadium(V) analogue [V(dtbc)$_2$(phen)][Sbf$_6$]. *J. Chem. Soc. Chem. Commun.*, pp. 17–18.

Karpishin, T. B., Dewey, T. M., and Raymond, K. M. (1993). The vanadium(IV) enterobactin complex: Structural, spectroscopic, and electrochemical characterization. *J. Am. Chem. Soc.* **115**, 1842–1851.

Kelm, H., and Krüger, H.-J. (1996). Synthesis structure, and spectroscopic characterization of a series of *N,N'*-dimethyl-2,11-diaza[3.3](2,6)pyridinophane vanadium(III), -(IV), and -(V) complexes. *Inorg. Chem.* **35**, 3533–3540.

Keramidas, A. D., Papaioannou, A. B., Vlahos, A., Kabanos, T. A., Bonas, G., Makriyannis, A., Rapropoulos, C. P., and Terzis, A. (1996). Model investigations for vanadium–protein interaction. Synthetic, structural, and physical studies of vanadium(III) and oxovanadium(IV/V) complexes with amidate ligands. *Inorg. Chem.* **35**, 357–367.

Kime-Hunt, E., Spartalian, K., DeRusha, M., Nunn, C. M., and Carrano, C. J. (1989). Synthesis, characterization, and molecular structures of a series of [(3,5-dimethyl-pyrazolyl)borato]vanadium(III) and -(IV) complexes. *Inorg. Chem.* **28**, 4392–4399.

Kondo, M., Minakoshi, S., Iwata, K., Shimizu, T., Matsuzaka, H., Kamigata, N., and Kitagawa, S. (1996). Crystal structure of a tris(dithiolene) vanadium(IV) complex having unprecedented D_{3h} symmetry. *Chem. Lett.*, pp. 489–490.

Kostrewa, D., Choe, H. W., Heinemann, U., and Saenger, W. (1989). Crystal structure of guanosine-free ribonuclease-T_1, complexed with vanadate(V), suggests conformational change upon substrate binding. *Biochemistry* **28**, 7592–7600.

Kustin, K., Robinson, W. E., Frankel, R. B., and Spartalian, K. (1996). Magnetic properties of tunicate blood cells. II. *Ascidia ceratodes*. *J. Inorg. Biochem.* **63**, 223–229.

Le Floc'h, C., Henderson, R. A., Hughes, D. L., and Richards R. L. (1993). Preparation and X-ray crystal structure of the triply hydrazide-bridged complex [NH₂Me₂]₂[(VCl₃)₂(μ-NNMe₂)₃]: A species derived from disproportionation of SiMe₃NHNMe₂. *J. Chem. Soc. Chem. Commun.*, pp. 175–176.

Leigh, G. J., Prieto-Alcón, R. and Sanders, J. R. (1991). The protonation of bridging dinitrogen to yield ammonia. *J. Chem. Soc. Chem. Commun.*, pp. 921–922.

Lin, M., Lee, M., Yue, K. T., and Marzilli, L. G. (1993). DNA–porphyrin adducts. Five-coordination of DNA-bound VOTMpyP(4) in an aqueous environment: New perspectives on the V=O stretching frequency and DNA intercalation. *Inorg. Chem.* **32**, 3217–3226.

Liu, Q., Yang, Y., Huang, L., Wu, D., Kang, B., Chen, C., Deng, Y., and Lu, J. (1995). Study on an assembly system including tetrathiovanadate. Synthesis and structural characterization of V₂Cu₂S₄ cubane-like clusters and VS₄Cu₄ bimetallic aggregates. *Inorg. Chem.* **34**, 1884–1893.

Maeda, H., Kanamori, K., Michibata, H., Tonno, T., Okamoto, K., and Hidaka, J. (1993). Preparation and properties of vanadium(III) complexes with L-cysteinate and D-penecillaminate. *Bull. Chem. Soc. Jpn.* **66**, 790–796.

Magill, C. P., Floriani, C., Chiesi-Villa, A., and Rizzoli, C. (1993). Vanadium(III)-α-amino acid homoleptic complexes from non-protic solutions: Reaction of [V(Mes)₃THF] with α-amino acids and the structures of tris(L-prolinato)vanadium-dimethyl sulfoxide and tris(D-prolinato)-vanadium-dimethyl sulfoxide. *Inorg. Chem* **32**, 2729–2735.

Malinak, S. M., Demadis, K. D., and Coucouvanis, D. (1995). Catalytic reduction of hydrazine to ammonia by the VFe₃S₄ cubanes. Further evidence for the direct involvement of the heterometal in the reduction of nitrogenase substrates and possible relevance to the vanadium nitrogenase. *J. Am. Chem. Soc.* **117**, 3126–3133.

Meisch, H.-U., Schmitt, J. A., and Reinle, W. (1978). Schwermetalle in Höheren Pilzen. III. Vanadium and Molybdän. *Z. Naturforsch. C* **33**, 1–6.

Mendz, G. L. (1991). Stimulation of mutases and isomerases by vanadium. *Arch. Biochem. Biophys.* **291**, 201–211.

Messerschmidt, A., and Wever, R. (1996). X-ray structure of a vanadium-containing enzyme: Chloroperoxidase from the fungus *Curvularia inaequalis*. *Proc. Natl. Acad. Sci. USA* **93**, 392–396.

Messerschmidt, A., Prade, L., and Wever, R. (1997). Implications for the catalytic mechanism of the vanadium-containing enzyme chloroperoxidase from the fungus *Curvularia inaequalis* by X-ray structures of the native and peroxide form. *Biol. Chem.* **378**, 309–315.

Mokry, L. M., and Carrano, C. J. (1993). Steric control of vanadium(V) coordination geometry: A mononuclear structural model for transition-state-analog RNase inhibitors. *Inorg. Chem.* **32**, 6119–6121.

Mondal, S., Rath, S. P., Dutta, S., and Chakravorty, A. (1996). Chemistry of oxovanadium(V) alkoxides: Synthesis and structure of mononuclear complexes incorporating ethane-1,2-diol. *J. Chem. Soc. Dalton Trans.*, pp. 99–103.

Murray, H. H., Novick, S. G., Armstrong, W. H., and Day, C. S. (1993). Synthesis, characterization and X-ray structure of $V_3(\mu_3\text{-}O)(\mu\text{-}Cl)Cl_6(\mu\text{-}\eta^1\text{-}\eta^1\text{-}PhNHNH_2)_2(THF)$: A new binding mode for phenylhydrazine. *J. Cluster Sci.* **4**, 439–451.

Nanda, K. K., Sinn, E., and Addison, A. W. (1996). The first oxovanadium(V)–thiolate complex, $[VO(SCH_2CH_2)_3N]$. *Inorg. Chem.* **35**, 1–2.

Otieno, T., Mokry, L. M., Bond, M. R., Carrano, C. J., and Dean, N. S. (1996). Non-template-centered, closed tetravanadium phosphate and phosphonate clusters. *Inorg. Chem.* **35**, 850–856.

Parry, D. L., Brand, S. G., and Kustin, K. (1992) Distribution of tunichromes in the ascidiae. *Bull. Mar. Sci.* **50**, 302–306.

Paul, P. C., Angus-Dunne, S. J., Batchelor, R. J., Einstein, F. W. B., and Tracey, A. S. (1997). Reactions of vanadate with N,N-dimethylhydroxylamine. *Can. J. Chem.* **75**, 429–440.

Percival, M. D., Doherty, K., and Gresser, M. J. (1990). Inhibition of phosphoglucomutase by vanadate. *Biochemistry* **29**, 2764–2769.

Pessoa, J. C., Silva, J. A. L., Vieira, A. L., Vilas-Boas, L., O'Brien, P., and Thornton, P. (1992). Salicylideneserinato complexes of vanadium. Crystal structure of the sodium salt of a complex of vanadium(-IV) and -(V). *J. Chem. Soc. Dalton Trans.*, pp. 1745–1749.

Plass, W. (1994). Structural and spectroscopic characterization of six-coordinate vanadium(V) complexes: A structural model for the active site of vanadium-dependent halo-peroxidases. *Z. Anorg. Allg. Chem.*, **620**, 1635–1644.

Rehder, D. (1988). Interaction of vanadate ($H_2VO_4^-$) with dipeptides investigated by ^{51}V NMR spectroscopy. *Inorg. Chem.* **27**, 4312–4316.

Rehder, D. (1991). The bioinorganic chemistry of vanadium. *Angew. Chem. Int. Ed. Engl.* **30**, 148–167.

Rehder, D. (1992). Structure and function of vanadium compounds in living organisms. *BioMetals* **5**, 3–12.

Rehder, D. (1995). Inorganic considerations of the function of vanadium in biological systems. In H. Sigel and A. Sigel (Eds.), *Vanadium and Its Role in Life. Metal Ions in Biological Systems*, Vol. 31. Marcel Dekker, New York, pp. 1–43.

Rehder, D., Vilter, H., Duch, A., Priebsch, W., and Weidemann, C. (1987). A vanadate(V)-dependent peroxidase from pigweed (*Ascophyllum nodosum*): The ^{51}V NMR study of an unusual enzyme and simple vanadate–peptide systems. *Recl. Trav. Chim. Pays-Bas* **106**, 408.

Rehder, D., Weidemann, C., Duch, A., and Priebsch, W. (1988). ^{51}V shielding in vanadium(V) complexes: A reference scale for vanadium binding sites in biomolecules. *Inorg. Chem.* **27**, 584–587.

Rehder, D., and Wieghardt, K. (1981). The ^{51}V NMR spectra of some oxo-, peroxo-, nitrosyl- and hydroxylamidovanadium complexes. *Z. Naturforsch. B* **36**, 1251–1254.

Root, C. A., Hoeschele, J. D., Comman, C. R., Kampf, J. W., and Pecoraro, V. L. (1993). Structural and spectroscopic characterization of dioxovanadium(V) complexes with asymmetric Schiff base ligands. *Inorg. Chem.* **32**, 3855–3861.

Saponja, J. A., and Vogel, H. J. (1996). Metal–Ion binding properties of the transferrins: A vanadium-51 NMR study. *J. Inorg. Biochem.* **62**, 253–270.

Schmidt, H., Bashirpoor, M., and Rehder, D. (1996). Structural characterization of possible intermediates in vanadium-catalysed sulfide oxidation. *J. Chem. Soc. Dalton Trans.*, pp. 3865–3870.

Schulz, D., Weyermüller, T., Wieghardt, K., and Nuber, B. (1995). The monofunctionalized 1,4,7-triazacyclononane derivatives 1,4,7-triazacyclononane-N-acetate (L^1) and N-(2-hydroxyben-zyl)-1,4,7-triazacyclononane (HL^2) and their complexes with vanadium(IV)/(V). Localized and delocalized electronic structures in compounds containing the mixed valent [OV^{IV}–O–V^VO]$^{3+}$ core. *Inorg. Chim. Acta* **240,** 217–229.

Schwendt, P., and Tyrselova, Pavelcic, F. (1995). Synthesis, vibrational spectra and single-crystal X-ray structure of the phosphato-bridged dinuclear peroxovanadate $(NH_4)_5[V_2O_2(O_2)_4$-$PO_4]\cdot H_2O$. *Inorg. Chem.* **34,** 1964–1966.

Sendlinger, S. C., Nicholson, J. R., Lobkovsky, E. B., Huffman, J. C., Rehder, D., and Christou, G. (1993). Reactivity studies of mononuclear and dinuclear vanadium–sulfide–thiolate compounds. *Inorg. Chem.* **32,** 204–210.

Shaver, A., Ng, J. B., Hall, D. A., Lum, B. S., and Posner, B. I. (1993). Insulin-mimetic peroxovana-dium complexes: Preparation and structure of potassium oxodiperoxo(pyridine-2-carboxylato)-vanadate(V), $K_2[VO(O_2)_2(C_5H_4 NCOO)]\cdot 2H_2O$, and potassium oxodiperoxo(3-hydroxypyri-dine-2-carboxylato)vanadate(V), $K_2[VO(O_2)_2(OHC_5H_3NCOO)]\cdot 3H_2O$, and their reactions with cysteine. *Inorg. Chem.* **32,** 3109–3113.

Simons, B. H., Barnett, P., Vollenbroek, E. G. M., Dekker, H. L., Muijsers, A. O., Messerschmidt, A., and Wever, R. (1995). Primary structure and characterization of the vanadium chloroperox-idase from the fungus *Curvularia inaequalis. Eur. J. Biochem* **229,** 566–574.

Smith, M. J. (1989). Vanadium biochemistry: The unknown role of vanadium-containing cells in ascidians (sea squirts). *Experientia* **45,** 452–457.

Song, J.-I., Berno, P., and Gambarotta, S. (1994). Dinitrogen fixation, ligand dehydrogenation, and cyclometalation in the chemistry of vanadium(III) amides. *J. Am. Chem. Soc.* **116,** 6927–6928.

Süssmilch, F., Olbrich, F., Gailus, H., Rodewald, D., and Rehder, D. (1994). σ-Alkenyl and halogeno complexes of vanadium(I and II). *J. Organomet. Chem.* **472,** 119–126.

Tasiopoulos, A. J., Vlahos, A. T., Keramidas, A. D., Kabanos, T. A., Deligiannakis, Y. G., Raptopoulos, C. P., and Terzis, A. (1996). Model investigations for oxovanadium(IV)–protein interactions: The first oxovanadium(IV) complexes with dipeptides. *Angew. Chem. Int. Ed. Engl.,* **35,** 2531–2533.

Taylor, S. W., Kammerer, B., and Bayer, E. (1997). New perspectives in the chemistry and biochemistry of the tunichromes and related compounds. *Chem. Rev.* **97,** 333–346.

Thorn, D. L., Harlow, R. L., and Herron, N. (1996). Phosphatovanadium chemistry: Behavior of phosphato groups covalently bound to only one vanadyl center. *Inorg. Chem.* **35,** 547–548.

Tracey, A. S., Gresser, M. J., and Parkinson, K. M. (1987). Vanadium(V) oxyanions. Interaction of vanadate with oxalate, lactate, and glycerate. *Inorg. Chem.* **26,** 629–638.

Tracey, A. S., Jaswal, J. S., Gresser, M. J., and Rehder, D. (1990). Condensation reactions of aqueous vanadate with the common nucleosides. *Inorg. Chem.* **29,** 4283–4288.

Tsagkalidis, W., and Rehder, D. (1996). Characterization of biorelated vanadium and zinc com-plexes containing tetradentate dithiolate-disulfide, -diamine and amine–amide ligands. *J. Biol. Inorg. Chem.* **1,** 507–514.

Tsagkalidis, W., Rodewald, D., and Rehder, D. (1995a). Coordination and oxidation of vana-daium(II) by 1,2-bis(2-sulfidophenylsulfanyl)ethane(2-) (S_4). *J. Chem. Soc. Chem. Commun.,* pp. 165–166.

Tsagkalidis, W., Rodewald, D., and Rehder, D. (1995b). Coordination and oxidation of *N,N'*-bis(*o*-mercaptophenyl)ethylenediamine (H*SNNS*H) by VO^{2+}: $\{V(^-SNNS^-)\}_4(\mu$-O)$_4\}$ and tetra-benzotetrathiatetraazacycloeicosane. *Inorg. Chem.* **34,** 1943–1945.

Tullius, T. D., Gillum, W. O., Carlson, R. M. K., and Hodgson, K. O. (1980). Structural study of the vanadium complex in living ascidian blood cells by X-ray absorption spectroscopy. *J. Am. Chem. Soc.* **102,** 5670–5676.

Vergopoulos, V., Jantzen, S., Rodewald, D., and Rehder, D. (1995). [Vanadium(salen)benzi-late]—A novel non-oxo vanadium(IV) complex. *J. Chem. Soc. Chem. Commun.,* pp. 377–378.

Vergopoulos, V., Priebsch, W., Fritzsche, M., and Rehder, D. (1993). Binding of L-histidine to vanadium. Structure of exo-[VO₂{N-(2-oxidonaphthal)-His }]. *Inorg. Chem.* **32**, 1844–1849.

Vilter, H. (1984). Peroxidases from *Phaeophyceae:* A vanadium(V)-dependent peroxidase from *Ascophyllum nodosum. Phytochemistry* **23**, 1387–1390.

Wang, W., Zeng, F.-L., Wang, X., and Tan, M.-Y. (1996). A Study of an oxovanadium(V)complex with a tridentate Schiff base ligand. Polyhedron, **15**, 1699–1703.

Wei, Y.-G., Zhang, S.-W., Huang, G.-Q., and Shao, M.-C. (1994). Synthesis, spectrum and single crystal structure of a peroxovanadium complex of nitrilotriacetate ligand without water of crystallization, K₂[VO(O₂)NTA]. *Polyhedron* **13**, 1587–1591.

Weidemann, C., Rehder, D., Kuetgens, U., Hormes, J., and Vilter, H. (1989). K-Edge X-ray absorption spectra of biomimetic oxovanadium coordination compounds. *Chem. Phys.* **136**, 405–412.

Weyand, M. (1996). Röntgenstrukturanalyse einer Vanadium-haltigen Haloperoxidase aus *Ascophyllum nodosum* bei 2.8 Å Auflösung. Dissertation, Gesellschaft für Biotechnologische Forschung, Braunschweig.

Weyand, M., Hecht, H.-J., Vilter, H., and Schomburg, D. (1996). Crystallization and preliminary X-ray analysis of a vanadium-dependent peroxidase from *Ascophyllum nodosum. Acta Crystallogr. D,* **52**, 864–865.

Wittenkeller, L., Abraha, A., Ramasamy, R., Mota de Freitas, D., Theisen, L. A., and Crans, D. C. (1991). Vanadate interaction with bovine Cu,Zn-superoxide dismutase as probed by ⁵¹V NMR spectroscopy. *J. Am. Chem. Soc.* **113**, 7872–7881.

Wright, D. W., Chang, R. T., Mandal, S. K., Armstrong, W. H., and Orme-Johnson, W. H. (1996). A novel vanadium(V) homocitrate complex: Synthesis, structure and biological relevance of [K₂(H₂O)₅][(VO₂)₂(R,S-homocitrate)₂]. *J. Bioinorg. Chem.* **1**, 143–151.

Wroblewski, J. T. (1988). Convenient synthesis of and additional characterization data for hydrogen phosphate tetrahydrate. *Inorg. Chem.* **27**, 946–948.

Xiaping, L., and Kangjing, Z. (1986). Crystal structure of tetraethylammonium L-histidinato(thiocyanato)vanadyl(IV) monohydrate. *J. Crystallogr. Spectrosc. Res.* **16**, 681–685.

Zhang, B., Zhang, S., and Wang, K. (1996). Synthesis, characterization and crystal structure of cyclic vanadate complexes with monosaccharide derivatives having a free adjacent diol system. *J. Chem. Soc. Dalton Trans.,* pp. 3257–3263.

Zhang, C., Markham, G. D., and LoBrutto, R. (1993). Coordination of vanadyl(IV) cation in complexes of S-adenosylmethionine synthase: Multifrequency electron spin echo envelope modulation study. *Biochemistry* **32**, 9866–9873.

Zhang, M., Zhou, M., van Etten, R. L., and Stauffacher, C. V. (1997). Crystal structure of bovine low molecular weight phosphotyrosyl phosphatase complexed with the transition state analog vanadate. *Biochemistry* **36**, 15–23.

Zhang, Y., and Holm, R. H. (1990). Vanadium-mediated oxygen atom transfer reactions. *Inorg. Chem.* **29**, 911–917.

Zhou, Z.-H., Yan, W.-B., Wan, H. L., Tsai, K.-R., Wang, J.-Z., and Hu, S.-Z. (1995). Metal-hydroxycarboxylate interactions: Syntheses and structures of K₂[VO₂(C₆H₆O₇)]₂·4H₂O and (NH₄)₂[VO₂(C₆H₆O₇)]·2H₂O. *J. Chem. Crystallogr.* **25**, 807–811.

12

VANADIUM IN ENZYMES

R. Wever and W. Hemrika

E. C. Slater Institute, Laboratory of Biochemistry, University of Amsterdam, Plantage Muidergracht 12, 1018 TV Amsterdam, The Netherlands

Vanadium in the Environment. Part 1: Chemistry and Biochemistry, Edited by Jerome O. Nriagu. ISBN 0-471-17778-4. © 1998 John Wiley & Sons, Inc.

1. INTRODUCTION

It is now well established that the element vanadium, which is universally distributed in the soil (Bertrand, 1950), plays a role in biological systems. The discovery that this metal is essential to the functioning of the bacterial nitrogen-fixation system and vanadium haloperoxidases from seaweeds and fungi has triggered research on the chemistry of model compounds for these biomolecules. As a result there is an exponential increase in publications, and several reviews have appeared dealing with the subject (Wever and Kustin, 1990; Wever and Krenn, 1990; Rehder, 1991; Butler and Walker, 1993; Vilter, 1995). Also two books (Chasteen, 1990; Sigel and Sigel, 1995) have appeared that cover several aspects, such as the role of the metal in the biosphere, the inorganic and redox chemistry of vanadium, and the insulin mimetic effect of vanadate. This chapter discusses the role of the metal in enzymes, with an emphasis on vanadium haloperoxidases. Vanadium nitrogenase will be discussed only very briefly (see Chapter 11). Vanadium has four oxidation states of which only vanadium(III), -(IV), and -(V) occur in biological systems (Chasteen, 1983). Of these states vanadium(V) and vanadium(IV) are the most common, and vanadium(III) has been detected only in tunicates (see, e.g., Wever and Kustin, 1990, Smith et al., 1995). At neutral pH vanadium is mostly found as vanadate but depending upon pH and concentration, vanadate oligomers, including the dimer, tetramer, and decamer forms, are found (Crans, 1994). At low concentrations such as in sea water the monomeric species is found, probably in the diprotonated state. Vanadate, by virtue of being a structural analogue of phosphate, has inhibitory, stimulatory, and regulatory effects on biochemical processes that in a number of cases are due to formation of enzyme–vanadate intermediates (Stankiewicz et al., 1995). Vanadium(IV) has one unpaired electron, which is strongly coupled to the ^{51}V nucleus ($I = 7/2$), giving rise to either eight or two sets of eight overlapping lines in axial symmetric complexes. This typical signal is easily detectable, and it is possible to observe relatively low concentrations of the vanadium peroxidases (De Boer et al., 1988a) by the electron paramagnetic resonance (EPR) technique.

2. VANADIUM NITROGENASE

The six-electron reduction of dinitrogen to ammonia is the central reaction of biological nitrogen fixation. The catalyst of this reaction is nitrogenase, a multicomponent metalloenzyme complex. The best-characterized nitrogenase is a Mo-dependent enzyme. Bortels (1936) was the first to report that vanadium could substitute for molybdenum as a trace element essential for nitrogen fixation in the bacterium *Azotobacter vinelandii*. Thirty years later it was demonstrated that *A. vinelandii* grown on nitrogen in media supplemented with vanadium instead of molybdenum synthesize an enzymatically active vanadium-containing nitrogenase (McKenna et al., 1970; Burns et al., 1971). As discussed in more detail by Cammack (1986), the significance of these findings was not appreciated at that time—in particular because one of the groups involved (Benemann et al, 1972) refuted the idea that vanadium itself was essential to the enzymic activity of the nitrogenase. Rather, they claimed that both molybdenum and vanadium were present in the vanadium nitrogenase and that only the molybdenum contained in the enzyme accounted for its observed activity. Later it was shown (Bishop, 1986; Chisnell et al., 1988) that *A. vinelandii* possesses three nitrogen fixation systems: the conventional molybdenum enzyme, a vanadium-containing enzyme, and a nitrogenase that lacks both metals. The vanadium nitrogenase is expressed only when the organism is grown under molybdenum limitations and is repressed in the presence of molybdenum. Robson et al. (1986) demonstrated that a mutant stain of *A. chroococcum* synthesized a vanadium nitrogenase when grown on vanadium. This strain lacks the structural genes for the conventional nitrogenase, which contains molybdenum at the active site. These findings were confirmed (Hales et al, 1986) and a vanadium nitrogenase was isolated that consisted of two proteins, one of which differs completely from the conventional component, that is, the MoFe protein. This molybdenum–iron protein has been studied in great detail and the crystal structures of these enzymes from both *A. vinelandii* and *Clostridium pasteurianum* are known at 2.2-Å and 3.0-Å resolution, respectively (Chan al., 1993; Kim et al., 1993), showing details of the FeMo cofactor. The vanadium in the vanadium nitrogenase is present in a similar cofactor, which consists of iron and acid-labile sulfur and has been studied by X-ray absorption spectroscopy and EPR (Morningstar and Hales, 1987; Harvey et al., 1990). However, since structural details of the VFe protein are still lacking, the role of vanadium in the reduction of nitrogen to ammonia and its possible ligation will not be discussed here. A detailed treatment and overview is given by Eady (1995).

3. VANADIUM HALOPEROXIDASES

3.1. Vanadium Bromo- and Iodoperoxidases

A striking array of halogenated products produced by natural sources (Gribble, 1992; Neidleman and Geigert, 1986) are found in the biosphere, the major part of which are probably formed by the enzymic activity of the haloperoxidases. The detailed mechanism by which haloperoxidases catalyze the insertion of a halogen into organic molecules is still a point of debate (Franssen, 1994). However, for some of these enzymes the reaction mechanism is simple and consists of a two-electron oxidation of the electron donor (Cl^-, Br^-, I^-) by H_2O_2 to hypohalous acids (eq 1):

$$H_2O_2 + X^- + H^+ \rightarrow HOX + H_2O \tag{1}$$

By definition enzymes that are able to oxidize chloride, bromide, and iodide are called chloroperoxidases and those able to oxidize bromide and iodide, bromoperoxidases. If a nucleophilic acceptor (AH) is present, a reaction will occur with HOX and halogenated compounds are produced (eq 2).

$$HOX + AH \rightarrow AX + H_2O \tag{2}$$

HOX will react with a broad range of acceptors (AH) to form a diversity of halogenated reaction products (AX). As has only recently been discovered, some of the haloperoxidases contain vanadium at the active site (Vilter, 1984; De Boer et al., 1986). These enzymes, which have been detected in a great variety of brown and red seaweed samples (Wever et al., 1991; Vilter, 1995) and also in marine diatoms (Moore et al., 1996), are widespread in the marine environment. Not all these enzymes contain vanadium; some are heme-containing peroxidases. Vanadium iodoperoxidases have also been identified (De Boer et al., 1986a) but they have hardly been studied. Bromoperoxidases have been found in seaweeds from the Great Barrier Reef in Australia, the Japanese Sea, the East Pacific, and the North Atlantic Ocean. Vanadium peroxidases are probably responsible for the formation of huge amounts of bromoform by seaweeds and phytoplankton (Wever et al., 1990, 1991). Some oceans are even saturated with this compound and other halogenated methanes, with the result that ventilation to the atmosphere occurs. There they play roles in the regulation of ozone in the stratosphere and perhaps in the Arctic troposphere at Polar Sunrise (Wever, 1988; Wever, 1993; Barrie et al., 1988). Clearly, the halogenated products of haloperoxidases have an unexpected and important effect on atmospheric processes. The reason why these algae produce these compounds is not clear. HOBr and some of the organohalogens formed have biocidal effects, and it is likely that formation of these compounds is part of the host defence system (Wever, 1993) and may prevent fouling by microorganisms or act as an antifeeding system.

3.2. Vanadium Chloroperoxidases

3.2.1. Discovery and Physiological Function

During the course of screening for haloperoxidases with tolerance to high pH environments and high temperatures, Hunter-Cevera and Sotos (1986) found that fungi isolated from plant material and soils collected in the Death Valley possesed haloperoxidase activity. The fungal species producing haloperoxidases belong to the group of dematiaceous hyphomycetes and these are quite ubiquitous (Ellis, 1971). A non-heme chloroperoxidase produced by one such hyphomycete, the fungus *Curvularia inaequalis* (Liu et al., 1987) was isolated and partially characterized. The enzyme was surprisingly stable towards oxidative attack. Iron and zinc were present and consequently these metals were thought to be involved in catalysis. In 1993 it was established (Van Schijndel et al., 1993a) that the prosthetic group in the enzyme was orthovanadate, as was also found in the vanadium bromoperoxidases. Later it was shown (Simons et al., 1995) that iron and zinc were present in less than stoichiometric amounts in the enzyme and that these metals were not involved in catalysis.

These peroxidases escape detection if classical peroxidase assay systems are used, and as a result it took a relatively long time to identify and characterize the peroxidases from hyphomycetes. Furthermore, when the fungus is grown in media to which no vanadate is added, the enzyme is secreted in an inactive apoform (Vollenbroek et al., 1995). Expression studies clearly showed (Simons et al., 1995) that the vanadium enzyme is produced in the secondary growth phase, the so-called idiophase. The expression appears to be mainly regulated at transcription level and is dependent on nutrient limitation of the fungus (Barnett et al., 1997). Several species in the group of the dematiaceous hyphomycetes produce haloperoxidases (Hunter-Cevera et al., 1990), and it was shown (Vollenbroek et al., 1995) that the haloperoxidases purified from some of these species (*Drechslera biseptata, D. subpapendorfii, Embellisia didymospora, and Ulocladium chartarum*) are vanadium chloroperoxidases. Thus, vanadium chloroperoxidases are in fact more widespread than previously thought. Some of these hyphomycetes are phytopathogenic and/or saprophytes, and this raises the question of the physiological function of the vanadium chloroperoxidases. The product of the enzyme is HOCl, which is a strong bactericidal and oxidizing agent known to degrade lignine (Kringstad and Lindström, 1994). Recent data (Barnett et al., 1997) show that the location of the enzyme is strongly dependent on culture conditions and that a considerable fraction of the enzyme remains bound on the surface of the fungal hyphae. Thus it may well be that this system is used by the fungus as an attack mechanism to oxidize the lignocellulose in the cell walls of the plant in order to facilitate penetration of the fungal hyphen into the host. The proposed mechanism of a freely diffusable agent involved in the breakdown of lignocellulose is reminiscent of the situation in lignin degradation. Ligninolytic fungi (Harvey et al., 1986) also produce extracellular peroxidases, which generate a veratryl alcohol radical cation, and the radical cation then catalyzes the oxidative depolymerization of lignine. It is likely that HOCl, which is generated by fungi in soils, also reacts with organic matter, resulting in the formation of these organohalogens. This may explain the presence of chlorinated compounds in soils (Asplund and Grimvall, 1991).

4. PROPERTIES OF HALOPEROXIDASES

4.1. Catalytic Activity

The vanadium peroxidases differ from the heme-containing peroxidases in their specificity for halides. Only chloride, bromide, iodide, and the pseudohalide thiocyanate (Walker and Butler, 1996) are oxidized by these vanadium enzymes. Classical organic electron donors such as guaiacol, *o*-dianisidine, and benzidine are not oxidized. Thus, the activity of the vanadium haloperoxidases can only be detected by using assay systems based on the formation of HOCl and HOBr. Routinely, the halogenating activity of these enzymes is determined by measuring the bromination or chlorination of the cyclic diketone monochlorodimedon (extinction coefficient of 20.2 $mM^{-1} \cdot cm^{-1}$ at 290 nm) to the dihalogenated compound (extinction coefficient of 0.2 $mM^{-1} \cdot cm^{-1}$ at 290 nm (De Boer et al., 1986b). At higher pH values and in particular at high concentrations of H_2O_2 a competing reaction between H_2O_2 and HOX occurs (eq. 3), resulting in singlet oxygen formation (Allen, 1975) and an apparent decrease in the rate of bromination or chlorination of monochlorodimedon.

$$H_2O_2 + HOX \rightarrow {}^1O_2 + H_2O + HX \qquad (3)$$

It is also possible to directly monitor the rate of dioxygen formation, using an oxygen electrode (Soedjak and Butler 1991) as a measure for enzyme activity. The dioxygen is formed by the secondary reaction between the H_2O_2 present and the HOBr produced (eq 3). At high pH, monitoring dioxygen formation is a more accurate measure of bromoperoxidase activity than measuring monochlorodimedon bromination. A drawback of this assay method is that, in particular at low pH values, the reaction between H_2O_2 and HOBr is slow and therefore initial rates cannot be obtained. Furthermore, if another nucleophlic acceptor is present, as is likely in partially purified preparations, the activity of the enzyme is underestimated. A convenient method for assaying the activity qualitatively is measuring the bromination of phenol red to bromophenol blue (De Boer et al., 1987). A marked color change takes place that allows screening for halogenating activity of a large number of samples by simple visual inspection. It should be noted, however, that HOBr has a low affinity for phenol red and this method should not be used in quantitative studies.

4.1.1. Kinetic Properties of Haloperoxidases

Extensive kinetic steady-state analyses of bromoperoxidase catalysis have been carried out (De Boer and Wever, 1988; Soedjak and Butler, 1990; Soedjak and Butler, 1991), the results of which show that hydrogen peroxide as a substrate first reacts with the enzyme and is followed by a reaction with a halide. The kinetic mechanisms for the bromoperoxidases from various sources are very similar. Slight inhibition of bromoperoxidase by high concentrations of H_2O_2 (5–10 mM) with stronger inhibition at higher pH has been observed. Bromide has been reported to be a competitive inhibitor with respect to H_2O_2, although Soedjak and Butler (1991) found that the data were more consistent with noncompetitive inhibition. In all cases inhibition by bromide is observed at pH values lower than the optimum. For the peroxidases studied the

Michaelis-Menten constant (K_m) for H_2O_2 increases with increasing H^+ concentration and the data indicate that although $V^{(V)}$ coordinated water may also be involved, an amino acid residue with a pK_a of 5.7–6.5 is important in the binding of peroxide to these enzymes. A steady-state kinetic analysis (Van Schijndel et al. 1994) showed that the kinetics of the chloroperoxidase resemble those of the vanadium bromoperoxidases (De Boer and Wever, 1988; Soedjak and Butler, 1991). The chloroperoxidase exhibits a pH profile similar to that of the vanadium bromoperoxidases, although the optimum pH of 5.0 is lower than that for the bromoperoxidases. At low pH, high chloride concentrations inhibit the enzyme in a competitive way, whereas at higher pH values the activity displays a normal Michaelis-Menten type of behavior. Both the K_m for chloride and the K_m for hydrogen peroxide are a function of pH. The log K_m for chloride increases linearly with pH, whereas that for hydrogen peroxide decreases with pH. These observations have led to the following simplified pingpong type of mechanism for the haloperoxidase (Fig. 1).

In the uninhibited reaction cycle, the enzyme first forms a complex with hydrogen peroxide, after which a halide ion and a proton react and an enzyme–hypohalous species is formed that decays rapidly to enzyme and free HOX. The linear dependency of the log K_m for hydrogen peroxide on pH suggests that a protonable group is involved in the binding of hydrogen peroxide. When this group is protonated, hydrogen peroxide is unable to bind. This phenomenon and the inhibition by chloride probably causes the decrease in enzymic activity at low pH. The decrease at higher pH value is due to both a decrease in the V_{max} and the pH dependence of the K_m for halide (Van Schijndel et al., 1994; De Boer and Wever, 1988). These effects account for the skewed curve of the activity versus pH seen for the bromoperoxidase from *A. nodosum* and the chloroperoxidase. It has been speculated that the group responsible for these effects in the bromoperoxidase is a histidine residue. Indeed, the X-ray structural data show (Messerschmidt and Wever,

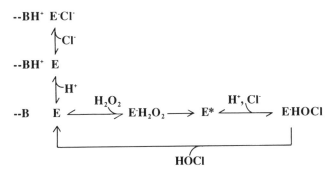

Figure 1. Simplified representation of the reaction of substrates with the vanadium haloperoxidase. For simplicity a number of protonated states have been omitted. The enzyme species E* may present an activated vanadium peroxo complex. B represents an acid–base group. When this group becomes protonated, hydrogen peroxide is unable to react with the enzyme.

1996) that two histidine residues are found in the active site of the fungal chloroperoxidase.

4.2. Properties of the Prosthetic Group

A common property of the vanadium peroxidases is that these enzymes lose their enzymic activity at low pH in buffers containing phosphate and EDTA. Also during purification procedures the peroxidases slowly inactivate if phosphate buffers are used. However, in the absence of phosphate-containing buffers the enzymes retain their activity for months (De Boer et al., 1987).

To obtain the apoenzyme and to inactivate the enzyme completely, the presence of phosphate in the procedure is essential (Wever and Krenn, 1990). Incubation of the apoenzyme at neutral pH with orthovanadate (VO_4^{3-}, which is the vanadium(V) oxyanion) results in recovery of the enzymic activity, as discovered originally by Vilter (1984) for the bromoperoxidase. Reconstitution of the bromoperoxidase by vanadium(IV) salts is also possible, but the effect is probably due to reoxidation of vanadium(IV) at neutral pH to vanadium(V) by oxygen. Reconstitution of apobromoperoxidase is inhibited competitively (Tromp et al., 1991) with an inhibition constant K_i of about 100 μM by phosphate (PO_4^{3-}), which is structurally and electronically analogous to vanadate. It should be noted, however, that the analogy may not always hold, since vanadate can interact with enzymes as a tetrahedral phosphate analogue or as a trigonal bipyrimidal phosphate transition state analogue (Crans, 1994). Also, arsenate (AsO_4^{3-}) or molybdate (MoO_4^{2-}) inhibit reconstitution, and in particular the tetrahedral compounds fluoroberyllate and fluoroaluminate, which have strong hydrogen-bonding properties, inhibit the reconstitution very strongly (K_i about 1 μM; Tromp et al., 1994). It is interesting that the apochloroperoxidase also binds WO_4^{2-} at the active site. This property was used (Messerschmidt and Wever, 1996) to obtain a heavy-metal derivative of the crystallized chloroperoxidase. The affinity of the apohaloperoxidases for vanadate is quite high. Values of 35–55 nM have been reported (Vilter, 1984; Tromp et al., 1990) for the bromoperoxidase from *Ascophyllum nodosum* that are close to the concentration of vanadate in sea water (about 50 nM; Chasteen, 1983), suggesting that for incorporation of vanadate in the apoenzyme *in vivo* no additional enzyme system is required. This affinity decreases rapidly at lower pH values, and the data suggest that protonation of a single group with a pK_a larger than 8.5 prevents the binding of vanadate to the apoenzyme. This group may be vanadate itself or some basic amino acid residues. The affinity of the apochloroperoxidase for vanadate is somewhat less (140 nM) and also increases at pH values lower than 7, suggesting that an amino acid residue with a pKa of 6–7 is involved in the binding of vanadate (Van Schijndel et al., 1993b). It is not likely that ionization of vanadate itself is responsible for this pH dependance, since the pKa values for vanadic acid (3.5, 7.8, and 12.5) are such that they do not correlate with the observed pH dependency. An obvious candidate for an amino acid with such a pKa value is histidine.

Interestingly, lysine, arginine, and histidine are directly involved in binding of the prosthetic group to the apoenzyme (Messerschmidt and Wever, 1996). The secreted apoform of the enzyme can easily be activated upon addition of vanadate (VO_4^{3-}) to the growth medium (Vollenbroek et al., 1995), also suggesting that these enzymes do not require an additional enzyme system for the incorporation of the vanadate but are capable of obtaining the metal directly from the environment. Also the recombinant inactive chloroperoxidase produced by yeast (Barnett et al., 1995; Hemrika et al., 1997) is easily reactivated by the addition of vanadate. In this respect it should be noted that in most soils very high concentrations (100 ppm) of vanadium are present (Bertrand, 1950), and the amount of vanadium in soils does not appear to represent a limiting factor in activating vanadium chloroperoxidase once produced by the fungi.

4.3. Stability of Haloperoxidases

Vanadium bromoperoxidases and in particular vanadium chloroperoxidase are thermostable and have a pronounced chemostability. When these enzymes are stored in organic solvents such as 40% (v/v) methanol, acetone, and dioxane, they remain stable for up to 6 weeks (Van Schijndel et al., 1994; Vollenbroek et al., 1995). Similarly, these enzymes are highly resistant towards elevated temperatures; the midpoint temperatures determined from thermal denaturation curves range from 82 to 90 °C. It is also possible to store the enzyme at elevated temperatures, and it has even been reported that incubation at 75 °C for 8 hours resulted in only a 20% loss of activity. The bromoperoxidase from A. *nodosum* was also shown to be stable under turnover conditions (De Boer et al., 1987). At room temperature the bromoperoxidase continues to catalyze bromination reactions for 3 weeks without loss of activity. The factors contributing to chemo- and thermostability are the compactness of the molecule and the large amounts of α-helical structure found for both the bromoperoxidase (Tromp et al., 1990) and the chloroperoxidase (Messerschmidt and Wever, 1996).

The enzymes are also resistant towards oxidative agents and unlike heme-containing haloperoxidases, the vanadium bromoperoxidases are just slightly inhibited by high concentrations of H_2O_2. Furthermore, the enzymic activity is hardly affected by HOBr, singlet oxygen, or oxidized bromine derivatives. However, at pH 4–5 an irreversble inactivation occurs by 100 mM hydrogen peroxide, which is not reversed by addition of vanadate. The inactivation is due to the formation of 2-oxohistidine (Meister Winter and Butler, 1996). There are two histidines in the vicinity of the active site, of which one ligates directly the vanadate and the other is suggested to play a role as an acid–base group. It is likely that one of these becomes modified during inactivation.

It was also shown (Van Schijndel et al., 1994) that the chloroperoxidase generates HOCl to concentrations as high as 4 mM without loss of enzymic activity. Liu et al. (1987) also showed the remarkably high resistance of the enzyme towards HOCl and very high concentrations of H_2O_2 (200 mM). Heme-containing peroxidases with their electron-rich aromatic ring system are easily oxidized and inactivated by these oxidative agents. In this respect vanadium chloroperoxidase is unique and the ability of the enzyme to handle the oxidative substrate and product is the consequence of the nature of the active site.

5. EVENTS DURING CATALYSIS

EPR studies (De Boer et al., 1988a) and K-edge X-ray studies (Arber et al., 1989; Kusthardt et al., 1993) on the vanadium bromoperoxidase from the brown seaweed *Ascophyllum nodosum* have shown that the oxidation state of the metal is vanadium(V) and that during turnover the redox state of the metal does not change. Since also the reduced enzyme is not active in bromination reactions (De Boer et al., 1988a; Wever et al., 1988), a model was proposed in which the vanadium(V) coordinates and activates hydrogen peroxide, after which the halide is able to react with this activated state of the enzyme to yield hypobromous acid. Vanadium may be considered to function as a Lewis acid, in which the metal acts as an effective withdrawer of electron density from the bound peroxo group in such a way that the halide is able to react with this activated state of the enzyme to yield hypohalous acid. In this mechanism, no redox changes occur in the vanadium metal center. Although the mechanisms formulated can be considered as a two-electron oxidation of the halide ion by the bound peroxointermediate, it may also be seen as a transfer of an oxygen atom from the bound peroxide or peroxo intermediate to the halide ion, directly yielding hypohalous acid. Such a mechanism has been shown to occur now in a number of heme peroxidases (Fu et al., 1992; Dexter et al., 1995), which are able to carry out stereoselective oxygen transfer reactions to organic sulfides and alkenes. The fungal chloroperoxidase has not yet been studied by detailed biophysical techniques, but since this enzyme shares so many properties with the vanadium bromoperoxidases, it is likely that for the oxidation of the halide the enzymes basically have the same catalytic mechanism.

According to De Boer and Wever (1988) Br_3^-, Br_2, or HOBr are the primary reaction products of the enzyme-mediated peroxidation of bromide. By using a rapid-scan spectrophotometer to circumvent the rapid reaction of these oxidized bromine species with H_2O_2 at neutral pH and at high bromide concentrations so as to favor detection of tribromide, Br_3^- formation is found. Measurements by Soedjak and Butler (1990) confirmed that these oxidized bromine species are formed by the enzyme. Peracetic acid was used as the oxidant, and since HOBr reacts only slowly with this oxidant, it was possible to monitor the formation of these oxidized bromine species in a more simple way. Similarly, the study by Van Schijndel et al. (1994) on the formation of oxidized chlorine species led to the conclusions that this enzyme produces free HOCl. Still, uncertainty exists as to which species is really produced enzymatically, owing to the rapid equilibria between OX^-, HOX, X_2, and X_3^-. In particular at neutral pH a further complication arises, since these oxidized halogen species react with hydrogen peroxide to yield singlet oxygen. It is likely, however, that primarily HOX is formed. However, contrary to these findings Tschirret-Guth and Butler (1994) obtained evidence that vanadium bromoperoxidase can bind indoles as organic substrates and that the active brominating species under these conditions is not enzyme-released bromine species. The data indicate that indole bromination is catalyzed by an enzyme-trapped brominating species. This is in contrast to the findings by De Boer and Wever (1988), who failed to demonstrate a ternary complex between the enzyme-halogenating intermediate and an organic substrate. Similarly, a study of the bromination of several barbituric acid derivatives (Franssen et al., 1988) suggested a mechanism by which the oxidized bromine species are released into the solution by the enzyme in a rate-determining step. However, it may well be that the vanadium bromoperoxidase has a specific binding site for the indole. This binding site is apparently not related to the known physiological role of the enzyme, since the seaweed from which the bromoperoxidase has been isolated produces only brominated methanes (Geschwend et al., 1985).

6. STRUCTURAL PROPERTIES OF HALOPEROXIDASES

6.1. X-ray Structure of Vanadium Chloroperoxidase and Details of the Active Site

Recently, the 2.1-Å crystal structure of chloroperoxidase was reported (Messerschmidt and Wever, 1996), revealing details of the structure and active site and also answering some of the unresolved questions. The enzyme molecule has an overall cylindrical shape (Fig. 2) with a length of about 80 Å and a diameter of 55 Å. The protein fold is mainly α-helical with two four-helix bundles as the main structural motives. The structure also contains some antiparallel β sheets. The very compact structure is different from other known protein structures, and it appears that the high stability, in line with the suggestion by Tromp et al. (1990) for vanadium bromoperoxidase, is due to compact packing of the helices, which reveals a strong stabilizing hydrophobic effect. The vanadium binding center is located on top of the second four-helix bundle, and the residues binding the prosthetic group span a length of approximately 150 residues in the primary structure.

X-ray structure analysis (Fig. 3) shows that vanadium is bound as orthovanadate(V) in which three oxygens form a plane and the fourth oxygen is found at the apex and is hydrogen-bonded to a water molecule.

The N^{e2} atom from a histidine (His496) ligating directly to the metal completes the trigonal bipyrimidal coordination. The negative charge of the vana-

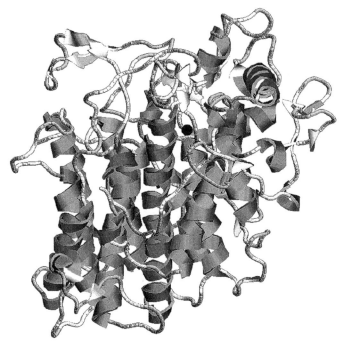

Figure 2. Ribbon-type presentation of the chloroperoxidase molecule at 0.21-nm resolution. The black dot on the top of the four-helix bundle represents the vanadium atom.

date group is compensated by hydrogen bonds to three positively charged residues (Arg 360, Arg 490, Lys 353). Further, the vanadate forms hydrogen bonds with a glycine (Gly 403) and a serine residue (Ser 402). There is also a histidine residue (His 404) close to the active site, which may act as an acid–base group in catalysis. Furthermore as in heme peroxidases (Poulos and Kraut, 1980), there is an arginine (Arg 490) nearby that may also have a function in catalysis. The pocket where vanadate is bound is at the end of a channel that supplies access and release of the small substrates and products of the reaction. Part of the surface of the channel is mainly hydrophobic with Pro 47, Pro 221, Trp 350, Phe 393, and Pro 395 as contributing side chains. The other half of the channels is predominantly polar, with several carbonyl oxygens and the ion pair Arg 490–Asp 292. It is interesting that the Arg 490 also directly binds the orthovanadate.

How does the real structure agree with some of the anticipated data that were obtained for the vanadium bromoperoxidase? The major difference is that the orthovanadate is five-coordinated rather than six-coordinated, as was originally proposed (Arber et al., 1989). However, using the bond valence sum analysis and a comparison of the intensity of the pre-edge feature in the extended X-ray fine structure analysis (EXAFS) spectra of the enzyme with that of vanadium model compounds, Carrano et al. (1994) arrived at the

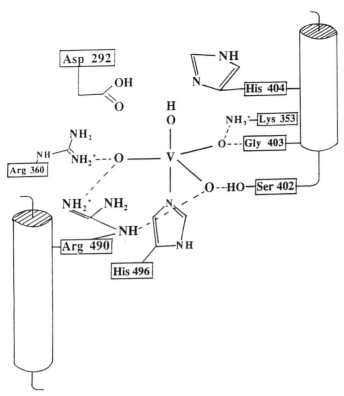

Figure 3. View of the active site of vanadium chloroperoxidase. Arg 490 and Gly 403, and His 404, are part of α-helix structures (residues 479–492 and residues 403–419, respectively). The other active site residues are in loops connecting the secondary structure elements. The helices are drawn as cylinders.

conclusion that the geometry was five-coordinate rather than a six-coordinate geometry. An EPR study on the reduced enzyme (De Boer et al., 1988a) demonstrated clearly that D_2O or $H_2^{17}O$ markedly affect the vanadium(IV) hyperfine line width, which is in line with the X-ray structure analysis showing that water is in the direct coordination sphere of the metal oxide. Similarly, an electron spine echo envelope modulation (ESEEM) study clearly showed (De Boer et al., 1988b) that the oxovanadium(IV) in the reduced enzyme is coupled to exchangeable protons.

Based upon EPR and EXAFS data, a short single terminal oxide with a short bond length was proposed to be present in the reduced enzyme, and this is clearly not observed in the structure of the native oxidized enzyme. However, significant structural changes were shown (Arber et al., 1989) to occur at the metal site upon reduction of the native enzyme, and a single terminal oxide ligand may well be present in the reduced enzyme, since structural details of reduced chloroperoxidase are not available yet. The X-ray structure clearly shows the presence of a histidine directly ligating the metal oxide. Indeed, the EXAFS data (Arber et al., 1989) clearly showed the presence of multiple scattering effects from outer atoms of a group corresponding to a histidine ligated to vanadium. Also the electron spin echo experiments (De Boer et al., 1988b) clearly indicated the presence of a histidine residue in the active site.

So what are the events that occur during catalysis? Figure 4 gives a hypothetical reaction pathway. The structural data suggest that, upon binding of peroxide to the vanadate structure, the histidine residue at the apical side may accept a proton from the peroxide and act as general acid–base catalyst facilitating the binding of the peroxide anion. The function of this histidine may therefore be analogous to the role of the distal histidine in heme-containing peroxidases (Messerschmidt and Wever, 1996). It is interesting that, also in model complexes that mimic vanadium haloperoxidases an acid–base group in the vicinity of the metal-bound peroxide seems to be essential (Colpas et al., 1996). Mainly because the structure of peroxide or peroxo intermediate is not known yet, the sequence of events in the enzyme following formation of a peroxovanadate intermediate is not clear. Numerous mono and diperoxo compounds have been characterized (see Wever and Kustin, 1990, for a overview), and the side-on peroxo complexes are the most frequently encountered yielding a seven-coordinate, distorted pentagonal bipyrimidal geometry. In the enzyme the peroxide may also be side-on bound as in the vanadium peroxo complexes. The vanadium atom by virtue of its Lewis acid properties will further polarize the bound peroxide, yielding an electronically asymmetrically bound peroxide. The negative charge on one of the oxygen atoms may be stabilized by a nearby positively charged residue (Arg 490, Arg 360 or Lys 353). The oxygen–oxygen bond now breaks, an oxygen atom is transferred to the bound halide, and the proton is donated back to the negatively charged

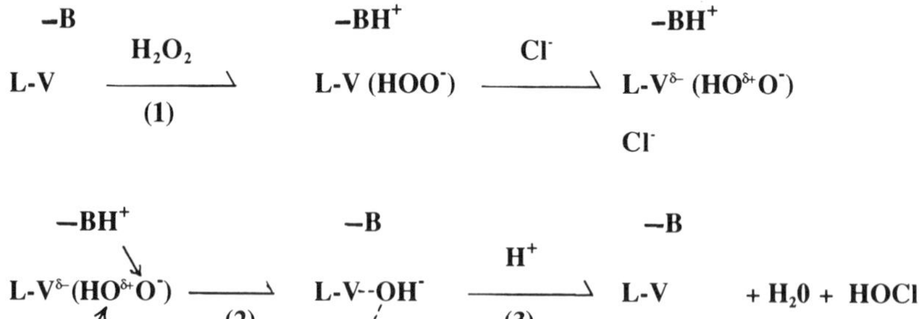

Figure 4. Hypothetical reaction pathway. In Step 1 hydrogen peroxide is bound and one of the protons is taken up by His 404. In Step 2 there is a nucleophilic attack of a chloride ion on one of the oxygen atoms, which is polarized by the vanadium. The oxygen–oxygen bond breaks and the complex takes up a proton (Step 3) to yield water, enzyme, and HOCl. Group B is His 404, the residue that acts as an acid–base group.

oxygen by the histidine, acting now as proton donor yielding OH^-. Finally, a proton reacts with the bound hydroxide and water and HOCl are formed.

The discovery of the vanadium haloperoxidases has given a considerable impetus to the development of transition metal peroxo complexes mimicking the activity of the vanadium haloperoxidases. These include the bis- and monoperoxo vanadium complexes (Colpas et al., 1996) and those of molybdenum(VI) and tungsten(VI) (Meister and Butler, 1994). Methylrhenium trioxide (Espenson et al., 1994) also catalyzes the oxidation of bromide in the presence of hydrogen peroxide. Some of these kinetic studies are consistent with an oxo-transfer mechanism for halide oxidation. The turnover of these complexes, which act as functional models for the haloperoxidases, is considerably less than that of the enzymes. The results demonstrate clearly that the local structural environment of the metal provided by the protein core tunes the reactivity of the bound vanadate and the vanadium peroxo intermediate in such a way that the enzyme-catalyzed reaction is able to proceed at neutral pH values and at a much higher rate. The fact that the nature of the site in these enzymes is now known in detail will provide the essential knowledge needed to synthesize catalysts that really resemble the vanadium peroxidases in both structural and kinetic aspects.

6.2. Sequence Homology of Vanadium Chloroperoxidases to Other Enzymes

The gene coding for the vanadium chloroperoxidase (Simons et al., 1995) has been cloned by reverse transcription of messenger RNA followed by amplification by the polymerase chain reaction. The deduced amino acid sequence predicts a protein of 609 residues with a molecular mass of 67,488 Da. This value is consistent with the estimated value of 67 kDa found by SDS/PAGE. Only a single gene encodes vanadium chloroperoxidase and there are no isoenzymes. Using the entire sequence and searching the Swiss-Prot database originally yielded hardly any sequence similarity to other enzymes. The only exception is that a correspondence is found with the bacterial nonheme haloperoxidases in one small region (residues 259–280). This region also has sequence similarity with other α/β-hydrolase-fold enzymes, a group to which the bacterial nonheme haloperoxidases belong (Hecht et al., 1994). This similarity is not in the region of the active-site residues of vanadium chloroperoxidase. Overall, however, the crystal structure of bacterial nonheme bromoperoxidase, which shows the general topology of the α/β-hydrolase fold, and that of vanadium chloroperoxidase, which is mainly α-helical, lack structural similarity.

Part of the primary structure of vanadium bromoperoxidase from the seaweed *Ascophyllum nodosum* is also available (Vilter, 1995). An amino acid sequence alignment (Messerschmidt and Wever, 1996) shows the conservation of residues involved in the active site of the vanadium CPO, though the overall sequence identity outside these regions is low. There is one exception; an asparagine residue forming hydrogen bonds to the vanadate that is present in bromoperoxidase is a lysine in the chloroperoxidase. It is unlikely that such a minor alteration plays any role in determining why the bromoperoxidase is able to oxidize only bromide at a significant rate, and chloroperoxidase is also able to oxidize chloride. A detailed structural comparison will probably reveal the factors that determine this difference.

Recently the very important and surprising discovery (Hemrika et al., 1997) was made that the amino acid residues contributing to the active site of the vanadate-containing haloperoxidases are conserved within at least three families of acid phosphatases that were previously considered unrelated. This was discovered by using small stretches of amino acid residues that form the active site of the chloroperoxidase. These three families are the glucose-6 phosphatases, the secreted nonspecific class A bacterial acid phosphatases, and a group of bacterial membrane-bound phosphatases.

Figure 5 shows the active site residues of vanadium chloroperoxidase which are conserved in the alignment of some of the phosphatases. It should be noted here that the loops connecting these residues are highly variable in length and also that there is very low similarity in these loops.

Is there direct evidence that the corresponding residues in the group of phosphatases form the active site in the phosphatases, in particular since structural data are not available for these enzymes? The data for glucose-6-phosphatase give a clear answer to this question. This enzyme is the key enzyme in gluconeogensis and glucose homeostasis, and its deficiency leads to a glycogen storage disease (Von Gierke disease). Mutagenesis studies (Lei et al., 1993; Lei et al., 1995) have shown that Arg 83 (see Fig. 5) is absolutely required for enzyme activity. In the alignment this residue corresponds to Arg 360 of the vanadium CPO, a residue directly hydrogen-bonded to the orthovanadate. Also His 119 turned out to be essential for phosphatase activity (Lei et al., 1995), and this histidine corresponds to His 404 in the vanadium chloroperoxidase, a residue involved in the uptake of a proton from hydrogen peroxide during catalysis (Fig. 2). There is also indirect evidence that the phosphate binding site and the vanadate binding site share similarities. As we have seen, the reconstitution of the chloroperoxidases is inhibited by phosphate, which is structurally and electronically analogous to vanadate.

$$-K^*-R^* P\text{------}P\ \ S^*G^*H^{**}\text{-------}S\ R^*\text{------}H^{\#}-$$
$$\quad 353 \quad\ 360\text{-}361 \qquad 401\text{-}402\text{-}403\text{-}404 \qquad\qquad 489\text{-}490 \qquad 496$$

Figure 5. The sequence similarity of vanadium chloroperoxidase and some phosphatases. $**$, His 404 acid–base group in catalysis. $^{\#}$, His 496 covalently linked to vanadate. The sequences differ in the loops connecting the active-site residues ($*$). These loops are highly variable in length and composition. The same residues are conserved in phophatidyl glycerophophate phophatase B, ATP-diphosphohydrolase, class A bacterial acid phophatase, and glucose-6-phosphatases from mammals. Some of the enzymes show more similarity with respect to each other (see Hemrika et al., 1997). In glucose-6-phosphatase, Arg 83 (Arg 360 in the chloroperoxidase) and His 119 (His 404 in the chloroperoxidase) are essential to the activity of the enzyme.

Furthermore, vanadate is a potent inhibitor of many different phosphatases (Stankiewicz et al., 1995), including mammalian glucose-6-phosphatase (Sing et al., 1981).

The conservation of active site residues strongly suggests that the binding pocket for vanadate in the peroxidases is similar to the phosphate binding site in the aligned phosphatases. This raised the question (Hemrika et al., 1997) whether the apochloroperoxidase from the fungus could act as a phosphatase. Indeed, activity measurements confirm that the apochloroperoxidase exhibits phosphatase activity. The phosphatase activity using p-nitrophenylphospate as a substrate for apochloroperoxidase is about a factor of 10^3 lower that of normal phosphatases. Surprisingly, the K_m for this substrate (50 μM) compares well with that for the phosphatases (Hemrika et al., 1997) and this means that the channel leading to the active site is large enough to accommodate this bulky substrate.

A common architecture of the active site has important implications for research in field of acid phosphatases. For this group of enzymes structural data are yet not available, and the structure of the chloroperoxidase can be used as a starting point to unravel structural details of these phosphatases. We are able now to predict which other residues in the class of these phosphatases are involved in catalytic activity. The data (Hemrika et al., 1997) indicate that the vanadium haloperoxidases and the phosphatases have divergently evolved from a common ancestor. We anticipate that the results of these studies will give important clues to research in the field of peroxidase as well as to research in the field of phosphatase. It is clear that the field of vanadium biochemistry is rapidly developing, and further discoveries and surprises are expected.

REFERENCES

Allen, R. C. (1975). Halide dependence of the myeloperoxidase mediated antimicrobial system of the polymorphonuclear leucocyte in the phenomenon of electronic excitation. *Biochem. Biophys. Res. Commun.* **63**, 675–683.

Arber, J. M., De Boer, E., Garner, C. D., Hasnain, S. S., and Wever, R. (1989). Vanadium K-edge X-ray absorption spectroscopy of bromoperoxidase from *Ascophyllum nodosum. Biochemistry* **28**, 7968–7973.

Asplund, G., and Grimvall, A. (1991). Organohalogens in nature; more widespread than previously assumed. *Environ. Science Technol.* **25**, 1347–1350.

Barnett, P., Hondmann, D. H., Simons, L. H., Ter Steeg, P. F., and Wever, R. (1995). Enzymatic antimicrobial compositions. Patent Cooperation Treaty WO95/27046.

Barnett, P., Kruitbosch, D. L., Hemrika, W., Dekker, H. L., and Wever, R. (1997). The regulation of the vanadium chloroperoxidase from *Curvularia inaequalis. Biochim. Biophys. Acta* **1352**, 73–84.

Barrie, L. A., Bottenheim, M. J. W., Schnell, R. C., Crutzen, P. J. and Rasmussen, R. A. (1988). Ozone destruction and photochemical reactions at polar sunrise in the lower Arctic atmosphere. *Nature* **318**, 550–553.

Benemann, J. R., McKenna, C. E., Lie, R. F., Traylor, T. G., and Kamen, M. D. (1972). The vanadium effect in nitrogen fixation by Azotobacter. *Biochim. Biophys. Acta* **264,** 25–38.

Bertrand, D. (1950). Survey of contempary knowledge of biogeochemistry. 2. The biochemistry of vanadium. *Bull. Am. Mus. Nat. Hist.* **94,** 403–456.

Bishop, P. (1986). A second nitrogen fixation system in *Azotobacter vinelandii. Trends Biochem. Sci.* **11,** 225–228.

Bortels, H. (1936) *Zentralbl. Bakteriol. Parasitenkd. Infectionskr. Hyg.* **Abt II 95** 193–218.

Burns, R. C., Fuchsman, W. H., and Hardy, R. W. F. (1971). Nitrogenase from vanadium-grown *Azotobacter:* Isolation, characteristics and mechanistic implications. *Biochem. Biophys. Res. Commun.* **42,** 353–358.

Butler, A., and Walker, J. V. (1993). Marine haloperoxidases. *Chem. Rev.* **93,** 1937–1944.

Cammack, R. (1986). Nitrogen fixation. A role for vanadium at least. *Nature* **322,** 312.

Carrano, C. J., Mohan, M., Holmes, S. M., De la Rosa, R., Butler, A., Charnock, J. M., and Garner, C. D. (1994). Oxovanadium(V) alkoxo-chloro complexes of the hydridotripyrazolylborates as models for the binding site in bromoperoxidase. *Inorg. Chem.* **33,** 646–655.

Chan, M. K., Kim, J., and Rees, D. C. (1993). The nitrogenase FeMo-Cofactor and P-cluster pair: 2.2 Å resolution structure. *Science* **260,** 792–794.

Chasteen, N. D. (1983). The biochemistry of vanadium. *Struct. Bonding* **53,** 105–138.

Chasteen, N. D. (1990). *Vanadium in Biological Systems.* Kluwer Academic Press, The Netherlands, 225 pp.

Chisnell, J. R., Premakumar, R., and Bishop, P. E. (1988). Purification of a second alternative nitrogenase from a *nifHDK* deletion strain of *Azotobacter vinelandii. J. Bacteriol.* **170,** 27–33.

Colpas, G. J., Hamstra, B. J., Kampf, J. W., and Pecoraro, V. L. (1996). Functional models for vanadium haloperoxidase: Reactivity and mechanisms of halide oxidation. *J. Am. Chem. Soc.* **118,** 3469–3478.

Crans, D. C. (1994). Aqueous chemistry of labile oxovanadates: Relevance to biological studies. *Comm. Inorg. Chem.* **16,** 1–33.

De Boer, E., Boon, K., and Wever, R. (1988a). Electron paramagnetic resonance studies on conformational states and metal ion exchange properties of vanadium bromoperoxidase. *Biochemistry* **27,** 1629–1635.

De Boer, E., Keijzers, C. P, Klaassen, A. A. K., Reijerse, E. J., Collison, D., Garner, C. D., and Wever, R. (1988b). ^{14}N-coordination to VO^{2+} in reduced vanadium bromoperoxidase, an electron spin echo study. *FEBS Lett.* **235,** 93–97.

De Boer, E., Plat, H., Tromp, M. G. M., Franssen, M. C. R., Van Der Plas, H. C., Meijer, E. M., Schoemaker, H. E., and Wever, R. (1987). Vanadium containing bromoperoxidase: An example of an oxidoreductase with high operational stability in aqueous and organic media. *Biotechnol. Bioeng.* **30,** 607–610.

De Boer, E., Van Kooyk, Y., Tromp, M. G. M., Plat, H., and Wever, R. (1986a). Bromoperoxidase from *Ascophyllum nodosum:* A novel class of enzymes containing vanadium as a prosthetic group? *Biochim. Biophys. Acta* **869,** 48–53.

De Boer, E., Tromp, M. G. M., Plat, H., Krenn, G. E., and Wever, R. (1986b). Vanadium(V) as an essential element for haloperoxidase activity in marine brown algae: Purification and characterisation of a vanadium(V)-containing bromoperoxidase from *Laminaria saccharina. Biochim. Biophys. Acta* **872,** 104–115.

De Boer, E., and Wever, R. (1988). The reaction mechanism of the novel vanadium-bromoperoxidase: A steady-state kinetic analysis. *J. Biol. Chem.* **263,** 12326–12332.

Dexter, A. F., Lakner, F. J., Campbell, R. A., and Hager, L. P. (1995). Highly enantioselective epoxidation of 1,1-disubstituted alkenes catalyzed by chloroperoxidase. *J. Am. Chem. Soc.* **117,** 6412–6413.

Eady, R. E. (1995). Vanadium nitrogenase of *Azotobacter*. In H. Sigel and A. Sigel (Eds.), *Metal Ions in Biological Systems. Vol. 31, Vanadium and Its Role in Life.* Marcel Dekker, New York, Basel, Hong Kong, pp. 363–405.

Ellis, M. B. (1971). *Dematiaceous hyphomycetes.* Commonwealth Mycologica Institute, Kew, Surrey, England, pp. 1–61.

Espenson, J. H., Pestovsky, O., Huston, P., and Staudt, S. (1994). Organometallic catalysis in aqueous solution-oxygen transfer to bromide. *J. Am. Chem. Soc.* **116**, 2869–2877.

Franssen, M. C. R. (1994). Halogenation and oxidation reactions with haloperoxidases. *Biocatalysis* **10**, 87–111.

Franssen, M. C. R., Jansma, J. D., Van der Plas, H. C., De Boer, E., and Wever, R. (1988). Enzymatic bromination of barbituric acid and some of its derivatives. *Bioorg. Chem.* **16**, 352–363.

Fu, H., Kondo, H., Ichiwata, Y., Look, G. C., and Wong, C.-H. (1992). Chloroperoxidase-catalyzed asymmetric synthesis and enantioselective reactions of chiral hydroperoxides with sulphide and bromohydration of glycals. *J. Org. Chem.* **57**, 7265–7270.

Geschwend, P. M., MacFarlane, J. K., and Newman, K. A. (1985). Volatile halogenated organic compounds released to seawater from temperate marine macroalgae. *Science* **227**, 1033–1036.

Gribble, G. W. (1992). Naturally occurring organohalogen compounds—A survey. *J. Nat. Products* **55**, 1353–1395.

Hales, B. J., Case, E. E., Morningstar, J. E., Dzeda, M. F., and Mauterer, L. A. (1986). *Biochemistry* **25**, 7251–7255.

Harvey, I., Arber, J. M., Eady, R. R., Smith, B. E., Garner, C. D., and Hasnain, S. S. (1990). Iron K-edge X-ray absorption spectroscopy of the iron-vanadium cofactor of the vanadium nitrogenase from *Azotobacter chroococcum*. *Biochem. J.* **266**, 929–931.

Harvey, P. J., Schoemaker, H. E., and Palmer, J. M. (1986). Veratryl alcohol as a mediator and the role of the radical cations in the lignin biodegradation by *Phanerochaete chrysosporium*. *FEBS Lett.* **195**, 242–246.

Hecht, M. J., Sobek, M., Maag, T., Pfeifer, O., and Van Pée, K.-H. (1994). The metal-ion free oxidoreductase from *Streptomyces aureofaciens* has an α/β-hydrolase fold. *Nature Struct. Biol.* **1**, 532–537.

Hemrika, H., Renirie, R. Dekker, H. L, Barnett, P., and Wever, R. (1997). From phosphatases to vanadium peroxidases: A similar architecture of the active site. *Proc. Natl. Ac. Sci. USA,* **94**, 2145–2149.

Hunter-Cevera, J. C., Belt, A., Sotos, L. S., and Fonda, M. E. (1990). Fungal chloroperoxidase method. United States Patent No. 4937192.

Hunter-Cevera, J. C., and Sotos, L. S. (1986). Screening for a new enzyme in nature: Haloperoxidase production by Death Valley dematiaceous hyphomycetes. *Microb. Ecol.* **12**, 121–127.

Kim, J., Woo, D., and Rees, D. C. (1993). X-ray crystal structure of the nitrogenase molybdenum–iron protein from *Clostridium pasteurianum* at 3.0-Å resolution. *Biochemistry* **32**, 7104–7115.

Kringstad, K. P., and Lindström K. (1994). Spent liquors from pulp bleaching. *Environ. Sci. Technol.* **18**, 236A–248A.

Kusthardt, U., Hedman, B., Hodgson, K. O., Hahn, R., and Vilter, H. (1993). High resolution XANES studies on vanadium-containing haloperoxidase; pH dependence and substrate binding. *FEBS Lett,* **329**, 5–8.

Lei, K. J., Pan, C. J., Liu, J. L, Shelly, L. L., and Yang Chou, J. (1995). Structure–function analysis of human glucose-6-phosphatase, the enzyme deficient in glycogen storage disease type 1a. *J. Biol. Chem.* **270**, 11882–11886.

Lei, K. J., Shelly, L. L., Pan, C. J., Sidbury, J. B., and Yang Chou, J. (1993). Mutations in the glucose-6-phosphatase gene that causes glycogen storage disease type 1a. *Science* **262**, 580–583.

Liu, T.-N., M'Timkulu, T., Geigert, J., Wolf, B., Neidleman, S. L., Silva, D., and Hunter-Cevera, J. C. (1987). Isolation and characterization of a novel non-heme chloroperoxidase. *Biochem. Biophys. Res. Commun.* **142,** 329–333.

McKenna, C. E., Benemann, J. R., and Taylor, T. G. (1970). A vanadium containing nitrogenase preparation: Implications for the role of molybdenum in nitrogen fixation. *Biochem. Biophys. Res. Commun.* **41,** 1501–1508.

Meister, G. E., and Butler, A. (1994). Molybdenum(VI) and tungsten(VI)-mediated biomimetic chemistry of vanadium bromoperoxidase. *Inorg. Chem.* **33,** 3269–3275.

Meister Winter, G. E., and Butler, A. (1996). Inactivation of vanadium bromoperoxidase: Formation of 2-oxohistidine. *Biochemistry* **35,** 11805–11811.

Messerschmidt, A., and Wever, R. (1996). X-ray structure of a vanadium-containing enzyme: Chloroperoxidase from the fungus *Curvularia inaequalis. Proc. Natl. Acad. Sci. USA* **93,** 392–396.

Moore, R. M., Webb, M., Tokarczyk, R., and Wever, R. (1996). Bromoperoxidase and iodoperoxidase enzymes and production of halogenated methanes in marine diatom cultures. *J. Geophys. Res.* **101,** 20,899–20,908.

Morningstar, J. E., and Hales, B. J. (1987). Electron paramagnetic resonance study of the vanadium iron protein of nitrogenase from *Azotobacter vinelandii. J. Am. Chem. Soc.* **107,** 6854–6855.

Neidleman, S. L., and Geigert, J. (1986). *Biohalogenation: Principles, Basic Rules and Application.* Ellis Horwood Limited, John Wiley and Sons, New York, 203 pp.

Poulos, T. L., and Kraut, J. (1980). The stereochemistry of peroxide catalysis. *J. Biol. Chem.* **255,** 8199–8205.

Rehder, D. (1991). The bioinorganic chemistry of vanadium. *Angew. Chem. Int. Ed. Engl.* **30,** 148–167.

Robson, R. L., Eady, R. R., Richardson, T. H., Miller, R. W., Hawkins, M., and Postgate, J. R. (1986). The alternative nitrogenase of *Azotobacter chroococcum* is a vanadium enzyme. *Nature* **322,** 388–390.

Sigel, H., and Sigel, A. (Eds.). (1995). *Metal Ions in Biological Systems.* Vol. 31, *Vanadium and Its Role in Life.* Marcel Dekker, New York, Basel, Hong Kong, 779 pp.

Simons, L. H., Barnett, P., Vollenbroek, E. G. M., Dekker, H. L., Muijsers, A. O., Messerschmidt, A., and Wever, R. (1995). Primary structure and characterization of the vanadium chloroperoxidase from the fungus *Curvularia inaequalis. Eur. J. Biochem.* **229,** 566–574.

Sing, J., Nordly, R. A., and Jorgenson, R. A. (1981). Vanadate; a potent inhibitor of multifunctional glucose-6 phosphatase. *Biochim. Biophys. Acta* **678,** 477–482.

Smith, M. J., Ryan, D. E., Nakanishi, K., Frank, P., and Hodgson, K. O. (1995). Vanadium in ascidians and the chemistry of tunichromes. In H. Sigel and A. Sigel (Eds.), *Metal Ions in Biological Systems. Vol. 31, Vanadium and Its Role in Life.* Marcel Dekker, New York, Basel, Hong Kong, pp. 423–490.

Soedjak, H. S., and Butler, A. (1990). Characterization of vanadium bromoperoxidase from *Macrocystis* and *Fucus:* Reactivity of vanadium bromoperoxidase towards acyl and alkylperoxides and bromination of amines. *Biochemistry* **29,** 7974–7981.

Soedjak, H. S., and Butler, A. (1991). Mechanism of dioxygen formation catalyzed by vanadium bromoperoxidase from *Macrocystis pyrifera* and *Fucus distichus:* Steady state kinetic analysis and comparison to the mechanism of bromination of V-BPO from *Ascophylum nodosum. Biochim. Biophys. Acta* **1079,** 1–7.

Stankiewicz, P. J. Tracey, A. S., and Crans, D. C. (1995). Inhibition of phosphate-metabolizing enzymes by oxovanadium(V) complexes. In H. Sigel and A. Sigel (Eds.), *Metal Ions in Biological Systems. Vol. 31, Vanadium and Its Role in Life.* Marcel Dekker, New York, Basel, Hong Kong, pp. 287–324.

Tromp, M. G. M., Olafsson, G., Krenn, B. E., and Wever, R. (1990). Some structural aspects of vanadium bromoperoxidase from *Ascophyllum nodosum. Biochim. Biophys. Acta* **1040,** 192–198.

Tromp, M. G. M., Van, T. T., and Wever, R. (1991). Reactivation of vanadium bromoperoxidase; inhibition by metallofluoric compounds. *Biochim. Biophys. Acta* **1079**, 53–56.

Tschirret-Guth, R. A., and Butler, A. (1994). Evidence for organic substrate binding to vanadium bromoperoxidase. *J. Am. Chem. Soc.* **116**, 411–412.

Van Schijndel, J. W. P. M, Barnett, P., Roelse, J., Vollenbroek, E. G. M., and Wever, R. (1994). The stability and steady-state kinetics of vanadium chloroperoxidase from the fungus *Curvularia inaequalis. Eur. J. Biochem.* **225**, 151–157.

Van Schijndel, J. W. P. M., Simons, L. H., Vollenbroek, E. G. M., and Wever, R. (1993a). The vanadium chloroperoxidase from the fungus *Curvularia inaequalis.* Evidence for the involvement of a histidine residue in the binding of vanadate. *FEBS Lett.* **336**, 239–242.

Van Schijndel, J. W. P. M, Vollenbroek, E. G. M., and Wever, R. (1993b). The chloroperoxidase from the fungus *Curvularia inaequalis:* A novel vanadium enzyme. *Biochim. Biophys. Acta* **1161**, 249–256.

Vilter, H. (1984). Peroxidases from Phaephycae: A vanadium(V) dependent peroxidase from *Ascophyllum nodosum. Phytochemistry* **23**, 1387–1390.

Vilter, H. (1995). Vanadium dependent haloperoxidases. In H. Sigel and A. Sigel (Eds.), *Metal Ions in Biological Systems.* Vol. 31, *Vanadium and Its Role in Life.* Marcel Dekker, New York, Basel, Hong Kong, pp. 325–362.

Vollenbroek, E. G. M., Simons, L. H., Van Schijndel, J. W. P. M., Barnett, P., Balzar, M., Dekker, H., Van der Linden, C., and Wever, R. (1995). Vanadium chloroperoxidases occur widely in nature. *Biochem. Soc. Trans.* **23**, 267–271.

Walker, J. V., and Butler, A. (1996). Vanadium bromoperoxidase-catalyzed oxidation of thiocyanate by hydrogen peroxide. *Inorg. Chem. Acta* **243**, 201–206.

Wever, R. (1988). Ozone destruction by algae in the Arctic atmosphere. *Nature* **335**, 501.

Wever, R. (1993). Sources and sinks of halogenated methanes in nature. In J. C. Murell and D. P. Kelly (Eds.), *Microbial Growth on CI Compounds.* Intercept Ltd., Andover, Hampshire, England. pp. 35–46.

Wever, R., De Boer, E., Krenn, B. E., Offenberg, H., and Plat, H. (1988). Structure and function of vanadium-containing bromoperoxidase. *Prog. Clin. Biol. Res.* **274**, 477–493.

Wever, R., and Krenn, B. E. (1990). Vanadium haloperoxidases. In N. D. Chasteen (Ed.), *Vanadium in Biological Systems.* Kluwer Academic Press, The Netherlands, pp. 81–97.

Wever, R., and Kustin, K. (1990). Vanadium: A biologically relevant element. *Adv. Inorg. Chem.* **35**, 103–137.

Wever, R., Tromp, M. G. M., Krenn, B. E., Marjani, A., and Van Tol, M. (1990). Formation of HOBr by seaweed: A novel defense mechanism? Impact on the biosphere. *Environ. Sci. Technol.* **25**, 446–449.

Wever, R., Tromp, M. G. M., Van Schijndel, J. W. P. M., Vollenbroek, E. G. M., Olssen, R. L., and Fogelqvist, E. (1991). Bromoperoxidases, their role in the formation of HOBr and bromoform by seaweeds. In R. S. Oremland (Ed.), *Biogeochemistry of Global Change.* Chapman and Hall, New York, pp. 811–824.

13

CATALYTIC EFFECTS OF VANADIUM ON PHOSPHORYL TRANSFER ENZYMES

George L. Mendz

School of Biochemistry and Molecular Genetics, The University of New South Wales, Sydney, NSW 2052, Australia

Vanadium in the Environment. Part 1: Chemistry and Biochemistry, Edited by Jerome O. Nriagu.
ISBN 0-471-17778-4.   © 1998 John Wiley & Sons, Inc.

307

1. INTRODUCTION

Vanadium is utilized in many processes in the petrochemical, mining, steel, glass, and utilities industries. For example, it is used as a catalyst in redox reactions to produce acids and in the combustion of lead-free petrol, and it is an industrial residue from power-generating plants (Chapter 1).

Although vanadium compounds exert potent effects in a wide variety of living systems, the recognition of its biological relevance and consequently its environmental importance does not have a long history. Some of its effects include (a) regulation of phosphate-metabolizing enzymes, (b) halogenation of organic compounds by vanadate-dependent non-heme peroxidases from seaweeds, (c) reductive protonation of nitrogen (nitrogen fixation) by alternative, that is, vanadium-containing, nitrogenases from N_2-fixing bacteria, (d) vanadium sequestering by sea squirts (ascidians), and (e) amavadine, a low-molecular-weight complex of V^{3+} accumulated in the fly agaric and related toadstools (Rehder, 1992).

Exposure to vanadium compounds could have noxious effects on humans. The metal can modulate immune responses and it induces structural, functional, and biochemical alterations in several types of immune cells in animal models. The carcinogenic effects of vanadium ions have been attributed to their ability to modify and cross-link DNA and to initiate lipid peroxidation; and their toxicity has been attributed to the effects of vanadyl and vanadate ions on phosphoryl transfer reactions. The wide utilization of vanadium and its potential impact on health make it an important pollutant; environmental and toxicological data indicate an increased exposure of inhabitants of industrial areas to this metal.

Several reviews describing interactions of vanadium with tissues, cells, and enzymes have appeared (Rehder, 1992; Dafnis and Sabatini, 1994; Leonard and Gerber, 1994; Zelikoff and Cohen, 1995) since the last major work on the effects of vanadium on biological systems (Chasteen, 1990). This chapter summarizes discoveries made during the last few years on the effects of vanadium on enzymes that catalyze phosphoryl transfer reactions, beginning with ion-translocating ATPases and ending with novel effects of vanadium ions on tissues and cells. To help in better understanding the significance of the work described, a brief outline of the biological background is given where appropriate.

2. ION-TRANSLOCATING ATPases

Many of the effects of vanadium ions on biological system occur through their action on ATPases. Considering that ATP hydrolysis is coupled to large numbers of cellular processes, for example transport, polypeptide synthesis, muscle contraction, proteolysis, energy production, and nitrogen fixation, it is easy to understand that vanadium ions would affect directly or indirectly many important cell functions.

Although the best-characterized interactions of vanadate and vanadyl ions with ATP-hydrolyzing enzymes are those with ion-translocating ATPases, new effects on this type of enzymes continue to be discovered. There are three families of ATPases that carry out active transport of ions across membranes: phosphorylated (P-type), vacuolar (V-type), and mitochondrial (F_0F_1-type) ATPases.

P-type ATPases form a covalent phosphorylated intermediate when the γ-phosphate of ATP reacts with an aspartic acid residue in the enzyme cycle. Examples are some Ca^{2+}-transporters, H^+-transporters of the plasma membrane of eukaryotes, gastric H^+/K^+-exchangers, Na^+/K^+-exchangers, and bacterial K^+-transporters (Pedersen and Carafoli, 1987). P-type ATPases are very sensitive to vanadate with IC_{50} of less than 10 μM (Forgac, 1989; Sachs et al., 1989). The inhibition of P-type ATPases does not require ADP, and the presence of Mg^{2+} is necessary and sufficient for vanadate inhibition.

Vanadate and vanadyl ions inhibit (Na^+, K^+)- and (Ca^{2+}, Mg^{2+})-ATPases from synaptosomal membranes of the parietal lobe of the human brain. The oxyanion is an effective inhibitor of both ATPases in vitro at concentrations above 50 μM, as is the oxoion at 10 nM concentrations. Vanadate seems to be an uncompetitive inhibitor of (Na^+, K^+)-ATPase ($K_i = 880$ nM) (Janiszewska et al., 1994).

The urine of salt-loaded healthy subjects contains at least three ouabain-like factors (OLF) with different polarities and cross-reactivities with a digoxin antibody. A more polar OLF-1 and a more apolar OLF-2 factor have been isolated recently and purified to single compounds that inhibit (Na^+, K^+)-ATPase in a dose-dependent manner (Kramer et al., 1995). Mass spectroscopy showed a molecular weight of around 400 and ^{1}H-NMR and IR spectroscopy suggested diascorbic acid salts. Vanadium diascorbates of molecular weight 403 have similar elution times from reverse phase HPLC as OLF. IC_{50} of 9 × 10^{-5} and 2 × 10^{-6} M were determined for V^{3+} diascorbate and V^{5+} diascorbate, respectively. Enzyme inhibition was noncompetitive with respect to sodium and Mg^{2+}-ATP; and assays with p-nitrophenylphosphate (p-NPP) showed strong inhibition in the E_2 configuration of the (Na^+,K^+)-ATPase. The results were interpreted as suggesting that vanadium diascorbates represent endogenous OLF excreted in human urine (Kramer et al., 1995).

V-type ATPases are associated with membrane-bound organelles other than mitochondria and sarcoplasmic and endoplasmic reticulums (Pedersen and Carafoli, 1987) and are involved in electrogenic H^+ translocation. They are found in yeast vacuoles and chromaffin granules (Beltran and Nelson, 1992), plant tonoplasts, and in vesicles such as those present in chicken osteoclasts (Chatterjee et al., 1992) and chicken kidney microsomes (David et al., 1996). V-type ATPases are much less sensitive to vanadate inhibition with IC_{50} of 100–1,000 μM. The mechanisms of inhibition of V-type ATPases appear to involve the formation of a vanadate complex with ADP and Mg^{2+} that may form a transition-state-like complex that interacts with the phosphate-binding region of the nucleotide-binding site of the enzyme, possibly at the catalytic site itself (David et al., 1996).

F-type ATPases are found in bacteria, chloroplasts, and mitochondria (Pedersen and Carafoli, 1987). They are the most complicated of the ATPases and comprise the F_1 moiety involved in catalytic activity, either ATP synthesis or hydrolysis, and the F_0 moiety involved in H^+ translocation. Their reaction cycle involves tight binding of ATP, but not the formation of a covalent phosphorylated intermediate. F-type ATPases are relatively insensitive to vanadate inhibition. For instance, plasma membranes from phototrophic cyanobacteria contain P-type ATPases, which are sensitive to orthovanadate and F-type ATPases, which are not inhibited by the oxyanion. Thylakoid membranes, on the other hand, contain only vanadate-insensitive F-type ATPases (Neisser et al., 1994).

However, the conventional classification of ion-translocating ATPases may require revision as knowledge of their properties increases. The cystic fibrosis transmembrane conductance regulator (CFTR) shares structural homology with the ABC (ATP-binding cassette) transporter superfamily of proteins that hydrolyze ATP to effect the transport of compounds across cell membranes. Some superfamily members are characterized as P-type ATPases. Vanadate does not alter the gating of CFTR when used at concentrations that completely inhibit the activity of other ABC transporters (1 mM). Higher concentrations of vanadate (10 mM) block the closing of CFTR but do not affect the opening of the channel. More selective P-type (ouabain), V-type (bafilomycin, SCN^-), and F-type (oligomycin) ATPase inhibitors do not affect either the opening or the closing of CFTR. Thus, CFTR does not share a pharmacological inhibition profile with other ATPases. Vanadate inhibition of channel closure might suggest that a hydrolytic step is involved, although the requirement of a high concentration raises the possibility of previously uncharacterized effects of the anion. This effect of vanadate demonstrates that the binding site of the oxyanion is considerably different than that of other ABC transporters (Schultz et al., 1996).

3. Ca²⁺-DEPENDENT ENZYMES

Calcium ions are required by many enzymic systems, including those responsible for the contractile properties of muscle, the transmission of the nerve impulse, the functioning of the blood-clotting mechanisms, and the modulation of hormonal action. Ca^{2+} ions regulate a number of metabolic processes in the cytosol and for this reason they are considered second messengers. Many cells respond to extracellular stimuli by altering their intracellular calcium concentrations, which in turn induce biochemical changes directly or through interactions with proteins such as calmodulin, calsequestrin, calcitonin, and regucalcin. Conversely, calcium levels are controlled to a large extent by the action of second messengers, in particular cyclic AMP. The metabolism of calcium is closely related to that of inorganic phosphate, from their initial dietary sources, through their metabolic behavior, to their excretory fates. For this reason it is expected that vanadium ions would have profound influence on metabolic processes involving Ca^{2+}. Another reason is that vanadyl ions are able to substitute Ca^{2+} and Mg^{2+} in binding interactions.

3.1. Calmodulin

Vanadyl (VO^{2+}) binds to calmodulin with an apparent stoichiometry of 4:1; the binding is competitive with calcium and magnesium cations (Nieves et al., 1987). In the presence of Ca^{2+}, VO^{2+} inhibits calmodulin-activated skeletal muscle myosin light chain kinase with an IC_{50} of 100 μM, presumably by forming VO^{2+}-calmodulin complexes (Parra-Diaz et al., 1995a).

3.2. Calcineurin

Calcineurin is a serine/threonine protein phosphatase 2B that exists as a heterodimer and requires metal ions and calmodulin for full enzymatic activity. The 2B protein phosphatases have an absolute requirement for Ca^{2+}, and very low phosphorylase phosphatase activity (Cohen, 1989). The catalytic site and binding region for calmodulin are contained in the A subunit of calcineurin. Four Ca^{2+}-binding sites are located in the B subunit in the form of "EF hands," which exhibit homology with calmodulin and troponin. The target of calcineurin is the nuclear factor of activated T cells, a phosphorylated transcriptional activator of the interleukin-2 gene. Ca^{2+} and vanadyl (VO^{2+}) bind to distinct sites in the phosphatase. There are two classes of VO^{2+} binding sites and this cation has greater affinity than Ca^{2+} for the first two binding sites in the B subunit. Enzyme activity is enhanced 100% in the presence of Ca^{2+} and calmodulin, and 160% in the presence of VO^{2+} and calmodulin. Vanadyl itself stimulates enzyme activity in the absence of Ca^{2+} and calmodulin (Parra-Diaz et al., 1995b).

3.3. Ca^{2+}-ATPases

Vanadate is an effective tool for studying the mechanisms of Ca^{2+}-ATPases because it can mimic phosphate in the formation of the phosphoenzyme intermediate during ATP hydrolysis.

The sarcoplasmic reticulum (SR) Ca^{2+}-ATPase constitutes a transmembrane transport system that accumulates Ca^{2+} at the expense of ATP hydrolysis. Initially it was thought that vanadate affected these Ca^{2+} pumps in the same way as other P-type ATPases (Pedersen and Carafoli, 1987), but recent studies show the existence of Ca^{2+}-ATPases with different sensitivities to vanadate. In SR vesicles coupled Ca^{2+} uptake and H^+ extrusion are inhibited by decavanadate (V_{10}) with an IC_{50} of 200 μM (2 mM total vanadate), but monovanadate (V_1) at concentrations up to 660 μM (2 mM total vanadate) has no effect. The inhibition of calcium translocation by V_{10} increases in the presence of Mg^{2+}, whereas the addition of 10 mM Mg^{2+} induces only a small stimulation of the pump in the presence of V_1. Hydrolysis of ATP is inhibited by both V_1 and V_{10} with an IC_{50} of approximately 3 and 1 mM, respectively. The effects of mono- and decavanadate are completely different in the uncoupled ATPase; V_1 does not inhibit ATP hydrolysis, and V_{10} does. Apparently the increase of Ca^{2+} concentration in the vesicles counteracts V_1 inhibition of the coupled SR Ca^{2+}-ATPase activity, but it does not affect V_{10} activity (Aureliano and Madeira, 1994).

The *Trypanosoma cruzi* plasma-membrane Ca^{2+}-ATPase partially purified from epimastigote plasma-membrane vesicles by calmodulin–agarose affinity chromatography has a molecular mass of 140 kDa. The purified enzyme is stimulated by *T. cruzi* or bovine brain calmodulin and is very sensitive to vanadate, indicating that it belongs to the P-type class of ionic pumps (Benaim et al., 1995).

Although vanadate inhibits with high affinity ($K_i = 3 \mu M$) the Ca^{2+}-ATPase activity of human red cell membranes, calmodulin-free membranes incubated with vanadate increase their Ca^{2+}-ATPase activity in media with Mg^{2+} and K^+. Pretreatment with Mg^{2+} and vanadate increases the apparent affinity of the Ca^{2+}-ATPase for Ca^{2+} and ATP, and elicits the appearance of a second (low-affinity) site for ATP (Romero, 1993). The effect is attributed to a change in the kinetic properties of the enzyme: an increase in the rate of phosphorylation and in turnover number. It is also suggested that the preincubation with Mg^{2+} and vanadate may stimulate the dephosphorylation and the conversion of conformer E_2 to E_1 that takes place during the reaction cycle of Ca^{2+}-ATPase (Romero and Rega, 1995).

3.4. Ecto-ATPases

Ecto-ATPases, which hydrolyze ATP to ADP, are present in almost all tissues. They differ from transport ATPases in having the ATP-hydrolyzing site located on the outside of the cell surface, in having broad nucleotide-hydrolyzing activity, and in being unaffected by several inhibitors of other ATPases, including vanadate. They are activated to similar extents by either Ca^{2+} or Mg^{2+}. Ecto-ATPases differ from nonspecific phosphatases in that the presence of other phosphorylated compounds such as β-glycerophosphate or p-NPP does not affect the hydrolysis of ATP (Meghji and Burnstock, 1995). Lack of sensitivity to vanadate is an intrinsic property of ecto-ATPases and is not tissue-dependent. For example, in rat myometrium there is a Ca^{2+}-ATPase involved in Ca^{2+} transport, and two Ca^{2+} ecto-ATPases; the transporter has an IC_{50} of 400 μM for vanadate (Enyedi et al., 1988), but the latter enzymes are unaffected by the oxyanion (Magocsi and Penniston, 1991).

3.5. Regucalcin and Ca^{2+}-Dependent Liver Enzymes

Regucalcin is a Ca^{2+}-binding protein found in the hepatic cytoplasm of rats. It increases the activity of the (Ca^{2+},Mg^{2+})-ATPase, which functions as a Ca^{2+} pump in the plasma membrane of rat liver cells. Administration of Ca^{2+} to rats stimulates the expression of regucalcin mRNA in rat liver (Shimokawa and Yamaguchi, 1992). Regucalcin has reversible effects on the activation or inhibition by Ca^{2+} and/or calmodulin of several liver enzymes, including protein kinase C, Ca^{2+}/calmodulin-dependent cyclic nucleotide phosphodiesterase, and deoxyuridine 5′-triphosphatase, suggesting that it plays an important role in the regulation of liver cell functions related to Ca^{2+}.

Vanadate reduces the (Ca^{2+},Mg^{2+})-ATPase activity of hepatic plasma membranes from normal rats with an IC_{50} greater than 200 μM, but it does not affect the increase in (Ca^{2+},Mg^{2+})-ATPase activity in liver plasma membranes of rats fed with $CaCl_2$. This activation is attributed to elevated levels of regucalcin and indicates that this protein is not involved in the Ca^{2+}-dependent phosphorylation of the enzyme but acts on the sulfhydryl groups at the active site of the (Ca^{2+},Mg^{2+})-ATPase (Takahashi and Yamaguchi, 1995).

4. HIGH-ENERGY PHOSPHATE PHOSPHORYL TRANSFER ENZYMES

The ability of vanadate to form mixed phosphate–vanadate anhydrides is an important factor in respect to the effects these ions have on the rates of phosphoryl transfer reactions involving sugar-phosphates and other high-energy phosphate metabolites.

It has been reported that a complex formed by glucose-1-phosphate (Glc1P) and inorganic vanadate (V_i) inhibits phosphoglucomutase. Both the inhibition of the steady state and the rate of approach to steady state are dependent on the concentrations of Glc1P and V_i (Percival et al., 1990). Inhibition is competitive versus glucose-1,6-bisphosphate and is ascribed to binding of the 6-vanadate ester of Glc1P to the dephospho form of phosphoglucomutase (Percival et al., 1990). Other studies of the effects of vanadate on the steady-state rates of exchange of phosphoryl groups and the reactions catalyzed by phosphoglucomutase and phosphofructokinase reported that the presence of vanadium stimulated the catalytic activity of the enzymes in vitro (Mendz, 1990, 1991). Stimulation of enzyme activity by vanadate and vanadyl complexes was also observed for the interconversion of aldehyde and keto groups catalyzed by the enzymes phosphomannose isomerase, phosphoribose isomerase, and phosphoglucose isomerase (Mendz, 1991). Addition of vanadate also increased the rate constants of the interconversion of glucose-6-phosphate and fructose-6-phosphate in hemolysates (Mendz, 1991). A significant difference between the two studies was the concentrations of vanadium ions used; phosphoglucomutase inhibition was observed at micromolar concentrations of the oxyanion (Percival et al., 1990), whereas enzyme activation was measured at millimolar concentrations of the oxyions and oxoions (Mendz, 1991). It remains to be elucidated whether this difference would account for the apparently contradictory results.

There have also been conflicting reports on the effects of vanadium ions on phosphoglycerate mutase (PGMase). The steady-state catalytic activity of phosphoglycerate mutase in the coupled system formed by PGMase and enolase is stimulated by vanadate, but enolase activity is not affected (Mendz, 1991). Studies on the formation of complexes of vanadate with 2-phosphoglycerate and 3-phosphoglycerate yielded equilibrium constants for the formation of the two 2,3-bisphosphate analogues of 2.5 M^{-1} for 2-vanadio-3-phosphoglycerate and 0.2 M^{-1} for 2-phospho-3-vanadioglycerate (Liu et al., 1992). The results of binding studies to the enzyme are consistent with noncooperativity in the binding of vanadiophosphoglycerate to the two active sites of PGMase, and they support the view that vanadate–phosphoglycerate complexes are transition state analogues for the phosphorylation of PGMase by 2,3-bisphosphoglycerate and are more potent inhibitors of the enzyme than either monomeric or dimeric vanadate (Liu et al., 1992).

The rates of vanadate-stimulated hydrolysis of 2,3-bisphosphoglycerate in red cells and hemolysates is attributed principally to the activation of the phosphatase activity of 2,3-bisphosphoglycerate synthase (Mendz et al., 1990). Redox reactions involving vanadium ions are important in establishing the final equilibrium concentrations of vanadate and vanadyl, but the activation of the enzyme results from direct action of the vanadium ions on the phosphatase activity and not as a consequence of the alteration in the equilibrium of intracellular oxidants and reductants (Mendz et al., 1990).

5. TYROSINE KINASES AND PHOSPHATASES

Vanadium ions have inhibitory effects on the activities of tyrosine kinases (PTKases) and phosphatases (PTPases). This property has been employed to characterize tyrosine phosphoryl transfer enzymes and to classify enzymes of unknown biological functions as PTKases or PTPases. Vanadate and vanadyl inhibit directly both the soluble and particulate alkaline phosphatase (ALPase) activity from UMR106 cells and from bovine intestinal ALPase (Cortizo et al., 1994). Peroxo- and hydroperoxovanadium compounds inhibit ALPase activity in the soluble fraction of osteoblasts, but not in the particulate fraction; nor do they inhibit bovine intestinal ALPase. By means of inhibitors of PTPases, the soluble ALPase was partially characterized as a tyrosine phosphatase. The major activity in the particulate fraction represents the bone-specific ALPase activity. This work demonstrated that different forms of vanadium are direct inhibitors of ALPase activity and that the effect of the ions depends on the enzymatic activity investigated and the origin of the ALPase (Cortizo et al., 1994).

· Vanadate was used to characterize a protein from *Streptomyces coelicolor A3(2)* of unknown biological function with a deduced molecular weight of 17,690 that contained significant amino acid sequence identity with mammalian and prokaryotic small, acidic phosphotyrosine protein phosphatases (Li and Strohl, 1996). The protein was expressed in *Escherichia coli* and was purified to homogeneity as a fusion protein containing five extra amino acids. The purified fusion enzyme catalyzes the removal of phosphate from *p*-NPP, phosphotyrosine, and a commercial phosphopeptide containing a single phosphotyrosine residue; but it does not cleave phosphoserine or phosphothreonine. Hydrolysis of *p*-NPP by the *S. coelicolor* enzyme is competitively inhibited by dephostatin with a K_i of 1.64 μM. Vanadate and the known PTPase inhibitors phenylarsine oxide (PAO) and iodoacetate also inhibit enzyme activity (Li and Strohl, 1996).

The roles of tyrosine kinases and phosphatases in cellular processes have been clarified by investigation of the effects of vanadium ions on these enzymes. The activation of cultured Raw 264.7 murine macrophages with interferon-γ (IFN-γ) and lipopolysaccharide results in the expression of inducible nitric oxide synthase (NOSase) and the subsequent production of nitric oxide. The NOSase expressed in activated cells was characterized for possible posttranslational protein modification by endogenous tyrosine protein kinases (Pan et al., 1996). Western blot analysis using phosphotyrosine antibodies revealed that nitric oxide synthase is phosphorylated on tyrosine residues and that this is an early event, coinciding with the appearance of newly synthesized NOSase. A brief exposure of activated cells to vanadate significantly increases the level of NOSase tyrosine phosphorylation, suggesting that tyrosine phosphatases are dynamically involved in the regulation of this process. Vanadate treatment of activated cells also results in a rapid increase in enzyme activity, occurring within 5 min of exposure. These results provide evidence that tyrosine kinases and phosphatases are involved in the posttranslational modification of NOSase and may potentially play a role in modulating the functional activity of the enzyme in macrophages (Pan et al., 1996).

New functions for intracellular pools of vanadium ions have been proposed through studies of their effects on PTKases. Vanadyl inhibits receptor tyrosine kinases such as the insulin receptor (IC_{50} = 23 ± 4 μM) and the insulin-like growth factor-I receptor (IC_{50} = 19 ± 3 μM). The inhibition is noncompetitive with respect to ATP, Mn^{2+}, and substrate concentrations. In contrast, staurosporine-sensitive cytosolic protein tyrosine kinases are poorly inhibited by vanadyl; and vanadate stimulated several PTKases by a factor of 2–6. Cytosolic protein tyrosine kinases derived from rat adipocytes, liver, and brain are activated, but PTKases from Nb2 lymphoma cells are not affected. PTKase extracted from insulin-responsive tissues is more sensitive to vanadate activation (ED_{50} = 3 ± 0.7 μM), than the brain enzyme (ED_{50} = 27 ± 3 μM) (Elberg et al., 1994). Tungstate, molybdate, and PAO also stimulate PTKase, suggesting that the effect of vanadate is secondary to inhibiting protein phosphotyrosine phosphatases. On the basis of these results the authors proposed a hypothesis implicating the intracellular vanadyl pool in modulating PTKase activity. Any physiological conditions that convert vanadyl to vanadate, for example, production of hydrogen peroxide, will activate PTKase; consequently there will be PTKase-dependent bioeffects (Elberg et al., 1994).

An interesting characteristic of PTKases and PTPases that is not well understood is their wide range of sensitivities to vanadium ions even for enzymes from the same source with similar biological functions. The 15-kDa fatty acid binding protein 422/aP2 is phosphorylated on Tyr19 both in vitro by the insulin receptor tyrosine kinase, and in intact 3T3-L1 adipocytes treated with insulin and phenylarsine oxide. Phosphorylated 422/aP2 accumulates in cells treated with insulin and PAO because the arsenical blocks turnover of the phosphoryl group of the protein. These findings suggest that a PAO-sensitive enzyme mediates turnover of the protein tyrosine phosphoryl group. Two membrane protein tyrosine phosphatases from 3T3-L1 adipocytes that catalyze hydrolysis of phospho-Tyr19 of 422/aP2 have been purified and characterized (Liao et al., 1991). These enzymes, designated PTPases HA1 and HA2, have molecular masses of approximately 60 and 38 kDa, respectively, and were shown to differ markedly in their sensitivity to both vanadate and phosphotyrosine. Vanadate at 250 μM concentration causes 100% inhibition of PTPase HA1 (K_i ~20 μM), and less than 10% inhibition of PTPase HA2 (K_i ~200 μM). Phosphotyrosine completely inhibits PTPase HA1 at 500 μM but has no detectable effect on PTPase HA2 activity. Both enzymes exhibit substrate preference for 422/aP2 when compared with other phosphotyrosine-containing protein substrates, and proteins containing phosphoserine and phosphothreonine do not serve as substrates for the enzymes. PTPase HA2 is expressed both in 3T3-L1 preadipocytes and adipocytes, whereas PTPase HA1 is expressed only in 3T3-L1 adipocytes (Liao et al., 1991).

When confluent 3T3-L1 preadipocytes are induced to differentiate into adipocytes, they undergo mitotic clonal expansion followed by growth arrest and then coordinate expression of adipocyte genes. During clonal expansion, expression of PTPase HA2 increases abruptly and then decreases concomitantly with the transcriptional activation of adipocyte genes. Constitutive expression of the PTPase by 3T3-L1 preadipocytes using a PTPase HA2 expression vector prevents adipocyte gene expression and differentiation into adipocytes. Exposure of transfected preadipocytes to vanadate, just as clonal expansion ceases, restores their capacity to differentiate. Treatment of transfected preadipocytes with vanadate prior to or during clonal expansion fails to reverse PTPase HA2-blocked differentiation, whereas treatment of untransfected preadipocytes during mitotic clonal expansion blocks differentiation. Vanadate added following clonal expansion has no effect on differentiation. Thus, the effects of vanadate on PTPase HA2 served to establish that critical tyrosine phosphorylation events occur between termination of clonal expansion and initiation of adipocyte gene expression, while critical tyrosine dephosphorylation events occur during clonal expansion (Liao and Lane, 1995). These results indicate that vanadate also affects cell differentiation.

6. RESPIRATORY ENZYMES

Many important physiological responses elicited by vanadium from biological systems have been studied in detail, but only recently has attention been given to the effects of vanadium ions on respiration at the subcellular level. Their effects on mitochondria of different organisms have been investigated (Henderson et al., 1989; Zychlinski and Byczkowski, 1990); and vanadate has been used as a probe to analyze respiratory status (Desautels and Dulos, 1993), metabolic control (Wisniewski et al., 1995), and calcium homeostasis (Gullapalli et al., 1989). However, few studies have been carried out on how vanadium affects individual components of respiratory systems.

The effects of vanadium on specific enzymes involved in the electron transfer of respiratory chains were investigated by means of ^{1}H-, ^{14}N-, and ^{31}P nuclear magnetic resonance spectroscopy and spectrophotometry. Vanadate produces opposite effects on the activities of enzymes operating at the beginning of aerobic and anaerobic respiratory chains; specifically, it enhanced aerobic NADH:FMN dehydrogenase activities, and inhibited anaerobic formate dehydrogenase activity (IC_{50} 30 $\pm$ 5 μM). Similarly NH_4VO_3 inhibits aerobic cytochrome c reductase only moderately, but inhibition of fumarate reductase depends linearly on oxyanion concentration (Mendz, 1996a).

The effects of vanadium on the activities of other electron carriers involved in both types of respiration are also qualitatively different. Vanadate enhances hydroxybutyrate dehydrogenase activity and does not affect the activity of *E. coli* transhydrogenase (Mendz, 1996b). On the other hand, anaerobic *E. coli* nitrate reductase activity decreases linearly with vanadate concentration (IC_{50} 148 $\pm$ 10 μM). The results suggest that the components of aerobic respiratory systems are less sensitive to oxidative and specific effects of vandate than the components of anaerobic respiration (Mendz, 1996b).

7. NEWLY DISCOVERED EFFECTS ON PHOSPHORYL TRANSFER ENZYMES

The number of enzymes found to be susceptible to vanadium ions has increased dramatically in the last few years. Vanadate and vanadyl have served also to provide insights into enzyme mechanisms by examination of their effects on enzyme activities.

7.1. Aldolase

Vanadate ions were employed as probes to investigate the interactions of oxoanions with aldolase by studying reductive, nonreductive, and photolytic effects of vanadate on rabbit fructose-1,6-bisphosphate aldolase (Crans et al., 1992a). The enzyme interacts strongly with these anions at low ionic strength and weakly at higher ionic strength. Oxoanions inhibit aldolase competitively with respect to fructose-1,6-biosphosphate, although the location of the oxo-anion binding site on aldolase remains unknown. Vanadate inhibits aldolase competitively in a manner analogous to arsenate, molybdate, phosphate, tung-state, and sulfate. Since vanadate solutions contain a mixture of vanadate anions, the nature of the inhibition was determined by a combination of enzyme kinetics and ^{51}V-NMR spectroscopy. Aldolase contains a significant number of thiol functional groups, and, as expected, vanadate undergoes redox chemistry with them, generating an irreversibly inhibited aldolase. This oxidative chemistry is attributed to the vandate tetramer, whereas the vanadate dimer is a reversible inhibitor. The vanadate monomer does not interact significantly with aldolase reversibly or irreversibly. Vanadyl cations have the lowest inhibition constant under the conditions of high ionic strength. Analyses of binding data suggest that phosphate, pyrophosphate, and sulfate bind to the same site on aldolase, whereas vanadate, arsenate, and molybdate bind to another site. Investigation of UV-light-induced photocleavage of aldolase by vanadate served to correlate the loss of aldolase activity with cleavage of the aldolase subunit. The study suggests that several sites on aldolase will accommodate oxoanions, and one of these sites also accommodates the vana-dyl cation (Crans et al., 1992a).

7.2. Krebs Cycle Enzymes

Vanadate inhibits ATP-dependent succinyl-CoA synthetase solubilized from rat brain mitochondria with an IC_{50} of 10 μM (Krivanek and Novakova, 1991a). This effect depends strongly on succinate concentration; inhibition of 50% of enzyme activity by vanadate anions takes place only at saturating concentrations (50 mM) of succinate. Vanadyl cations induce a less potent inhibition of ATP-dependent succinyl-CoA synthetase with only 20% inhibi-tion at 100 μM concentration (Krivanek and Novakova, 1991a). The effect is attributed to the inhibition by vanadate and vanadyl ions of the phosphoryla-tion of a histidine residue in the α-subunit of the enzyme.

ATP-citrate lyase also contains a histidyl as the only residue with an acid-labile phosphate bond. In the presence of 1 mM vanadate the endogenous phosphorylation of this histidine residue is inhibited by 67%. In contrast, the phosphorylation of serine and threonine residues is not affected by vanadate (Krivanek and Novakova, 1991b).

Monovanadate and decavanadate weakly inhibit rat liver ATP citrate lyase; the inhibition was only 37% at 1 mM concentration of each of the metallates (Krivanek, 1994). Vanadyl also inhibits the enzyme with 32% and 66% reduction in the activity at concentrations of 1 mM and 10 mM, respectively. Kinetic data show that the vanadates inhibit ATP-citrate lyase competitively with respect to ATP and citrate; and vanadyl inhibits the enzyme noncompetitively with respect to both substrates. All three species of vanadium ions inhibit ATP-citrate lyase noncompetitively with respect to CoA (Krivanek, 1994).

The data on the effects of vanadium ions on succinyl-CoA synthetase and ATP-citrate lyase support the hypothesis that inhibition of histidyl phosphorylation at the catalytic site may be a common mechanism by which vanadium ions suppress the activity of histidyl containing enzymes that catalyze phosphoryl transfer reactions (Krivanek, 1994).

7.3. Lipid Metabolism Enzymes

A wide range of signal transduction pathways involve activation of phospholipases. *Streptomyces chromofuscus* phospholipase D (PPLaseD) catalyzes the formation of cyclic lysophosphatidic acid from lysophosphatidylcholine or lysophosphatidylethanolamine. The cyclic intermediate is hydrolyzed further to lysophosphatidic acid by the phosphatase activity of PPLaseD. The ring–opening reaction of synthetic cyclic lysophosphatidic acid is partially inhibited by addition of 5–10 mM sodium vanadate. This effect of the oxyanion could increase significantly the accumulation of cyclic lysophosphatidic acid in cells, which may act as a dormant configuration of the physiologically active lysophosphatidic acid or as a modulator of cell function in its own right (Friedman et al., 1996).

7.4. Phosphatidylinositol Phosphatases

The major 5-phosphatase activity of phosphatidylinositol 3,4,5-triphosphate in the cytosol of rat brain tissue is a 145-kDa Mg^{2+}-dependent enzyme. The enzyme was purified to near homogeneity and shown to be a magnesium-dependent 5-phosphatase that hydrolyzes phosphatidylinositol 4,5-bisphosphate and phosphatidylinositol 3,4,5-triphosphate, and is inhibited by vanadate with an IC_{50} of ca. 700 μM (Woscholski et al., 1995).

7.5. Proteolysis

ATP regulates the activity of intracellular proteolytic enzymes and the modification of proteins that will make them susceptible to proteolytic enzyme systems. In these processes the nucleotide is hydrolyzed, but its specific role is not clear. The 26 S proteasome complex is a heteromultimeric complex containing the multicatalytic proteinase 20 S proteasome as a core enzyme; it is responsible for the ATP-dependent degradation of protein–ubiquitin conjugates and nonubiquinated substrates such as ornithine decarboxylase. Vanadate does not reduce the peptide-hydrolyzing activity of the 26 S proteasome complex significantly, nor does it prevent its four to five-fold activation by ATP. For this reason ATP does not appear to serve as a donor of free energy, but to serve more as an allosteric effector that activates the complex by inducing conformational changes (Dahlmann et al., 1995).

7.6. Transporters

Many systems that translocate metabolites across cell membranes are nucleotide-dependent, and the linkage of these transport processes to phosphoryl transfer reactions may make them susceptible to vanadium effects.

PDR5 and SNQ2 are 160-kDa multidrug transporter proteins of *Saccharomyces cerevisiae* that belong to the ABC superfamily of the ATP-binding cassettes. They are located in the plasma membrane-enriched fraction and have nucleotide triphosphate phosphatase activities. At pH 6.3 their UTPase activity is inhibited by vanadate with an IC_{50} of 3–5 μM (Decottignies et al., 1995)

The overexpression of the P-glycoprotein MDR1, the multidrug resistance gene product, has been linked to the development of resistance to multiple cytotoxic natural-product anticancer drugs in certain cancers and cell lines derived from tumors. P-glycoprotein, a member of the ABC superfamily of transporters, mediates ATP-dependent vanadate-sensitive drug efflux when reconstituted in plasma vesicles (Horio et al., 1988). In vivo P-glycoprotein appears to function as a pump with broad specificity for chemically unrelated hydrophobic compounds. Recent studies on purified and reconstituted P-glycoprotein have been carried out to elucidate the mechanism of drug transport. P-glycoprotein from the human carcinoma multidrug-resistant cell line KB-V1 was purified and reconstituted into proteoliposomes that exhibit high levels of drug-stimulated ATPase activity as well as ATP-dependent [^{3}H]vinblastine accumulation. Both the ATPase and vinblastine transport activities of the reconstituted P-glycoprotein are inhibited by vanadate. In addition, vinblastine transport is inhibited by verapamil and daunorubicin. These studies provide strong evidence that human P-glycoprotein functions as an ATP-dependent drug transporter (Ambudkar, 1995).

7.7. Uteroferrin

Uteroferrin is the purple acid phosphatase from porcine uterine fluid. The enzyme is inhibited noncompetitively by vanadate in a time-dependent manner under both aerobic and anaerobic conditions (Crans et al., 1992b). This time-dependent inhibition is observed only with the FeFe enzyme and is absent when the FeZn enzyme is used. These observations are attributed to the sequential formation of two uteroferrin–vanadium complexes. The first complex forms rapidly and reversibly, whereas the second complex forms slowly and results in the production of catalytically inactive oxidized uteroferrin and vanadyl, which is observable by EPR. The redox reaction can be reversed by treatment of the oxidized enzyme first with vanadyl and then with EDTA to generate a catalytically active uteroferrin. Multiple inhibition kinetics suggests that vanadate is mutually exclusive with molybdate, tungstate, and vanadyl cation. The binding site for each of these anions is distinct from the site to which the competitive inhibitors phosphate and arsenate bind. The time-dependent inhibition by vanadate of uteroferrin containing the diiron core represents a new type of mechanism by which vanadium can interact with proteins and gives additional insight into the binding of anions to uteroferrin (Crans et al., 1992b).

8. NOVEL EFFECTS OF VANADIUM ON TISSUES AND CELLS

Since it was established that vanadium as a trace metal is diuretic and natridiuretic, causes vasoconstriction (Nechay, 1984), and increases hexose uptake and glucose metabolism like insulin (Shechter et al., 1990), new effects of vanadium on tissues and cells have been discovered. Vanadium ions are also a very useful tool for understanding complex cellular functions in vivo. Owing to a number of studies in this area and to the importance of their findings, the scope of this section of the review has been widened to also include some investigations not directly related to the effects of vanadium ions on phosphoryl transfer enzymes.

8.1. Anticarcinogenesis

The effect of vanadium on diethylnitrosamine-induced (DENA-induced) hepatocarcinogenesis was examined in male Sprague-Dawley rats. Rats were divided into four groups: Group 1 served as normal group, and hepatocarcinogenesis was initiated in the other three by a single intraperitoneal injection of DENA (200 mg/kg body weight). Supplementary ammonium vanadate, at the levels of 0.2 and 0.5 ppm added to the drinking water, was given ad libitum to Groups 3 and 4, respectively, 4 weeks before DENA administration and continued for 8 or 16 weeks after the carcinogenic insult. The rats were sacrificed at the 12th or 20th week after the experiment began. There were no significant differences in food and water intakes or in growth rates between the groups. Supplementation with vanadium (Groups 3 and 4) for 12 or 20 weeks reduced the incidence, total number, and multiplicity and altered the size distribution of visible persistent (neoplastic) hepatocyte nodules in a dose-dependent manner relative to DENA controls (Group 2). Mean nodular volume and nodular volume as percentage of liver volume were also inhibited upon vanadium supplementation. Morphometric analysis of preneoplastic focal lesions revealed the occurrence in the vanadium-supplemented groups (Groups 3 and 4) of a decrease in the number of altered liver cell foci per square centimeter and in the average focal area, coupled with a decrement in the percentage area of liver parenchyma occupied by foci. These results are more pronounced at the 20th week than at the 12th. The findings of the study suggest that vanadium at a dose of 0.5 ppm may have a potential anticarcinogenic property against chemically induced hepatic neoplasia without any apparent signs of toxicity. The authors propose that this property may open new vistas in cancer chemotherapy (Bishayee and Chatterjee, 1995).

Among vanadium complexes that show inhibition of cell growth for human nasopharyngeal carcinoma KB cells, 1:1 vanadyl-1,10-phenanthroline complex, $VO(Phen)^{2+}$, cleaves supercoiled plasmid Col E1 DNA effectively when hydrogen peroxide is added; but vanadyl ions, VO^{2+}, are less effective. Lineweaver-Burk plots of the complex binding to calf thymus DNA indicate that $VO(Phen)^{2+}$ has a high affinity to DNA, as supported by CD spectral measurements. ESR spin trapping was performed to examine the active species for DNA cleavage by the complex; it was found that hydroxyl radicals are generated in a pH-dependent manner in the $VO(phen)^{2+}$–H_2O_2 system, the optimal pH region being 8.5–9.5. In contrast, no optimum pH was observed in the VO^{2+}–H_2O_2 system. Thus, the $VO(Phen)^{2+}$ complex is proposed to bind DNA and cleave it when hydrogen peroxide is present (Sakurai et al., 1995).

The effect of oral administration of vanadate at 100-, 200-, and 400-nM concentrations for 30 days on the activity of the detoxifying enzyme system glutathione S-transferase (GSTase) in rat liver and in several extrahepatic tissues was examined. Vanadate shows high activity as GSTase inducer in liver and in small intestine mucosa, followed by large intestine mucosa and kidney in a dose-dependent manner. No significant alterations in GSTase activity are observed in forestomach and lung tissues after vanadate administration. Vanadate treatment that results in an enhancement of GSTase activity impairs neither hepatic nor renal function, as evidenced by serum glutamic oxaloacetic transaminase, glutamic pyruvic transaminase, sorbitol dehydrogenase, urea, and creatinine. Since the ability of anticarcinogenic agents to induce an increase of detoxifying enzyme activity is correlated with their activity in the inhibition of tumorigenesis, the trace element vanadium might be considered a potential cancer chemopreventive agent (Bishayee and Chatterjee, 1993).

8.2. Cytotoxicity

Cytotoxicity and morphological transformation have been studied in BALB/3T3 Cl A31-1-1 mouse embryo cells exposed to ammonium vanadate and vanadyl sulfate alone or in combination with diethylmaleate (DEM), a cellular glutathione (GSH)-depleting agent (Sabbioni et al., 1993). Cells exposed for

24 h to 10^{-5} M NH_4 VO_3 alone or in combination with 3×10^{-6} M DEM showed the characteristic hyperfine EPR signal of $VOSO_4$, which was more obvious in the case of exposure to vanadate alone. This suggests that the amount of VO_3^- reduced to VO^{2+} decreased on GSH-depleted cells. While $VOSO_4$ at concentrations of 3×10^{-6} M and 10^{-5} M does not have transforming effects in the cells, NH_4VO_3 showed neoplastic transforming activity in comparison to cells not exposed to vanadium ions. Cytotoxicity and morphological transformation in cells exposed to NH_4VO_3 in combination with 3×10^{-6} M DEM were significantly more intensive compared to the corresponding values observed in cells exposed to vanadate alone. This suggests that the final transforming activity response is dependent on the intracellular GSH-mediated mechanism of reduction of vanadate to vanadyl, because (1) the intensity of the observed neoplastic action is determined by the extent to which VO_3^- is bioreduced to less-toxic VO^{2+} via intracellular GSH; and (2) the carcinogenic potential of VO_3^- is strictly dependent on its intracellular persistence, owing a to lack of GSH-mediated reduction (Sabbioni et al., 1993).

8.3. Glucose Homeostasis

The effect of pervanadate as an insulinomimetic agent that inhibits insulin receptor dephosphorylation was assessed in vivo. A single intraperitoneal administration of pervanadate at concentrations as low as 700 μg vanadium per kilogram body weight markedly lowered blood glucose levels in streptozotocin-induced diabetic rats from 430 ± 28 to 212 ± 30 mg/100 ml within 3 h (Shisheva et al., 1994) A decrease was already observed a half hour after treatment; it continued in accelerating fashion until the third hour and persisted for at least 24 h. The initial hyperglycemia reoccurred on the second day and remained thereafter. In a comparable fashion, pervanadate decreased the blood glucose levels of control healthy rats, treated identically. Within this period body weight was not significantly altered in either group. The data indicate that rapid and efficient management of glucose homeostasis is achieved via inhibiting receptor dephosphorylation. This observation may lead to a new therapeutic approach of protein tyrosine phosphatase inhibition for future treatment of diabetes in general, and in insulin-resistant states in particular (Shisheva et al., 1994).

8.4. Lipid Peroxidation

It has been known for decades that redox-active metal ions induce lipid peroxidation in biological systems, but it is not known whether this effect plays an important role in the toxicity of metals. The ability of various redox-active metal ions to induce lipid peroxidation was investigated in normal and α-linolenic acid-loaded (LNA-loaded) cultured rat hepatocytes (Furuno et al., 1996). At low-metal-ion concentrations induction of lipid peroxidation is highest with ferrous ions (Fe^{2+}), whereas at high concentrations, vanadyl ions

(V^{3+}) and copper ions (Cu^{2+}) have the greatest effect on both groups of hepatocytes. The extent of lipid peroxidation in the presence of chloride or sulfate salts of any one of the three metal ions at concentrations up to 1 mM is several times greater in LNA-loaded cells than in normal cells. In addition, upon the addition of Fe^{2+} or V^{3+}, LNA-loaded hepatocytes are injured, whereas normal cells are not. The addition of Cu^{2+} causes substantial cell injury in normal hepatocytes and even greater injury in LNA-loaded cells. The prevention of lipid peroxidation in LNA-loaded hepatocytes by addition of an antioxidant like N,N'-diphenyl-p-phenylene-diamine (DPPD) prevents Fe^{2+}-and V^{3+}-induced cell injury almost completely and reduces Cu^{2+}-induced cell injury. α-Tocopherol behaves in a similar way but is less effective than DPPD. OH$\cdot$ radical scavengers such as mannitol and dimethyl sulfoxide (DMSO) have no effect on lipid peroxidation induced by any metal ions in LNA-loaded hepatocytes. Addition of cadmium ions (Cd^{2+}), which require the lowest concentration to cause cell injury, induces a slight increase in lipid peroxidation in normal hepatocytes but does not induce lipid peroxidation to the same extent as seen in LNA-loaded cells treated with any of the three metal ions mentioned (Furuno et al., 1996).

Two mechanisms have been proposed to explain the induction of lipid peroxidation by metal ions. One is the production of hydroxyl radicals, which initiate lipid peroxidation; radicals may be produced via Fenton reactions in the reduction of hydrogen peroxide by metal ions in low-valency states. The other mechanism is chain propagation of lipid peroxidation; reactive alkoxyl radicals may be produced by decomposition of lipid peroxides by low-valence-state metal ions and initiate radical propagation by abstracting hydrogens from other lipids. The observed lipid peroxidation induced by metal ions in LNA-loaded hepatocytes suggests that the likely mechanism is the formation of alkoxyl radicals (Furuno et al., 1996).

8.5. Muscle Contraction

The toxic effects of vanadate on rat myocardium were evaluated by measuring the changes in calcium metabolism and contractile force of Sprague-Dawley rats administered intragastric orthovanadate. Left ventricular pressure of hearts exposed to vanadium was 23% lower than in control hearts, and the first derivative was 36% lower, indicating that the myocardial contractile force decreased. The cellular content of exchangeable calcium in both rested and excited ventricles of vanadium-treated rats is considerably higher than that of the control group, suggesting that the effects of vanadate on contractile force are probably due to accumulation of Ca^{2+} and cell damage (Pytkowski and Jagodzinska-Hamnn, 1996).

8.6. Proliferation and Differentiation

The effects of different vanadium compounds on proliferation and differentiation were examined in osteoblast-like UMR106 cells (Cortizo and Etcheverry,

1995). Vanadate increased cell growth in a biphasic manner, the higher doses inhibiting cell progression. Much like vanadate, pervanadate increased osteoblast-like cell proliferation in a biphasic manner but no inhibition of growth was observed. Vanadyl stimulated cell proliferation in a dose-response manner. Vanadyl and pervanadate were stronger stimulators of cell growth than vanadate, but only vanadate was able to regulate cell differentiation as measured by cell alkaline phosphatase activity. These results suggest that vanadium derivatives behave like growth factors on osteoblast-like cells and are potential pharmacological tools in the control of cell growth (Cortizo and Etcheverry, 1995).

8.7. Receptors and Transporters

Transmembrane signaling through cell surface Fc receptors (FcR) is controlled by aggregation (Metzger, 1992). The cross-linking of high-affinity IgG Fc receptors (FcγRI) or IgA Fc receptors (FcαR) on U937 monocytes results in the rapid generation of oxygen radicals and tyrosine phosphorylation of their respective γ chains. The low-affinity IgG Fc receptors (FcγRII) lack γ chains, and cross-linking of these receptors results in the phosphorylation of tyrosyl residues in its cytoplasmic domain.

Aggregation of the FcγRI receptors of U937 cells results in the transient tyrosine phosphorylation of FcγRI γ chain, but not in the phosphorylation of γ chains associated with nonaggregated FcαR on the same cells, indicating that tyrosine phosphorylation of γ chains is limited normally to FcR in aggregates (Pfefferkorn and Swink, 1996). In contrast, aggregation of FcγRI in the presence of vanadate, besides inducing a sustained tyrosine phosphorylation of FcγRI γ chains, also induces rapid and extensive phosphorylation of nonaggregated FcαR γ chains and of low-affinity FcγRII receptors. This global phosphorylation of motifs on nonaggregated FcR is detected also upon aggregation of FcαR or FcγRII, which induces the phosphorylation of nonaggregated FcγRI γ chains. Vanadate prevents dephosphorylation of proteins and increases kinase activity in stimulated cells. There is no evidence to support alternative explanations such as acquisition of phospho-γ through subunit exchange or a coalescence of nonaggregated with aggregated FcR. It is likely, therefore, that activated kinases interact with nonaggregated FcR in stimulated cells. Pervanadate induced the tyrosine phosphorylation of γ chains in the absence of FcR cross-linking, indicating that the kinases could be activated by phosphatase inhibition and could react with nonaggregated substrates. Pfefferkorn and Swink (1996) concluded that under normal conditions there is a vanadate-sensitive mechanism that prevents tyrosine phosphorylation of nonaggregated FcR γ-chain motifs in activated cells, restricting their phosphorylation to aggregates.

Vanadyl ions and vanadate ions stimulate furosemide-sensitive electrogenic Cl$^-$ secretion in isolated epithelia of rabbit descending colon (Plass et al., 1992). This effect is associated with an increased release of prostaglandin E2 from the tissue. Inhibitors of phospholipase A2 or cyclooxygenase abolish both vanadium-induced release of prostaglandin E2 and Cl^- secretion. Neuronal mechanisms are not likely to be involved, as tetrodotoxin does not affect the vanadate-induced Cl^- secretion. Although vanadate is known to inhibit (Na^+, K^+)-ATPase activity, no inhibition of active Na^+ transport was observed in intact colonic epithelia, suggesting a rapid intracellular reduction of vanadate ions to vanadyl ions, which have no inhibitory effect on the Na^+,K^+-ATPase. The findings indicated that vanadate-stimulated colonic Cl^- secretion involves intracellular conversion of vanadate to vanadyl and release of prostaglandin E2 (Plass et al., 1992).

8.8. Signaling Cascades

Extracellular polypeptides such as cytokines and growth factors transmit their signals from cell surface to nucleus through a number of protein messengers (Ihle et al., 1994). Physical and chemical changes in these messengers propagate signals down different pathways. Interaction of interferon-γ (IFN-γ) with its receptor on the plasma membrane leads to structural changes in the receptor complex, which in turn triggers tyrosyl phosphorylation of the Janus kinases Jak1, Jak2, and Tyk2. The activated kinases then phosphorylate a unique tyrosyl residue on the cytosolic tail of the receptor that serves as the docking site for signal transducers and activators of transcription (Stat) proteins which take the message to the nucleus and bind cognate DNA elements, thus coupling ligand binding and activation of gene expression. Stat proteins appear to be the substrate of Janus kinases and have a conserved tyrosine residue near the carboxyl terminus (Tyr-701) that is phosphorylated in the signaling cascade and is essential for function (Ihle et al., 1994). Stat1α is a specific DNA-binding protein phosphorylated in the IFN-γ pathway.

Permeabilized HeLa S3 cells treated with 200–500 μM peroxo derivatives of vanadium, molybdenum, and tungsten accumulate constitutively phosphorylated Stat1α molecules. In contrast, permeabilized cells treated with orthovanadate, vanadyl sulfate, molybdate, and tungstate in the same range of concentrations do not accumulate activated Stat1α molecules in the absence of ligand. These compounds, however, inhibit the inactivation of IFN-γ-induced DNA-binding activity of Stat1α (Haque et al., 1995). A 4- to 6-h exposure of permeabilized cells to orthovanadate, molybdate, and tungstate, but not vanadyl sulfate, results in a ligand-independent activation of Stat1α, which is blocked by the inhibition or depletion of NADPH oxidase activity in the cells, indicating that superoxide formation catalyzed by NADPH oxidase is required for the bioconversion of these metal oxides to the corresponding peroxo compounds. Interestingly, ligand-independent Stat1α activation by peroxo derivatives of these transition metals does not require Jak1, Jak2, or Tyk2 kinase activity, suggesting that other kinases can phosphorylate Stat1α on tyrosine 701 (Haque et al., 1995).

The reorganization of cytoskeletal actin underlies several essential cellular responses, such as mitosis, adhesion, and cell aggregation, that are induced by external stimuli. Studies of cytoskeletal actin organization in fibroblasts have shown that an increase in protein tyrosine phosphorylation is accompanied by the formation of stress fibers (SF) and focal adhesions (FA). Rho proteins are ras-like small GTP binding proteins involved in SF and FA formation. Stress fiber formation is activated by tyrosine kinase-linked receptors and inhibited by PTKase inhibitors. From these observations it has been concluded that tyrosine phosphorylation is involved in the transduction of signals that lead to cytoskeletal actin organization. Vanadate stimulates SF formation in Madin-Darby canine kidney cells (Volberg et al., 1992) and mouse Swiss 3T3 fibroblasts (Barry and Critchley, 1994), but the site of action is unclear.

The mechanisms of vanadate-induced actin reorganization were investigated in cultured astrocytes (Koyama et al., 1996). Treatment of protoplasmic astrocytes with 0.5 mM dibutyryl cAMP caused the disappearance of stress fibers and focal adhesions accompanied by cellular stellation. Addition of 1 mM orthovanadate reorganized SF and FA in dibutyryl cAMP-treated cells, and the newly formed FA showed increased phosphotyrosine levels. Orthovanadate also reorganized SF and FA in stellate astrocytes induced by 5 μM cytochalasin B, 50 μM ML-9 (a myosin light-chain kinase inhibitor), and 20 μM W-7 (a calmodulin inhibitor). Cytoplasmic microinjection of 20 μg/ml C3 ADP-ribosyltransferase of *Clostridium botulinum*, which specifically inactivates rho proteins, caused disappearance of SF. The effect of C3 enzyme on SF was not reversed by a subsequent addition of VO_4^{3-}. These results suggest that rho proteins are involved in vanadate-induced reorganization of cytoskeletal actin (Koyama et al., 1996).

Zinc, vanadate, and selenate are termed insulin-mimetics because they mimic the effects of insulin on several processes. Recent studies have compared the effects of zinc, vanadate, and selenate on insulin-sensitive processes in an attempt to probe the mechanism of insulin action. Insulin regulates the expression of genes involved in a variety of metabolic processes. In chick-embryo hepatocytes in culture, insulin amplifies the triiodothyronine-induced (T3-induced) enzyme activity, and the level and rate of transcription of mRNA for both fatty acid synthase (FASase) and decarboxylating malate dehydrogenase (DMDase, "malic enzyme"). Insulin alone, however, has little or no effect on the expression of these genes. In chick embryo hepatocytes, the mechanism by which insulin regulates the expression of these or other genes is not known. Investigation of the effects of zinc, vanadate, and selenate on the T3-induced expression of both FASase and DMDase in chick embryo hepatocytes in culture showed that, like insulin, these agents had little or no effect on the basal activities of the enzymes for 48 h. Unlike insulin, however, zinc, vanadate, and selenate inhibited the T3-induced activities and mRNA levels of both FASase and DMDase. Maximal inhibition was achieved at concentrations of 50 μM zinc or vanadate, or 20 μM selenate. Zinc and vanadate also inhibited the T3-induced transcription of the FASase and DMDase genes. Although the mechanism of this inhibition is unknown, the results indicate that it is not mediated through inhibition of binding of T3 to its nuclear receptor nor through a general toxic effect. Thus zinc, vanadate, and selenate are not insulin mimetics under all conditions, and their effects on other insulin-sensitive processes may be fortuitous and unrelated to actions or components of the insulin signaling pathway (Zhu et al., 1994).

9. CONCLUSIONS

The hundreds of studies on the effects of vanadate and vanadyl ions on living systems that have appeared in the last few years reflect the complexity of these interactions and indicate the interest that vanadium ions have for their environmental importance and potential medical applications.

Space restrictions precluded a more detailed review of the work carried out on the physiological effects of vanadium, even within the limits of systems involving phosphoryl transfer reactions.

Although discoveries continue to be made in well-established areas as the interactions of vanadate and vanadyl with ATPases, two particular trends are noticeable in the recent literature: (a) the utilization of vanadium ions as probes for understanding biochemical mechanisms, and (b) the investigation of their interactions with biological systems of increasing complexity. The discoveries emerging from the latter are opening new vistas on the environmental and medical relevance of these metal ions.

REFERENCES

Ambudkar, S. V. (1995). Purification and reconstitution of functional human P-glycoprotein. *J. Bioenerg. Biomembr.* **27,** 23–29.

Aureliano, M., and Madeira, V. M. (1994). Interactions of vanadate oligomers with sarcoplasmic reticulum Ca^{2+}-ATPase. *Biochim. Biophys. Acta* **1221,** 259–271.

Barry, S. T., and Critchley, D. R. (1994). The RhoA-dependent assembly of focal adhesions in Swiss 3T3 cells is associated with increased phosphorylation and the recruitment of both pp 125FAK and protein kinase C-δ to focal adhesions. *J. Cell Sci.* **107,** 2033–2045.

Beltran, C., and Nelson, N. (1992) The membrane sector of vacuolar H^+-ATPase by itself is impermeable to protons. *Acta Physiol. Scand. Suppl.* **607,** 41–47.

Benaim, G., Moreno, S. N., Hutchinson, G., Cervino, V., Hermoso, T., Romero, P. J., Ruiz, F., de Souza, W., and Docampo, R. (1995). Characterization of the plasma-membrane calcium pump from *Trypanosoma cruzi. Biochem. J.* **306,** 299–303.

Bishayee, A., and Chatterjee M. (1993). Selective enhancement of glutathione S-transferase activity in liver and extrahepatic tissues of rat following oral administration of vanadate. *Acta Physiol. Pharmacol. Bulg.* **19,** 83–89.

Bishayee, A., and Chatterjee, M. (1995). Inhibition of altered liver cell foci and persistent nodule growth by vanadium during diethylnitrosamine-induced hepatocarcinogenesis in rats. *Anticancer Res.* **15,** 455–61.

Chasteen, N. D. (1990). *Vanadium in Biological Systems.* Kluwer Academic Publishers, Dordrecht, 222 pp.

Chatterjee, D., Chakraborty, M., Leit, M., Neff, L., Jamsa-Kellokumpu, S., Fuchs, R., and Baron, R. (1992). Sensitivity to vanadate and isoforms of subunits A and B distinguish the osteoclast proton pump from other vacuolar H$^+$-ATPases. *Proc. Natl. Acad. Sci. USA* **89,** 6257–6261.

Cohen, P. (1989). The Structure and Regulation of Protein Phosphatases. *Annu. Rev. Biochem.* **58,** 453–508.

Cortizo, A. M., and Etcheverry, S. B. (1995). Vanadium derivatives act as growth factor-mimetic compounds upon differentiation and proliferation of osteoblast-like UMR106 cells. *Mol. Cell. Biochem.* **145,** 97–102.

Cortizo, A. M., Salice, V. C., and Etcheverry S. B. (1994). Vanadium compounds. Their action on alkaline phosphatase activity. *Biol. Trace Elem. Res.* **41,** 331–339.

Crans, D. C., Sudhakar, K., and Zamborelli T. J. (1992a). Interaction of rabbit muscle aldolase at high ionic strengths with vanadate and other oxoanions. *Biochemistry* **31,** 6812–6821.

Crans, D. C., Simone, C. M., Holz, R. C., and Que, L. Jr. (1992b). Interaction of porcine uterine fluid purple acid phosphatase with vanadate and vanadyl cation. *Biochemistry* **31,** 11731–11739.

Dafnis, E., and Sabatini S. (1994). Biochemistry and pathophysiology of vanadium. *Nephron* **67,** 133–143.

Dahlmann, B., Kuehn, L., and Reinauer, H. (1995). Studies on the activation by ATP of the 26 S proteasome complex from rat skeletal muscle. *Biochem. J.* **309,** 195–202.

David, P., Horne, W. C., and Baron R. (1996). Vanadate inhibits vacuolar H$^+$-ATPase-mediated proton transport in chicken kidney microsomes by an ADP-dependent mechanism. *Biochim. Biophys. Acta* **1280,** 155–160.

Decottignies, A., Lambert, L., Catty, P., Degand, H., Epping, A., Moye-Rowley, W. S., Balzi, E., and Goffeau, A. (1995). Identification and characterization of SNQ2, a new multidrug ATP binding cassette transporter of the yeast plasma membrane. *J. Biol. Chem.* **270,** 18150–18157.

Desautels, M., and Dulos., R. A. (1993). Nucleotide requirement and effects of fatty acids on protein synthesis and degradation in brown adipose tissue mitochondria. *Can. J. Physiol. Pharmacol.* **71,** 17–25.

Domingo, J. L. (1995). Prevention by chelating agents of metal-induced developmental toxicity. *Reprod. Toxicol.* **9,** 105–113.

Elberg, G., Li, J., and Shechter, Y. (1994). Vanadium activates or inhibits receptor and non-receptor protein tyrosine kinases in cell-free experiments, depending on its oxidation state. Possible role of endogenous vanadium in controlling cellular protein tyrosine kinase activity. *J. Biol. Chem.* **269,** 9521–9527.

Enyedi, A., Minami, J., Caride, A. J., and Penniston, J. T. (1988). Characteristics of the Ca^{2+} pump and Ca^{2+}-ATPase in the plasma membrane of rat myometrium. *Biochem. J.* **252,** 215–220.

Forgac, M. (1989). Structure and function of vacuolar class of ATP-driven proton pumps. *Physiol. Rev.* **69,** 765–796.

Friedman, P., Haimovitz, R., Markman, O., Roberts, M. F., and Shinitzky, M. (1996). Conversion of lysophospholipids to cyclic lysophosphatidic acid by phospholipase D. *J. Biol. Chem.* **271,** 953–957.

Fuchs, R., and Baron, R. (1992). Sensitivity to vanadate and isoforms of subunits A and B distinguish the osteoclast proton pump from other vacuolar H$^+$ ATPases. *Proc. Natl. Acad. Sci. USA* **89,** 6257–6261.

Furuno, K., Suetsuga, T., and Sugihara, N. (1996). Effects of metal ions on lipid peroxidation in cultured rat hepatocytes loaded with α-linolenic acid. *J. Toxicol. Environ. Health* **48,** 121–129.

Gullapalli, S., Shivaswamy, V., Ramasarma, T., and Kurup, C. K. (1989). Redistribution of subcellular calcium rat liver on administration of vanadate. *Mol. Cell Biochem.* **90,** 155–164.

Haque, S. J., Flati, V., Deb, A., and Williams, B. R. (1995). Roles of protein–tyrosine phosphatases in Stat1α-mediated cell signalling. *J. Biol. Chem.* **270**, 25709–25714.

Henderson, G. E., Evans, I. H., and Bruce, I. J. (1989). Vanadate inhibition of mitochondrial respiration and H$^+$-ATPase activity in *Saccharomyces cerevisiae. Yeast* **5**, 73–77.

Horio et al. (1988). ATP-dependent transport of vinblastine in vesicles from human multidrug-resistant cells. *Proc. Natl. Acad. Sci. USA* **85**, 3580–3584.

Ihle, J. N., Witthuhn, B. A., Quelle, F. W., Yamamoto, K., Thierfielder, W. E., Kreider, B., and Silvennoinen, O. (1994). Signalling by the cytokine receptor superfamily: JAKs and STATs. *Trends Biochem. Sci.* **19**, 222–227.

Janiszewska, G., Lachowicz, L., Jaskolski, D., and Gromadzinska E. (1994). Vanadium inhibition of human parietal lobe ATPases. *Int. J. Biochem.* **26**, 551–553.

Koyama, Y., Fukuda, T. and Baba, A. (1996). Inhibition of vanadate-induced astrocytic stress fiber formation by C3 ADP-ribosyltransferase. *Biochem. Biophys. Res. Commun.* **218**, 331–336.

Kramer, H. J., Krampitz, G., Backer, A., Krampitz, G. Jr., and Meyer-Lehnert H. (1995). Vanadium-diascorbates are strong candidates for endogenous ouabain-like factors in human urine: Effects on (Na$^+$,K$^+$)-ATPase enzyme kinetics. *Biochem. Biophys. Res. Commun.* **213**, 289–294.

Krivanek, J. (1994). Effect of vanadium ions on ATP citrate lyase. *Gen. Physiol. Biophys.* **13**, 43–55.

Krivanek, J., and Novakova, L. (1991a). A novel effect of vanadium ions: Inhibition of succinyl-Coa synthetase. *Gen. Physiol. Biophys.* **10**, 71–82.

Krivanek, J., and Novakova, L. (1991b). ATP-citrate lyase is another enzyme the histidine phosphorylation of which is inhibited by vanadate. *FEBS Lett.* **282**, 32–34.

Leonard, A., and Gerber, G. B. (1994). Mutagenicity, carcinogenicity and teratogenicity of vanadium compounds. *Mutat. Res.* **317**, 81–87.

Li, Y., and Strohl, W. R. (1996). Cloning, purification, and properties of a phosphotyrosine protein phosphatase from *Streptomyces coelicolor A3(2). J. Bacteriol.* **178**, 136–142.

Liao, K., Hoffman, R. D., and Lane, M. D. (1991). Phosphotyrosyl turnover in insulin signalling. Characterization of two membrane-bound pp15 protein tyrosine phosphatases from 3T3-L1 adipocytes. *J. Biol. Chem.* **266**, 6544–6553.

Liao, K., and Lane, M. D. (1995). The blockade of preadipocyte differentiation by protein-tyrosine phosphatase HA2 is reversed by vanadate. *J. Biol. Chem.* **270**, 12123–12132.

Liu, S., Gresser, M. J., and Tracey, A. S. (1992). ^{1}H and ^{51}V NMR studies of the interaction of vanadate and 2-vanadio-3-phosphoglycerate with phosphoglycerate mutase. *Biochemistry* **31**, 2677–2685.

Magocsi, M., and Penniston, J. T. (1991). Ca^{2+} or Mg^{2+} nucleotide phosphohydrolases in myometrium: Two ecto enzymes. *Biochim. Biophys. Acta* **1070**, 163–172.

Meghji, P., and Burnstock, G. (1995). Inhibition of extracellular ATP degradation in endothelial cells. *Life Sci.* **57**, 763–771.

Mendz, G. L. (1990). Effects of vanadium on the activities of kinases, mutases and phosphatases. In P. Collery, L. A. Poirier, M. Manfait, and J. C. Etienne (Eds.), *Metal Ions in Biology and Medicine.* John Libbey, Paris, Vol. 1, pp. 309–311.

Mendz, G. L. (1991). Stimulation of mutases and isomerases by vanadium. *Arch. Biochem. Biophys.* **291**, 201–211.

Mendz, G. L. (1996a). Effects of vanadium on aerobic and anaerobic respiratory enzymes. In P. Collery, J. Corbella, J. L. Domingo, J. C. Etienne, and J. M. Llobet (Eds.), *Metal Ions in Biology and Medicine.* John Libbey, Paris, Vol. 4, pp. 126–129.

Mendz, G. L. (1996b). Oxidative and specific effects of vanadium ions on respiratory enzymes. *Biochim. Biophys. Acta* (submitted).

Mendz, G. L., Hyslop, S. J., and Kuchel, P. W. (1990). Stimulation of human erythrocyte 2,3-bisphosphoglycerate phosphatase by vanadate. *Arch. Biochem. Biophys.* **291**, 201–211.

Metzger, H. (1992). Transmembrane signalling: The joy of aggregation. *J. Immunol.* **149,** 1477–1487.

Nechay, B. R. (1984). Mechanism of action of vanadium. *Annu. Rev Pharmacol.* **24,** 501–524.

Neisser, A., Fromwald, S., Schmatzberger, A., and Peschek, G. A. (1994). Immunological and functional localization of both F-type and P-type ATPases in cyanobacterial plasma membranes. *Biochem. Biophys. Res. Commun.* **200,** 884–892.

Nieves, J., Lim, L., Puett, D., and Echegoyen, L. (1987). Electron spin resonance of calmodulin–vanadyl complexes. *Biochemistry* **26,** 4523–4527.

Pan, J., Burgher, K. L., Szczepanik, A. M., and Ringheim GE. (1996). Tyrosine phosphorylation of inducible nitric oxide synthase: Implications for potential posttranslational regulation. *Biochem. J.* **314,** 889–894.

Parra-Diaz, D., Echegoyen, L., Zott, H. G., and Puett, D. (1995a). Vanadium(IV) inhibits calmodulin-stimulated skeletal muscle myosin light chain activity. *Biofactors* **5,** 25–28.

Parra-Diaz, D., Wei, Q., Lee, E. Y., Echegoyen, L., and Puett, D. (1995b). Binding of vanadium(IV) to the phosphatase calcineurin. *FEBS Lett.* **376,** 58–60.

Pedersen, P. L., and Carafoli, E. (1987). Ion motive ATPases. I. Ubiquity, properties, and significance to cell function. *Trends Biochem. Sci.* **12,** 146–150.

Percival, M. D., Doherty, K., and Gresser, M. J. (1990). Inhibition of phosphoglucomutase by vanadate. *Biochemistry* **29,** 2764–2769.

Pfefferkorn, L. C., and Swink, S. L. (1996). Intracluster restriction of Fc receptor γ-chain tyrosine phosphorylation subverted by a protein–tyrosine phosphatase inhibitor. *J. Biol. Chem.* **271,** 11099–11105.

Plass H., Roden M., Wiener, H., and Turnheim, K. (1992). Vanadium-induced Cl⁻-secretion in rabbit descending colon is mediated by prostaglandins. *Biochim. Biophys. Acta* **1107,** 139–142.

Pytkowski, B., and Jagodzinska-Hamnn, L. (1996). Effects of *in vivo* vanadate administration on calcium exchange and contractile force of rat ventricular myocardium. *Toxicol. Lett.* **84,** 167–173.

Rehder, D. (1992) Structure and function of vanadium compounds in living organisms. *Biometals* **5,** 3–12.

Romero, P. J. (1993). Synergistic activation of the human red cell calcium ATPase by magnesium and vanadate. *Biochim. Biophys. Acta* **1143,** 43–50.

Romero, P. J., and Rega, A. F. (1995). Effects of magnesium plus vanadate on partial reactions of Ca^{2+}-ATPase from the human red cell membranes. *Biochim. Biophys. Acta* **1235,** 155–157.

Sabbioni, E., Pozzi, G., Devos, S., Pintar, A., Casella, L., and Fischbach, M. (1993). The intensity of vanadium(V)-induced cytotoxicity and morphological transformation in BALB/3T3 cells is dependent on glutathione-mediated bioreduction to vanadium(IV). *Carcinogenesis* **14,** 2565–2568.

Sachs, G., Munson, K., Balaji, V. N., Aures-Fischer, D., Hersey, S. J., and Hall, K. (1989). *J. Bioenerg. Biomembr.* **21,** 573–578.

Sakurai, H., Tamura, H., and Okatani, K. (1995). Mechanism for a new antitumour vanadium complex: Hydroxyl radical-dependent DNA cleavage by 1,10-phenanthroline-vanadyl complex in the presence of hydrogen peroxide. *Biochem. Biophys. Res. Commun.* **206,** 133–137.

Schultz, B. D., Bridges, R. J., and Frizzell, R. A. (1996). Lack of conventional ATPase properties in CFTR chloride channel gating. *J. Memb. Biol.* **151,** 63–75.

Shechter, Y., Meyerovitch, J., Farfel, Z., Sack, J., Bruck, R., Bar-Meir, S., Amir, S., Deganis, H., and Karlish, S. J. D. (1990). Insulin mimetic effects of vanadium. In N. D. Chasteen, (Ed.), *Vanadium in Biological Systems.* Kluwer Academic Publishers, Dordrecht, pp. 129–142.

Shimokawa, N., and Yamaguchi, M. (1992). Calcium administration stimulates the expression of calcium-binding protein regucalcin mRNA in rat liver. *FEBS Lett.* **305,** 151–154.

Shisheva, A., Ikonomov, O., and Shechter Y. (1994). The protein tyrosine phosphatase inhibitor, pervanadate, is a powerful antidiabetic agent in streptozotocin-treated diabetic rats. *Endocrinology* **134,** 507–510.

Takahashi, H., and Yamaguchi, M. (1995). Increase of the (Ca^{2+}-Mg^{2+})-ATPase activity in hepatic plasma membranes of rats administered orally calcium: The endogenous role of regucalcin. *Mol. Cell Biochem.* **144,** 1–6.

Volberg, T., Zick, Y., Dror, R., Sabanay, I., Gilon, C., Levitzki, A., and Geiger, B. (1992). The effect of tyrosine-specific protein phosphorylation on the assembly of adherens-type junctions. *EMBO J.* **11,** 1733–1742.

Wisniewski, E., Gellerich, F. N., and Kunz, W. S. (1995). Distribution of flux control among the enzymes of mitochondrial oxidative phosphorylation in calcium-activated saponin-skinned rat *musculus soleus* fibers. *Eur. J. Biochem.* **230,** 549–554.

Woscholski, R., Waterfield, M. D., and Parker, P. J. (1995). Purification and biochemical characterization of a mammalian phosphatidylinositol 3,4,5-triphosphate 5-phosphatase. *J. Biol. Chem.* **270,** 31001–31007.

Zelikoff, J. T., and Cohen, M. D. (1995). Immunotoxicity of inorganic metal compounds. In R. J. Smialowicz and M. P. Holsapple (Eds.), *Experimental Immunotoxicology.* CRC Press, Boca Raton, pp. 189–228.

Zhu, Y, Goodridge, A. G., and Stapleton, S. R. (1994). Zinc, vanadate and selenate inhibit the tri-iodothyronine-induced expression of fatty acid synthase and malic enzyme in chick-embryo hepatocytes in culture. *Biochem. J.* **303** (Pt 1), 213–216.

Zychlinski, L., and Byczkowski, J. Z. (1990). Inhibitory effects of vanadium pentoxide on respiration of rat liver mitochondria. *Arch. Environ. Contam. Toxicol.* **19,** 138–142.

14

ENERGY TRANSDUCTION MECHANISMS AS AFFECTED BY VANADIUM(V) SPECIES: Ca²⁺-PUMPING IN SARCOPLASMIC RETICULUM

Manuel Aureliano*

Centro de Neurociências de Coimbra, Departamento de Zoologia, Universidade de Coimbra, 3000 Coimbra, Portugal

Vítor M. C. Madeira

Departamento de Bioquímica, Universidade de Coimbra, 3000 Coimbra, Portugal

* Present address: Instituto Superior de Ciências da Saúde—Norte Cidadela Universitária, 4580 Paredes, Portugal.

Vanadium in the Environment. Part 1: Chemistry and Biochemistry, Edited by Jerome O. Nriagu.
ISBN 0-471-17778-4. © 1998 John Wiley & Sons, Inc.

1. INTRODUCTION

Twenty years ago vanadium was found in commercial ATP and rediscovered for biology as an inhibitor in muscle and Na$^+$,K$^+$-ATPase inhibitor (Josephson and Cantley, 1977; Cantley et al., 1977; Beaugé and Glynn, 1978). Vanadium is currently used as inhibitor of E1-E2 ion transport ATPases (Pedersen and Carafoli, 1987; Beffagna et al., 1993; Larsson et al., 1994; Yoshikawa et al., 1995), for example, the sarcoplasmic reticulum (SR) Ca^{2+} pump, although the role of vanadium in these transport systems is not yet clearly understood. Recent data on vanadium association with membrane-bound transport systems (Willsky et al., 1984; Bode et al., 1990; Gabbay-Azaria et al., 1994; Uyama et al., 1994) involved in energy transduction mechanisms (Aureliano and Madeira, 1994a, 1994b) suggest an effective participation of vanadium in the energetic coupling of transport ATPases.

Vanadium is present in a variety of natural products, mainly in crude oil, coal and gasoline. The metal and derivatives used in industry as components of alloys and catalysts are known for environmental and biological impacts (Chasteen, 1983; Erdmann et al., 1984; Sadiq et al., 1992; French and Jones, 1993; Sadiq and Mian, 1994). Most of the biological activity of vanadium is associated with the +5 oxidation state (vanadate) which mimicks the chemistry of phosphorus in ortophosphate (Chasteen, 1983, Stankiewicz et al., 1995). It has been reported that vanadate reverses drug resistance (Colin et al., 1994; Desoize et al., 1994), increases glucose metabolism (Shechter, 1990), and influences, by either inhibition or stimulation, the activity of several enzymes (Simons, 1980; Macara, 1980; Chasteen, 1983; Nour-Eldeen et al., 1985; Nechay et al., 1986; Stankiewicz et al., 1987).

In vanadium(V) solutions, different n-meric ($n = 1$–10) vanadate species can occur simultaneously in equilibrium, for example, monomeric (V1), dimeric (V2), tetrameric (V4), and decameric (V10) and, in some cases, with different states of protonation and conformations (Rossoti and Rossoti, 1956; Ingri and Brito, 1959; Howarth and Jarrold, 1978; Baes and Mesmer, 1978; Heath and Howarth, 1981; Harnung et al., 1992; Amado et al., 1993). Often the species variety is not accounted for in most biological studies, although it is recognized that the individual species may differently influence enzyme activities (DeMaster and Mitchell, 1973; Choate and Mansour, 1979; Chasteen, 1983; Crans, 1993a; Aureliano and Madeira, 1994a; Stankiewicz et al., 1995). Furthermore, the chemical equilibrium between the vanadate oligomers, in the time scale of the millisecond (Crans et al., 1990a), makes difficult the identification of the vanadate species responsible for the effects promoted in biological systems. The effects can be conveniently appraised by combining biochemical kinetic studies with ^{51}V NMR spectroscopy (Crans et al., 1990b; Crans 1993a; Kalyani and Ramasarma, 1992; Hochman et al., 1993; Aureliano and Madeira, 1994a, 1994b). It has been possible to identify inhibitory effects correlated with specific vanadate species (Stankiewicz et al., 1987; Crans et al., 1990b; Crans and Schelble, 1990; Aureliano and Madeira, 1994a). Therefore, it is of primary importance to precisely characterize the vanadate species and the interactions with the system before attempting to understand the promoted effects (Chapter 7). Allegorically, vanadate studies in biological systems compare to iceberg phenomena: There is an invisible part, probably not the most interesting, but certainly the major part. With the sarcoplasmic reticulum calcium pump, experiments can be designed to detect different interactions and different effects for several oligomeric vanadate species (Aureliano and Madeira, 1994a, 1994b). Some of the interactions, (e.g., decameric species) disrupt the energetic coupling and the enzyme turnover. Other interactions of vanadium (e.g., monomeric species) may be without effect or even improve the coupling of Ca^{2+} pumping.

2. CHEMICAL SPECIES IN NOMINAL "MONOVANADATE" AND "DECAVANADATE" SOLUTIONS

Distinct methods for the preparation of vanadate solutions containing only monomeric or decameric species have been developed to test the effects promoted by these species in SR Ca^{2+}-ATPase (Csermely et al., 1985a; Varga et al., 1985; Csermely et al., 1985b; Coan et al., 1986; Vegh et al., 1990; Molnar et al., 1991; Aureliano and Madeira, 1994a; 1994b). Although it is possible to obtain a sole vanadate species exclusively in a stock solution, the denominated "monovanadate" solution (containing presumably only monomeric species) or the "decavanadate" solution (decameric species) is often contaminated with other species, as indicated by ^{51}V NMR spectroscopy (Fig. 1). This spectroscopic technique is very powerful for evaluating the composition of solutions as a function of vanadate concentration, pH, ionic strength, and the presence of compounds of biological interest (Howarth and Jarrold, 1978; Heath and Howarth, 1981; Howarth, 1991; Amado et al., 1993; Aureliano and

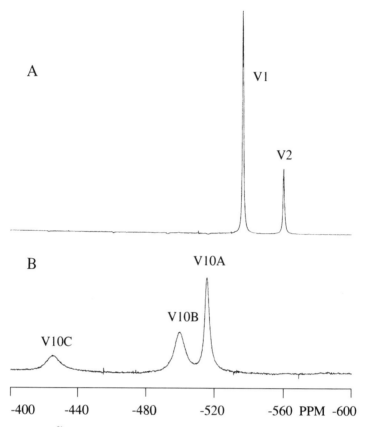

Figure 1. 52.6 MHz ^{51}V NMR spectra, at 22 °C, of a vanadate stock solution 100 mM in vanadium at pH 12.3 (A) and pH 4.0 (B). Vertical scale in B is amplified 4×. V1 and V2 NMR signals correspond, respectively, to monomeric (VO_4^{3-}, HVO_4^{2-} and $H_2VO_4^-$) and dimeric ($HV_2O_7^{3-}$, and $H_2V_2O_7^{2-}$) vanadate species regardless of the protonation state. V10A, V10B, and V10C are signals of vanadium atoms from the decameric ($V_{10}O_{28}^{6-}$) species. The concentrations of distinct vanadate oligomers were calculated from the fractions of the total integrated areas observed in spectra, as described elsewhere (Aureliano and Madeira, 1994a).

Madeira, 1994a; Crans, 1995). As expected, when the stock vanadate solutions are diluted into a reaction medium, several vanadate species appear with time. For instance, 1 h after dilution of nominal "decavanadate" into a reaction medium monomeric, dimeric, and tetrameric vanadate species appear in solution (Aureliano and Madeira, 1994a). Whenever possible, experimental conditions should be optimized to obtain n-meric vanadate with minimal contamination from other species.

2.1. Effect of pH and Total Vanadium Concentration

In the millimolar range of vanadium(V) concentrations several vanadate species occur simultaneously in equilibrium: monomeric (VO_4^{3-}, HVO_4^{2-}, and

$H_2VO_4^-$), dimeric ($HV_2O_7^{3-}$ and $H_2V_2O_7^{2-}$), tetrameric ($V_4O_{12}^{4-}$), pentameric ($V_5O_{15}^{5-}$), and even decameric vanadates ($V_{10}O_{28}^{6-}$) may occur upon adicidification, and several structural species may eventually occur (Rossoti and Rossoti, 1956; Ingri and Brito, 1959; Howarth and Jarrold, 1978; Habayed and Hileman, 1980; Heath and Howarth, 1981; Petterson et al., 1985; Harnung et al., 1992; Amado et al., 1993; Crans 1993b). Most of the vanadate species can be detected by [51]V NMR (O'Donnel and Pope, 1976; Howarth and Jarrold, 1978; Habayed and Hileman, 1980, Heath and Howarth, 1981; Amado et al., 1993). The spectra of "decavanadate" stock solution, 100 mM in total vanadium, contain only 10 mM of decameric species, since only signals from decameric vanadate species are observed: V10A at -515 ppm, V10B at -500 ppm, and V10C at -424 ppm (Fig. 1B). The concentrations of each vanadate oligomer were obtained by integrating the respective areas of NMR spectra and were calculated by using an algorithm described elsewhere (Aureliano and Madeira, 1994a). When the pH is adjusted to 7.0, small amounts of monomeric (V1) at -560 ppm, dimeric, (V2) at -574 ppm and tetrameric (V4) species at -578 ppm were detected (Aureliano and Madeira, 1994a). After dilution of this "decavanadate" solution into the reaction medium in the concentration range normally used in kinetic studies (up to 5 mM in vanadium), the concentration of decameric vanadate increases linearly with total vanadate, whereas monomeric vanadate remains constant (Aureliano and Madeira, 1994a), in agreement with data of other researchers (Csermely et al., 1985a).

"Monovanadate" stock solutions, at 100 mM total vanadium concentration, contain monomeric (70 mM) and often a significant amount of dimeric (15 mM) species (Fig. 1A). If the pH is adjusted to 7.0, the concentrations of these species are much lower, whereas tetrameric becomes the major species (23 mM) (Aureliano and Madeira, 1994a). Upon dilution of this "monovanadate" solution into the reaction medium, in the concentration range normally used in biochemical studies, different concentrations of mono- (V1), di- (V2), tetra- (V4), and pentameric (V5) species are present, these being the higher molecular vanadate oligomers favored by increasing total vanadate concentration. A series of [51]V NMR spectra of "monovanadate" solutions were scanned for concentrations up to 5 mM in a medium containing 5 mM $MgCl_2$, 5 mM HEPES, pH 7.0, at 22 °C (Fig. 2). The concentration of each vanadate oligomer, calculated by integrating the respective areas of the NMR spectra as a function of total vanadate, exhibits different profiles according to the species (Aureliano and Madeira, 1994a). For example, nominal 2 mM "monovanadate," in addition to 662 ± 46 μM monomeric (V1), also contains 143 ± 9 μM dimeric (V2), and 252 ± 16 μM tetrameric (V4) species.

2.2. Stability of Vanadate Solutions

In contrast to "monovanadate," the "decavanadate" solution is unstable in the assay medium. When "decavanadate" is diluted at room temperature into the reaction medium at low concentrations (2 mM total vanadium equivalent

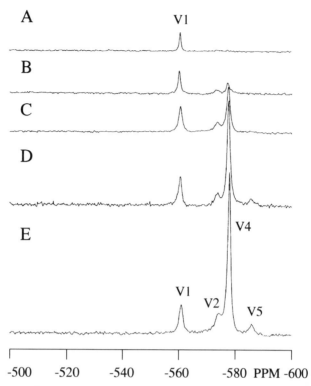

Figure 2. 52.6 MHz ^{51}V NMR spectra, at 22 °C, of different concentrations of a nominal "mono-vanadate" solution in a medium containing 25 mM HEPES, pH 7.0, and 0.1 M KCl: 0.25 mM (A); 0.75 mM (B); 2 mM (C); 3.5 mM (D); and 5 mM (E) total vanadium. V1 and V2 NMR signals correspond, respectively, to monomeric (VO_4^{3-}, HVO_4^{2-}, and $H_2VO_4^{-}$) and dimeric ($HV_2O_7^{3-}$ and $H_2V_2O_7^{2-}$) vanadate species regardless of the protonation state, whereas V4 and V5 correspond to cyclic tetrameric ($V_4O_{12}^{4-}$) and pentameric ($V_5O_{15}^{5-}$) vanadate species.

to 200 μM decameric species), ^{51}V NMR spectra show an additional signal from monomeric vanadate. Thus, after 17 min the solution contains 190 μM decameric vanadate and about 100 μM monomeric vanadate. After 34 min the monomeric concentration rises to 193 μM. Also, the monomeric concentration present in decavanadate solutions increases in the presence of 2.5 mM ATP (up to 177 and 250 μM after 17 and 34 min, respectively), and the presence of SR (1.25 mg/ml) prevents the appearance of monomeric species significantly (up to 50%) (Aureliano and Madeira, 1994a). Therefore, at room temperature, the rate of depolymerization is slow, making it possible to study the effect of the decameric species on the activity of sarcoplasmic reticulum Ca^{2+} pump. However, at 37 °C, the disappearance of the decameric vanadate is faster and can be easily followed by UV/Vis spectroscopy (Fig. 3A). The degree of depolymerization of decameric species, under several experimental conditions, can be evaluated by the second derivative of the UV/Vis spectra of "decavana-

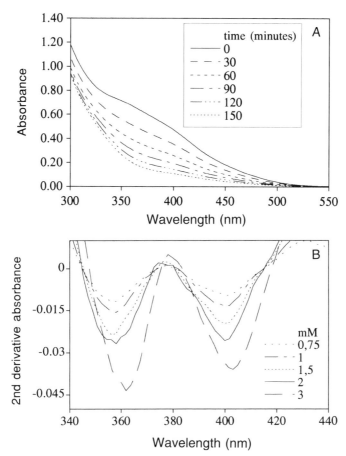

Figure 3. UV/Vis spectra of "decavanadate" in the assay medium containing 0.1 M KCl, 5 mM MgCl$_2$, 5 mM HEPES, pH 7.0 A. Spectra of "decavanadate" solution (1 mM total vanadium), at 37 °C, obtained at different times (0, 30, 60, 90, 120, and 150 min). B. Second derivative UV/Vis spectra, at 25 °C, of increasing concentrations of nominal "decavanadate" solution (0.75, 1, 1.5, 2, and 3 mM total vanadium).

date" solutions (Fig. 3B). As observed in Figure 3B, "decavanadate" concentration is proportional to the increase in the negative bands at 360 nm and 400 nm ascribed to decameric species, as described elsewhere (Aureliano, 1995).

3.   VANADATE COMPLEXES

During the course of experiments with the SR Ca^{2+} pump, the presence of several compounds in the reaction medium induces changes to the NMR spectra of vanadate solutions, affecting the composition in oligomeric species. The effects of the addition of acetylphosphate (ACP), pyruvate, sucrose, and phosphoenolpyruvate (PEP) to "monovanadate" solutions were studied by ^{51}V NMR spectroscopy, which has been widely used in the characterization of vanadate interactions with several compounds of biological interest (Rheder, 1991; Howarth, 1991; Crans, 1995). These studies are particularly interesting because the biochemical effects of vanadates are mediated by esterification with phosphate or hydroxyl group of biological molecules (Lindquist et al., 1973; Lopez et al., 1976; Crans, 1995; Stankiewicz et al., 1995). For instance, the inhibition of myosin is due to the formation of a ternary complex between vanadate, ADP, and the protein (Goodno, 1982). In the case of the membrane-bound E1-E2 transport ATPase, the inhibition is promoted by the binding to the conformation of the protein, which is phosphorylated by orthophosphate, the E2 conformation (Pick, 1982). The relevance of ACP is related to the catalytic site of sarcoplasmic reticulum Ca^{2+}-ATPase, which contains an aspartyl residue that is phosphorylated by ATP during the catalytic cycle, forming an acyl phosphate anhydride (de Meis and Vianna, 1979). Phosphoenolpyruvate is often used in studies when regeneration of ATP is required (Galina and de Meis, 1991), and sucrose is often added to preparations to be frozen (Eletr and Inesi, 1972; Da Costa and Madeira, 1986).

3.1. Evidence of New Bound Vanadate NMR Signals

3.1.1. *Sucrose*

Two new vanadate signals were clearly identified in the ^{51}V NMR spectra of vanadate solutions in the presence of SR Ca^{2+}-ATPase (Aureliano and Madeira, 1994a). These bound vanadate signals are due to the presence of sucrose in the SR samples, since they are observed upon addition of sucrose. Two signals are observed at -554 ppm (C1) and at -540 ppm (C2), whereas the signal ascribed to tetrameric vanadate decreases in intensity as a function of sucrose concentration (Fig. 4). The half line widths of these signals are similar (300 ± 50 Hz), and the relative areas gave populations of C1 amounting to twice those of C2. Probably C1 and C2 are complexes of vanadate with geometric isomers of sucrose. Although scarce, studies about the interaction between vanadate and saccharides (Tracey and Gresser, 1988; Sreedhara et al., 1994) are important in biology, since it has been found that vanadate influences enzymatic reactions involving these compounds (Nour-Eldeen et al., 1985; Shechter, 1990). Sucrose is routinely added during the isolation of E1-E2 membrane-bound ATPases. As shown in Figure 4, the formation of vanadate:sucrose complexes induces the disappearance of tetrameric vanadate. If this vanadate species is the one responsible for promoting a specific effect, the absence of tetrameric species may in part explain the lack of sensitivity to vanadate of E1-E2 enzymes isolated in the presence of sucrose, as described elsewhere (Kasai and Sawada, 1994).

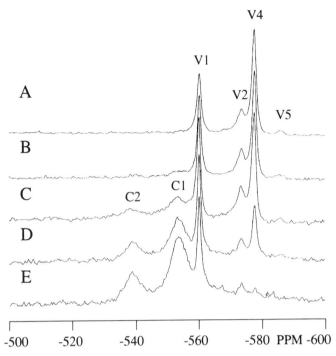

Figure 4. 52.6-MHz ^{51}V NMR spectra of 2.0 mM nominal "monovanadate" solution in a medium containing 0.1 M KCl, 5 mM MgCl$_2$, 5 mM HEPES, pH 7.0, at 22 °C, in the absence (A) or in the presence of sucrose: 25 (B); 50 (C); 100 (D); 200 mM (E). New vanadate NMR signals, C1 and C2, are observed. The area of C1 is approximately twice as large as the area of C2.

3.1.2. Acetylphosphate, Phosphoenolpyruvate, and Pyruvate

In the presence of phosphoenolpyruvate or acetylphosphate, the monomeric vanadate ^{51}V NMR signal shifts and suffers a large broadening (Aureliano et al., 1994). This behavior has been observed in studies of vanadate–orthophosphate interactions (Gresser et al., 1986) and may be related to the formation of a complex between vanadate and the phosphate group. Broadening of monomeric signal, with no up-field shift, was observed with simultaneous decrease in intensity of all the vanadate signals, upon the addition of ADP (Aureliano et al., 1994). Compared to phosphoenolpyruvate, acetyl-phosphate shifted up-field and broadened the monomeric signal in a larger extension, but decreased the intensities of the monomeric as well as the tetrameric vanadate signals, probably owing to a different type of interaction with vanadate. The cleavage of phosphoenolpyruvate by vanadate with libera-tion of orthophosphate in the medium may be due to a special interaction of vanadate with phosphoenolpyruvate (Aureliano et al., 1994). The mechanism of phosphoenolpyruvate cleavage excludes the formation of pyruvate as a product, since the ^{51}V NMR spectra of the vanadate solution, in the presence of pyruvate, displays a decrease of the monomeric resonance signal without broadening, whereas dimeric and tetrameric species almost disappear (Aureliano et al., 1994). Several products of vanadate and pyruvate can be formed and three main bound vanadate signals were observed with chemical shifts at -490, -518, and -546 ppm (not shown), which were assigned to octahedral and trigonal bipyramidal vanadate, as observed for oxalate, succinate, and lactate (Tracey et al., 1987; Tracey et al., 1990). Probably, upon the cleavage of the $C-O$ bond of phosphoenolpyruvate by vanadate, vanadoenolpyruvate complex and orthophosphate are formed (Aureliano et al., 1994; Aureliano, 1995). The mechanism would be similar to phosphorolysis, except that the cleavage would be promoted by vanadate instead of phosphate (Cori et al., 1938).

4. VANADATE INTERACTIONS WITH SARCOPLASMIC RETICULUM Ca^{2+}-ATPase

Widely known for the toxic effects, vanadium is vestigial in muscles and other tissues that are considered an essential oligoelement for humans, although its role is far from clearly identified (French and Jones, 1993; Harland and Harden-Williams, 1994). Recently, the discovery of vanadium-containing enzymes in microorganisms (e.g., nitrogenases) increased the interest on vanadate interactions with proteins (Hales et al., 1986; Kreen et al., 1989; Messerschmidt and Wever, 1996). The pioneer ^{51}V NMR studies on the interaction between oligovanadates and proteins were performed with SR Ca^{2+}-ATPase. These and other studies summarized in Table 1 have been useful in determining the vanadate species potentially responsible for the effects promoted in enzyme activities. With SR Ca^{2+}-ATPase, vanadate interacts particularly with the E2 conformation of the enzyme (Medda and Hasselbach, 1983; Csermely

Table 1 ^{51}V RMN Studies on Vanadate Oligomer Interactions with Proteins

Protein	RMN Signals More Affected	Reference
SR ATPase	V10 and V4	Csermely et al., 1985a
Ribonuclease A	V1	Borah et al., 1985
Phosphoglycerate mutase	V2	Stankiewicz et al., 1987
Apotransferrin	V1	Butler and Eckert, 1989
Myosin	V1 and V4	Ringel et al., 1990
Superoxide dismutase	V4	Wittenkeller et al., 1991
Myosin	V10 and V4	Aureliano, 1991
Phosphoglycerate mutase	V2	Liu et al., 1992
SR ATPase	V10; V4 and V1[a]	Aureliano and Madeira, 1994a

[a] Signals differently affected according to the conformation of the enzyme.

et al., 1985a), inducing crystallization of SR Ca^{2+}-ATPase dimers (Dux and Martonosi, 1983; Maurer and Fleischer, 1984), mediating photocleavage (Vegh et al., 1990; Molnar et al., 1991), and inhibiting enzyme activity (Pick, 1982; Csermely et al., 1985a; Molnar et al., 1991). The E1 conformation of the enzyme may also be favorable to interaction with vanadate (Vegh et al., 1990; Molnar et al., 1991). Recently it has been suggested that the interaction of monomeric vanadate with sarcoplasmic reticulum ATPase is favored during enzyme phosphorylation by ATP, probably by forming an ATP intermediate adduct or by condensation with the aspartylphosphate group (Aureliano et al., 1994; Aureliano and Madeira, 1994a).

4.1 Interactions of Vanadate Oligomers with Different SR Ca²⁺-ATPase Conformations

A strong affinity for SR ATPase in the E2 conformation was reported for several oligomeric species (e.g., decameric and tetrameric), as described previously (Csermely et al., 1985a). However, monomeric and other oligovanadates may also interact with other forms of the ATPase than the E2 conformation (Andersen and Moller, 1985; Coan et al., 1986; Vegh et al., 1990; Vaschencko et al., 1991; Molnar et al., 1991). With the aid of NMR spectroscopy, experiments can be designed to identify specific interactions of several vanadate oligomers with different forms of the SR ATPase (Aureliano and Madeira, 1994a). For instance, the presence of ATP in the medium induces important changes on the ^{51}V NMR spectra of vanadate, upon addition of SR. These changes in the NMR spectra are possibly the consequence of interactions of the different vanadate species with the Ca^{2+}-ATPase. The presence of natural ligands of the enzyme may induce ATPase conformations that are particularly specific for certain vanadate oligomers. The relative order of interaction of the vanadate oligomers with the SR ATPase is V4 > V10 > V1 = V2 = 1 and V4 = V10 = V1 > V2 = 1 in the absence and in the presence of ATP, respectively, whereas no changes were observed for dimeric vanadate (Aureliano and Madeira, 1994a). Therefore, we assign a value of 1 to the dimeric species. A deeper appraisal of this order of interaction of oligovanadates present in "monovanadate" solutions can be followed in Figure 5. The interaction of monomeric vanadate with the enzyme is significantly potentiated by the presence of ATP, whereas the interaction with tetrameric vanadate is decreased. Furthermore, other studies indicate that the affinity of tetrameric vanadate is stronger for the E2 than for the E1 conformation of the enzyme, and that the affinity of tetrameric vanadate for both forms of the ATPase decreases upon phosphorylation by orthophosphate or by ATP (Aureliano and Madeira, 1994a). In studies with myosin, similar findings were reported for tetrameric and monomeric species, after addition of ATP (Ringel et al., 1990). However, the broadening effects are at least two to three times larger, depending on the vanadate oligomers, upon addition of SR ATPase, as compared with myosin (Aureliano 1991; Aureliano, 1995).

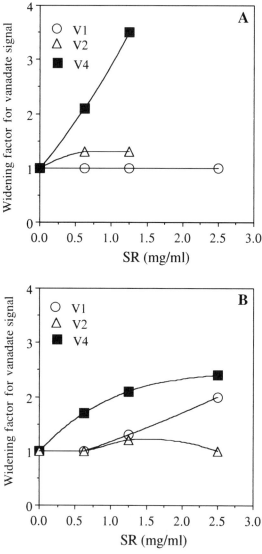

Figure 5. Effect of SR addition (up to 2.5 mg/ml), on the half-width of ^{51}V NMR signals of vanadate oligomers present in 2 mM "monovanadate" solution in a medium containing 0.1 M KCl, 5 mM MgCl$_2$, 5 mM HEPES, pH 7.0, at 22 °C, in the absence (A) or in the presence (B) of ATP 2.5 mM. Notice that the interaction of tetrameric species with SR is decreased by ATP, but the interaction of monomeric species is increased.

4.2. Binding Sites of Vanadate Oligomers

Decameric, tetrameric, and dimeric vanadates have been shown to have a stronger affinity for the SR ATPase relative to the monomeric species (Csermely et al., 1985a). It was initially reported that monomeric and decameric species may bind to the ATPase at the vicinity of the phosphate acceptor (351 aspartyl residue), thus blocking the active site by preventing formation of the phosphoenzyme intermediate (Varga et al., 1985). The decameric vanadate binding site to the enzyme may be the ATP binding site (Varga et al., 1985) or as very close to it (Molnar et al., 1991) in addition to being binding site for monomeric species (Varga et al., 1985; Csermely et al., 1985b). In these studies, the "decavanadate" solutions also contained monomeric species, and may account for the mimicking of monomeric species by interacting with the enzyme at the phosphate binding site. Therefore, it is always very difficult to ascertain if the effect is due to decameric vanadate (Boyd et al., 1985). The same reasoning can be argued in respect to "monovanadate" solutions owing to the presence of dimeric and tetrameric species. It has been recently suggested that with use of small concentrations of "decavanadate" and experimental conditions that minimize the contamination from others species, the decameric species may bind to the SR ATPase after ATP binding, whereas the appearance of monomeric species induces decameric vanadate to compete with ATP (Fig. 6) (Aureliano, 1995). Data in Figure 6 indicate that two different inhibition mechanisms for "decavanadate" solution may occur. At low "decavanadate" concentrations, incompetitive or noncompetitive inhibition is observed, whereas for 1.6 mM in total vanadium, probably containing about 155 μM decameric species and 50 μM monomeric species, the inhibition is competitive regarding ATP. It is suggested that low "decavanadate" has insufficient monomeric vanadate to promote the "opening" of the ATP binding site to decameric species.

The interaction of monomeric vanadate with the SR ATPase that is observed only when phosphorylation of the enzyme by ATP may occur (Aureliano and Madeira, 1994a) is probably a consequence of increased affinity of this species for the protein (Andersen and Moller, 1985; Markus et al., 1986), suggesting that monomeric species may bind with high or low affinity depending on the phosphorylation state of the enzyme, as described elsewhere (Yamasaki and Yamamoto, 1992). A possible mechanism of monomeric vanadate interaction may comprise the formation of ADP.V or ATP.V intermediate complexes, as has been suggested for the inhibition of myosin ATPase (Goodno, 1982). Condensation with the phosphorylation domain (Aureliano et al., 1994) may be also considered as an alternative hypothesis.

As in the case of monomeric vanadate the affinity of tetrameric species changes with phosphorylation, but the affinity of tetrameric vanadate decreases. The affinity of tetrameric vanadate to SR decreases after phosphorylation of the enzyme by ATP or by orthophosphate (Aureliano and Madeira, 1994a). This phosphate-analog-like behavior of the tetrameric vanadate was

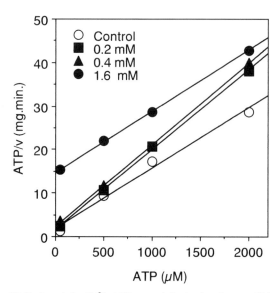

Figure 6. Hanes–Wolf plot of the Ca^{2+}-ATPase activity in the absence (○) and in the presence of "decavanadate": 0.2 (■), 0.4 (▲), and 1.6 (●) mM total vanadium. Reactions were performed at 25 °C with a pH electrode in 2 ml assay medium containing 0.5 mM HEPES, pH 7.0, 0.1 M KCl, 5 mM $MgCl_2$, and SR 0.18 mg/ml. The reactions were initiated by the addition of 84 μM Mg-ATP and the Ca^{2+}-ATPase activity was stimulated by the addition of 50 μM $CaCl_2$, 90 s after addition of ATP. Two mechanisms of inhibition are observed for "decavanadate" solutions. High concentration of "decavanadate" changed the inhibition from incompetitive to competitive in respect to ATP, probably owing to the presence of monomeric species.

first ascertained for monomeric vanadate, since its binding to SR ATPase is prevented after enzyme phosphorylation by inorganic phosphate (Pick, 1982). The interaction between tetrameric vanadate and the SR ATPase is particularly favorable when the enzyme is in the E2 conformation, probably by vanadate binding to enzyme sites with positive charges, namely, residues of lysine and of arginine (Varga et al., 1985; Aureliano and Madeira, 1994a). Nevertheless, the different cyclic structure of tetrameric vanadate (Fuchs et al., 1976; Heath and Howarth, 1981; Amado et al., 1993) may lead to a specific interaction with the SR ATPase, when compared with the interactions between decameric or monomeric species with the enzyme.

5. EFFECTS OF VANADATE ON THE ACTIVITY OF SARCOPLASMIC RETICULUM Ca^{2+} PUMP

Several studies have been carried out using vanadate for the characterization of SR ATPase, following the original studies reporting inhibition of the enzyme (Wang et al., 1979; Pick, 1982; Dupont and Bennett, 1982; Dux and Martonosi,

1983; Inesi et al., 1984; Ortiz et al., 1984; Csermely et al., 1985a; Coan et al., 1986; Molnar et al., 1991; Bigelow and Inesi, 1992). In our laboratory there is a long tradition of study of the molecular mechanisms of Ca^{2+} translocation by the SR calcium pump and proteins involved in muscle contraction (Carvalho and Leo, 1967; Vale and Carvalho, 1973; Pires and Perry, 1977; Madeira, 1980, 1982; Da Costa and Madeira, 1986; Aureliano et al., 1995; Madeira, 1995) Studies on the interaction of species present in vanadate solutions with macro-molecular assemblies involved in energy transduction (e.g., myosin ATPase and SR Ca^{2+}-ATPase) have been performed (Aureliano et al., 1987; Aureliano et al., 1988; Malva et al., 1988; Teixeira-Dias et al., 1991; Aureliano, 1991; Aureliano and Madeira, 1994a). Original findings that decameric and tetra-meric vanadate also affect myosin ATPase activity and that the inhibitory effects are restricted to some of the species were reported (Aureliano et al., 1987; Aureliano et al., 1991; Aureliano, 1991). Also significant additional data were obtained with SR ATPase (Aureliano and Madeira, 1994a; Aureliano and Madeira, 1994b).

Inhibition of the SR Ca^{2+}-ATPase was observed when the SR vesicles were incubated with vanadate in the presence of a Ca^{2+} ionophore (Pick, 1982). In these conditions, ATP hydrolysis is uncoupled from Ca^{2+} accumulation. Conversely, in intact SR native vesicles, 200 μM vanadate did not inhibit Ca^{2+} uptake or Ca^{2+}-dependent ATP hydrolysis; only a lag of Ca^{2+} uptake caused by preincubation with vanadate was observed (Pick, 1982). In these studies, EGTA is normally present in the medium to promote the E2 conformation of enzyme that is favorable for vanadate inhibition (Pick, 1982). The lack of effect is probably a consequence of the formation of a 1:1 vanadate complex with EGTA, affecting not only the concentration of monomeric species but also, and to a greater extent, the concentration of tetrameric vanadate. Actu-ally, in the interaction between decameric species and SR ATPase, EGTA can be used to prevent the appearance of monomeric species (Aureliano and Madeira, unpublished observations). Furthermore, the complexation of vanadate by EGTA also affects the concentration of free Ca^{2+} in the medium (Aureliano and Madeira, unpublished observations).

In the absence of EGTA, monomeric vanadate can normally be observed upon dilution of "decavanadate" in the medium. At room temperature, the rate of depolymerization of decameric into monomeric species is slow, allowing study of the effect of the decameric species on the activity of the sarcoplasmic reticulum Ca^{2+} pump. As observed in Section 2.2, after 17 min of dilution, 2 mM nominal "decavanadate" contains about 190 μM decameric vanadate and at least 100 μM monomeric species, depending on medium conditions. Kinetic studies can be performed in conditions such that the elapsed time before beginning the reactions is never longer than 5 min and usually shorter than 1 min, so that the conversion into monomeric vanadate is always less than 5% of total vanadium (Aureliano and Madeira, 1994a; Aureliano and Madeira, 1994b).

5.1.   Calcium Uptake and ATP Usage

5.1.1.   Coupled SR Ca²⁺ Pump

Kinetic studies performed by electrometry and colorimetry indicate that decameric vanadate clearly differs from vanadate oligomers present in "monovanadate" solutions in preventing the accumulation of Ca^{2+} by sarcoplasmic reticulum (SR) vesicles coupled to ATP hydrolysis. A nominal "decavanadate" solution, 2 mM in total vanadium, containing about 190–200 μM decameric and less than 100 μM monomeric species, depresses the rate of Ca^{2+} uptake by 50%, whereas a nominal 2 mM "monovanadate" solution, containing about 662 μM monomeric, 143 μM dimeric, and 252 μM tetrameric species had no effect on the rate of Ca^{2+} accumulation. However, the same "decavanadate" concentration (2 mM total vanadium) inhibits by 75% the SR Ca^{2+}-ATPase activity (Aureliano and Madeira, 1994a). An equivalent "monovanadate" concentration also inhibits the ATP hydrolysis by 50% without corresponding inhibition of Ca^{2+} accumulation by SR. Apparently both vanadate solutions promote important deviations from the orthodox Ca^{2+}:ATP stoichiometry (2:1). Deviations from classically accepted stoichiometry have also been reported previously under the effect of Mg^{2+} (Da Costa and Madeira, 1986) or other conditions (Gafni and Boyer, 1985; Galina and de Meis, 1991). For the case of the "monovanadate" concentration that inhibits by 50% the Ca^{2+}-ATPase activity (2 mM total vanadium), a Ca^{2+}:ATP ratio of 4 was obtained, meaning a significant increase of the efficiency of the pump (Aureliano and Madeira, 1994a).

In the presence of oxalate, inhibition of SR Ca^{2+}-ATPase activity by "decavanadate" and "monovanadate" solutions is enhanced to 97% and 86%, respectively, whereas in the presence of the ionophore lasalocid the inhibitions were 87% and 19% for these solutions (2 mM in vanadium), These results suggest that the uncoupling of ATP hydrolysis and Ca^{2+} accumulation by SR may prevent the effects promoted by the vanadate species present in "monovanadate," while not significantly affecting "decavanadate" inhibition. This behavior of "monovanadate" may be related to the observation that the rate of crystallization induced by "monovanadate" solutions, probably by tetrameric vanadate, is inhibited by the presence of the ionophore A23187 (Varga et al., 1985). However, as discussed above, 2 mM "monovanadate" contains other vanadate species. Therefore, it is not possible to clearly define the effects promoted by each species.

5.1.2.   Uncoupled SR Ca²⁺ Pump

The first work using "monovanadate" and "decavanadate" solutions as inhibitors of the SR Ca^{2+} pump did not report any differences between the two vanadate solutions (Csermely et al., 1985a; Molnar et al., 1991). Recent results point out that decameric vanadate is the only species able to depress the rate of Ca^{2+} uptake by SR and to strongly inhibit ATP hydrolysis uncoupled from Ca^{2+} accumulation (Aureliano and Madeira, 1994a). Conversely, two distinct effects for "monovanadate" solutions were observed when ATP hydrolysis is uncoupled from Ca^{2+} accumulation by the ionophore lasalocid. "Monovanadate" does not inhibit the ATP hydrolysis of the uncoupled Ca^{2+} pump until 3.5 mM is reached, at which the rate of ATP hydrolysis is depressed, probably owing to a significant presence of tetravanadate species at this "monovanadate" concentration (Aureliano and Madeira, 1994a). Tetrameric vanadate, recently reported to be responsible for the inhibition of several enzymes (Crans et al., 1990b), may also be responsible for depressing the rate of ATP hydrolysis by sarcoplasmic reticulum Ca^{2+}-ATPase, as observed for decameric species. New indications of the interaction of tetrameric vanadate with the SR calcium pump will be pointed out and discussed below, in Section 5.2.

5.2. Proton Ejection

In parallel with the inhibition of Ca^{2+} uptake, "decavanadate" also prevents the associated ejection of H^+ (Aureliano and Madeira, 1994b). Ejection of H^+, an early coupling event of Ca^{2+} pumping, is strongly inhibited by "decavanadate," whereas limited effects were observed for "monovanadate" solutions (Aureliano and Madeira, 1994b). Decameric vanadate 0.2 mM (2 mM total vanadium) delays the beginning of the accumulation of Ca^{2+} and decreases by 48% the rate of Ca^{2+} uptake, while the H^+ ejection is strongly affected (75%). An equivalent vanadium concentration in the form of "monovanadate" (2 mM) inhibits the ejection of H^+ by only 25%, whereas no additional inhibition was observed at higher concentrations. Rather, it increases the uptake of Ca^{2+} and only at 4 mM is a slight inhibition observed. Furthermore, "monovanadate" solutions synergistically increase the Ca^{2+} uptake inhibition promoted by decameric species, since the presence of 2 mM "monovanadate," which per se does not inhibit Ca^{2+} uptake, promotes an increase in the inhibitory effect of 2 mM "decavanadate" from 48% to 81%.

A linear decrease of the H^+/Ca^{2+} ratio as a function of "decavanadate" concentration indicates that the decameric species primarily affects the H^+ ejection associated with Ca^{2+} transport. "Monovanadate," although not effective at Ca^{2+} uptake, affects to a limited extent the H^+/Ca^{2+} ratio in concentrations up to 1.2 mM (Fig. 7). The effect fades out as the concentration increases beyond this limit, and essentially no effect is detected at 4 mM. This surprising "monovanadate" effect may be due to the presence of tetrameric vanadate, since it increases linearly and significantly in relation to other oligomeric species in solution when the nominal "monovanadate" concentration ranges from 1 to 5 mM (Aureliano and Madeira, 1994a).

The effect of "monovanadate" in depressing 25% of H^+ ejection at 1.2 mM indicates that some of the ejected protons are not directly coupled to Ca^{2+} transport. This interpretation is corroborated by the fact that "decavanadate" (2 mM total vanadium) decreases H^+ ejection by 75% but Ca^{2+} uptake by only 50% (Aureliano and Madeira, 1994a; Aureliano and Madeira, 1994b). Therefore, we are tempted to suggest that the fraction of released protons

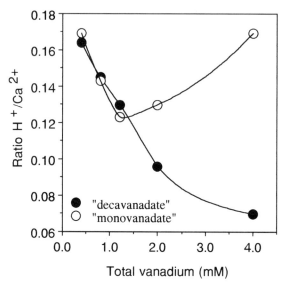

Figure 7. H$^+$/Ca^{2+} ratio as a function of "monovanadate" ($\bigcirc$) and "decavanadate" ($\bullet$) concentrations. The rates of H$^+$ ejection and Ca^{2+} uptake mean the amount of Ca^{2+} taken or the H$^+$ ejected per minute per milligram protein, after addition of Mg-ATP. The different behavior observed for "monovanadate" solution suggests an effective participation of tetrameric species in the events promoted by the SR Ca^{2+} pump.

not associated with the Ca^{2+} transport is possibly related to a partial energetic uncoupling of the pump in the sense that some ejected protons result from energy lost with no benefit to Ca^{2+} transport.

5.3. Calcium Efflux

When SR vesicles are loaded with radioactive Ca^{2+}, sedimented by centrifugation, and diluted in media containing EGTA, Ca^{2+} passively leaks through the ATPase. The presence of "monovanadate" or "decavanadate" inhibits the Ca^{2+} efflux through the pump, the effect being prevented when the Ca^{2+} concentration in the medium increases from 3 to 39 μM (Aureliano, 1995). Blockage of Ca^{2+} efflux was also reported for natural ligands of ATPase, namely, Ca^{2+}, Mg^{2+}, ADP, and orthophosphate (de Meis et al., 1990; de Meis, 1991) and for specific inhibitors of ATPase (de Meis, 1991). In the presence of ADP and orthophosphate, ATPase-reversed activity (ATP synthesis) is coupled with Ca^{2+} efflux and, in this condition, only "decavanadate" solutions strongly affect Ca^{2+} efflux from the vesicles (Aureliano, 1995). Probably only the binding of decameric vanadate stabilizes the ATPase channel in a closed state, and the blockage of the ATPase channel by decameric species is a consequence of the interaction between decameric species and the SR Ca^{2+}-ATPase ATP binding site, as suggested in Figure 6. The lack of effect observed for "monovanadate" may be due to the presence of Ca^{2+} in the medium and also to orthophosphate, since it prevents the binding of "monovanadate" species to SR Ca^{2+}-ATPase, as described elsewhere (Aureliano and Madeira, 1994a; Aureliano, 1995). Thus, the inhibition of Ca^{2+} uptake and Ca^{2+} efflux observed in the presence of "decavanadate" may result from the blockage of the Ca^{2+} binding sites located in the transmembranous portion of the enzyme (MacLennan and Toyofuku, 1992). It may ultimately be due to a protein conformation change in the region of the calcium binding sites (Cheng and Lepock, 1992) promoted by decameric species binding to the active site. According to other reports, the decameric binding to SR ATPase may induce changes in protein–protein interactions that promote dimerization of the AT-Pase (Dux and Martonosi, 1983; Maurer and Fleischer, 1984; Coan et al., 1986). Although the physiological role of the association between the ATPase monomers in the translocation of Ca^{2+} by the SR calcium pump is not yet understood (Boldyrev and Quinn, 1994), the formation of ATPase dimers induced by decameric vanadate may be related with the above effects promoted by "decavanadate" solutions.

6. SUMMARY

The effects of vanadium(V) solutions containing several vanadate oligomeric species on the activity of the Ca^{2+} pump of sarcoplasmic reticulum of rabbit skeletal muscle are appraised by combining biochemical kinetic studies with ^{51}V NMR and UV/Vis spectroscopy. The complexity of the system, which contains several vanadate species, interacts with several compounds, and has multiple sites on the SR Ca^{2+}-ATPase, in several conformations, hampers the clear understanding of the effects of each vanadate species present in solutions trivially denominated "decavanadate" and "monovanadate." With the sarcoplasmic reticulum Ca^{2+} pump, experiments can be designed for observing different interactions and different effects for several oligomeric species of vanadate. Additionally, new bound vanadate ^{51}V NMR signals were detected upon addition of compounds often used in experiments with SR ATPase—namely, sucrose, phosphoenolpyruvate, acetylphosphate, and pyruvate. These studies are important owing to the capacity of vanadate to condense with itself and with several compounds that are of biological interest and therefore may be useful in understanding the interactions of vanadium(V) in biological systems.

Changes in the interaction of several vanadate oligomers with different SR ATPase forms are detected by ^{51}V NMR spectroscopy. ATP changes the affinity of the enzyme for tetrameric and monomeric vanadates, the interaction with the tetrameric species being prevented by the presence of orthophosphate, whereas the interaction of monomeric species is observed only when ATP is present. The relative order of interaction of the different vanadate oligomers with SR ATPase is V4 > V10 > V1 = V2 = 1 and V1 = V10 =

V4 > V2 = 1 in the absence and in the presence of ATP, respectively, with no changes observed for dimeric vanadate (arbitrarily assigned a value of 1). Furthermore, tetrameric vanadate affinity is stronger for the E2 than for the E1 conformation of SR ATPase and it decreases upon phosphorylation by orthophosphate or ATP.

Kinetic studies performed by electrometry, colorimetry, and radiometry indicate that decameric vanadate clearly differs from other oligomeric species in inhibiting Ca^{2+} uptake by SR coupled to ATP hydrolysis, Ca^{2+} efflux coupled with ATPase-reversed activity (ATP synthesis), and H^+ ejection promoted by the SR ATPase. Also, the inhibition of these activities that is caused by some species can be prevented, or the activities can even be stimulated, by other vanadate oligomers, for example, tetrameric vanadate. The Ca^{2+}/ATP ratio of Ca^{2+} transport is affected by vanadate solutions and in particularly enhanced above the classic stoichiometry by "monovanadate." It is concluded that Ca^{2+} uptake is tightly coupled to proton ejection and ATP hydrolysis through molecular events that are sensitive to the interaction of vanadate species. Apparently, the Ca^{2+}/ATP and H^+/Ca^{2+} stoichiometries are variable and modulated by molecular events that are involved in vanadate interaction, suggesting alterations in the energetic coupling associated with Ca^{2+} translocation.

REFERENCES

Amado, A. M., Aureliano, M., Ribeiro-Claro, P. J., and Teixeira-Dias, J. J. C. (1993). Vanadium(V) oligomerization in aqueous alkaline solutions: A combined raman and [51]V NMR spectroscopic study. *J. Raman Spectrosc.* **24,** 669–703.

Andersen, J. P., and Moller, J. V. (1985). The role of Mg^{2+} and Ca^{2+} in the simultaneous binding of vanadate and ATP at the phosphorylation site of sarcoplasmic reticulum Ca^{2+}-ATPase. *Biochim. Biophys. Acta* **815,** 9–15.

Aureliano, M. (1991). Effect of myosin light chain phosphorylation and vanadate oligomers on myosin and actomyosin ATPase activity (Portuguese). Master's thesis, University of Coimbra, Portugal.

Aureliano, M. (1995). Energy transduction in sarcoplasmic reticulum Ca^{2+}-pump: Interactions of vanadates (Portuguese). Ph.D. thesis, University of Coimbra, Portugal.

Aureliano, M., Geraldes, C. F. G. C., and Pires, E. M. V. (1991). [51]V-analysis of the interaction of vanadium oligoanions with myosin. Abstract, Fourth Portuguese–Spanish Biochemistry Congress, Póvoa de Varzim, Portugal.

Aureliano, M., Leta, J., Madeira, V. M. C., and de Meis, L. (1994). The cleavage of phosphoenolpyruvate by vanadate. *Biochem. Biophys. Res. Commun.* **201,** 155–159.

Aureliano, M., Lima, M. C. P., Carvalho, A. P., and Pires, E. M. V. (1995) Effect of myosin phosphorylation on actomyosin ATPase activity—A flow microcalorimetric study. *Thermochim. Acta* **258,** 59–66.

Aureliano, M., Lima, M. C. P., and Pires, E. M. V. (1988). Inhibition of actomyosin MgATPase by decavanadate: A flow microcalorimetric study. Abstract, Third Portuguese–Spanish Biochemistry Congress, Santiago de Compostela, Spain.

Aureliano, M., and Madeira, V. M. C. (1994a). Interactions of vanadate oligomers with sarcoplasmic reticulum Ca^{2+}-ATPase. *Biochim. Biophys. Acta* **1221,** 259–271.

Aureliano, M., and Madeira, V. M. C. (1994b). Vanadate oligoanions interact with the proton ejection by Ca^{2+} pump of sarcoplosmic reticulum. *Biochem. Biophys. Res. Commun.* **205,** 161–167.

Aureliano, M., Silva, P. C., and Pires, E. M. V. (1987). Inhibition of Myosin ATPase by differente vanadate molecular species (Portuguese). Abstract, Tenth Portuguese Chemistry Society Meeting, Porto, Portugal.

Baes, C. F., and Mesmer, R. E. (1976). *The Hydrolysis of Cations.* John Wiley and Sons, New York.

Beaugé, L. A., and Glynn, I. M. (1978). Commercial ATP containing traces of vanadate alters the response of (Na^+,K^+) ATPase to external potassium. *Nature* **272,** 551–552.

Beffagna, N., Romani, G., and Lovadina, S. (1993). Inhibition of malate synthesis by vanadate: Implications for the cytosolic alkalinization induced by vanadate concentrations partially inhibiting the plasmalemma H^+ pump. *J. Exp. Botany* **267,** 1535–1542.

Bigelow, D. J., and Inesi, G. (1992). Contribution of chemical derivatization and spectroscopy studies to the characterization of the Ca^{2+} transport ATPase of sarcoplasmic reticulum. *Biochim. Biophys. Acta* **113,** 323–338.

Bode, H.-P., Friebel, C., and Fuhrmann, G. F. (1990). Vanadium uptake by yeast cells. *Biochim. Biophys. Acta* **1022,** 163–170.

Boldyrev, A. A., and Quinn, P. J. (1994). E1/E2 type cation transport ATPases: Evidence for transient associations between protomers. *Int. J. Biochem* **12,** 1323–1331.

Borah, B., Chen, C.-W., Egan, W., Miller, M., Wlodawer, A., and Cohen, J. S. (1985). Nuclear magnetic resonance and neutron diffraction studies of the complex of ribonuclease A with uridine vanadate, a transition-state analogue. *Biochemistry* **24,** 2058–2067.

Boyd, D. W., Kustin, K., and Niwa, M. (1985). Do vanadate polyanions inhibit phosphotransferase enzymes? *Biochim. Biophys. Acta* **827,** 472–475.

Butler, A., and Eckert, H. (1989). ^{51}V NMR as a probe of vanadium (V) coordination to human apotransferrin. *J. Am. Chem. Soc.* **111,** 2802–2809.

Cantley Jr., L. C., Josephson, L., Warner, R., Yanagisawa, M., Lechene, C., and Guidotti, G. (1977). Vanadate is a potent (Na^+,K^+)-ATPase inhibitor found in ATP derived from muscle. *J. Biol. Chem.* **252,** 7421–7423.

Carvalho, A. P., and Leo, B. (1967). Effects of ATP on the interaction of Ca^{2+}, Mg^{2+}, and K^+ with fragmented sarcoplasmic reticulum isolated from skeletal muscle. *J. Gen. Physiol.* **50,** 1327–1331.

Chasteen, N. D. (1983). The biochemistry of vanadium. *Struct. Bonding* **53,** 105–138.

Cheng, K. H., and Lepock, J. R. (1992). Inactivation of calcium by EGTA is due to an irreversible thermotropic conformational change in the calcium binding domain of the Ca^{2+}-ATPase. *Biochemistry* **31,** 4074–4080.

Choate, G., and Mansour, T. E. (1979). Studies on heart phosphofructokinase. *J. Biol. Chem.* **254,** 11457–11462.

Coan, C., Scales, D. J., and Murphy, A. J. (1986). Oligovanadate binding to sarcoplasmic reticulum ATPase. *J. Biol. Chem.* **261,** 10394–10403.

Colin, M., Madoulet, C., Baccard, N., Arsac, F., and Jardillier, J. C. (1994). Study of sodium orthovanadate as a reverser of multidrug resistance on lymphoblastic leukemic CEM/VLB$_{100}$ cells. *Anticancer Res.* **14,** 2383–2388.

Cori, G. T., Colowick, S. P., and Cori, C. F. (1938). The formation of glucose-1-phosphoric acid in extracts of mammalian tissues and of yeast. *J. Biol. Chem.* **123,** 375–380.

Crans, D. C. (1993a). Enzyme interactions with labile oxovanadates and other polyoxometalates. *Comments Inorg. Chem.* **16,** 35–76.

Crans, D. C. (1993b). Aqueous chemistry of labile oxovanadates: Relevance to biological studies. *Comments Inorg. Chem.* **16**, 1–33.

Crans, D. C. (1995). Interaction of vanadate with biogenic ligands. In H. Sigel and A. Sigel (Eds.), Metal Ions in Biological Systems, Vol. 31, pp. 147–209. Marcel Dekker, New York.

Crans, D. C., Rithner, C. D., and Theisen, L. A. (1990a). Application of time-resolved ^{51}V 2D NMR for quantitation of kinetic exchange pathways between vanadate monomer, dimer, tetramer, and pentamer. *J. Am. Chem. Soc.* **112**, 2901–2908.

Crans, D. C., and Schelble, S. M. (1990). Vanadate dimer and tetramer both inhibit glucose-6-phosphate dehydrogenase from *Leuconostoc mesenteroides*. *Biochemistry* **29**, 6698–6706.

Crans, D. C., Willging, E. M., and Butler, S. R. (1990b). Vanadate tetramer as the inhibiting species in enzyme reactions in vitro and in vivo. *J. Am. Chem. Soc.* **112**, 427–432.

Csermely, P., Martonosi, A., Levy, G. C., and Ejchart, A. J. (1985a). ^{51}V NMR analysis of the binding of vanadium(V) oligoanions to sarcoplasmic reticulum. *Biochem. J.* **230**, 807–815.

Csermely, P., Varga, S., and Martonosi, A. (1985b). Competition between decavanadate and fluorescein isothiocyanate on the Ca^{2+}-ATPase of sarcoplasmic reticulum. *Eur. J. Biochem.* **150**, 455–460.

Da Costa, G., and Madeira, V. M. C. (1986). Magnesium and manganese ions modulate Ca^{2+} uptake and its energetic coupling in sarcoplasmic reticulum. *Arch. Biochem. Biophys.* **249**, 199–206.

DeMaster, E. G., and Mitchell, R. A. (1973). A comparison of arsenate and vanadate as inhibitors or uncouplers of mitochondrial and glycolitic energy metabolism. *Biochemistry* **12**, 2616–3621.

De Meis, L. (1991). Fast efflux of Ca^{2+} mediated by the sarcoplasmic reticulum Ca^{2+} ATPase. *J. Biol. Chem.* **266**, 5736–5742.

De Meis, L., Suzano, V. A., and Inesi, G. (1990). Functional interactions of catalytic site and transmembrane channel in the sarcoplasmic reticulum ATPase. *J. Biol. Chem.* **265**, 18848–18851.

De Meis, L., and Vianna, A. L. (1979). Energy interconversion by the Ca^{2+}-dependent ATPase of the sarcoplasmic reticulum. *Annu. Rev. Biochem.* **48**, 275–292.

Desoize, B., Briois, F., and Carpentier, Y. (1994). Reversion of resistance to adriamycin by vanadate on four Friend cell sublines. *Int. J. Oncol.* **5**, 87–91.

Dupont, Y., and Bennett, N. (1982). Vanadate inhibition of the Ca^{2+}-dependent conformational change of the SR Ca^{2+}-ATPase. *FEBS Lett.* **139**, 237–240.

Dux, L., and Martonosi, A. (1983). Two-dimensional arrays of proteins in sarcoplasmic reticulum and purified Ca^{2+}-ATPase vesicles treated with vanadate. *J. Biol. Chem.* **258**, 2599–2603.

Eletr, S., and Inesi, G. (1972). Phospholipid orientation in sarcoplasmic reticulum membranes: Spin label ESR and proton NMR studies. *Biochim. Biophys. Acta* **282**, 174–179.

Erdmann, E., Werdan, K., Krawietz, W., Schmitz, W., and Scholz, H. (1984). Vanadate and its significance in biochemistry and pharmacology. *Biochem. Pharmacol.* **33**, 945–950.

French, R. J., and Jones, P. J. H. (1993). Role of vanadium in nutrition: Metabolism, essentiality and dietary considerations. *Life Sci.* **52**, 339–346.

Fuchs, J., Mahjour, S., and Pickardt, J. (1976). Structure of the true metavanadate ion. *Angew. Chem. Int. Ed.* **15**, 374–375.

Gabbay-Azaria, R., Pick, U., Ben-Hayyim, G., and Tel-Or, E. (1994). The involvement of a vanadate-sensitive ATPase in plasma membranes of a salt tolerant cyanobacterium. *Physiol. Plant.* **90**, 692–698.

Gafni, A., and Boyer, P. D. (1985). Modulation of stoichiometry of the sarcoplasmic reticulum calcium pump may enhance thermodynamic efficiency. *Proc. Natl. Acad. Sci. USA* **82**, 98–101.

Galina, A., and de Meis, L. (1991). Ca^{2+} translocation and catalytic activity of the sarcoplasmic reticulum ATPase. *J. Biol. Chem.* **266**, 17978–17982.

Goodno, C. C. (1982). Myosin active-site trapping with vanadate ion. *Methods Enzymol.* **85,** 116–123.

Gresser, M. J., Tracey, A. S., and Parkinson, K. M. (1986). Vanadium(V) oxyanions: The interaction of vanadate with pyrophosphate, phosphate, and arsenate. *J. Am. Chem. Soc.* **108,** 6229–6234.

Habayed, M. A., and Hileman Jr., O. E. (1980). ^{51}V FT-nmr investigations of metavanadate ions in aqueous solutions. *Can. J. Chem.* **58,** 2255–2261.

Hales, B. J., Case, E. E., Morningstar, J. E., Dezeda, M. F., and Mauterer, L. A. (1986). Isolation of a new vanadium-containing nitrogenase from *Azotobacter vinelandii. Biochemistry* **25,** 7251–7255.

Harland, B. F., and Harden-Williams, B. A. (1994). Is vanadium of human nutricional importance yet? *J. Am. Diet. Assoc.* **94,** 891–894.

Harnung, S. E., Larsen, E., and Pedersen, E. J. (1992). Structure of monovanadates in aqueous solution. *Acta Chim. Scand.* **47,** 674–682.

Heath, E., and Howarth, O. W. (1981). Vanadium-51 and oxygen-17 nuclear magnetic resonance study of vanadate(V) equilibria and kinetics. *J. Chem. Soc. Dalton Trans.*, pp. 1105–1110.

Hochman, Y., Carmeli, S., and Carmeli, C. (1993). Vanadate, a transition state inhibitor of chloroplast CF1-ATPase. *J. Biol. Chem.* **268,** 12373–12379.

Howarth, O. W. (1991). Vanadium-51 NMR. *Prog. NMR Spectrosc.* **22,** 453–485.

Howarth, O. W., and Jarrold, M. (1978). Protonation of the decavanadate(6-) ion: A vanadium-51 nuclear magnetic resonance study. *J. Chem. Soc. Dalton. Trans.*, pp. 503–506.

Inesi, G., Lewis, D., and Murphy, A. J. (1984). Interdependence of H$^+$, Ca^{2+}, and Pi (or vanadate) sites in sarcoplasmic reticulum ATPase. *J. Biol. Chem.* **259,** 996–1003.

Ingri, N., and Brito, F. (1959). Equilibrium studies of polyanions. VI. Polyvanadates in alkaline Na(Cl) medium. *Acta Chim. Scand.* **13,** 1971–11996.

Josephson, L., and Cantley Jr., L. C. (1977). Isolation of a potent (Na$^+$-K$^+$) ATPase inhibitor from striated muscle. *Biochemistry* **16,** 4572–4578.

Kalyani, P., and Ramasarma, T. (1992). Polyvanadate-stimulated NADH oxidation by plasma membranes—The need for a mixture of deca and meta forms of vanadate. *Arch. Biochem. Biophys.* **297,** 244–252.

Kasai, M., and Sawada, S. (1994). Evidence for decrease in vanadate-sensitive Mg^{2+} ATPase activity of higher plant membrane preparations in sucrose solution. *Plant Cell Physiol.* **35,** 697–700.

Kreen, B. E., Izumi, Y., Yamada, H., and Weber, R. (1989). A comparison of a different (vanadium) bromo peroxidases—The bromoperoxidase from *Corallina-pillulifera* is also a vanadium enzyme. *Biochim. Biophys. Acta* **998,** 63–68.

Larsson, C., Sommarin, M., and Widell, S. (1994). Isolation of highly purified plant plasma membranes and separation of inside-out and right-side-out vesicles. *Methods Enzymol.* **228,** 451–469.

Lindquist, R. N., Lynn, J. L., and Lienhard, G. E. (1973). Possible transition state analogs for ribonuclease the complexes of uridine with oxovanadium(IV) ion and vanadium(V) ion. *J. Am. Chem. Soc.* **95,** 8762–8768.

Liu, S., Gresser, M. J., and Tracey, A. S. (1992). ^{1}H and ^{51}V NMR studies of the interaction of vanadate and 2-vanadio-3-phosphoglycerate with phosphoglycerate mutase. *Biochemistry* **31,** 2677–2685.

Lopez, V., Stevens, T., and Lindquist, R. N. (1976). Vanadium ion inhibition of alkaline phosphatase-catalyzed phosphate ester hydrolysis. *Arch. Biochem. Biophys.* **175,** 31–38.

Macara, I. G. (1980). Vanadium—An element in search of a role. *Trends Biochem. Sci.* April, pp. 92–94.

MacLennan, D. H., and Toyofuku, T. (1992). Structure–function relationship in the Ca^{2+} pump of the sarcoplasmic reticulum. *Biochem. Soc. Trans.* **20**, 559–562.

Madeira, V. M. C. (1980). Proton movements across the membranes of sarcoplasmic reticulum during the uptake of calcium ions. *Arch. Biochem. Biophys.* **200**, 319–325.

Madeira, V. M. C. (1982). Oxalate transfer across the membranes of sarcoplasmic reticulum during the uptake of Ca^{2+}. *Cell Calcium* **3**, 67–79.

Madeira, V. M. C. (1995). Acid–base reactions as driving force in metabolism, ATP synthases and ion-motive ATP-ases: The role of the oxide anion. *Bioelectrochem. Bioenerg.* **39**, 45–52.

Malva, J., Aureliano, M., and Vale, M. G. P. (1988). Effect of vanadium oligoanions on the Ca^{2+}-ATPase activity of synaptic plasma membranes isolated from sheep brain. Abstract, Third Portuguese–Spanish Biochemistry Congress, Santiago de Compostela, Spain.

Markus, S., Priel, Z., and Chipman, D. M. (1986). Simultaneous binding of calcium and vanadate to the Ca^{2+}-ATPase of sarcoplasmic reticulum. *Biochim. Biophys. Acta* **874**, 128–135.

Maurer, A., and Fleischer, S. (1984). Decavanadate is responsible for vanadate-induced two-dimensional crystals in sarcoplasmic reticulum. *J. Bioenerg. Biomembr.* **16**, 491–505.

Medda, P., and Hasselbach, W. (1983). The vanadate complex of the calcium-transport ATPase of the sarcoplasmic reticulum, its formation and dissociation. *Eur. J. Biochem.* **137**, 7–14.

Messerschmidt, A., and Wever, R. (1996). X-ray structure of a vanadium-containing enzyme: Chloroperoxidase from the fungus *Curvularia inaequalis*. *Proc. Natl. Acad. Sci. USA* **93**, 392–396.

Molnar, E., Varga, S., and Martonosi, A. (1991). Differences in the susceptibility of various cation transport ATPases to vanadate-catalyzed photocleavage. *Biochim. Biophys. Acta* **1068**, 17–26.

Nechay, B. R., Nanninga, L. B., Nechay, P. S. E., Post, R. L., Grantham, J. J., Macara, I. G., Kubena, L. F., Philipps, T. D., and Nielsen, F. H. (1986). Role of vanadium in biology. *Fed. Proc.* **45**, 123–132.

Nour-Eldeen, A. F., Craig, M. M., and Gresser, M. J. (1985). Interaction of inorganic vanadate with glucose-6-phosphate dehydrogenase. Non enzymic formation of glucose-6-vanadate. *J. Biol. Chem.* **260**, 6836–6842.

O'Donnel, S. E., and Pope, M. T. (1976). Applications of vanadium-51 and phosphorus-31 nuclear magnetic resonance spectroscopy to the study of iso- and hetero-polyvanadates. *J. Chem. Soc. Dalton Trans.*, pp. 2290–2297.

Ortiz, A., García-Carmona, F., García-Cánovas, F., and Gómez-Fernández, J. C. (1984). A kinetic study of the interaction of vanadate with the Ca^{2+} Mg^{2+}-dependent ATPase from sarcoplasmic reticulum. *Biochem. J.* **221**, 213–222.

Pedersen, P. L., and Carafoli, E. (1987). Ion-motive ATPases. 1. Ubiquity, properties and significance to cell function. *Trends Biochem. Sci.* **12**, 146–150.

Petterson, L., Andersson, I., and Hedman, B. (1985). Multicomponent polyanions. *Chemica Scripta* **25**, 309–317.

Pick, U. (1982). The interaction of vanadate ions with the Ca-ATPase from sarcoplasmic reticulum. *J. Biol. Chem.* **257**, 6111–6119.

Pires, E. M. V., and Perry, S. V. (1977). Purification and properties of myosin light-chain kinase from fast skeletal muscle. *Biochem. J.* **167**, 137–146.

Rehder, D. (1991). The bioinorganic chemistry of vanadium. *Angew. Chem. Int. Ed. Engl.* **30**, 148–167.

Ringel, I., Peyser, Y. M., and Muhlrad, A. (1990). [51]V NMR study of vanadate binding to myosin and its subfragments. *Biochemistry* **29**, 9091–9096.

Rossoti, F. J. C., and Rossotti, H. (1956). Equilibrium studies of polyanions. I. Isopolyvanadates in acidic medium. *Acta Chim. Scand.* **10**, 957–984.

Sadiq, M., Al-Thagafi, K. M., and Mian, A. A. (1992). Preliminary evaluation of metal contamination of soils from the gulf war activities. *Bull. Environ. Contam. Toxicol.* **49**, 633–639.

Sadiq, M., and Mian, A. A. (1994). Nickel and vanadium in air particulates at Dhahran (Saudi Arabia) during and after the Kwait oil fires. *Atmos. Environ.* **28,** 2249–2253.

Shechter Y. (1990). Perspective in diabetes: Insulin—Mimetic effects of vanadate. Possible implications for future treatment of diabetes. *Diabetes* **39,** 1–5.

Simons, T. J. B. (1980). Vanadate—A new tool for biologists. *Nature* **281,** 337–338.

Sreedhara, A., Srinivasa, R., and Rao, C. P. (1994) Transition metal–saccharide interactions: Synthesis and characterization of vanadyl saccharides. *Carbohyd. Res.* **264,** 227–235.

Stankiewicz, P. J., Gresser, M. J., Tracey, A. S., and Hass, L. F. (1987). 2,3-Diphosphoglycerate phosphatase activity of phosphoglycerate mutase. Stimulation by vanadate and phosphate. *Biochemistry* **26,** 1264–1269.

Stankiewicz, P. J., Tracey, A. S., and Crans, D. C. (1995). Inhibition of phosphate metabolizing enzymes by oxovanadium(V) complexes. In H. Sigel and A. Sigel (Eds.), *Metal Ions in Biological Systems,* Vol. 31, pp. 287–323. Marcel Dekker, New York.

Teixeira-Dias, J. J. C., Pires, E. M. V., Ribeiro-Claro, P. J. A., Batista de Carvalho, L. A. E., Aureliano, M., and Amado, A. M. (1991). *Vibrational Raman Spectroscopy Study of Myosin and Myosin–Vanadate Interactions. Cellular Regulation by Protein Phosphorylation.* NATO-ASI Series, H56, pp. 29–33. Springer-Verlag, Heidelberg.

Tracey, A. S., and Gresser, M. J. (1988). Vanadium(V) oxyanions. Interactions of vanadate with cyclic diols and monosaccharides. *Inorg. Chem.* **27,** 2695–2702.

Tracey, A. S., Gresser, M. J., and Parkinson, K. M. (1987). Vanadium(V) oxyanions. Interactions of vanadate with oxalate, lactate and glycerate. *Inorg. Chem.* **26,** 629–638.

Tracey, A. S., Li, H., and Gresser, M. T. (1990). Interactions of vanadate with mono and dicarboxylic acids. *Inorg. Chem.* **29,** 2267–2271.

Uyama, T., Moriyama, Y., Futai, M., and Michibata, H. (1994). Immunological detection of a vacuolar-type H^+-ATPase in vanadocytes of the ascidian *Ascidia sydneiensis samea. J. Exp. Zool.* **270,** 148–154.

Vale, M. G. P., and Carvalho, A. P. (1973). Effect of ruthenium red on Ca^{2+} uptake and ATPase of sarcoplasmic reticulum of skeletal muscle. *Biochim. Biophys. Acta* **325,** 29–37.

Varga, S., Csermely, P., and Martonosi, A. (1985). The binding of vanadium(V) oligoanions to sarcoplasmic reticulum. *Eur. J. Biochem.* **148,** 119–126.

Vaschencko, V. I., Utegalieva, R. S., and Esyrev, O. V. (1991). Vanadate inhibition of ATP and p-nitrophenyl phosphate hydrolysis in the fragmented sarcoplasmic reticulum. *Biochim. Biophys. Acta* **1079,** 8–14.

Vegh, M., Molnar, E., and Martonosi, A. (1990). Vanadate-catalyzed, conformationally specific photocleavage of the Ca^{2+}-ATPase of sarcoplasmic reticulum. *Biochim. Biophys. Acta* **1023,** 168–183.

Wang, T., Tsai, L.-T., Solaro, R. J., De Gende, A. O. G., and Schwartz, A. (1979). Effects of potassium on vanadate inhibition of sarcoplasmic reticulum Ca^{2+}-ATPase from dog cardiac and rabbit skeletal muscle. *Biochem. Biophys. Res. Commun.* **91,** 356–361.

Willsky, G. R., White D. A., and McCabe, B. (1984). Metabolism of added orthovanadate to vanadyl and high-molecular-weight vanadates by *Saccharomyces cerevisiae. J. Biol. Chem.* **259,** 13273–13281.

Wittenkeller, L., Abraha, A., Ramasamy, R., Freitas, D. M., and Crans, D. C. (1991). Vanadate interactions with bovine Cu, Zn-superoxide dismutase as probed by ^{51}V NMR spectroscopy. *J. Am. Chem. Soc.* **113,** 7872–7881.

Yamasaki, K., and Yamamoto, T. (1992). Inhibition of phosphoenzyme formation from phosphate and sarcoplasmic reticulum Ca^{2+}-ATPase by vanadate binding to high- or low-affinity site on the enzyme. *J. Biochem.* **112,** 658–664.

Yoshikawa, S., Chikara, K. I., Hashimoto, H., Mitsui, N., Shimosaka, M., and Okasaki, M. (1995). Isolation and characterization of *Zygosaccharomyces rouxii* mutants defective in proton pumpout activity and salt tolerance. *J. Ferment. Bioeng.* **79,** 6–10.

15

BIOACTIVITY OF VANADIUM COMPOUNDS ON CELLS IN CULTURE

Susana B. Etcheverry and Ana M. Cortizo

Cátedra de Bioquímica Patológica (S. B. E., A. M. C.) and CEQUINOR (S. B. E.), Facultad de Ciencias Exactas, Universidad Nacional de La Plata, Calle 47 y 115, 1900 La Plata, Argentina

Vanadium in the Environment. Part 1: Chemistry and Biochemistry, Edited by Jerome O. Nriagu.
ISBN 0-471-17778-4. © 1998 John Wiley & Sons, Inc.

1. INTRODUCTION

1.1. Sources and Biologically Relevant Chemical Forms

The Spanish mineralogist Andrés Del Río discovered vanadium in 1802–1803 in a mineral (vanadinite) present in a lead ore in Mexico (Hoppe et al., 1990). The element was rediscovered in 1830 by N. G. Sleftröm, a Swedish chemist who called the element vanadium in honor of Vanadis, the goddess of beauty in Northern mythology (Chapter 1).

The metallic vanadium was first obtained pure by H. E. Roscoe, who was one of the greatest pioneers in vanadium chemistry (Rehder, 1990; Hoppe et al., 1990; Baran, 1996). Vanadium as a metal is not found in the environment. There are more than 70 minerals containing vanadium, among which vanadinite, descloizite, carnotite, roscoelite, and patronite are the most important for mining (Chapter 1).

Many countries extracted vanadium from fossil fuels like crude oils (4% vanadium), coal, tars, bitumens, and asphaltites (Leighton et al. 1991). In these fuels, vanadium is mainly present as vanadyl porphyrins. Vanadyl porphyrins are also found in vivo (Cantley and Aisen, 1979; Macara et al., 1980). These compounds are converted into V_2O_5 in furnaces and emitted into the atmosphere. In this way they may cause health damage and also catalyze the conversion of SO_2 into SO_3, which produces acid rain. Vanadium is very broadly used in the chemical industry, especially in H_2SO_4 and plastic production, in the production of several types of steel and in atomic energy and space technology.

In sea water, vanadium is found as a salt and only about 10% (20–35 nM) in soluble form. The principal species in this form is $H_2VO_4^-$ and, to a minor degree, HVO_4^{2-} and $NaHVO_4^-$. Highly mineralized waters in Argentina contain vanadium up to 0.3–10 μg/L (Trelles et al., 1970). These concentrations often occur together with high levels of arsenic and/or fluorides. The abundance of vanadium in the earth crust is 0.014%, similar to the percentage of zinc.

Vanadyl $[VO(H_2O)_4OH]^+$ and vanadate $[H_2VO_4]^-$ are the major species of vanadium(IV) and -(V), respectively that are present in aqueous solution under physiological conditions: total concentration of vanadium in human tissues, 0.1 μM (Rohder, 1991). These vanadium species interact with a variety of ligands such as carboxyl, phosphate, and amino groups (Nechay et al., 1986; Crans et al., 1989; Crans, 1994a; Baran, 1995). In BALB/3T3 fibroblasts, Sabbioni et al. (1991) found ~ 20% of vanadium bound to proteins, which is in agreement with the previous estimation of Nechay et al. (1986). They had found that in the intracellular environment only 1% of VO^{2+} would be free; less than 16% would be bound to proteins, and the rest coordinated to ATP, ADP, AMP, and inorganic phosphates.

Other vanadium species, like VO_2^+, VO^{3+}, VO^{2+}, V^{4+}, V^{3+}, are found at neutral pH only as coordination compounds (Rehder, 1992). The free ions $[VO(H_2O)_5]^{2+}$ and $[V(H_2O)_6]^{3+}$ undergo hydrolysis reactions to mono- and dinuclear hydroxy species (Gmelins Handbuch der Anorganische Chemie: Vanadium, 1968) and readily precipitate the hydroxides.

$[VO_2(H_2O)_4]^+$ is a species present only under pH = 2. Above this pH it converts into vanadate. Vanadate and vanadyl are linked by the redox reactions:

$$H_2VO_4^- + 4H^+ + e^- \Leftrightarrow VO^{2+} + 3H_2O$$

Vanadate and vanadyl interconvert easily under physiological conditions (Rehder, 1995).

Vanadate seems to be more toxic for living systems, and its conversion to vanadyl may be a detoxification mechanism (Sabbioni et al., 1991, 1993).

As a phosphate analogue (Gresser and Tracey, 1990), vanadate and vanadate derivatives inhibit and stimulate several enzymes of phosphate metabolism (Cantley et al., 1977; Borah et al., 1985; Stankiewicz et al., 1987; Stankiewicz and Gresser, 1988; Mendz et al., 1990; Percival et al., 1990; Ray and Puvathingal, 1990; Ray et al., 1990; Crans and Schelble, 1990; Crans, 1994b; Cortizo et al., 1994; Vescina et al., 1996).

The phosphate (PO_4^{3-}) and vanadate (VO_4^{3-}) anions are very similar in some chemical and biological aspects, but they also show striking differences. In fact, the major vanadate species under physiological conditions, $H_2VO_4^-$, has one negative charge and the phosphate form under similar conditions, HPO_4^{2-}, carries two negative charges. Vanadate, but not phosphate, tends to form oligomers at concentrations higher than 0.1 mM. Vanadium, as a transition metal, shows a more flexible coordination sphere than phosphate. Vanadate has available d orbitals, through which it could have coordination numbers of 5, 6, 7, and 8; but the biologically active phosphate compounds only have a coordination number of 4 (tetrahedral). A coordination number of 5 may be shown by phosphorous in the transition state of some enzymatic reactions (Ray and Puvathingal, 1990; Crans et al., 1991). On the other hand, vanadate, but not phosphate, undergoes reduction under physiological conditions (Macara et al., 1980; Willsky and Dosch, 1986; Rehder, 1992).

The possible coordination geometry for vanadate species VO_4^{3-} and HVO_4^{2-} are tetrahedral (Td) or a trigonal bipyramid which has been determined by different spectroscopies, as shown in Figure 1, (a) and (b), respectively (Howarth, 1990; Rehder, 1990, 1991, 1995). The first one of those structures (Td) supports the similarity between vanadate and phosphate. The trigonal bipyramid is an analogue structure for the transition state of several enzymes involved in phosphate metabolism.

It is important to point out that what is called metavanadate, obtained by dissolving $NaVO_3$ in H_2O, is a mixture of mono- and oligovanadates, the composition of which depends on the total vanadium concentration, pH, and ionic strength of the medium (Chapter 14). Monovanadates, divana-dates, and linear tetra- and decavanadates may be protonated, the degree of

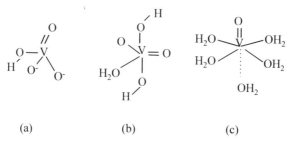

(a)   (b)   (c)

Figure 1. Structure of vanadate in tetrahedral (a) and bipyramid trigonal (b) coordinations, and vanadyl cation structure(c).

protonation being dependent on pH values, as determined by potentiometric techniques and ^{17}O and ^{51}V NMR spectroscopy (Howarth, 1990; Rehder, 1990, 1991; Elvingson et al., 1994).

The oligomer species of vanadate, although they are not present under physiological conditions, may be present in toxic vanadium concentrations or in special cellular compartments where vanadium accumulation takes place.

These oligomers have shown inhibitory effects on different enzyme activities (Soman et al., 1983; Crans and Simone, 1991; Wittenkeller et al., 1991; Crans et al., 1992).

On the other hand, the most relevant vanadium(IV) species, the vanadyl cation VO^{2+}, exists as $[VO(H_2O)_5]^{2+}$ at acidic pH (Fig. 1c). This form was determined by its characteristic EPR spectrum (Chasteen, 1981). Unless $[VO(H_2O)_5]^{2+}$ cation is coordinated, it hydrolyzes at neutral pH, forming polymeric species of $VO(OH)_2$. It can be seen from model studies that vanadyl cation forms different types of complexes with various biologically interesting ligands that have been characterized by physicochemical techniques (Muschietti et al., 1982; Baran et al., 1987; Etcheverry et al., 1989; Williams et al., 1993, 1994, 1996, 1997).

Bone is an active vanadium accumulator and vanadium apparently plays a role in the first steps of mineralization. Especially interesting are model studies on the interactions of vanadyl(IV) cation and the inorganic phase of hard tissues (hydroxylapatite) (Etcheverry et al., 1984; Narda et al., 1992). One of us has already investigated the interaction of vanadyl(IV) cation with some components of extracellular matrix such as chondroitin sulfate A (Etcheverry et al., 1994) and with the moities of this mucopolysaccharide (Etcheverry et al., 1996a, 1996b).

Other groups of vanadium compounds that are relevant from the biological point of view are the peroxo compounds. They show specific characteristic signals in ^{51}V NMR (Howarth and Hunt, 1979; Howarth, 1990).

Intracellular levels of peroxides may be high, although living organisms have several enzymes in charge of decreasing hydrogen peroxide and related compounds inside cells (e.g. catalase, peroxidase, superoxidodismutase). Many peroxo compounds have catalytic properties and they also interact with a number of enzymes, in this way influencing biological response.

1.2. Environmental Levels and Metabolism

The main source of vanadium in the general population is food (range of concentration 0.1–10 μg/kg of wet tissue), with a typical concentration of 1 μg/kg. From this source, less than 30 μg vanadium enters the organism daily. Drinking water has an average vanadium level of 5 μg/L (for detailed daily vanadium intake in a diet see Myron et al., 1978, and Byrne and Kosta 1978). The vanadium content in food varies greatly among different types of food (Nielsen, 1995).

The vanadium content of air varies over a broad range of concentrations, depending on the relative presence of furnaces, electric power plants, metallurgical industries, (Chapter 3) and highly populated cities. In contrast, in rural areas the vanadium level in air is significantly lower. It can be estimated that from this source (50 ng V/m^3 of air as average), 1 μg vanadium enters the respiratory tract per day (World Health Organization, 1988).

The absorption and distribution of vanadium compounds depend on the rate of entrance and the solubility of vanadium derivatives in biological body fluids. Absorption of vanadium takes place through different pathways: inhalation and the gastrointestinal tract (major routes), and the skin (minor or negligible entrance in human beings) (World Health Organization, 1988).

Once absorbed, vanadate is reduced to vanadyl by the gluthatione (GSH) of erythrocytes or by ascorbic acid, catecholamines, or cysteine in plasma. VO^{2+} is transported by albumin or transferrin. As a consequence of the O_2 tension, part of vanadyl is reoxidized to vanadate and is transported by transferrin (Sabbioni et al., 1979). The chemical form of vanadium and the amount bound to albumin and transferrin depend on several factors (Chasteen et al., 1986a, 1986b).

Vanadium is specially stored in organs rich in ferritin such as the liver. Other sites for vanadium retention are the spleen, testes, and bones. Experiments with animals have supported the following retention order: bone > kidney > liver > spleen > bowel > stomach > blood > lung > brain (Chasteen, 1983). Nielsen (1995) provided the mean vanadium concentrations in human fluids and organs.

By means of very sensitive and reliable techniques, it has been determined that very little vanadium is retained in the body under normal physiological conditions. The total amount of vanadium is about 100 μg, with most tissues containing less than 10 ng/g wet tissue (Parr et al., 1991; Cornelis et al., 1981; Vrbic et al., 1987; Ishida et al., 1989).

Most of the vanadium ingested is then excreted in feces. It seems that bile facilitates the vanadium excretion especially when vanadium levels in plasma are high (Sabbioni et al., 1981). If vanadium compounds have been administered parenterally, the main route of excretion is urine (Nielsen, 1995). In general, most studies on human beings and higher animals support the fact that vanadium absorption through the gastrointestinal tract is quite low. The oxidation state of vanadium and the presence of other dietary components influence the absorption of vanadium by the gastrointestinal tract (Hill, 1990; Nielsen, 1995).

1.3.   Biological Effects of Vanadium Compounds

The essential nature of vanadium in higher animals was assumed on the basis of deprivation signs (impaired reproductive performance, failures in growth, elevation of hematocrit and plasma cholesterol, decreased bone development). These deprivation signs were not present in control animals fed diets supplemented with vanadium. Because the final levels of vanadium in those diets were much higher than those normally found in deprived diets, doubts arose about deficiency characteristics in animals. In fact, many of these signs are overcome with vanadium supplementation, but it may be assumed that it is due to the high pharmacological activity of vanadium compounds. In the 1980s and 1990s, newer research work was carried out on animals in order to elucidate the uncertainty about the essentiality of vanadium (Anke et al., 1989; Uthus and Nielsen, 1989; Nielsen and Uthus, 1990). At present it is accepted that alterations in both bone structure and development, and changes in plasma cholesterol and reproductive performance, are signs of vanadium deficiency. It is thought that the reversal of some deficiency signs, especially those of bone alteration, cannot be attributable to a pharmacological effect of the small amount of vanadium (0.5 μg/g) present in the diets in recent studies (Nielsen, 1995).

Many studies developed in vitro or in experimental animals have shown that vanadium has insulin-mimetic properties: mitogenic effects, stimulatory or inhibitory action on cell differentiation, and numerous metabolic effects. In vitro studies in cell-free systems have shown the actions of vanadium compounds on the activity of many enzymes, especially those related to phosphate reactions.

Heyliger et al. (1985) determined that the addition of vanadate to the drinking water of streptozotocin-diabetic rats alleviated the diabetic symptoms (normalization of glycemia and improvement of cardiac performance) without any change in insulin plasma levels. That study also showed the lack of vanadate effect on blood glucose levels and on hepatic enzyme activities in normal rats. Meyerovitch et al. (1987) established the conditions required to maintain normoglycemia for long periods in streptozotocin-diabetic rats orally fed low and high doses of vanadate in drinking water. Low vanadate doses normalized the glycemia, lipogenesis, and insulin binding and did not cause toxic effects. Brichard et al. (1993) was able to demonstrate that vanadate treatment of streptozotocin-diabetic rats produced a change in the relative ratio of gluconeogenesis and glycolysis. It was shown that vanadate acts on the expression of genes involved in liver glycolytic and gluconeogenic pathways. Orally administered vanadate partially restored glucokinase mRNA and glucokinase activity and totally restored L-type pyruvate kinase (L-PK) parameters. In contrast, vanadate treatment decreased the activity and mRNA level of the gluconeogenic enzyme phosphoenolpyruvate carboxykinase (PEPCK). This treatment also corrected the levels of liver glucose transporter (GLUT2) mRNA and protein, which are altered in diabetes.

Oral vanadate administration decreased blood glucose in ob/ob and db/db mice, two rodent models for type II diabetes (Brichard et al., 1990; Meyerovitch et al., 1991). Ferber et al. (1994) examined the effect of vanadate in liver carbohydrate metabolism (mRNA expression for key enzymes and transport proteins).

McNeill and coworkers have studied the properties of vanadium derivatives in oxidation state 4 as hypoglycemic agents. In fact, in a recent study they tested this property for vanadyl sulfate and the organic compound bis(maltolato) oxovanadium(IV) in streptozotocin-diabetic rats (Yuen et al., 1995). Although the normoglycemic effects of vanadium are independent of its oxidation state (4 or 5), the toxicity of vanadium(V) has been reported to be higher than that of vanadium(IV) (Sabbioni et al., 1993).

Vanadyl cation is less toxic than vanadate but it has limited solubility under physiological conditions. In order to increase its solubility, coordination compounds with different kinds of ligands have been used (McNeil et al., 1992; Shechter et al., 1992; Cortizo et al., 1996b; Etcheverry et al., 1997). Moreover, vanadate, vanadyl sulfate, and bis(maltolato) oxovanadium(IV) produced no significant alterations in the hematological parameters, suggesting lack of toxicity in rats (Dai et al., 1995). Vanadate proved to be an active agent on cardiovascular function in rabbits (Carmignani et al., 1996). It reduced the synthesis and/or the release of nitric oxide, the endothelium derived vasodilating agent, and increased the plasma cathecholamine levels, so it behaves like a probable environmental hazard for cardiovascular homeostasis.

Since the nineteenth century, vanadium compounds have been used for the treatment of diseases such as diabetes, tuberculosis, and anemia. Lyonnel and Martin (1899) reported a study carried out on two patients with diabetes mellitus treated with vanadate. A decrease in their glucosuria was observed during treatment. Two decades later, the discovery of insulin diminished the initial interest in insulin substitutes for several years. Nevertheless, in the period 1975–1980, experiments on different types of cells and on cell-free systems renewed interest in vanadium compounds as therapeutic drugs for diabetes treatment (Shechter, 1990; Shechter and Shisheva, 1993). Recently, Cohen et al. (1995) have shown that vanadyl sulfate, given orally to six non-insulin-dependent diabetes mellitus (NIDDM) patients, improved hepatic and peripheral insulin sensitivity. Basal glycemia control was poor and improved after treatment. This effect can be attributed to the stimulation of glucose uptake and glycogen synthase activation. In contrast with vanadate, vanadyl sulfate has been previously shown not to be associated with toxic side effects (Pedersøn et al., 1985; Ramanadham et al., 1989; Dai and McNeill, 1994).

Goldfine et al. (1994) tested the effects of a 2-week vanadate treatment in NIDDM patients without significant amelioration in diabetes markers after treatment.

Vanadium has been shown to possess insulin-mimetic properties in a variety of systems, as we will describe later. Different forms of vanadium also behave like growth factors, including effects on cell growth and oncogen expression of cells in cultures.

Among its insulin-mimetic effects, vanadium has been shown to stimulate glucose uptake and oxidation, as well as glycogen synthesis both in adipose cells and skeletal muscle (Dubyak and Kleinzeller, 1980; Shechter and Karlish, 1980; Clausen et al., 1981; Degani et al., 1981). It was also determined that vanadate, like insulin, inhibited isoproterenol- or ACTH-stimulated lipolysis in rat adipocytes (Degani et al., 1981). Vanadate and vanadyl cation stimulated lipogenesis in the same cell types, this effect being greatly increased by vanadyl complexes (Shechter et al., 1992). Vanadate, vanadyl, and some derivatives with organic ligands have also been shown to increase glucose consumption and protein content in two osteoblast-like cell lines (Etcheverry et al., 1997).

On the other hand, it has recently been shown that vanadate, vanadyl, and pervanadate act as growth factor-mimetic compounds on osteoblast-like cells in culture (Cortizo and Etcheverry, 1995).

In spite of the many studies carried out in vivo and in vitro, up to now it has not been possible to find a specific function of vanadium in human beings and higher animals. However, vanadium compounds and their possible biological roles have recently been described in lower forms of life (ascidians, bacteria, fungi, algaes, worms) (Smith et al., 1995; Eady, 1995; Bayer, 1995; Vilter, 1995; Ishii et al., 1995).

Several biochemical and physiological functions for vanadium compounds have been suggested on the basis of numerous in vitro and pharmacological studies. Nevertheless, it cannot be established that vanadium is the direct stimulator or inhibitor of any enzyme in higher forms of life. However, the fact that certain haloperoxidases in lower forms of life require vanadium for their activity suggests a possible role of vanadium compounds in the thyroid haloperoxidase in higher animals. On the other hand, in vivo and in vitro studies also suggest bone and connective tissue as target sites for vanadium biological actions (Nielsen, 1995).

2. ACTIONS OF VANADIUM ON METABOLIC EVENTS

Vanadium, like insulin, produces a number of cellular events including glucose transport and metabolism, ion and aminoacid transport, lipid metabolism, glycogen synthesis, gene transcription, and protein synthesis. These effects may be characterized according to their induction period as shown in Table 1. Vanadium induces early events such as metabolite and ion transport, as well as receptor and protein phosphorylation, which are observed after some

Table 1   Insulin-Like Effects of Vanadium

Early (minutes)	Insulin receptor tyrosine phosphorylation
	Protein phosphorylation
	Glucose transport
	Plasma membrane Ca-ATPase inhibition
	Na, K-ATPase inhibition
Intermediate (minutes–hours)	Glucose oxidation
	Lipogenesis
	mRNA synthesis
	Protein synthesis
Prolonged (hours–days)	DNA synthesis
	Growth promotion
	Cell differentiation
	Oncogene expression
	Protein/enzyme induction

seconds or minutes. Following these effects, the most important of which occurs at the plasma membrane level, vanadium also regulates the intermediate metabolism, such as glucose oxidation and glycogen synthesis, lipogenesis, and mRNA synthesis. Finally, vanadium may also directly affect DNA synthesis, oncogen expression, growth promotion, and cell differentiation. In this context, it has been suggested that vanadium acts not only as an insulin-mimetic factor but as a growth factor as well (Stern et al., 1993; Cortizo and Etcheverry, 1995), a concept that we will redefine later.

Vanadium's effects, and its mechanisms of action, are specific to species and tissues. For instance, unlike rat adipocytes, human adipocytes are found to be unresponsive to vanadate when glucose uptake and antilipolytic effects are studied (Lönnroth et al., 1993). Furthermore, human beings and rats responded differently to pervanadate, since the dose-response curve for peroxovanadate was biphasic in human fat cell, but that was not observed in rat adipocytes (Duckworth et al., 1988; Shisheva and Shechter, 1993).

It was observed in early in vitro studies that vanadate strongly inhibits Na^+, K^+-ATPase, Ca^{2+}-ATPase, acid and alkaline phosphatases, ribonuclease, phosphodiesterase, phosphotyrosyl-phosphatase, and other enzymes in vitro (Stern et al., 1993). In addition, it stimulates tyrosine kinase phosphorylase and adenylate cyclase (Tamura et al., 1984; Bernier et al., 1988).

Vanadium stimulates tyrosine kinase autophosphorylation, which is associated with insulin receptor in intact rat adipocytes (Tamura et al., 1984; Bernier et al., 1988). Vanadate stimulates glucose transport in insulin-sensitive and insulin-resistant tissues presumably by inhibiting tyrosine phosphatases (Carey et al., 1995). Thus, vanadate can stimulate glucose transport by overcoming a defect in insulin-resistant diabetic tissues.

Leighton et al. (1991) showed that pervanadate increases glycogen synthesis, lactate formation, and glucose oxidation in incubated rat soleus muscle.

Vanadate increases insulin sensitivity by enhancing insulin binding capacity and affinity in insulin rat adipocytes but not in human fat cells (Fantus et al., 1990; Eriksson, et al., 1992; Lönnroth et al., 1993).

In pancreatic islet cells, vanadate favors insulin release (Fagin et al., 1987). In addition, vanadate exhibits dose-dependent inhibition of islet Na^+, K^+-ATPase activity. Nevertheless, the vanadate mechanism of action in pancreatic cells remains unknown. Furthermore, vanadate potentiates the glucose effect on insulin release, an effect that seems to be due to changes in the membrane potential, phosphoinositide metabolism, and calcium mobilization (Zhang et al., 1991).

Gay's group has characterized a Ca^{2+}-ATPase in the plasma membrane of osteoblast- and osteoclast-like cells (Akisaka et al., 1988; Bekker and Gay, 1990; Lloyd et al., 1995). This enzyme seems to regulate intracellular calcium by extruding it into the extracellular media. This Ca pump is polarly distributed and it has been proved, by histochemical and biochemical methods, to be inhibited by vanadium. These results suggest that vanadate could modulate bone turnover by regulating membrane Ca^{2+}-ATPase and, consequently, the cytosolic calcium levels in bone cells.

We have recently shown that an alkaline phosphatase (ALP) from UMR106 osteoblast-like cells is directly inhibited by vanadium derivatives in vitro (Cortizo et al., 1994). The cytosolic ALP was partially characterized as a protein tyrosine phosphatase (PTPase).

The UMR106 osteoblastic line has a glucose transport system that is sensitive to insulin (Thomas et al., 1995). These cells express GLUT1 but not the insulin-responsive glucose transporter GLUT4, even though insulin is able to increase glucose transport by about 50% when acting acutely. However, long-term regulation of glucose transport is observed after a 24-h incubation. Using this cell line, we have observed that different vanadium compounds are able to increase glucose consumption after 24 h in culture (Etcheverry et al., 1997). The effects were similar for all the tested vanadium derivatives (vanadate, complexes with oxalate and citrate, vanadyl and complexes with tartrate, and peroxovanadium complexes). V-NTA, a peroxovanadium derivative, was more potent than vanadate. The latter result is in agreement with previous observations either in vivo or in vitro, suggesting that peroxovanadium is more potent than vanadate in different systems (Fantus et al., 1989; Lönnroth et al., 1993; Shisheva and Shechter, 1993; Yale et al., 1995).

Thus, vanadium compounds seem to regulate osteoblast metabolism via multiple mechanisms, such as calcium pump, glucose transporter, and different phosphatases.

3. MITOGENIC AND DIFFERENTIATIVE EFFECTS OF VANADIUM

In spite of the early events regulated by vanadium, recent evidence indicates that the growth of a number of cells is affected by vanadate. Vanadium induction of cell proliferation is similar to that produced by growth factors. This effect has required 24–48 h, depending on the cell line studied, and is relatively weak without serum. In addition, the growth promoter effect of vanadium is induced in a narrow range of concentrations. This last characteristic contrasts with the growth factor effects that are common in a range that goes from 10^{-10} to 10^{-7} M in culture systems, and the growth factor actions are shown in a dose-response manner. Additionally, growth factors act by binding to a cognate-specific receptor, which in turn becomes autophosphorylated. This phosphorylation process induces a cascade of signal transduction mechanisms that finally stimulates gene transcription and cell progression. In contrast, vanadium compounds seem to act by inhibiting PTPases, which act at different or distal levels in the phosphorylation cascade, inducing cell proliferation. Recently, Pardey et al. (1995), using deficient CHO cells in insulin receptor protein kinase, have suggested that the mitogenic effect of vanadate is independent of insulin receptor phosphorylation. This study supports the above concepts of the vanadium mechanism of action.

Vanadate stimulates thymidine incorporation into DNA and cell numbers and cell growth in fibroblasts (Carpenter, 1981; Smith, 1983; Jamieson et al., 1988; Cortizo et al., 1996a); in Leydig cells (Sato et al., 1987); in bone cells (Canalis, 1985; Kato et al., 1987; Lau et al., 1988; Davidai et al., 1992; Cortizo and Etcheverry, 1995). In most cases, the vanadate effect on cell proliferation was biphasic, being cytotoxic for cells over a concentration range of 50 to 100 μM.

We have also addressed the issue of the fate of vanadium mitogenicity by using different vanadium derivatives in fibroblast- and osteoblast-like cells in culture (Cortizo and Etcheverry, 1995; Cortizo et al. 1996a, 1996b; Etchevery et al., 1997). In these culture systems, the proliferative effects of different vanadium compounds were biphasic, the maximal response occurring in the concentration range 10–25 μM (Fig. 2). The potency of vanadium was similar for vanadium(IV) (vanadyl, vanadium(IV)-tartrate, malto-vanadyl) and vanadium(V) (vanadate, vanadium(V) complexes with citrate and oxalate, and maltol), with the exception of the peroxo vanadium(V) derivatives (vanadium-NTA, peroxovanadate), which were less effective, and usually cytotoxic, for both types of cells. Vanadium derivatives inhibit cell proliferation and induce morphological transformation within only 12 h at a 50 μM concentration. Vanadium(IV) compounds were good mitogens after a 24-h incubation, and they did not show any cytotoxicity up to 100 μM; below that level a weak effect was observed in fibroblasts.

During the last decade, studies on vanadium bioactivity have begun to address another growth factor mimetic property of these compounds, that is, the ability to regulate differentiated phenotypes in a variety of cells. In most of these studies, it is evident that this behavior is related to vanadium's property of inhibiting PTPases. We will describe some of these studies so as to point out the role of phosphorylation/dephosphorylation events in cell differentiation.

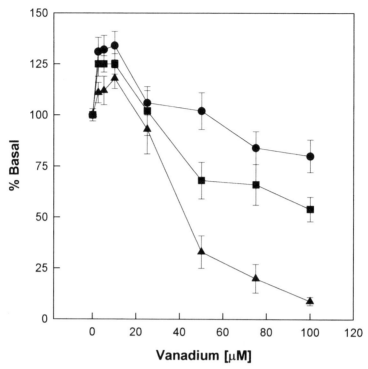

Figure 2. Effect of vanadium compounds on cell proliferation. Swiss 3T3 fibroblasts were incubated in DMEM plus vanadate (■), vanadyl (●), or pervanadate (▲) at the indicated doses. After an incubation of 24 h at 37°C, cells were fixed and stained with crystal violet. Dye was extracted and absorbance read at 540 nm. Results are expressed as % basal, $X \pm SEM$, $n = 6$.

Kachinskas et al. (1994) observed that human keratinocyte differentiation, as assessed by expression of the specific marker involucrin, is inhibited by vanadate, arsenate, molybdate, and tungstate. It is known that all these agents strongly inhibit PTPases. This effect was dose-dependent, with half-maximum inhibition at 3 μM after 10 days of vanadate treatment.

HL-60 cells exposed to 12-O-tetradecanoyl phorbol-13 acetate (TPA) differentiate into monocyte/macrophage-like cells (Wei and Yung, 1995). Vanadate enhanced the extent of monocyte differentiation induced by low doses of TPA. This effect seems to be mediated by the inhibition of the PTPases in this cell line.

Preadipocyte 3T3-L1 in culture could be induced to differentiate by expressing adipocyte gene after mitogenic expansion and growth arrest (Liao and Lane, 1995). This transformation is accompanied by an increase in PTPases during clonal expansion and a decrease in gene activation. Addition of vanadate during mitotic expansion prevents differentiation; vanadate added after clonal expansion has no effect on differentiation. Thus, these studies suggest that an essential tyrosine phosphorylation event(s) occurs during a specific step in the differentiation program, suggesting that vanadate may control cell maturation through the phosphorylation pathway.

There is some evidence that shows that vanadium can play a role in the metabolism and growth of the skeleton. Vanadate increases the incorporation of [^{3}H]-proline into type I collagen in cultures of 21-day-old fetal rat calvariae (Canalis, 1985). This effect was evident after 24 h of incubation, even though a high concentration of vanadate inhibited collagen synthesis. In addition, the treatment with 1 mM or 10 μM vanadate for 24 and 96 h, respectively, inhibited alkaline phosphatase activity. Thus, high doses of vanadate show an irreversible inhibitory effect on these parameters, indicating that vanadate has cytotoxic effects.

The action of vanadate has been also tested on the regulation of chondrocyte phenotype expression (Kato et al., 1987). Results indicate that vanadate (0.1–6 μM) stimulates the synthesis of cartilage-matrix proteoglycans of rabbit costal chondrocyte cultures. These authors also found that 2–6 μM vanadate induced morphological differentiation of fibroblastic cells to spherical chondrocytes after a 24-h incubation. However, the transformed phenotype was expressed by cells exposed to 60 μM vanadate. In additional experiments it was found that 6–60 μM vanadate increased the phosphotyrosine level in chondrocyte proteins (Owada et al., 1989), suggesting a tyrosine phosphorylation role in the regulation of chondrocyte differentiation.

Lau et al. (1988) also showed that vanadate stimulates bone collagen synthesis in alkaline phosphatase calvaria cells in culture. This last discovery is in contrast with other observations of the effects of bone growth factors (Wergedal et al., 1988; Cortizo and Etcheverry, 1995), which showed alkaline phosphatase inhibition in osteoblast-like cells in culture. Additionally, vanadate inhibited osteoblast acid phosphatase/phosphotyrosyl protein phosphatase activity, an effect that could lead to stimulation of bone cell proliferation and bone collagen synthesis.

In recent studies we have shown that vanadium regulates phenotype characteristics of osteoblast-like cells, as assessed by the levels of alkaline phosphatase activity (Cortizo and Etcheverry, 1995; Cortizo et al., 1996a; Etcheverry et al., 1997). For these studies we used the rat osteosarcoma cell line UMR106, exposed to different vanadium(V) and vanadium(IV) complexes as well as peroxo vanadium(V) derivatives. The peroxo vanadate compounds were the strongest inhibitors of cell differentiation, after 24-h incubation at mitogenic doses (5–25 μM). Vanadate, as well as malto-vanadyl and the complexes of vanadium(V) with oxalate and citrate, equally inhibited alkaline phosphatase activity. On the other hand vanadyl cation, and vanadium(IV)-tartrate and malto-vanadate, were not inhibitors of the osteoblast marker ALP, or they weakly inhibited this activity. The oxidation state of vanadium compounds is not the only property determining the growth-promoting effect of these agents. We used malto-vanadium(V) and malto-vanadium(IV), two complexes of vanadium(V) and vanadium(IV) with the sugar maltol (McNeill et al., 1992;

Elvingson et al., 1996) to explore the factors determining vanadium mitogenic behavior in UMR106 osteoblast-like cells (Cortizo et al., 1996b). In a dose-response study (Fig. 3), vanadate strongly inhibited ALP activity after a 24-h incubation. Under similar conditions, vanadyl was only a weak inhibitor of cell differentiation at doses of 10–50 μM. Contrary to the expected response based only on the oxidation state, malto-vanadate, the vanadium(V) complex, induced an inhibition of ALP activity as weak as vanadyl did. The coordination geometry of this derivative (Elvingson et al., 1996) is octahedral and it may be reduced to a vanadium(IV) derivative by the cells. On the other hand, upon testing of the malto-vanadium(IV) derivative, a dose-response similar to that of vanadate was observed (Fig. 3). This complex shows a square pyramid structure (McNeill et al., 1992).

All these experiments together suggest that the coordination number and the stability of vanadium complexes under physiological conditions, as well as the oxidation state, play a role in determining the growth-promoting and cell differentiation effects of these compounds.

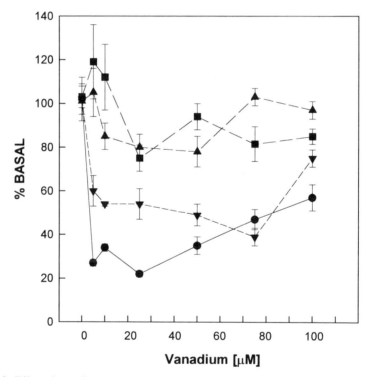

Figure 3. Effect of vanadium compounds on osteoblast differentiation. UMR106 osteoblast-like cells were incubated in DMEM plus vanadate (●), maltovanadyl (▼), vanadyl (■), and malto vanadate (▲), at 37°C/24 h. After extraction with triton X-100, cell differentiation was assessed by alkaline phosphatase specific activity. Results are expressed as % basal, $X \pm SEM$, $n = 9$.

The fact that vanadium compounds are not only insulin-mimetics but also behave like growth factor has become evident through these studies. These factors may be defined as agents acting locally and playing a role in the mechanism controlling cell growth. They behave throughout like specific receptors that in turn induce a phosphorylation cascade and then biological effects. Several laboratories are investigating whether vanadium acts via growth factor receptor phosphorylation or by additional mechanisms.

4. MORPHOLOGICAL TRANSFORMATIONS INDUCED BY VANADIUM

In addition to its insulin-mimetic actions and mitogenic effects, it was early observed that vanadate causes neoplastic transformations in different cells (Klarlund, 1985). This behavior is similar to that induced by many growth factors and various tumorigenic agents (Ullrich and Schlessinger, 1990). Thus, it has been reported that vanadate leads to morphological transformation in kidney and fibroblast cells (Klarlund, 1985; Mountjoy and Flier, 1990; Cortizo et al., 1996a), BALB/3T3 cells (Sabbioni et al., 1991, 1993), hamster embryo cells (Rivedal et al., 1990), bovine papilloma virus DNA-transfected C3H10T1/2 line (Kowalski et al., 1992), and MC3T3E1 osteoblast-like cells (Etcheverry et al., 1997).

It has been suggested that morphological changes are associated with an increase of tyrosine-phosphorylation of numerous intracellular proteins. This effect seems to be mediated by the ability of vanadate to inhibit PTPases (Gresser et al., 1987), although this hypothesis was not directly assessed. For example, pp60scr, a tyrosine kinase, is stimulated by vanadate in Rous sarcoma virus transformed cells, but not in normal fibroblasts (Brown and Gordon, 1984; Ryden and Gordon, 1987), suggesting that vanadate may regulate the activity of pp60scr differently in normal and tumor cells.

In TE-2R cells, a clone of human esophageal cancer cell line that expresses a considerable amount of epidermal growth factor (EGF) receptor, EGF induces a transformation in the cell shape from round to fibroblastic and in the colony formation from compact to sparse (Shiozaki et al., 1995). Vanadate potentiates the EGF response in this model. These striking changes were not evident until after 24 h in culture, a time in which E-cadherin adhesion molecules have appeared distributed over the whole cell surface. Shiozaki et al. (1995) have suggested that Tyr-phosphorylation of the E-cadherin system is responsible for the morphological changes described. They have also observed morphological changes and phosphorylation of β-caterin, an associated cytoplasmic protein to cell–cell adhesion molecules, in other squamous cell carcinomas. The above study has been supported by Gavrilivic et al. (1990), who have reported that a fibroblast-like phenotype in vitro is induced by the transforming growth factor alpha (TGF-α), an EGF receptor ligand.

The idea of a relationship between the shape changes of cells and the tyrosine-phosphorylation state of certain proteins is also supported by studies in human erythrocytes (Bordin et al., 1995). Different environment effects together with certain drugs may produce a change in the normal red cell shape. Bordin et al. (1995), using agents that specifically inhibit Ser/Tyr-phosphatases (okadaic acid) or Tyr-phosphorylation (vanadate and pervanadate), have shown that Tyr-phosphorylation of the major membrane-spanning band 3 is related to certain morphological changes in human erythrocytes. By means of scanning electron microscopy, they observed that vanadate and pervanadate but not okadaic acid induced a transformation from the normal smooth biconcave disk to the spinning form. These morphological transformations were accompanied by changes in the Tyr-phosphorylation pattern. Thus, the phosphorylation state of cytoskeletal proteins induced by vanadium compounds leads to alterations in the membrane structure and the cell morphology of human erythrocytes.

In BALB/3T3 cells, Sabbioni et al. (1993) have developed studies on vanadate and vanadyl cytotoxicity and morphological transformation. They found that whereas vanadyl was not a transformer at 3–10 μM, vanadate induced neoplastic changes under similar conditions in this cell line.

In recent studies, we have examined mitogenic responses in Swiss 3T3 fibroblasts after they had been exposed to different vanadium compounds such as vanadate, vanadyl cation, and pervanadate (Cortizo et al., 1996a). These three derivatives regulate cell growth in a biphasic manner, being toxic for the cells at high doses (Fig. 4). The potency order for this effect was pervanadate > vanadate > vanadyl. Morphological transformations were also observed and they were directly correlated with the ability of vanadium derivatives to inhibit cell proliferation. Pervanadate, the strongest cytotoxic agent, induced condensation of cytoplasm and membranes as well as a loss of cellular processes. These alterations were observed in as little as 12-h of culture under pervanadate exposure. Similar changes were induced by vanadate at a 24-h incubation, whereas a slight transformation was detected after cell incubation with vanadyl. Thus, these experiments suggest that cytotoxic effects and morphological changes are associated with and dependent on the oxidation state and the coordination geometry of the studied vanadium compounds.

In addition, we have investigated the cell growth and morphological changes induced by a new series of vanadium compounds in the mouse calvaria MC3T3E1 cell line (Etcheverry et al., 1997). In osteoblast-like cells exposed to vanadyl and vanadyl-tartrate complex, we observed an increase in cell proliferation with slight cell shape alterations after a 24-h incubation at a 25 μM concentration. In contrast, vanadium(V) compounds (vanadate and vanadium(V) complexes with citrate and oxalate) caused striking morphological changes: Cells became spindle-like and condensed, whereas cell processes were lost. These changes were accompanied by induction of cell proliferation and the inhibition of osteoblast differentiation as evaluated by the ALP activity. Furthermore, it was found that a peroxo-derived vanadium(V) compound

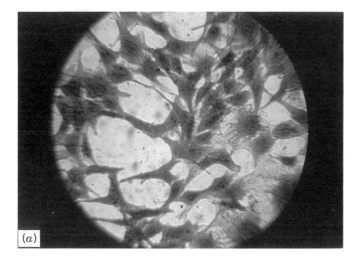

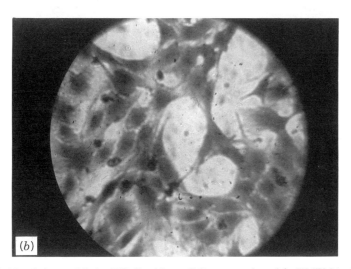

Figure 4. Morphology of Swiss 3T3 fibroblasts. Cells were cultured in DMEM at 37°C/24 h without vanadate (A), and with 10 μM vanadate (B), 25 μM vanadate (C), and 75 μM vanadate (D). After this incubation period, cells were fixed and stained with crystal violet, washed, and observed. Control fibroblasts show a polyhedral morphology with processes among neighboring cells. Fibroblasts incubated with 10 μM vanadate show slight differences from control cells. Exposure to 25 μM vanadate show fusiform-shaped cells, with condensed cytoplasm and loss of cell connections. Cells incubated with 75 μM vanadate show many cells with the transformed phenotype and fewer cells per camp in comparison with the control cultures. Magnification × 40.

Figure 4 continues

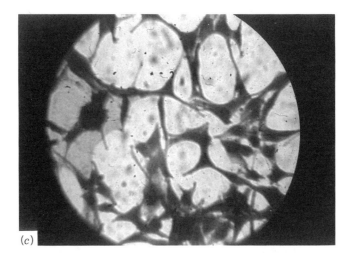

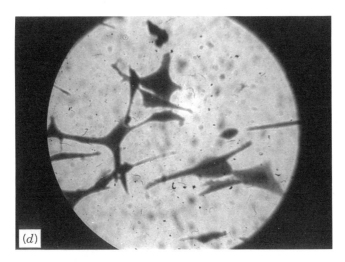

Figure 4. *(Continued)*

(V-NTA) was the strongest morphological transformer. The fact that this agent did not stimulate cell proliferation in this cell line suggests some degree of cytotoxicity.

All these results together indicate different cell sensitivities to different vanadium compounds. Accordingly, studies might indicate that the ability of vanadium compounds to induce morphological transformation would depend on (a) the tissue or cell culture system under study, (b) the ability of each cell to take up and metabolize each vanadium derivative (which we will discuss later), and (c) particular properties such as vanadium stability under physiological conditions, the oxidation state, and/or the geometrical coordination. The mechanism by which vanadium compounds induce neoplastic transformation deserves further investigation.

5. UPTAKE AND METABOLISM OF VANADIUM COMPOUNDS BY CELLS IN CULTURE

The most widely investigated vanadium compound, vanadate, is usually reduced to vanadium(IV) under physiological conditions (Rehder, 1995). Even though many biological papers refer to this compound as vanadyl cation, it has been demonstrated that this is only one of the forms of vanadium(IV), but not the most important, because at physiological pH vanadyl cation undergoes hydrolytic processes with the formation of different hydroxy complexes (Chasteen, 1981).

Under physiological conditions vanadate(V) exists in several forms, depending on concentration, pH, and other environmental factors (presence of reductans, ligands, etc.). As there is not much detailed information, we will refer to "vanadate" as representative of all these vanadate-derived species present under assay conditions (Crans, 1994a).

The reduction of vanadate to vanadium(IV) has been previously reported in yeast (Willsky et al., 1984), red blood cells (Heinz et al., 1982), fibroblast-like cells (Sabbioni et al., 1991), and rat adipocytes (Sakurai et al., 1980; Degani et al., 1981). It has been demonstrated in *Saccharomyces cerevisiae* (Willsky et al., 1984) cultured in the presence of orthovanadate (VO_4^{3-}) or vanadyl (VO^{2+}) that both ions were transported into the cells and they inhibited plasma membrane Mg^{2+}-ATPase. ESR spectroscopy is a useful tool for measuring cell-associated paramagnetic vanadyl cation. [51]V NMR spectroscopy is adequate for detecting cell-associated diamagnetic vanadium (e.g., vanadium(V)). When the yeast *S. cerevisiae* was exposed to both toxic (5 mM) and nontoxic (1 mM) vanadate concentrations in the culture medium, they were able to associate vanadate and to reduce it to vanadyl, as demonstrated by ESR measurement. Vanadyl was then accumulated in the cell culture medium. Cells exposed to the toxic vanadate concentration accumulated dimeric vanadate and decavenadate forms, as could be determined by [51]V NMR spectroscopy. It was proved that the inhibitory effects of vanadium(V) usually observed in in vitro experiments is due to vanadate in one or more of its derived forms. The same data support the idea that the stimulatory species of vanadium present in experiments with whole cells is the vanadyl ion or one or more of its derivatives.

Working on different strains of *Saccharomyces cerevisiae*, Willsky et al. (1984) showed by EPR and [51]V NMR spectroscopies that the vanadate-resistant strains contained more cell-associated vanadyl than the parental strains. They were also able to determine, using [51]V NMR spectroscopy, that the resonances associated with cell toxicity were not observed in the resistant strains. It was also shown that these cells were able to accumulate phosphate, vanadate, and vanadyl. The amount of remaining vanadate in the culture media after growth was lower in the resistant strains than in the sensitive ones. On the other hand, the resistant cells showed more cell-associated vanadyl and less of the cell-associated ^{51}V NMR resonance previously correlated with vanadate cell toxicity. Nevertheless, vanadate resistance could not be explained by the inability to transport vanadate or vanadyl into the cells. Although some monomeric vanadate could be bound to certain enzymes, forming a transition state analogue in the pathway of protein phosphorylation, the structure alterations of these enzymes were not enough to explain the observed vanadate resistance. The results with *S. cerevisiae* showed that vanadate resistance is correlated with the decrease in the accumulation of a toxic form of vanadate by the cells. On the other hand, with 1 mM vanadate in the growth media, vanadate entered into the yeast probably by the anion transport system, and then it was reduced to vanadyl by glutathione (Macara et al., 1980) or other reductants such as catechols. After that, vanadyl accumulated in the medium (Cantley et al., 1978). In the presence of a higher vanadate concentration, the above mechanisms, which prevented vanadate from accumulating into the cells, seemed to be overloaded. A new ^{51}V NMR vanadate peak associated with toxicity was accumulated into cells in the sensitive, but not in the resistant strains (Willsky et al., 1985). In another set of experiments Willsky and Dosch (1986) showed the relevance of mitochondrial function (oxidative phosphorylation) in vanadium metabolism, since yeast respiratory-deficient strains accumulated higher amounts of a cell associated vanadate compound.

The uptake and metabolism of vanadate and vanadyl in human erythrocytes has also been tested (Heinz et al., 1982). At the pH used in these studies, free monomeric vanadium(V) was mainly in the metavanadate (VO_3^-) form. The red cells in acid solution (below pH 2) produced rapid reduction of vanadium(V) to vanadium(IV), reducing agents such as glutathione and NADPH being released from the cells. The rapid formation of vanadium(IV) was measured by quantitative EPR spectroscopy (Chasteen, 1983).

In order to investigate the uptake and transport processes of vanadium(V) and -(IV), Heinz et al. (1982) used a combination of EPR and ^{48}V-traced compounds. They found a rapid uptake of vanadium(V) from the medium (t1/2 = 1.9 min in Tris buffer), followed by a slow intracellular reduction to vanadium(IV) (t1/2 = 17 min). The composition of the media (presence or absence of plasma proteins, different buffers, oxygen tension) affected the transport process. In fact, the plasma contains a variety of proteins that may interact with vanadate in the medium, reducing the free concentration of the anion in the solution. In that study, it was demonstrated that only about 17% of the total vanadate in the plasma medium was free. Buffer composition also showed a great effect on the uptake process. The third variable taken into account was the O_2 tension in the media. The kinetic parameters determined in that study showed that high O_2 tension increased the influx rate constant and decreased the rate of reduction of vanadium(V) to vanadium(IV). The rates and equilibrium constants were essentially the same for 1 and 10 μM vanadium(V) concentrations. These constants were changed by a 100 μM concentration of vanadium(V), suggesting that high levels of vanadium(V) might change the transport uptake properties of the erythrocytes.

The accumulation process of vanadium inside the cells proved to be pH-dependent (inverse relationship). For the red cells, kinetic analysis showed that vanadium(V) influx and efflux decreased above pH 7.2, the effect being stronger on influx. The elevation of pH also decreased the conversion of vanadium(V) to vanadium(IV) inside the cells.

On the other hand, erythrocytes were able to take up vanadium(IV) from the incubation medium. VO^{2+} could be transported by red cells via a divalent cation transporter in the membrane. Heinz et al. (1982) showed that the rate of vanadium(IV) transport and accumulation was not dependent on calcium concentration in the external medium. As to erythrocytes, vanadium(IV) is transported at a slower rate than vanadium(V) into the cells. Macara et al. (1980) had also determined that vanadate was converted almost quantitatively to VO^{2+} via cytoplasmatic GSH in red cells, and that the vanadyl cation was bound to hemoglobin.

In rat adipocytes, Degani et al. (1981) demonstrated that vanadate was reduced to vanadyl and the cation was complexed by glutathione inside the cells. In their study it was shown that the addition of sodium metavanadate (200 μM) and D-glucose (100 mM) to a suspension of fat cells resulted in the gradual appearance of characteristic EPR signals of vanadyl ion complexes. Structural and bonding features of vanadyl complexes inside the adipocytes were determined by EPR studies and computer simulations. The EPR signals were exclusively associated with the cell fraction.

Sabbioni et al. (1991) established that BALB/3T3 fibroblasts incubated with [48]V-vanadate or vanadyl showed radioactivity associated with the cells, in a dose-dependent manner. By means of gel filtration experiments, these authors were able to determine the intracellular distribution of vanadium(V). During the first hours of incubation, vanadate was found in the cytosol. Then cytosolic vanadate decreased and this fact matched with a higher nuclear association. The distribution of vanadium in the cells proved to be a dynamic process. In another series of experiments, Sabbioni et al. (1993) showed, by EPR spectroscopy, that the vanadate(V) taken up by the BALB/3T3 fibroblasts in culture is reduced to vanadium(IV) in a GSH-dependent metabolic pathway. On the other hand, vanadium(IV) did not change its oxidation state in these cells. Once again, cells try to reduce more toxic forms of vanadium(V) to less toxic species of vanadium(IV), probably as a detoxification mechanism. The same authors had previously established that vanadate acted as an inhibitor of some mammalian DNA polymerase, suggesting in that way that vanadate may affect the processes of DNA synthesis and repair. In this sense, vanadyl(IV) was less effective (Sabbioni et al., 1983).

In experiments carried out in collaboration with Dr. D. C. Crans, we demonstrated that UMR-106 osteoblast-like cells, grown in DMEM medium with vanadate or vanadyl cation, were able to take up both types of vanadium derivatives (unpublished results). In order to understand the uptake and metabolism of vanadium compounds in bone-related cells, we have used a combination of EPR and ^{51}V NMR spectroscopies. Cells, conditioned media (supernatant), and washes were studied by means of these two techniques. ^{51}V NMR spectroscopy has provided information about the speciation of vanadium(V) derivatives in the cell, supernatant, and wash fractions. After 1-h incubation of the osteoblast-like cells with 1 mM of sodium orthovanadate, the supernatant fraction showed the characteristic ^{51}V NMR pattern of vanadate (monomer, dimer, and tetramer peaks could be seen). An identical spectrum was shown by the washes and by a control sample (medium plus vanadate; final concentration 1 mM). The cells presented only the peak corresponding to the monomer species (-557 ppm). These results suggest that the monomer (phosphate analogue) is taken up by the osteoblasts. Vanadyl EPR signals were present in the cells. This fact has demonstrated that these cells reduced the vanadate to vanadium(IV). The EPR signals were absent in the supernatant fractions, and a weak vanadyl signal could be observed in the washes, suggesting a release of vanadyl cation from the cells.

It was also demonstrated in these experiments that the vanadate-to-vanadyl reduction depends on the vanadate concentration in the incubation medium. In order to determine if osteoblast-like cells are able to take up vanadyl cation, cells were incubated with 1 mM vanadyl at 37 °C in periods of 1–4 h. EPR spectra of the cells showed that vanadyl cation enters into cells in a short time (1 h) and that EPR signals decrease after a 4-h incubation. On the other hand, the supernatant fractions showed an increase in the EPR signal over a period of 1 to 4 h. These results have allowed us to suggest that UMR-106 osteoblast-like cells take up vanadyl cation very quickly and then release it into the incubation medium in a slower process.

Another possibility to be considered is that the vanadyl cation could be complexed by different intracellular ligands, being transformed into EPR-silent species. However, in this case the increase in the EPR signal showed by the incubation medium (supernatant fractions) as a function of the incubation time cannot be explained.

These experiments have contributed to clarifying which is the active vanadium species in the proliferation and differentiation studies developed in Section 4 of this chapter. Because of the complexity of the aqueous chemistry of vanadium, it is important to use specific spectroscopic techniques to establish the speciation of active vanadium forms relevant to biological studies.

Our results have shown, by means of ^{51}V NMR and EPR spectroscopy, that vanadate in the monomer form is transported into cells, probably through the anion channel of the plasma membrane. Vanadate was reduced to vanadyl inside the cells, as was previously observed in other cellular systems. Bone-related cells (UMR106) were also able to take up vanadium(IV) from the incubation media. The vanadyl cation might be transformed to vanadium(V), at least partially, inside the cells, even though we were not able to detect it by ^{51}V NMR.

Finally, several cell types have been proved to take up vanadium(V) and -(IV) derivatives. Vanadium(V) was reduced to vanadium(IV), possibly as part of a detoxification mechanism. The uptake and transport processes depend on pH, oxygen tension, vanadium concentration, and the presence of different ligands in the culture media. The vanadate(V) forms possibly enter cells by the anion channel, especially the lower-molecular-weight forms. Use of ^{48}V derivatives and appropriate spectroscopic methods (EPR and ^{51}V NMR) are useful tools for determining the active species of vanadium involved in biological effects.

6. MOLECULAR MECHANISM OF VANADIUM ACTION

In spite of the amount of information accumulated about vanadium, the mechanisms of action of this element are still partially known. For many years it was observed that vanadium promotes a number of cellular events, as insulin and other growth factors do. Most of the evidence suggests that vanadate acts through the insulin receptor pathway. We will briefly present the mechanisms involved in the action of insulin and the evidence pointing out the pathway of the action of vanadium.

Insulin, like other peptide growth factors, regulates cellular processes via specific cell-surface receptors (Ulrich and Schlessinger, 1990; Cheatham and Kahn, 1995). After insulin is bound to its receptor, autophosphorylation of the insulin receptor occurs on tyrosine residues of the insulin receptor β-subunit. In addition, serine and threonine residues are also phosphorylated. The receptor acquires protein tyrosine kinase activity with specificity for tyrosine residues of different cellular proteins. They include the major substrate for insulin receptor (IRS-1), mitogen-activated protein kinase (MAPK), and members of the ribosomal S6 kinases, pp70 S6 and pp90 S6 kinase. Evidence now suggests that this tyrosine kinase activity is required for insulin action.

Insulin activates a phosphorylation cascade in intact cells, as summarized in Figure 5. This insulin signaling network plays an important role in the regulation of kinases and phosphatases involved in the intermediate metabolism. Thus, it seems that insulin regulates the state of phosphorylation of key regulatory enzymes. However, insulin has many other effects that do not seem to involve protein phosphorylation/dephosphorylation directly. One of them is the stimulation of glucose transport. Insulin causes the translocation of the facilitative glucose transporter to the plasma membrane, resulting in an increase in the rate of glucose uptake. Even though the mechanism of this stimulation is unclear, recent evidence suggests that it may involve insulin-stimulated PI 3 kinase (Berger et al., 1994).

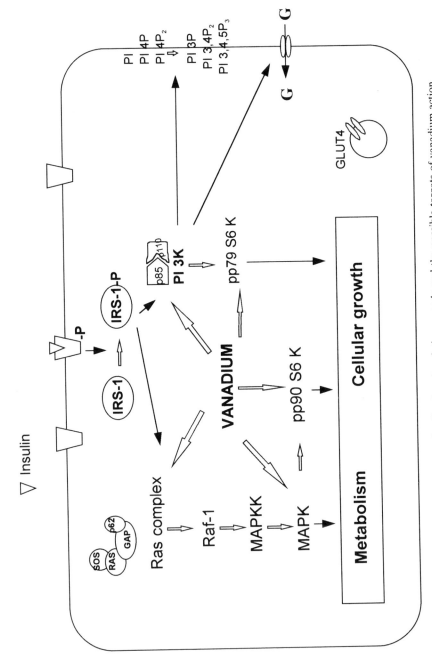

Figure 5. Model showing the insulin phosphorylation cascade and the possible targets of vanadium action.

In addition, insulin causes changes in the expression of the genes for many proteins. The effects on the expression of certain proto-oncogenes such as *c-ras, c-fos,* and *c-myc* are particularly interesting. For instance, insulin induces the expression of *c-Ha-ras* in 3T3 fibroblasts in culture (Lu et al., 1989). Insulin enhances the *c-fos* and *c-myc* mRNA in Reuber H35 rat hepatoma cells (Taub et al., 1987) and in transformed rat fibroblasts (Burgering et al., 1989).

More than 20 years ago it was discovered that vanadate mimics insulin actions in isolated adipocytes (Dubyak and Kleinzeller, 1980; Shechter and Karlish, 1980). Thereafter, studies demonstrated that sodium vanadate is a potent phosphotyrosine phosphatase inhibitor (Swarup et al., 1982). Consequently, the effects of vanadium on different cell systems have been explained by its ability to regulate the phosphorylation/dephosphorylation processes. However, other mechanisms could also be involved. We will present studies in support of these ideas.

Vanadate downregulates the number of insulin receptors in primary culture adipocytes, as insulin does (Marshall and Monzon, 1987). The characteristics of this effect were similar to that of this hormone and consistent with its insulin-mimetic action in these cells. Inhibition of protein synthesis with cycloheximide markedly blunted vanadate-induced receptor loss. Even though the mechanism by which vanadate induced insulin receptor downregulation is unknown, these authors hypothesized that a regulatory protein was involved in the mechanism of vanadium action.

Initially, Tamura et al. (1984) discovered that vanadate stimulates insulin receptor function, although little increase in autophosphorylation was usually observed (Fantus et al., 1989; Mooney et al., 1989; Strout et al., 1989). On the other hand, the mechanism of action of vanadium could depend on the form of compound under investigation. For instance, it was reported that pervanadate, a product of vanadate peroxidation, seems to act via the insulin receptor. Pervanadate stimulated the insulin receptor tyrosine kinase (Kadota et al., 1987; Fantus et al., 1989) and enhanced the autophosphorylation of this receptor (Shisheva and Shechter, 1993).

Lately, the effects of vanadate as well as insulin on the phosphorylation of downstream targets of insulin receptor autophosphorylation have been investigated (Fig. 5). For example, $p21^{ras}$ is a G-protein involved in the signal transduction pathway of different tyrosine kinases. This protein has been shown to play a major role in the control of cell growth and tumor formation (Der, 1989; Macara, 1991; Chardin, 1991). Its activity is regulated by binding to GTP and its hydrolysis, the latter being stimulated by a cytosolic $p21^{ras}$ GTPase-activating protein (GAP) (Medema and Bos, 1993). Indeed, GAP function was associated with other phosphoproteins, termed p62 and p190. It has been shown that pervanadate, like insulin or activated EGF receptor, phosphorylated p62 protein (Pronk et al., 1993a, 1993b).

Several studies have been performed to explore the effect of vanadium on tyrosine phosphorylation of proteins in the insulin signaling pathway. Vanadate was found to induce the tyrosine phosphorylation of 210-kDa protein, the 95-kDa band corresponding to the β-subunit of the insulin receptor and the IRS-1 (185-kDa band) in H4 hepatoma cells (Yurkow and Kim, 1995).

Liao and Lane (1995) discovered that a protein tyrosine phosphatase HA2 is involved in the induction of the differentiation program of 3T3-L1 preadipocytes. Upon induction of differentiation, endogenous PTPase levels reach a maximum during clonal expansion (days 1 and 2) and then decrease as clonal expansion ceases and adipocyte gene expression begins (days 3 and 4). Incubation with vanadate extends the half-life of the phosphoryl pp15, a substrate of protein tyrosine phosphatase HA2. Exposure to vanadate during the mitotic phase completely inhibited cell differentiation. Contrarily, when 3T3-L1 preadipocytes transfected with protein tyrosine phosphatase HA2 were exposed to vanadate at the beginning of adipocyte cell expression, the inhibition of differentiation previously observed was reversed. These experiments suggest that vanadate, regulating phosphotyrosine dephosphorylation events, can control the expression of specific genes and, in doing so, control preadipocyte differentiation as well.

In ovarian granulosa cells stimulated by cAMP, vanadate induced a massive tyrosine phosphorylation of intracellular proteins and enhanced steroidogenesis (Aharoni et al., 1993). In addition, vanadate induced pronounced changes in cell shape and a collapse of the actin network as revealed by rhodamine-phalloidin staining. Electron microscopy showed the breakdown of thin filament cables in cells stimulated with vanadate, as well as the clustering of lipid droplets and mitochondria. These authors suggest that these changes may be due to the phosphorylation of cytoskeleton or associated proteins.

As we previously mentioned, vanadate stimulates cell proliferation and transforms normal cells phenotypically. It also strongly inhibits tyrosine phosphorylation in different cell systems, leading to an increase in tyrosine phosphorylation of intracellular proteins. Thus, it is believed that stimulation of cell growth and transformation processes by vanadium are mediated mainly by the elevation of tyrosine phosphorylation as a consequence of the inhibition of different phosphatases (Klarlund, 1985; Aharoni et al., 1993). However, this hypothesis is not directly tested and other possibilities may be considered, since the mechanism of action of vanadium is not completely known.

Another pathway involved in the insulin action is the activation of phosphatidylinositol (PI) 3-kinase subsequent to the phosphorylation of IRS-1 by insulin receptor activation (Fig. 5) (Cheathan and Kahn, 1995). Consequently, polyphosphoinositides became phosphorylated in the D-3 position. These compounds (PI 3-P, PI 3,4-P_2, PI 3,4,5-P_3) are shown to play a role in the regulation of the cytoskeleton. In addition, the glucose transport stimulated by insulin seems to be mediated by the activation of PI 3-kinase in adipocytes and myotubes (Okada et al., 1994; Cheathan et al., 1994; Berger et al., 1994). Berger et al. (1994) showed that, in L6 myotubes, insulin and vanadate stimulate glucose transport. However, only sixty percent of the vanadate effect was inhibited by worthmanin, an inhibitor of PI 3-kinase, in contrast to the almost complete inhibition of the insulin effect. Thus, it was concluded that insulin activation of PI 3-kinase is necessary to stimulate glucose transport into L6 muscle cells. In contrast, vanadate seems to increase glucose transport by PI 3-kinase-dependent and -independent pathways.

Vanadate not only interacts with the signal transduction pathway of the tyrosine-kinase receptor family but also with receptors associated with GTP-binding proteins. For instance, vanadate increased phosphotyrosine in several proteins and stimulated prostaglandin PGE_2 production in rat Kupffer cells in culture (Chao et al., 1992). It also decreases the surface expression of platelet-activating factor (AGEPC) receptors. Consequently, tyrosine phosphorylation of the AGEPC receptor protein, or other cellular factors regulating that receptor, was proposed as a possible molecular mechanism of vanadium action.

Other studies dealing with the involvement of protein phosphorylation in vanadium-induced cell proliferation and transformation have used bone-related cells. Our laboratory has demonstrated that cell growth, cytotoxicity, and transformed phenotype in fibroblasts and osteoblast-like cells induced by vanadium are dependent on the oxidation state and the coordination geometry of the compound under investigation (Cortizo et al., 1996a, 1996b; Etcheverry et al., 1997). Studies on tyrosine phosphorylation of intracellular proteins after exposure of these cells to different vanadium compounds could give more information about the relationship between the mechanism of action and the effect of these compounds. Preliminary results in our laboratory suggest that the magnitude and the pattern of tyrosine-phosphorylated proteins are dependent on the oxidation state as well as on the geometry of the vanadium compounds. They suggest a possible correlation between tyrosine protein phosphorylation and the growth promotion and transformation effects of vanadium derivatives.

In addition to the induction of tyrosine phosphorylation events in the way that growth factors do, evidence has been accumulated to suggest that vanadium may enhance protooncogene expression (Stern et al., 1993). The products of these genes are involved in the control of cell proliferation. Thus, growth factors as well as agents with tumorigenic properties seem to increase cell proliferation by induction of different oncogenes. It is for this reason that scientists have been concerned with the potential carcinogenic properties of vanadium compounds (Sabbioni et al., 1993; Stern et al., 1993).

The evidence presented points out the pleiothropic effects exerted by vanadium and suggests that multiple pathways may be involved in its mechanism of action.

7. SUMMARY AND CONCLUSIONS

Vanadium is a trace transition element. At present, it has been established that vanadium plays biological functions in lower forms of life. The essentiality of vanadium for higher animals and human beings, although not yet accurately demonstrated, has been accepted. Vanadium compounds have been shown to be active in experimental animals, different cell types, and cell-free systems. Vanadium derivatives show biological and pharmacological effects in different systems, in which they behave like insulin- and growth-factor-mimetic compounds. This point has renewed scientific interest in the synthesis of new vanadium derivatives because of their possible therapeutic use.

Vanadate, pervanadate, vanadyl, and different coordination compounds of vanadium(V) and -(IV) have been tested in several cells in culture. Vanadium derivatives induce metabolic events, promote proliferative processes, and act upon cellular differentiation. Some vanadium compounds produce morphological changes of cells in culture, the peroxo vanadium derivatives being the strongest transformer agents. The potencies of the biological, pharmacological, and cytotoxic effects seem to depend on the oxidation state and coordination geometry of vanadium.

Vanadium derivatives are taken up by cells in culture, and it has been shown that vanadate is reduced to vanadyl inside the cells, probably as a detoxification mechanism.

Owing to the complex aqueous chemistry of vanadium, it is necessary to use appropriate techniques to identify the active species responsible for the biological and pharmacological actions of vanadium.

Although further investigations are required in order to know the molecular mechanism of action of vanadium compounds, some of the observed effects can be explained on the basis of a vanadium–phosphate analogy. In this sense, some experimental evidence points to the fact that different forms of vanadium can act on the phosphorylation/dephosphorylation cascade, producing cellular events through different pathways.

ACKNOWLEDGMENTS

The authors greatfully acknowledge the contribution of their esteemed colleagues and collaborators whose names appear in the references. S.B.E and A.M.C. are especially indebted to Dr. E. J. Baran for his encouragement to develop this subject at the Cátedra de Bioquímica Patológica, Facultad de Ciencias Exactas, UNLP. S.B.E and A.M.C. wish to thank Prof. M. C. Bernal for her kind language revision, and Ms. A. M. Martinez for preparation of the manuscript. The authors would like to express their gratitude to Dr. G. R. Vasta for his assistance in providing many reference materials. S.B.E. is a member of the Carrera del Investigador Científico (CONICET, Argentina), and A.M.C. is a member of the Carrera del Investigador (CICPBA, Argentina). The laboratory work reported herein has been supported by the Facultad de Ciencias Exactas (UNLP), Universidad Nacional de La Plata (Argentina), and the Third World Academy of Sciences.

REFERENCES

Aharoni, D., Dantes, A., and Amsterdam, A. (1993). Cross-talk between adenylate cyclase activation and tyrosine phosphorylation leads to modulation of the actin cytoskeleton and to acute progesterone secretion in ovarian granulosa cells. *Endocrinology* **133,** 1426–1436.

Akisaka, T., Yamamoto, T., and Gay, C. V. (1988). Ultracytochemical investigation of calcium-activated adenosine triphosphatase (Ca^{++}-ATPase) in chick tibia. *J. Bone Miner. Res.*, **3,** 19–25.

Anke, M., Groppel, B., Gruhn, K., Langer, M., and Arnhold, W. (1989). The essentiality of vanadium for animals. In M. Anke, W. Baumann, H. Bräunlich, C. Brücker, B. Groppel, and M. Grün (Eds.), *Sixth International Trace Element Symposium 1989*. Friedrich-Schiller Universität, Jena, pp. 17–27.

Baran, E. J. (1995). Vanadyl (IV) complexes of nucleotides. In: H. Sigel and A. Sigel (Eds.), *Metal Ions in Biological Systems*. Marcel Dekker, New York, Vol. 31, pp. 129–146.

Baran, E. J. (1996). [Some contributions to the bioinorganic chemistry of vanadium]. [In Spanish]. *Anal. Acad. Nac. Cs. Fis. Nat. Buenos Aires* **46,** 35–43.

Baran, E. J., Etcheverry, S. B., and Haiek, D. S. M. (1987). Two new vanadyl (IV) complexes containing biuret. *Polyhedron* **6,** 841–844.

Bayer, E. (1995). Amavadin, the vanadium compound of amanitae. In H. Sigel and A. Sigel (Eds.), *Metal Ions in Biological Systems*. Marcel Dekker, New York, Vol. 31, pp. 407–427.

Bekker, P., and Gay, C. V. (1990). Characterization of a Ca^{++}-ATPase in osteoclast plasma membrane. *J. Bone Miner. Res.* **5,** 557–567.

Berger, J., Hayes, N., Szalkowski, D. M., and Zhang, B. (1994). PI 3-kinase activation is required for insulin stimulation of glucose transport into L6 myotubes. *Biochem. Biophys. Res. Commun.* **205,** 570–576.

Bernier, M., Laird, D. M., and Lane, M. D. (1988). Effect of vanadate on the cellular accumulation of pp15, an apparent product of insulin receptor tyrosine kinase action. *J. Biol. Chem.* **263,** 13626–13634.

Borah, B., Chen, C., Egan, W., Miller, M., Whochaner, A., and Cohen, J. S. (1985). Nuclear magnetic resonance and neutron diffraction studies of the complex of ribonuclease A with uridine vanadate, a transition state analog. *Biochemistry* **24,** 2058–2067.

Bordin, L., Clari, G., Moro, I., Vecchia, F. D., and Moret, V. (1995). Functional link between phosphorylation state of membrane proteins and morphological changes of human erythrocytes. *Biochem. Biophys. Res. Commun.* **213,** 249–257.

Brichard, S. M., Bailey, C. J., and Henquin, J. C. (1990). Marked improvement of glucose homeostasis in diabetic ob/ob mice given oral vanadate. *Diabetes* **39,** 1326–1332.

Brichard, S. M., Desbuquois, B., and Girard, J. (1993). Vanadate treatment of diabetic rats reverses the impaired expression of genes involved in hepatic glucose metabolism: Effects on glycolytic and gluconeogenic enzymes, and on glucose transporter GLUT 2. *Mol. Cell. Endocrinol.* **91,** 91–97.

Brown, D. J., and Gordon, J. A. (1984). The stimulation of pp60v-src kinase activity by vanadate in intact cells accompanies a new phosphorylation state of the enzyme. *J. Biol Chem.* **259,** 9580–9586.

Burgering, B. M., Snijders, A. J., Maassen, J. A., van der Eb, A. J., and Bos, J. L. (1989). Possible involvement of normal p21 H-ras in the insulin/insulin-like growth factor I signal transduction pathway. *Mol. Cell Biol.* **9,** 4312–4322.

Byrne, A. R., and Kosta, L. (1978). Vanadium in foods and human body fluids and tissues. *Sci. Total Environ.* **10,** 17–30.

Canalis, E. (1985). Effect of sodium-vanadate on deoxyribonucleic acid and protein synthesis in cultured rat calvariae. *Endocrinology* **116,** 855–862.

Cantley, L. C. Jr., and Aisen, P. (1979). The fate of cytoplasmic vanadium: Implications on (Na,K)-ATPase inhibition. *J. Biol. Chem.* **254,** 1781–1784.

Cantley, L. C. Jr., Ferguson, J. H., and Kustin, K. (1978). Norepinephrine complexes and reduces vanadium (V) to reverse vanadate inhibition of the (Na,K)-ATPase. *J. Am. Chem. Soc.* **100,** 5210–5212.

Cantley, L. C. Jr., Josephson, L., and Warner R. (1977). Vanadate is a potent (Na+,K+)-ATPase inhibitor found in ATP derived from muscle. *J. Biol. Chem.* **252,** 7421–7423.

Carey, J. O., Azeved, J. L., Morris, P. G., Pories, W. J., and Dhon, G. L. (1995). Okadaic acid, vanadate, and phenylarsine oxide stimulate 2-deoxyglucose transport in insulin-resistant human skeletal muscle. *Diabetes* **44,** 682–688.

Carmignani, M., Volpe, A. R., Masci, O., Boscolo, P., Di Giacomo, F., Grilli, A., Del Rosso, G., and Felaco, M. (1996). Vanadate as factor of cardiovascular regulation by interactions with the cathecholamine and nitric oxide system. *Biol. Trace Element Res.* **51,** 1–12.

Carpenter, G. (1981). Vanadate, epidermal growth factor and the stimulation of DNA synthesis. *Biochem. Biophys. Res. Commun.* **102,** 1115–1121.

Chao, W., Liu, H., Hanahan, D. J., and Olson, M. S. (1992). Protein tyrosine phosphorylation and regulation of the receptor for platelet-activating factor in rat Kupffer cells. *Biochem. J.* **288,** 777–784.

Chardin, P. (1991). Small GTP-binding proteins of the *ras* family: A conserved functional mechanism? *Cancer Cells* **3,** 117–126.

Chasteen, N. D. (1981). Vanadyl (IV) EPR spin probes: Inorganic and biochemical aspects. In J. Reuber and L. Berliner (Eds.), *Biological Magnetic Resonance.* Plenum Press, New York, pp. 53–119.

Chasteen, N. D. (1983). The biochemistry of vanadium. *Struct. Bonding* **53,** 105–138.

Chasteen, N. D., Grady, J. K., and Holloway, C. E. (1986a). Characterization of the binding, kinetics, and redox stability of vanadium(IV) and vanadium(V) protein complexes in serum. *Inorg. Chem.* **25,** 2754–2760.

Chasteen, N. D., Lord, E. M., Thompson, H. J., and Grady, J. K. (1986b). Vanadium complexes of transferrin and ferritin in the rat. *Biochem. Biophys. Acta* **884,** 84–92.

Cheatham, B., and Kahn, C. R. (1995). Insulin action and the insulin signaling network. *Endocrine Rev.* **16,** 117–142.

Cheathman, B., Vlahos, C. J., Wang, L., Blenis, J., and Kahn, C. R. (1994). Phosphatidylinositol 3-kinase activation is required for insulin stimulation of pp70 S6 kinase, DNA synthesis and glucose transporter translocation. *Mol. Cell Biol.* **14,** 4902–4911.

Clausen, T., Andersen, T. L., Sturup-Johansen, M., and Petkova, O. (1981). The relationship between the transport of glucose and cations across cell membranes in isolated tissues: The effect of vanadate in the ^{45}Ca-efflux and sugar transport in adipose tissue and in skeletal muscle. *Biochem. Biophys. Acta* **646,** 261–267.

Cohen, N., Halberstam, M., Shlimovich, P., Chang, C. J., Shamoon, H., and Rossetti, L. (1995). Oral vanadyl sulfate improves hepatic and peripheral insulin sensitivity in patients with non-insulin-dependent diabetes mellitus. *J. Clin. Invest.* **95,** 2501–2509.

Cornelis, R., Versieck, J., Mees, L., Hoste, J., and Barbieri, F. (1981). The ultratrace element vanadium in human serum. *Biol. Trace Element Res.* **3,** 257–263.

Cortizo, A. M., and Etcheverry, S. B. (1995). Vanadium derivatives act as growth factor-mimetic compounds upon differentiation and proliferation of osteoblast-like UMR106 cells. *Mol. Cell. Biochem.* **145,** 97–102.

Cortizo, A. M., Salice, V. C., and Etcheverry, S. B. (1994). Vanadium compounds: Their action on alkaline phosphatase activity. *Biol. Trace Element Res.* **41,** 331–339.

Cortizo, A. M., Salice, V. C., Vescina, C. M., and Etcheverry, S. B. (1996a). Proliferative and morphological changes induced by vanadium compounds in Swiss 3T3 fibroblasts. *Biometals* **10,** 127–133.

Cortizo, A. M., Barrio, D. A., and Etcheverry, S. B. (1996b). Vanadium compounds alter cell proliferation and differentiation on cultured osteoblast-like cells. In: P. Collery, J. Corbella, J. L. Domingo, J. C. Etienne, and J. M. Llobet (Eds), *Metal Ions in Biology and Medicine.* John Libbey Eurotext, Paris, Vol. 4, pp. 294–297.

Crans, D. C. (1994a). Aqueous chemistry of labile oxovanadates of relevance to biological studies. *Comm. Inorg. Chem.* **16,** 1–33.

Crans, D. C. (1994b). Enzyme interactions with labile oxovanadates and other oxometalates. *Comm. Inorg. Chem.* **16,** 35–76.

Crans, D. C., Bunch, R. L., and Theisen, L. A. (1989). Interactions of trace levels of vanadium (IV) and vanadium (V) in biological system. *J. Am. Chem. Soc.* **111,** 7597–7607.

Crans, D. C., Felty, R. A., and Miller, M. M. (1991). Cyclic vanadium (V) alkoxide: An analogue of the ribonuclease inhibitors. *J. Am. Chem. Soc.* **113,** 265–268.

Crans, D. C., and Schelble, S. M. (1990). Vanadate dimer and tetramer both inhibit glucose-6-phosphate dehydrogenase from *Leuconostoc mesenteroids*. *Biochemistry* **29,** 6998–6706.

Crans, D. C., and Simone, C. M. (1991). Nonreductive interaction of vanadate with an enzyme containing a thiol group in the active site: Glycerol-3-phosphate dehydrogenase. *Biochemistry* **30,** 6734–6741.

Crans, D. C., Sudhakar, K., and Zamborelli, T. (1992). Interaction of rabbit muscle aldolase at high ionic strengths with vanadate and other oxoanions. *Biochemistry* **31,** 6812–6821.

Dai, S., and McNeill, J. H. (1994). One-year treatment of non-diabetic and streptozotocin-diabetic rats with vanadyl sulfate did not alter blood pressure or haematological indices. *Pharmacol. J. Toxicol.* **74,** 110–115.

Dai, S., Vera, E., and McNeill, J. H. (1995). Lack of hematological effect of oral vanadium treatment in rats. *Pharmacol. Toxicol.* **76,** 263–268.

Davidai, G., Lee, A., Schuartz, Y., and Hazum, E. (1992). PDGF induces tyrosine phosphorylation in osteoblast-like cells: Relevance to mitogenesis. *Am. J. Physiol.* **263,** E205–E209.

Degani, H., Gochin, M., Karlish, S. J. D., and Shechter, Y. (1981). Electron paramagnetic studies and insulin-like effects of vanadium in rat adipocytes. *Biochemistry* **20,** 5795–5799.

Der, C. J. (1989). The *ras* family of oncogenes. *Cancer Treat. Res.* **47,** 73–119.

Dubyak, G. R., and Kleinzeller, A. (1980). The insulin-mimetic effects of vanadate in isolated rat adipocytes. Dissociation from effects of vanadate as a (Na+-K+) ATPase inhibitor. *J. Biol. Chem.* **255,** 5306–5312.

Duckworth, W. C., Solomon, S. S., Liepnieks, J., Hamel, F. G., Hand, S., and Peavy, D. E. (1988). Insulin-like effects of vanadate in isolate rat adipocytes. *Endocrinology* **122,** 2285–2289.

Eady, R. R. (1995). Vanadium nitrogenases of *Azotobacter*. In H. Sigel and A. Sigel (Eds.) *Metal Ions in Biological Systems*. Marcel Dekker, New York, Vol. 31, pp. 363–406.

Elvingson, K., Fritzsche, M., Rehder, D., and Pettersson, L. (1994). Speciation in vanadium bioinorganic systems: 1.A. Potentiometric and ^{51}VNMR study of aqueous equilibria in the H$^+$-vanadate (V)-L-α-alanyl-L-histidine system. *Acta Chem. Scand.* **48,** 878–885.

Elvinson, K., González Baró, A., and Pettersson, L. (1996). Speciation in vanadium bioinorganic systems. 3. An NMR ESR and potentiometric study of the aqueous H$^+$-vanadate-maltol system. *Inorg. Chem.* **35,** 3388–3393.

Eriksson, J. W., Lönnroth, P., and Smith, U. (1992). Vanadate increases cell surface insulin binding and improves insulin sensitivity in both normal and insulin-resistant rat adipocytes. *Diabetologia* **35,** 510–516.

Etcheverry, S. B., Apella, M. C., and Baran, E. J. (1984). A model study of the incorporation of vanadium in bone. *J. Inorg. Biochem.* **20,** 269–274.

Etcheverry, S. B., Ferrer, E. G., and Baran, E. J. (1989). Nene Beweise über die Wechselwirkung des Vanadyl (IV)-Kations mit Nucleotides: ^{31}P-NMR- und IR-Messungen in Lössung. *Z. Naturforsch.* **44b,** 1355–1358.

Etcheverry, S. B., Williams, P. A. M., and Baran, E. J. (1994). The interaction of the vanadyl (IV) cation with chondroitin sulfate A. *Biol. Trace Element Res.* **42,** 43–52.

Etcheverry, S. B., Williams, P. A. M., and Baran, E. J. (1996a). A spectroscopic study of the interaction of the VO^{2+} cation with the two components of chondroitin sulfate. *Biol. Trace Element Res.* **51,** 169–176.

Etcheverry, S. B., Williams, P. A. M., and Baran, E. J. (1996b). Synthesis and characterization of a solid vanadyl (IV) complex of D-glucuronic acid. *J. Inorg. Biochem.* **63,** 285–289.

Etcheverry, S. B., Crans, D. C., Keramidas, A. D., and Cortizo, A. M. (1997). Insulin-mimetic action of vanadium compounds on osteoblast-like cells in culture. *Arch. Biochem. Biphys.* **338,** 7–14.

Fagin, J. A., Ikejiri, K., and Levin, S. R. (1987). Insulinotropic effects of vanadate. *Diabetes* **36,** 1448–1452.

Fantus, I. G., Ahmad, F., and Deragon, G. (1990). Vanadate augments insulin binding and prolongs insulin action in rat adipocytes. *Endocrinology* **127,** 2716–2725.

Fantus, I. G., Kadota, S., Deragon, G., Foster, B., and Posner, B. I. (1989). Pervanadate [Peroxide(s) of vanadate] mimics insulin action in rat adipocytes *via* activation of the insulin receptor tyrosine kinase. *Biochemistry* **28,** 8864–8871.

Ferber, S., Meyerovitch, J., Kriauciaunas, K. M., and Kahn, C. R. (1994). Vanadate normalizes hyperglycemic and phosphoenolpyruvate carboxykinase mRNA levels in ob/ob mice. *Metabolism* **43**, 1346–1354.

Gavrilivic, J., Moens, G., Thiery, J. P., and Jouanneau, J. (1990). Expression of transfected transforming growth factor alpha induces a motile fibroblastlike phenotype with extracellular matrix-degrading potential in a rat bladder carcinoma cell line. *Cell. Regulation* **1**, 1003–1014.

Gmelins Handbuch der Anorganischen Chemie. (1968). *Vanadium, Part A.* Verlag Chemie, Weinheim, p. 529.

Goldfine, A. B., Folli, F., Patti, M. E., Simonson, D. C., and Kahn, C. R. (1994). Effect of sodium vanadate on *in vivo* and *in vitro* insulin action in diabetes. *Clin. Res.* **42**, 116A.

Gresser, M. J., and Tracey, A. S. (1990). Vanadate as phosphate analogs in biochemistry. In N. D. Chasteen (Ed.), *Vanadium in Biological Systems.* Kluver Academic Publishers, The Netherlands, pp. 63–79.

Gresser, M. J., Tracey, A. S., and Stankiewic, P. J. (1987). The interaction of vanadate with tyrosine kinases and phosphatases. *Adv. Prot. Phosphatases* **4**, 35–57.

Heinz, A., Rubbinson, K. A., and Gruntham, J. J. (1982). The transport and accumulation of oxyvanadium compounds in human erythrocytes *in vitro. J. Lab. Clin. Med.* **100**, 593–612.

Heyliger, C. E., Tahiliani, A. G., and McNeil, J. H. (1985). Effect of vanadate on elevated blood glucose and depressed cardiac performance of diabetic rats. *Science* **227**, 1474–1477.

Hill, C. H. (1990). Interaction of vanadate and chloride in chicks. *Biol. Trace Element Res.* **23**, 1–10.

Hoppe, G., Siemroth, J., and Damaschun, F. (1990). Alexander von Humboldt and the discovery of vanadium. *Chem. Erde* **50**, 81–94.

Howarth, O. W. (1990). Vanadium ^{51}NMR. *Prog. Nucl. Mgn. Reson. Spectrosc.* **22**, 463–485.

Howarth, O. W., and Hunt, J. R. (1979). Peroxocomplexes of vanadium (V): a vanadium-51 nuclear magnetic resonance study. *J. Chem. Soc. Dalton.* **9**, 1388–1391.

Ishida, O., Kihira, K., Tsukamoto, Y., and Marumo, F. (1989). Improved determination of vanadium in biological fluids by electrothermal atomic absorption spectrometry. *Clin. Chem.* **35**, 127–130.

Ishii, T., Nakai, I., and Koshi, K. (1995). Biochemical significance of vanadium in a polychaete worm. In H. Sigel and A. Sigel (Eds.), *Metal Ions in Biological Systems.* Marcel Dekker, New York, Vol. 31, pp. 491–510.

Jamieson, G. A. Jr., Etscheid, B. G., Muldoon, L. L., and Villareal, M. L. (1988). Effects of phorbol ester on mitogen and orthovanadate stimulated responses of cultured human fibroblasts. *J. Cell Physiol.* **134**, 220–228.

Kachinskas, D. J., Phillips, M. A., Qin, Q., Stokes, J. D., and Rice, R. H. (1994). Arsenate perturbation of human keratinocyte differentiation. *Cell Growth Differen.* **5**, 1235–1241.

Kadota, S., Fantus, I. G., Deragon, G., Guy, H. J., Hersh, B., and Posner, B. I. (1987). Peroxide(s) of vanadium: A novel and potent insulin-mimetic agent which activates the insulin receptor kinase. *Biochem. Biophys Res. Commun.* **147**, 259–266.

Kato, Y., Iwamoto, M., Koike, T., and Suzuki, F. (1987). Effect of vanadate on cartilage-matrix proteoglycan synthesis in rabbit costal chondrocyte cultures. *J. Cell Biol.* **104**, 311–319.

Klarlund, J. K. (1985). Transformation of cell by an inhibitor of phosphatases acting on phosphotyrosine in proteins. *Cell* **41**, 707–711.

Kowalski, L. A., Tsang, S.-S., and Davison, A. J. (1992). Vanadate enhances transformation of bovine papillomavirus DNA-transformated C3H/10T½ cells. *Cancer Lett. (Shannon, Irel.)* **64**, 83–90.

Lau, K.-H. W., Tanimoto, H., and Baylink, D. J. (1988). Vanadate stimulates bone cell proliferation and bone collagen synthesis *in vitro. Endocrinology* **123**, 2858–2867.

Leighton, B., Cooper, G. J. S., DaCosta, C., and Foot, E. (1991). Peroxovanadates have full insulin-like effects on glycogen synthesis in normal and insulin-resistant skeletal muscle. *Biochem. J.* **276**, 289–292.

Liao, K., and Lane, M. D. (1995). The blockade of preadipocyte differentiation by protein-tyrosine phosphatase HA2 is reversed by vanadate. *J. Biol. Chem.* **270**, 12123–12132.

Lloyd, Q. P., Kuhn, M. A., and Gay, C. V. (1995). Characterization of calcium translocation across the plasma membrane of primary osteoblasts using a lipophilic calcium-sensitive fluorescent-dye, calcium green C_{18}. *J. Biol. Chem.* **270**, 22445–22451.

Lönnroth, P., Eriksson, J. W., Posner, B. I., and Smith, U. (1993). Peroxovanadate but not vanadate exerts insulin-like effects in human adipocytes. *Diabetologia* **36**, 113–116.

Lu, K., Levine, R. A., and Campos, J. (1989). c-ras-Ha gene expression is regulated by insulin or insulin like growth factor and by epidermal growth factor in murine fibroblasts. *Mol. Cell Biol.* **8**, 3411–3417.

Lyonnet, B. M., and Martin, E. (1899). L'emploi therapeutique des derives du vanadium. *Presse Med.* **1**, 191–192.

Macara, I. G. (1991). The ras superfamily of molecular switches. *Cell Signal.* **3**, 179–187.

Macara, I. G., Kustin, K., and Cantley, L. C. Jr. (1980). Glutathione reduces cytoplasmic vanadate: Mechanism and physiological implications. *Biochem. Biophys. Acta* **629**, 95–106.

Marshall, S., and Monzon, R. (1987). Down-regulation of cell surface insulin receptors in primary cultured rat adipocytes by sodium vanadate. *Endocrinology* **121**, 1116–1122.

McNeill, J. H., Youen, V. G., Hoveyda, H. R., and Orvig, C. (1992). Bis (maltolato) oxovanadium (IV) is a potent insulin mimetic. *J. Med. Chem.* **35**, 1489–1491.

Medema, R. H., and Bos. J. L. (1993). The role of p21^{ras} in receptor tyrosine kinase signaling. *Crit. Rev. Oncogenesis* **4**, 615–661.

Mendz, G. L., Hylop, S. J., and Kuchel, P. W. (1990). Stimulation of human erythrocyte 2,3-biphosphoglycerate phosphatase by vanadate. *Arch. Biochem. Biophys.* **276**, 160–171.

Meyerovitch, J., Farfel, Z., Sack, J., and Shechter, Y. (1987). Oral administration of vanadate normalizes blood glucose levels in streptozotocin-treated rats. *J. Biol. Chem.* **262**, 6658–6661.

Meyerovitch, J., Rothenberg, P., Shechter, Y., Weir, A., and Kahn, C. R. (1991). Vanadate normalized hyperglycemia in two mouse models of non-insulin dependent diabetes mellitus. *J. Clin. Invest.* **87**, 1286–1294.

Mooney, R. A., Bordwell, K. L., Luhowskyj, S., and Casnellie, J. E. (1989). The insulin-like effect of vanadate on lipolysis in rat adipocytes is not accompanied by an insulin-like effect on tyrosine phosphorylation. *Endocrinology* **124**, 422–429.

Mountjoy, K. G., and Flier, J. S. (1990). Vanadate regulates glucose transporter (Glut-1) expression in NIH3T3 mouse fibroblasts. *Endocrinology* **127**, 2025–2034.

Muschietti, L. I., Etcheverry, S. B., and Baran, E. J. (1982). Zur Stöchiometric der Vanadyl-Diphosphat-Komplexe. *Monatsh. Chem.* **113**, 1399–1401.

Myron, D. R., Zimmerman, T. J., Shuler, T. R., Klevay, L. M., Lee, D. E., and Nielsen, F. H. (1978). Intake of nickel and vanadium by humans: A survey of selected diets. *Am. J. Clin. Nutr.* **31**, 527–531.

Narda, G. E., Vega, E. D., Pedregosa, J. C., Etcheverry, S. B., and Baran, E. J. (1992). Über die Wechselwirkung des Vanadyl (IV)-Kations mit Calcium-Hydroxylapatit. *Z. Naturforsch.* **47b**, 395–398.

Nechay, B. R., Nanninga, L. B., and Nechay, P. S. E. (1986). Vanadyl(IV) and vanadate(V) binding to selected endogenous phosphate, carboxyl and amino ligands: Calculation of cellular vanadium species distribution. *Arch. Biochem. Biophys.* **251**, 128–138.

Nielsen, F. H. (1995). Vanadium in mammalian physiology and nutrition. In H. Sigel and A. Sigel (Eds.), *Metal Ions in Biological Systems,* Vol. 31: *Vanadium and Its Role in Life.* Marcel Dekker, New York, pp. 543–573.

Nielsen, F. H., and Uthus, E. O. (1990). The essentially and metabolism of vanadium. In N. D. Chasteen (Ed.), *Vanadium in Biological Systems.* Kluwer, Dordrecht, The Netherlands, pp. 51–62.

Okada, T., Kawaro, Y., Sakakibara, T., Hazeki, O., and Ui, M. (1994). Essential role of phospatidyl-inositol 3-kinase in insulin-induced glucose transport and antilipolysis in rat adipocytes. *J. Biol. Chem.* **269**, 3568–3573.

Owada, M. K., Iwamoto, M., Koike, T., and Kato, K. (1989). Effects of vanadate on tyrosine phosphorylation and the pattern of glycosaminoglycan synthesis in rabbit chondrocytes in culture. *J. Cell Physiol.* **138**, 484–492.

Pardey, S. K., Chiasson, J.-L., and Srivastava, A. K. (1995). Vanadium salts stimulate mitogen activated (MAP) kinases and ribosomal S6 kinases. *Mol. Cell. Biochem.* **153,** 69–78.

Parr, R. M., DeMaeyer, E. M., Iyengar, V. G., Byrne, A. R., Kirkbright, G. F., Schöch, G., Ninistö, L., Pineda, O., Vis, H. L., Hofvander, Y., and Omololu, A. (1991). Minor and trace elements in human milk from Guatemala, Hungary, Nigeria, Philippines, Sweden, and Zaire. *Biol. Trace Element Res.* **29,** 51–75.

Pedersøn, R. A., Ramanadham, S., Buchan, A. M. J., and McNeill, J. H. (1985). Long-term effects of vanadyl treatment on streptozotocin induced diabetes in rats. *Diabetes* **38,** 1390–1395.

Percival, M. D., Doherty, K., and Gresser, M. J. (1990). Inhibition of phosphoglucomutase by vanadate. *Biochemistry* **29,** 2764–2769.

Pronk, G. J., Medema, R. H., Burgering, B. M., Clark, R., McCormick, F., and Bos, J. L. (1993a). Interaction between the p21^{ras} GTPase activating protein and the insulin receptor. *J. Biol. Chem.* **267,** 24058–24063.

Pronk, G. J., de Vries-Smits, A. M. M., Ellis, C., and Bos, J. L., (1993b). Complex formation between the p21^{ras} GTPase-activating protein and phosphoproteins p62 and p190 is independent of p21^{ras} signalling. *Oncogene* **8,** 2773–2780.

Ramanadham, S., Mongold, J. J., Brownsey, R. W., Cros, G. H., and McNeill, J. H. (1989). Oral vanadyl sulfate in treatment of diabetes mellitus in rats. *Am. J. Physiol.* **257,** H904–H911.

Ray, W. J. Jr., Burgner, J. W., II, and Post, C. V. (1990). Characterization of vanadate-based transition state analogue complexes of phosphoglucomutase by spectral and NMR techniques. *Biochemistry* **29,** 2770–2778.

Ray, W. J. Jr., and Puvathingal, J. M. (1990). Characterization of vanadate based transition in state analogue complex of phosphoglucomutase by kinetic and equilibrium binding structures: Mechanistic implications. *Biochemistry* **20,** 2790–2801.

Rehder, D. (1990). Biological applications of ^{51}VNMR spectroscopy. In N. D. Chasteen (Ed.), *Vanadium in Biological Systems.* Kluwer Academic Publishers, Dordrecht, The Netherlands, pp. 173–197.

Rehder, D. (1991). The bioinorganic chemistry of vanadium. *Angew. Chem. Int. Ed. Engl.* **30,** 148–167.

Rehder, D. (1992). Structure and function of vanadium compounds in living organisms. *BioMetals* **5,** 3–12.

Rehder, D. (1995). Inorganic considerations on the function of vanadium in biological systems. In H. Sigel and A. Sigel (Eds.), *Metal Ions in Biological Systems.* Marcel Dekker, New York, Vol. 31, pp. 1–43.

Rivedal, E., Roseng, L. E., and Sanner, T. (1990). Vanadium compounds promote the induction of morphological transformation of hamster embryo cells with no effect on gap junctional communication. *Cell Biol. Toxicol.* **6,** 303–314.

Ryden, J. W., and Gordon, J. A. (1987). In vivo effect of sodium orthovanadate on pp60c-src kinase. *Mol. Cell. Biol.* **7,** 1139–1147.

Sabbioni, E., Clerici, L., and Branzzelli, A. (1983). Different effects of vanadium on some DNA-metabolizing enzymes. *J. Toxicol. Environ. Health* **12,** 737–748.

Sabbioni, E., Marafante, E., Pietra, R., Gómez, L., Girardi, F., and Orvini, E. (1979). The association of vanadium with the iron transport system in human blood as determined by gel filtration and neutron activation analysis. In *Proceedings of a Symposium on Nuclear Activation Techniques in the Life Sciences, Vienna, 22–26 May, 1978.* International Atomic Energy Agency, Vienna, pp. 179–192.

Sabbioni, E., Pozzi, G., Devos, S., Pintar, A., Casella, L., and Fischbach, M. (1993). The intensity of vanadium (V)-induced biotoxicity and morphological transformation in BALB/3T3 cells independent on gluthatione-mediated bioreduction to vanadium (IV). *Carcinogenesis* **14,** 2565–2568.

Sabbioni, E., Pozzi, G., Pintar, A., Casella, L., and Garattini, S. (1991). Cellular retention, cytoxicity and morphological transformation by vanadium(IV), and vanadium(V) in BALB/3T3 cell lines. *Carcinogenesis* **12,** 47–52.

Sabbioni, E., Rada, J., Gregotti, C., Di Nacci, A., and Manzo, L. (1981). Biliary excretion of vanadium in rats. *Toxicol. Eur. Res.* **3**, 93–98.

Sakurai, H., Shimomura, S., Fukuzawa, K., and Ishizu, K. (1980). Detection of oxovanadium (IV) and characterization of its ligand environment in subcellular fractions of the liver of rat treated with pentavalent vanadium(V). *Biochem. Biophys. Res. Commun.* **96**, 293–298.

Sato, B., Miyashita, Y., Maeda, Y., Noma, K., Kishimoto, S., and Matsumoto, K. (1987). Effects of estrogen and vanadate on the proliferation of newly established transformed mouse Leydig cell line *in vitro*. *Endocrinology (Baltimore)* **120**, 1112–1120.

Shechter, Y. (1990). Perspectives in diabetes: Insulin mimetic effects of vanadate, possible implications for future treatment of diabetes. *Diabetes* **39**, 1–5.

Shechter, Y., and Karlish, S. J. D. (1980). Insulin-like stimulation of glucose oxidation in rat adipocytes by vanadyl(IV) ions. *Nature* **284**, 556–558.

Shechter, Y., and Shisheva, A. (1993). Vanadium salts and the future treatment of diabetes. *Endeavour* (new ser.) **17**, 27–31.

Shechter, Y., Shisheva, A., Lazar, R., Libman, J., and Shanzer A. (1992). Hydrophobic carriers of vanadyl ions augment the insulinomimetic actions of vanadyl ions in rat adipocytes. *Biochemistry* **31**, 2063–2068.

Shiozaki, H., Kadowaki, T., Doki, Y., Inove, M., Tamura, S., Oka, H., Iwazawa, T., Matsui, S., Simaya, K., Takeichi, M., and Mori, T. (1995). Effect of epidermal growth factor on cadherin-mediated adhesion in a human oesophageal cancer cell line. *Br. J. Cancer* **71**, 250–258.

Shisheva, A., and Shechter, Y. (1993). Mechanism of pervanadate stimulation and potentiation of insulin-activated glucose transport in rat adipocytes: Dissociation from vanadate effect. *Endocrinology* **133**, 1562–1568.

Smith, J. B. (1983). Vanadium ions stimulate DNA synthesis in Swiss mouse 3T3 and 3T6 cells. *Proc. Natl. Acad. Sci. USA* **80**, 6162–6166.

Smith, M. J., Ryan, D. E., Nakaninshi, K., Frank, P., and Hodgson, K. O. (1995). Vanadium in ascidians and the chemistry of tunichromes. In H. Sigel and A. Sigel (Eds.), *Metal Ions in Biological Systems, Vol. 31: Vanadium and Its Role in Life,* pp. 423–490.

Soman, G., Chang, Y. C., and Graves, D. J. (1983). Effect of oxyanions of early transition metals on rabbit skeletal muscle phosphorylase. *Biochemistry* **22**, 4994–5000.

Stankiewicz, P. J., and Gresser, M. J. (1988). Inhibition of phosphatase and sulfatase by transition-state analogues. *Biochemistry* **27**, 206–212.

Stankiewicz, P. J., Gresser, M. J., Tracey, A. S., and Has, L. F. (1987). 2,3-Diphosphoglycerate phosphatase activity of phosphoglycerate mutase: Stimulation by vanadate and phosphate. *Biochemistry* **26**, 1264–1269.

Stern, A., Yin, X., Tsang, S.-S., Davison, A., and Moon, J. (1993). Vanadium as a modulator of cellular regulatory cascades and oncogene expression. *Biochem. Cell Biol.* **71**, 103–112.

Strout, H. V., Vicario, P. P., Suprstein, R., and Slater, E. E. (1989). The insulin-mimetic effect of vanadate is not correlated with insulin receptor tyrosine kinase activity nor phosphorylation in mouse diaphragm *in vivo*. *Endocrinology* **124**, 1918–1924.

Swarup, G., Cohen, S., and Garbers, D. L. (1982). Inhibition of membrane phosphotyrosyl-protein phosphatase activity by vanadate. *Biochem. Biophys. Res. Commun.* **107**, 1104–1109.

Tamura, S., Brown, T. A., Whipple, J. H., Fujita-Yamaguchi, Y., Dubler, R. E., Cheng, K., and Larner, J. (1984). A novel mechanism for the insulin-like effects of vanadate on glycogen synthase in rat adipocytes. *J. Biol. Chem.* **259**, 6650–6658.

Taub, R., Roy, A., Dieter, R., and Koontz, J. (1987). Insulin as a growth factor in rat hepatoma cells. Stimulation of protooncogene expression. *J. Biol. Chem.* **262**, 10893–10897.

Thomas, D. M., Rogers, S. D., Sleeman, M. W., Pasquin, G. M., Bringhurst, F. R., Ng, K. W., Zajac, J. D., and Best, J. D. (1995). Modulation of glucose transport by parathyroid hormone and insulin in UMR 106-01, a clonal rat osteogenic sarcoma cell line. *J. Endocrinol.* **14**, 263–275.

Trelles, R. A., Larghi, L. A., and Paez, L. J. P. (1970). The health problem of human drinking water with high contents of arsenic, vanadium and fluorine. [In Spanish.] *Saneamiento* **34**, 31–80.

Ulrich, A., and Schlessinger, J. (1990). Signal transduction by receptors with tyrosine kinase activity. *Cell* **61**, 203–212.

Uthus, E. D., and Nielsen, F. H. (1989). The effect of vanadium, iodine and their interaction on thyroid status indices. In M. Anke, W. Baumann, H. Bräunlich, C. Brücker, B. Groppel, and M. Grün, (Eds.), *Sixth International Trace Element Symposium, 1989.* Friedrich-Schiller Universität, Jena, pp. 44–49.

Vescina, C. M., Sálice, V. C., Cortizo, A. M., and Etcheverry, S. B. (1996). Effect of vanadium compounds on acid phosphatase activity. *Biol. Trace Element Res.* **53,** 185–191.

Vilter, H. (1995). Vanadium-dependent haloperoxidases. In H. Sigel and A. Sigel (Eds.), *Metal Ions in Biological Systems.* Marcel Dekker, New York, Vol. 31, pp. 325–362.

Vrbic, V., Stupar, J., and Byrne, A. R. (1987). Trace element content of primary and permanent tooth enamel. *Caries Res.* **21,** 37–39.

Wei, L., and Yung, B. Y.-M. (1995). Effects of okadaic acid and vanadate on TPA-induced monocytic differentiation in human promyelocytic leukemia cell line HL-60. *Cancer Lett.* **90,** 199–205.

Wergedal, J. E., Lau, K. H. W., and Baylink, D. J. (1988). Fluoride and bovine bone extract influence cell proliferation and phosphatase activities in human bone cell cultures. *Clin. Orthop. Res.* **233,** 274–279.

Williams, P. A. M., Etcheverry, S. B., and Baran, E. J. (1993). Uber die Wechselwikung des Vanadyl (IV) Kations mit Nucleobasen. *Z. Naturforsch.* **48b,** 1845–1847.

Williams, P. A. M., Etcheverry, S. B., and Baran, E. J. (1994). Interaction of the vanadyl (IV) cation with nucleosides. *An. Asoc. Quím. Argent.* **82,** 13–17.

Williams, P. A. M., Etcheverry, S. B., and Baran, E. J. (1996). A spectrophotometric study of the interactions of VO^{2+} with cytosine in nucleotides. *J. Inorg. Biochem.* **61,** 285–289.

Williams, P. A. M., Etcheverry, S. B., and Baran, E. J. (1997). Synthesis and characterization of solid vanadyl (IV) complexes of D-ribose and D-ribose-5-phosphate. *J. Inorg. Biochem.* **65,** 133–136.

Willsky, G. R., and Dosch, F. (1986). Vanadium metabolism in wild type and respiratory defficient strains of *S. cerevisiae. Yeast* **2,** 77–85.

Willsky, G. R., Leung, J. O., Offermann, P. V., Plotnik, E. K., and Dosch, F. (1985). Isolation and characterization of vanadate mutants of Saccharomyces cerevisiae. *J. Bacteriol.* **164,** 611–617.

Willsky, G. R., White, D. A., and McCabe, B. C. (1984). Metabolism of added orthovanadate to vanadyl and high-molecular weight vanadate by *Saccharomyces cerevisiae. J. Biol. Chem.* **259,** 13273–13281.

Wittenkeller, L., Abraha, A., Ramasanny, R., Defreitas, D. M., Theisen, L. A., and Crans, D. C. (1991). Vanadate interactions with bovine Cu, Zn-superoxide dismutase as probed by [51]VNMR spectroscopy. *J. Am. Chem. Soc.* **113,** 7872–7881.

World Health Organization. (1988). *Environmental Health Criteria,* 81: *Vanadium.* WHO, Geneva, p. 13.

Yale, J. F., Lachance, D., Bevan, A. P., Vigeant, C., Shaver, A. and Posner, B. I. (1995). Hypoglycemic effects of peroxovanadium compounds in Sprague-Dawley and diabetic BB rats. *Diabetes* **44,** 1274–1279.

Yuen, V. G., Orvig, C., and McNeill (1995). Comparison of the glucose-lowering properties of vanadyl sulfate and bis (maltolato) oxovanadium (IV) following acute and chronic administration. *Can. J. Physiol. Pharmacol.* **73,** 55–64.

Yurkow, E. J., and Kim, G. (1995). Effects of chromium on basal and insulin-induced tyrosine phosphorylation in H4 hepatoma cells: Comparison with phorbol-12-myristate-13-acetate and sodium orthovanadate. *Mol. Pharmacol.* **47,** 686–695.

Zhang, A., Gao, Z.-Y., Gilon, P., Nenquin, M., Drews, G., and Henquin, J. C. (1991). Vanadate stimulation of insulin release in normal mouse islets. *J. Biol. Chem.* **266,** 21649–21656.

INDEX